S0-CAH-091

Phase Equilibria and Fluid Properties in the Chemical Industry

Estimation and Correlation

Truman S. Storvick, EDITOR
University of Missouri, Columbia

Stanley I. Sandler, EDITOR
University of Delaware

A symposium co-sponsored

by the Engineering Foundation,

the American Institute

of Chemical Engineers,

and the National

Science Foundation at the

Asilomar Conference Grounds

Pacific Grove, CA,

January 16–21, 1977

ACS SYMPOSIUM SERIES 60

Library of Congress CIP Data

Phase equilibria and fluid properties in the chemical industry.
(ACS symposium series; 60 ISSN 0097-6156)

Includes bibliographical references and index.

1. Phase rule and equilibrium—Congresses. 2. Thermodynamics—Congresses. 3. Liquids—Congresses.
I. Storvick, Truman S., 1928- . II. Sandler, Stanley I., 1940- . III. Engineering Foundation, New York. IV. Series: American Chemical Society. ACS symposium series; 60.

QD501.P384 660.2'9'63 77-13804
ISBN 0-8412-0393-8 ACSMC8 60 1-537 (1977)

PRINTED IN THE UNITED STATES OF AMERICA

FOREWORD

The ACS SYMPOSIUM SERIES was founded in 1974 to provide a medium for publishing symposia quickly in book form. The format of the SERIES parallels that of the continuing ADVANCES IN CHEMISTRY SERIES except that in order to save time the papers are not typeset but are reproduced as they are submitted by the authors in camera-ready form. As a further means of saving time, the papers are not edited or reviewed except by the symposium chairman, who becomes editor of the book. Papers published in the ACS SYMPOSIUM SERIES are original contributions not published elsewhere in whole or major part and include reports of research as well as reviews since symposia may embrace both types of presentation.

DEDICATION

This work is dedicated to the memory of three men who contributed to our understanding of fluid properties.

Ping L. Chueh
Shell Development Co.
Houston, TX

Gerald A. Ratcliff
McGill University
Montreal, Quebec, Canada

Thomas M. Reed
University of Florida
Gainesville, FL

Illness and accident cut short their careers in 1976 and have left us with their last contribution.

CONTENTS

PREFACE

We had two goals in organizing this conference. The first was to provide a forum for state-of-the-art reviews of an area of chemical engineering often referred to as "thermodynamics and physical properties." The reviews should represent the work of both the academic researcher and the industrial practitioner. This we thought was both necessary and timely because there were obvious dislocations between the current needs of the industrial chemical engineer and the research being done at universities, on the one hand, and the slow acceptance of new theoretical tools by the industrial people on the other.

Our second objective was, through these reviews and the ensuing discussion, to develop a collection of research objectives for the next decade. We asked the session reporters to try to identify the important research problems that were suggested in the presentations and discussions of the sessions, as well as to set down their thoughts in this regard. In this way, the major papers in this volume summarize the current state of research and industrial practice, while the reporter's summaries provide a listing of important questions and research areas that need attention now.

The conference was attended by 135 engineers and scientists from North America, Europe, Asia and Africa. They represented, in almost equal numbers, the industrial and academic sectors. Recognized authorities, presently active in physical properties work, were chosen to be speakers, panel members, session reporters and session chairmen. The conference was held at the Asilomar Conference Grounds on the Monterey Penninsula of California, the beautiful setting matched by idyllic weather. We have tried to give an accurate account of the material presented at the conference sessions, but the printed word cannot reflect the friendships that were established nor the extent of the academic–industrial dialogue which was initiated. Similarly, the unusual enthusiasm of the conference is not reflected here. Indeed, this enthusiasm was so great that there were six ad-hoc sessions, continuations of scheduled sessions and meetings packed into the four sunny afternoons of the meeting.

Many important areas of work were identified as needing further attention during the next decade. Several obvious to us (in no special order) are listed below:

- It was generally agreed that nine out of 10 requests for data by design engineers were for vapor–liquid equilbrium or mixture enthalpy data. Reduction to field-level practice of either data banks or estimating procedures to supply this information would be very useful.

• Significant progress has been made on the group contribution methods for estimating phase equilibrium data. Further development of these procedures is clearly justified.

• Perturbation methods based on theory from physics and chemistry, electronic computer simulation studies, and careful comparisons with real fluid behavior are moving quickly toward producing an effective equation of state for liquids. These efforts are in the hands of the theoretician today, but further development and reduction to practice should be explored.

• Fluid transport properties were not the primary concern at this conference, but progress was also shown here. Remarkable agreement between prediction and experiment for viscosities and thermal conductivities of gaseous mixtures was reported. Clearly, much work needs to be done, especially for liquids.

• Real difficulties remain when attempts are made to predict, to extrapolate, or even to interpolate data for multicomponent mixtures containing hydrocarbons, alcohols, acids, etc. Such systems were affectionately identified as a "Krolikowski mess" at the conference. Multicomponent mixtures of this kind may include more than one liquid and/or solid phase and with components that "commit chemistry" as well as physically distribute between the phases are commonly encountered in industrial practice. The goal for the future is to reduce these problems from nightmare to headache proportions in industrial applications, though they may continue to remain an enigma for the theoretician.

• Cries for more experimental data were often heard. Special needs include high pressure vapor–liquid equilibrium data; data on several properties for mixtures with very light, volatile components in heavy hydrocarbon mixtures; ionic solutions; acid gases in hydrocarbons; and certainly more emphasis on mixtures containing aromatic hydrocarbons. Data with intrinsic value for design work and accurate enough for discriminating theoretical comparisons should have high priority. Significantly, several conferees stated that their primary sources of new experimental data are rapidly shifting to laboratories outside the United States.

An important measure of the success of a conference is its long-term impact. It remains to be seen whether this conference results in any permanent interchange of ideas between academic and industrial engineers and whether the ideas expressed influence research in the coming years.

Acknowledgments

This volume is based on the Engineering Foundation Conference, "The Estimation and Correlation of Phase Equilibria and Fluid Properties in the Chemical Industry," convened at the Asilomar Conference Grounds, Pacific Grove, CA, on Jan. 16–21, 1977. The views presented here are not necessarily those of the Engineering Foundation, 345 East 47th St., New York, N.Y. 10017. The advice, financial and moral support, and the concern for local arrangements, publicity, registration by Sandford Cole, Harold Commerer, Dean Benson and their staff permitted us to concentrate on the technical aspects of the meeting. Manuscript typing was done by the University of Missouri, Stenographic Services Department.

Major funding for the conference by the National Science Foundation was a key ingredient in its success. These funds made it possible for many American and European academicians to attend who would have been otherwise unable to participate. The interest and support of Marshall Lih and William Weigand of the National Science Foundation were especially appreciated.

The American Institute of Chemical Engineers made important contributions by co-sponsoring and publicizing the conference.

We also thank the members of the Organizing Committee: Stanley Adler, Pullman-Kellogg Co.; Howard Hanley of the National Bureau of Standards; Robert Reid of the Massachusetts Institute of Technology; and Lyman Yarborough of the Amoco Production Co. They brought focus and structure to the general concept of the conference we brought to them.

Finally, and most important we thank the speakers, session reporters, and chairman who did their work diligently and in the best scientific tradition; and the conferees for their enthusiastic participation and important discussion contributions that made this conference special.

T. S. Storvick
University of Missouri—Columbia
June 1977

Stanley I. Sandler
University of Delaware

1

Origin of the Acentric Factor

KENNETH S. PITZER

University of California, Berkeley, Calif. 94720

It was a pleasure to accept Dr. Sandler's invitation to open this conference by reviewing the ideas and general point of view which led me to propose the acentric factor in 1955. Although I had followed some of the work in which others have used the acentric factor, the preparation of this paper provided the incentive to review these applications more extensively, and I was most pleased to find that so much has been done. I want to acknowledge at once my debt to John Prausnitz for suggestions in this review of recent work as well as in many discussions through the years.

Beginning in 1937, I had been very much interested in the thermodynamic properties of various hydrocarbon molecules and hence of those substances in the ideal gas state. This arose out of work with Kemp in 1936 on the entropy of ethane (1) which led to the determination of the potential barrier restricting internal rotation. With the concept of restricted internal rotation and some advances in the pertinent statistical mechanics it became possible to calculate rather accurately the entropies of various light hydrocarbons (2). Fred Rossini and I collaborated in bringing together his heat of formation data and my entropy and enthalpy values to provide a complete coverage of the thermodynamics of these hydrocarbons in the ideal gas state (3). As an aside I cite the recent paper of Scott (4) who presents the best current results on this topic.

But real industrial processes often involve liquids or gases at high pressures rather than ideal gases. Hence it was a logical extension of this work on the ideal gases to seek methods of obtaining the differences in properties of real fluids from the respective ideal gases without extensive experimental studies of each substance.

My first step in this direction came in 1939 when I was able to provide a rigorous theory of corresponding states (5) on the basis of intermolecular forces for the restricted group of substances, argon, kryptron, xenon, and in good approximation also methane. This pattern of behavior came to be called that of a simple fluid. It is the reference pattern from which the acentric factor measures the departure. Possibly we should recall the key ideas. The intermolecular potential must be given by a universal function with scale factors of energy and distance for each substance. By then it was well-known that the dominant attractive force followed an inverse sixth-power potential for all of these substances. Also the repulsive forces were known to be very sudden. Thus the inverse sixth power term will dominate the shape of the potential curve at longer distances. Even without detailed theoretical reasons for exact similarity of shorter-range terms, one could expect that a universal function might be a good approximation. In addition one assumed spherical symmetry (approximate for methane), the validity of classical statistical mechanics, and that the total energy was determined entirely by the various intermolecular distances.

I should recall that it was not feasible in 1939 to calculate the actual equation of state from this model. One could only show that it yielded corresponding states, i.e., a universal equation of state in terms of the reduced variables of temperature, volume, and pressure.

One could postulate other models which would yield a corresponding-states behavior but different from that of the simple fluid. However, most such molecular models were special and did not yield a single family of equations. Rowlinson (6) found a somewhat more general case; he showed that for certain types of angularly dependent attractive forces the net effect was a temperature dependent change in the repulsive term. From this a single family of functions arose.

I had observed empirically, however, that the family relationship of equations of state was much broader even than would follow from Rowlinson's model. It included globular and effectively spherical molecules such as tetramethylmethane (neopentane), where no appreciable angular dependence was expected for the intermolecular potential, and for elongated molecules such as carbon dioxide the angular dependence of the repulsive forces seemed likely to be at least as important as that of the attractive forces. Thus the core model of Kihara (7) appealed to me; he assumed that the Lennard-Jones 6-12 potential applied to the shortest distance between cores instead of the distance between molecular centers. He was able to calculate the second virial coefficient for various shapes of core. And I was able to show that one obtained in good approximation a single family of reduced second virial coefficient functions for cores of all reasonable shapes. By a single family I mean that one additional parameter sufficed to define the equation for any particular case. While this did not prove that all of the complete equations of state would fall into a single family, it gave me enough encouragement to go ahead with the numerical work--or more accurately to persuade several students to undertake the numerical work.

Let me emphasize the importance of fitting globular molecules into the system. If these molecules are assumed to be spherical in good approximation, they are easy to treat theoretically. Why aren't they simple fluids? Many theoretical papers ignore this question. In fluid properties neopentane departs from the simple fluid pattern much more than propane and almost as much as n-butane. But propane is much less spherical than neopentane. The explanation lies in the narrower attractive potential well. The inverse-sixth-power attractive potential now operates between each part of the molecule rather than between molecular centers. Thus the attractive term is steeper than inverse sixth power in terms of the distance between molecular centers. This is shown in Figure 1, taken from my paper (8) in 1955. We need not bother with the differences between the models yielding the dotted and dashed curves for the globular molecule. The important feature is the narrowness of the potential well for either of these curves as compared to the solid curve for the molecules of a simple fluid.

It was easy to show that the intermolecular potential curves for spherical molecules would yield a single family of reduced equations of state. If one takes the Kihara model with spherical cores, then the relative core size can be taken as the third parameter in addition to the energy and distance scale factors in the theoretical equation of state.

With an adequate understanding of globular molecule behavior, I then showed as far as was feasible that the properties of other non-polar or weakly polar molecules would fall into the same family. It was practical at that time only to consider the second virial coefficient. The Kihara model was used for nonpolar molecules of all shapes while Rowlinson's work provided the basis for discussion of polar molecules. Figure 2 shows the reduced second virial coefficient for several cases. Curves labeled a/ρ_o refer to spherical-core molecules with a indicating the core size, correspondingly ℓ/ρ_o indicates a linear molecule of core length ℓ, while y refers to a dipolar molecule with $y = \mu^4/\varepsilon_o{}^2 r_o{}^6$ where μ is the dipole moment. The non-polar potential is

$$\varepsilon = \varepsilon_o\left[\left(\frac{\rho_o}{\rho}\right)^{12} - 2\left(\frac{\rho_o}{\rho}\right)^{6}\right] \qquad (1)$$

where ρ is the shortest distance between cores. For the polar molecules I omitted the core, thus $\rho = r$.

While the curves in Figure 2 appear to fall into a single family, this is investigated more rigorously in Figure 3 where the reduced second virial coefficient at one reduced temperature is compared with the same quantity at another temperature. T_B is the Boyle temperature which is a convenient reference temperature for second virial coefficients. One sees that the non-polar core molecules fall accurately on a single curve (indeed a straight line). While the polar molecules deviate, the difference is only 1% at $y = 0.7$ which I took as a reasonable standard of accuracy at that time. For comparison the y values of chloroform, ethylchloride, and ammonia are 0.04, 0.16, and 4, respectively. Thus the first two fall well below the 0.7 value for agreement of polar with non-polar effects while ammonia is beyond that value.

The next question was the choice of the experimental basis for the third parameter. The vapor pressure is the property most sensitive to this third parameter; also it is one of the properties most

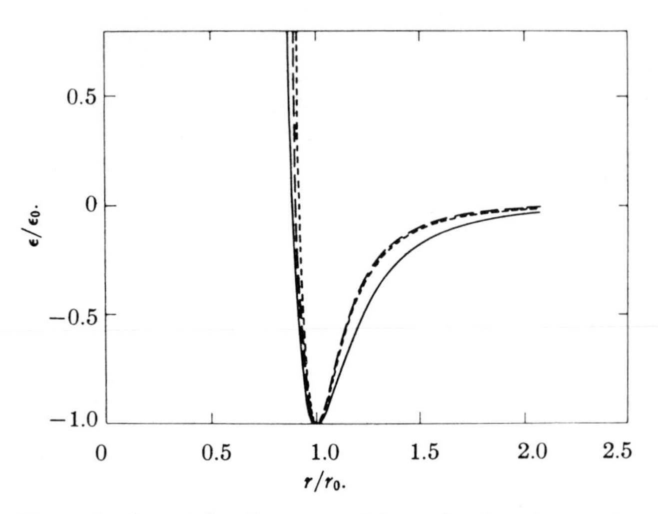

Figure 1. Intermolecular potential for molecules of a simple fluid, solid line; and for globular molecules such as $C(CH_3)_4$, dashed lines

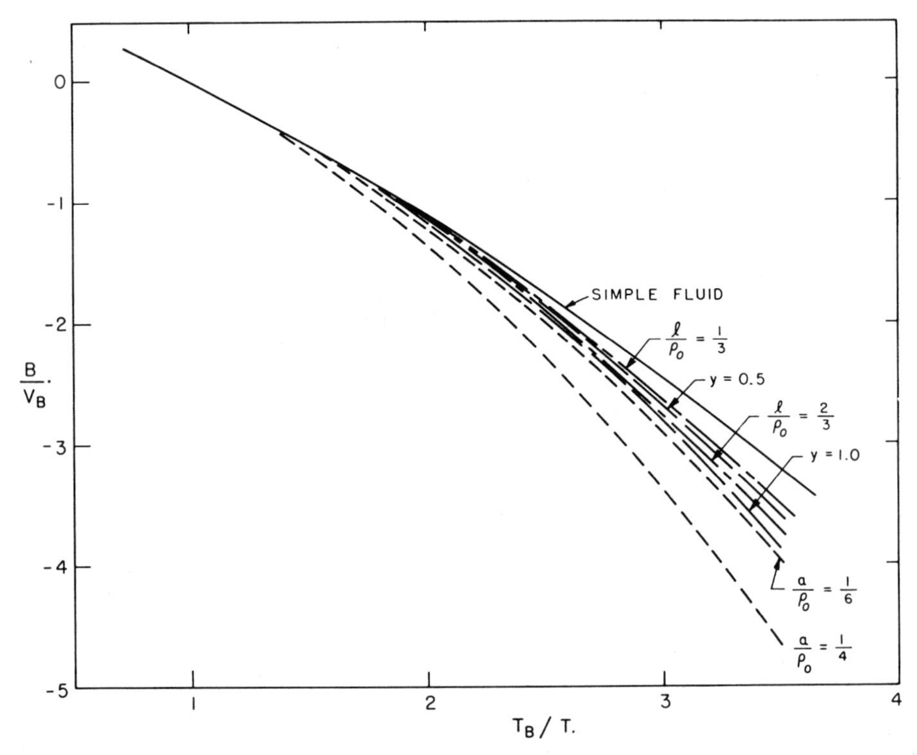

Figure 2. Reduced second virial coefficients for several models: solid curve, simple fluid; curves labeled by a/ρ_0, *spherical cores of radius* a; *curves labeled by* l/ρ_0, *linear cores of length* l; *curves labeled by* y, *molecules with dipoles*

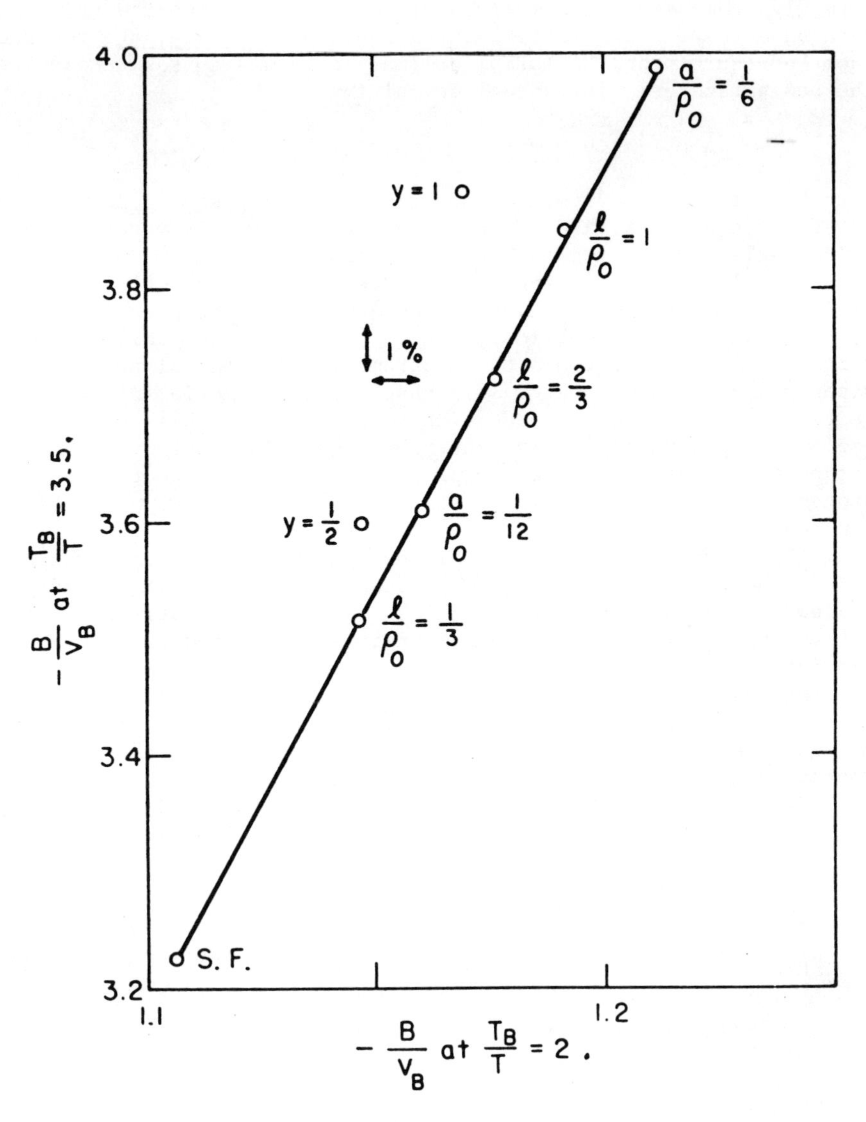

Figure 3. Check on family relationship of curves of Figure 2. Comparison of deviations from simple fluid at $(T_B/T) = 3.5$ *with that at* $(T_B/T) = 2.0$

widely measured at least near the normal boiling point. Thus both the availability of data and the accuracy of the data for the purpose strongly indicated a vapor pressure criterion. Since the critical data have to be known for a reduced equation of state, the reduced vapor pressure near the normal boiling point was an easy choice for the new parameter. The actual definition

$$\omega = -\log P_r - 1.000 \tag{2}$$

with P_r the reduced vapor pressure at $T_r = 0.700$ seemed convenient, but the actual determination of ω can be made from any vapor pressure value well-removed from the critical point.

Here I should note the work of Riedel (9) which was substantially simultaneous with mine but whose first paper preceded slightly. His work was purely empirical, but was excellent and fully complementary. He chose for his third parameter also the slope of the vapor pressure curve, but in his case the differential slope at the critical point. That seemed to me to be less reliable and accurate, empirically, although equivalent otherwise. Fortunately Riedel and I chose to emphasize different properties as our respective programs proceeded; hence the full area was covered more quickly with little duplication of effort.

Also I needed a name for this new parameter, and that was difficult. The term "acentric factor" was suggested by some friendly reviewer, possibly by a referee; I had made a less satisfactory choice initially. The conceptual basis is indicated in Figure 4. The intermolecular forces between complex molecules follow a simple expression in terms of the distances between the various portions of the molecule. Since these forces between non-central portions of the molecules must be considered, the term "acentric factor" seemed appropriate.

It is assumed that the compressibility factor and other properties can be expressed in power series in the acentric factor and that a linear expression will usually suffice.

$$\frac{pv}{RT} = z = z^{(0)} + \omega z^{(1)} + \ldots$$

$$z^{(0)} = z^{(0)}(T_r, P_r)$$

$$z^{(1)} = z^{(1)}(T_r, P_r)$$

The preference of P_r over V_r as the second independent variable is purely empirical; the critical pressure is much more accurately measurable than the critical volume.

The empirical effectiveness of this system was first tested with volumetric data as shown on Figure 5. Here pv/RT at a particular

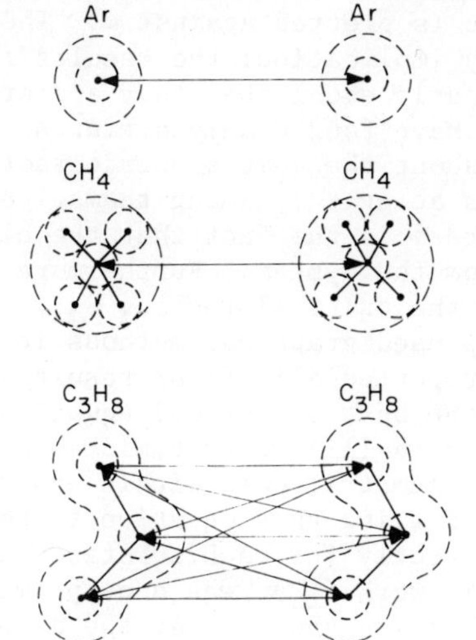

Figure 4. Intermolecular forces operate between the centers of regions of substantial electron density. These centers are the molecular centers for Ar and (approximately) for CH_4, but are best approximated by the separate CH_3 and CH_2 groups in C_3H_8—hence the name acentric factor for the forces arising from points other than molecular centers.

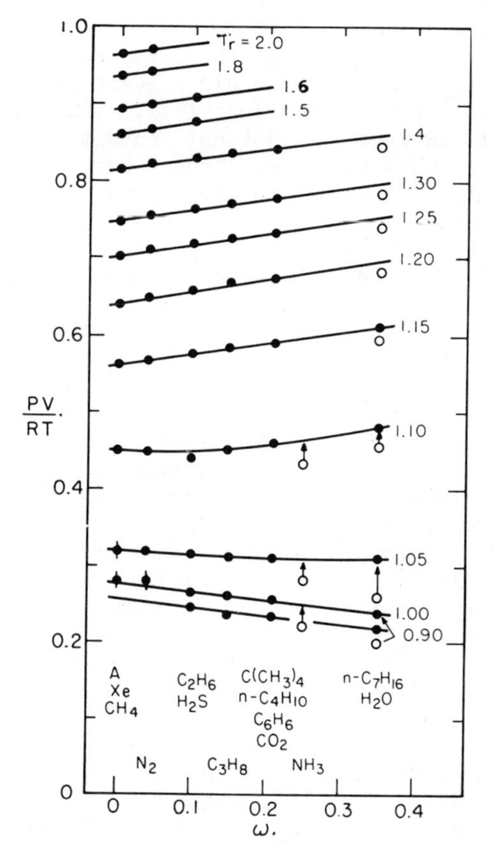

Figure 5. Compressibility factor as a function of acentric factor for reduced pressure 1.6 and reduced temperature shown for each line. Where several substances have approximately the same acentric factor, the individual points are indistinguishable except for n-C_7H_{16} (●) and H_2O (○).

reduced temperature and pressure is plotted against ω. The most important result appears only by implication; the results for $C(CH_4)_4$, n-C_4H_{10}, C_6H_6, and CO_2 are so nearly equal that they appear as single points on these plots. Here we have four widely different shapes of molecules which happen to have about the same acentric factor, and they follow corresponding states accurately among themselves.

Also to be noted from Figure 5 is the fact that the highly polar molecules NH_3 and H_2O depart from the system. Furthermore the dependence on ω is linear except for the critical region.

My immediate research group used graphical methods in dealing with the experimental data and reported all of our results in numerical tables (10). At that time the best analytical equation of state was that of Benedict, Webb and Rubin (11) which employed eight parameters and still failed to fit volumetric data within experimental accuracy. Bruce Sage suggested fitting this equation to the data for the normal paraffins both directly for each substance and within the acentric factor system. This work (12) was done primarily by J. B. Opfell at Cal Tech. The results showed that the acentric factor system was a great advance over the simple postulate of corresponding states, but the final agreement was inferior to that obtained by graphical and numerical methods.

Thus we continued with numerical methods for the fugacity, entropy, and enthalpy functions (13), although we did present an empirical equation for the second virial coefficient (14). This work was done by Bob Curl; he did an excellent job but found the almost interminable graphical work very tiresome. Thus I was pleased that the British Institution of Mechanical Engineers included Curl in the award of their Clayton Prize for this work. A fifth paper with Hultgren (15) treated mixtures on a pseudocritical basis, and a sixth with Danon (16) related Kihara core sizes to the acentric factor.

Naturally, I am very pleased to note that others have extended the accuracy and range of our tables and equations with consideration of more recent experimental results. Of particularly broad importance is the 1975 paper by Lee and Kessler (17) which presents both improved tables and analytical equations for all of the major functions including vapor pressures, volumetric properties, enthalpies, entropies, fugacities, and heat capacities. Their equation is an extension of that of Benedict, Webb, and Rubin now containing twelve parameters. They considered more recent experimental data as well as a number of papers which had already extended my earlier work in particular areas. I refer to their bibliography (17) for most of this more detailed work, but I do want to note the improved equation of Tsonopoulos (18) for the second virial coefficient. This equation deals also with effects of electrical polarity.

In addition to references cited by Lee and Kessler there is the work Lyckman, Eckert, and Prausnitz (19) dealing with liquid volumes; they found it necessary to use a quadratic expression in ω. Also Barner and Quinlan (20) treated mixtures at high temperatures and pressures, and Chueh and Prausnitz (21) treated the compressibility of liquids. Reid and Sherwood (22) give an extensive table including acentric factors as well as critical constants for many substances.

On the theoretical side, one great advance has been in the development of perturbation theories of a generalized van der Waals type. Here one assumes that the molecular distribution is determined primarily by repulsive forces which can be approximated by hard cores. Then both the softness of the cores and the atractive forces are treated by perturbation methods. Barker and Henderson (23) have recently reviewed theoretical advances including their own outstanding work. Rigby (24) applied these modern Van der Waals' methods to non-spherical molecules which represent one type of molecules with non-zero acentric factors. In a somewhat similar manner Beret and Prausnitz (25) developed equations applicable even to high polymers and related the initial departures from simple fluids to the acentric factor.

But in my view the most effective approach would concentrate first on globular molecules. These could be modeled by Kihara potentials with spherical cores or by other potentials allowing the well to be narrowed. The great advantage would be the retention of spherical symmetry and its theoretical simplicity. Rogers and Prausnitz (26) made an important beginning in this area with calculations based on Kihara models appropriate for argon, methane, and neopentane with excellent agreement for the properties studied. While they do not discuss these results in terms of the acentric factor, the transformation of spherical core radius to acentric factor is well established (16, 27), Rogers and Prausnitz were also able to treat mixtures very successfully although those calculations were burdensome even with modern computers. I believe further theoretical work using spherical models for globular molecules would be fruitful.

The move to an analytical equation by Lee and Kessler was undoubtedly a wise one in view of the marvelous capacity of modern computers to deal with complex equations. I would expect future work to yield still better equations.

There remains the question of the ultimate accuracy of the acentric factor concept. How accurately do molecules of different shapes but with the same acentric factor really follow corresponding states? Apparently this accuracy is within experimental error for most, if not all, present data. Thus the acentric factor system certainly meets engineering needs, and it is primarily a matter of scientific curiosity whether deviations are presently measurable.

It has been a pleasure to review these aspects of the "acentric factor" with you and I look forward to your discussion of recent advances in these and other areas.

Literature Cited

1. Kemp, J. D. and Pitzer, K. S., J. Chem. Phys., (1936) 4, 749; J. Am. Chem. Soc. (1937) 59, 276.

2. Pitzer, K. S., J. Chem. Phys., (1937) 5, 469, 473, 752; (1940) 8, 711; Chem. Rev. (1940) 27, 39.
3. Rossini, F. D., Pitzer, K. S., Arnett, R. L., Braun, R. M. and Pimentel, G. C., "Selected Values of the Physical and Thermodynamic Properties of Hydrocarbons and Related Compounds," Carnegie Press, Pittsburgh (1953).
4. Scott, D. W., J. Chem. Phys. (1974) 60, 3144.
5. Pitzer, K. S., J. Chem. Phys. (1939) 7, 583.
6. Rowlinson, J. S., Trans. Faraday Soc. (1954) 50, 647; "Liquids and Liquid Mixtures," 2nd ed. Chapter 8, Butterworth, London (1969).
7. Kihara, T., Rev. Mod. Phys. (1953) 25, 831 and papers there cited.
8. Pitzer, K. S., J. Am. Chem. Soc. (1955) 77, 3427.
9. Riedel, L., Chem. Ing. Tech. (1954) 26, 83, 259, 679; (1955) 27, 209, 475; (1956) 28, 557.
10. Pitzer, K. S., Lippman, D. Z., Curl, Jr., R. F., Huggins, C. M. and Petersen, D. E., J. Am. Chem. Soc. (1955) 77, 3433.
11. Benedict, M., Webb, G. B. and Rubin, L. C., J. Chem. Phys. (1940) 8, 334.
12. Opfell, J. B., Sage, B. H. and Pitzer, K. S., Ind. Eng. Chem. (1956) 48, 2069.
13. Curl, Jr., R. F., and Pitzer, K. S., Ind. Eng. Chem. (1958) 50, 265.
14. Pitzer, K. S. and Curl, Jr., R. F., J. Am. Chem. Soc., (1957) 79, 2369.
15. Pitzer, K. S. and Hultgren, G. O., J. Am. Chem. Soc. (1958) 80, 4793.
16. Danon, F. and Pitzer, K. S., J. Chem. Phys. (1962) 36, 425.
17. Lee, B. I. and Kesler, M. G., A.I.Ch.E. Journal (1975) 21, 510.
18. Tsonopoulos, C., A.I.Ch.E. Journal (1974) 20, 263.
19. Lyckman, E. W., Eckert, C. A. and Prausnitz, J. M., Chem. Engr. Sci. (1965) 20, 703.
20. Barner, H. E. and Quinlan, C. W., I. and E.C. Proc. Des. Dev. (1969) 8, 407.
21. Chueh, P. L. and Prausnitz, J. M., A.I.Ch.E. Journal (1969) 15, 471.
22. Reid, R. C. and Sherwood, T. K., "The Properties of Gases and Liquids," 2nd ed., McGraw-Hill Book Co., New York (1966).
23. Barker, J. A. and Henderson, D., Rev. Mod. Phys. (1976) 48, 587.
24. Rigby, M., J. Phys. Chem (1972) 76, 2014.
25. Beret, S. and Prausnitz, J. M., A.I.Ch.E. Journal (1975) 21, 1123.
26. Rogers, B. L. and Prausnitz, J. M., Trans. Faraday Soc. (1971) 67, 3474.
27. Tee, L. S., Gotoh, S., and Stewart, W. E., Ind. Eng. Chem. Fund. (1966) 5, 363.

2

State-of-the-Art Review of Phase Equilibria

J. M. PRAUSNITZ

University of California, Berkeley, Calif. 94720

I welcome the opportunity to discuss the state of the art for calculating phase equilibria in chemical engineering first, because I consider it a high honor to have been chosen for this important assignment and second, because it may give me a chance to influence the direction of future research in this field. When I mentioned these two reasons to one of my more candid coworkers, he said "What you really mean is, that you enjoy the opportunity to go on an ego trip and that you are glad to have an audience which you can subject to your prejudices."

While this restatement of my feelings is needlessly unkind, I must confess that it bears an element of truth. The assignment that Professor Sandler has given me--to review applied phase equilibrium in an hour or two--is totally impossible and it follows that in choosing material for this presentation, I must be highly selective. Since time is limited, I must omit many items which others, in exercising their judgment, might have included. At the outset, therefore, I want to apologize to all in the audience who may feel that some publications, notably their own, have received inadequate attention.

While I have tried to be objective and critical in my selection, it is human nature to give preference to that work with which one is most familiar and that, all too often, tends to be one's own. Nevertheless, I shall try to present as balanced a picture as I can. After more than 20 years, I have developed a certain point of view conditioned by my particular experience and I expect that it is pervasive in what I have to say. However, I want very much to assure this audience that I present my point of view without dogmatic intent; it is only a personal statement, a point of departure for what I hope will be vigorous discussion during the days ahead. My aim in attending this conference is the same as yours: at the end of the week I want to be a little wiser than I am now, at the beginning.

Thermodynamics: Not Magic but a Tool

All too often, when I talk with chemical engineers from industry who have little experience in thermodynamics, I obtain the impression that they look upon me as a medicine man, a magician who is supposed to incant obscure formulas and, in effect, produce something out of nothing. This audience knows better but nevertheless, we must remind ourselves that thermodynamics is not magic, that it is only a useful tool for efficient organization of knowledge. Thermodynamics alone never tells us the value of a desired equilibrium property; instead, it tells us how the desired equilibrium property is related to some other equilibrium property. Thus thermodynamics provides us with a time-saving bookkeeping system: we do not have to measure all the equilibrium properties; we measure only some and then we can calculate others. Thus, from an engineer's point of view, the main advantage of thermodynamics is that it reduces experimental effort: e.g., if we know how the Gibbs energy of mixing varies with temperature, we need not measure the enthalpy of mixing since we can calculate it using the Gibbs-Helmholtz equation, or, in a binary system, if we know how the activity coefficient of one component varies with composition, we can use the Gibbs-Duhem equation to calculate the other. We must keep reminding ourselves and others as to just what thermodynamics can and cannot do. False expectations often lead to costly disappointments.

While the limitations of classical thermodynamics are clear enough, the potentially vast possibilities opened by statistical thermodynamics are still far from realized. Just what modern physics can do for us will be discussed later in the week; for now, I just want to say that even at this early stage, simple molecular ideas can do much to stretch the range of application of thermodynamics. When thermodynamics is coupled with the molecular theory of matter, we can construct useful models; while these only roughly approximate true molecular behavior, they nevertheless enable us to interpolate and extrapolate with some confidence, thereby reducing further the experimental effort required for reliable results. When my nontechnical friends ask me what I, a molecular thermodynamicist do, I answer with a naive but essentially accurate analogy: I am a greedy tax collector. From the smallest possible capital, I try to extract the largest possible revenue.

Keeping in mind that thermodynamics is no more than an efficient tool for organizing knowledge toward useful ends, I find that, for phase-equalibrium work, thermodynamics provides us with two procedures, as shown in Figure 1. Our aim is to calculate fugacities and we can do so either using method (a), based entirely on an equation of state applicable to both phases α and β, or using method (b), which uses an equation of state only for calculating the vapor-phase fugacity and a completely different method, expressed by the activity coefficient, for calculating condensed-phase fugacities.

I now want to examine these two methods because they are the ones which have been used in essentially all applied phase-equilibrium work

FOR EVERY COMPONENT i, IN PHASES α AND β

$$f_i^{\alpha} = f_i^{\beta} \qquad f = \text{FUGACITY}$$

EITHER

(a) $$\ln f_i = \frac{1}{RT}\int_V^{\infty}\left[\left(\frac{\partial P}{\partial n_i}\right)_{T,V,n_j} - \frac{RT}{V}\right]dV - \ln \frac{V}{n_i RT}$$

n_i = MOLES OF i; V = TOTAL VOLUME

OR

(b) $$f_i^V = \phi_i y_i P \quad \text{AND} \quad f_i^L = \gamma_i x_i f_i^o$$

y,x = COMPOSITION; ^{o} = STANDARD STATE

ϕ = FUGACITY COEFFICIENT (FROM EQUATION OF STATE)

γ = ACTIVITY COEFFICIENT

Figure 1. Two thermodynamic methods for calculation of fluid-phase equilibria

METHOD	ADVANTAGES	DISADVANTAGES
(a)	1. NO STANDARD STATES. 2. P-V-T-X DATA ARE SUFFICIENT; IN PRINCIPLE, NO PHASE EQUILIBRIUM DATA NEEDED. 3. EASILY UTILIZES THEOREM OF CORRESPONDING STATES. 4. CAN BE APPLIED TO CRITICAL REGION.	1. NO REALLY GOOD EQUATION OF STATE AVAILABLE FOR ALL DENSITIES. 2. OFTEN VERY SENSITIVE TO MIXING RULES. 3. DIFFICULT TO APPLY TO POLAR COMPOUNDS, LARGE MOLECULES, OR ELECTROLYTES.
(b)	1. SIMPLE LIQUID-MIXTURE MODELS ARE OFTEN SATISFACTORY. 2. EFFECT OF TEMPERATURE IS PRIMARILY IN f^o, NOT γ. 3. APPLICABLE TO WIDE VARIETY OF MIXTURES, INCLUDING POLYMERS AND ELECTROLYTES.	1. NEED SEPARATE METHOD TO FIND $\overline{V}$. 2. CUMBERSOME FOR SUPER-CRITICAL COMPONENTS. 3. DIFFICULT TO APPLY IN CRITICAL REGION.

Figure 2. Brief comparison of methods (a) and (b)

When encountering a particular phase-equilibrium problem, the very first decision is to decide which of these methods is most suitable for the particular problem. It is therefore important to review the relative advantages and disadvantages of both methods; these are summarized in Figure 2.

The state of the art today is such that for mixtures of simple, or what Pitzer has called "normal" fluids, we can often calculate vapor-liquid equilibria, even at high pressures, with good success using some empirical equation of state. However, for mixtures including one or more strongly polar or hydrogen-bonding component, we must resort to the use of activity coefficients and standard-state fugacities.

As indicated in Figure 2, an equation of state for all fluid phases has many advantages because one very troublesome feature, viz. specifying a standard state, is avoided. This feature is troublesome because we frequently are concerned with multicomponent mixtures where at least one component is supercritical. In that event, the choice of a properly defined activity coefficient and standard state introduces formal difficulties which are often mathematically inconvenient and, for practical implementation, require parameters from experimental data that are only rarely available.

For liquid-phase mixtures, polar or nonpolar, including polymers and electrolytes, at low or moderate pressures, the activity coefficient provides the most convenient tool we have but our fundamental knowledge about it is sparse. Thermodynamics gives us little help; we have three well-known relations: first, the Gibbs-Duhem equation which relates the activity coefficient of one component in a solution to those of the others, second, the Gibbs-Helmholtz equation which relates the effect of temperature on the activity coefficient to the enthalpy of mixing and finally, an equation which relates the partial molar volume to the effect of pressure on the activity coefficient. These illustrate what I said earlier, viz. that classical thermodynamics is little more than an efficient organization of knowledge, relating some equilibrium properties to others, thereby reducing experimental work. But the practical applications of these classical thermodynamic relations for activity coefficients are limited, in contrast to the more powerful thermodynamic relations which enable us to calculate fugacities using only volumetric properties. From a strictly thermodynamic point of view, using an equation of state is more efficient than using activity coefficients. If we have an equation of state applicable to all phases of interest, we can calculate not only the fugacities from volumetric data but also all the other configurational properties such as the enthalpy, entropy and volume change on mixing.

Our inability to use equations of state for many practical situations follows from our inadequate understanding of fluid structure and intermolecular forces. Only for simple situations do we have theoretical information on structure and forces for establishing an equation of state with a theoretical basis and only for the more common fluids do we have sufficient experimental information to establish reliable empirical equations of state. Thanks to corresponding states, we can extend the available empirical basis to a much wider class of fluids but again, we are limited here because corresponding states cannot easily be extended to polar or hydrogen-bonding materials. Our biggest bottleneck is that we have not been able to establish a useful statistical mechanical treatment for such fluids nor even to characterize the intermolecular forces between their molecules. At liquid-like densities, the dipole moment is not good enough and the strength of a hydrogen bond depends not only on particular conditions like density and temperature but, what is worse, also on the method used to measure it. Later in the week, when we discuss the contribution of theory, we shall hopefully return to some of these problems.

Equations of State for Both Fluid Phases

Let us now see what kind of practical phase-equilibrium problems we can handle using nothing beyond one of the many currently available equations of state. For relatively simple mixtures, e.g., those found in processing of natural gas and light petroleum fractions, we do well with one of the many modifications of the Benedict-Webb-Rubin equation; in its original version, this equation had eight constants for each fluid but in later versions this number had increased, sometimes considerably so. To illustrate, Figure 3 shows calculated and observed K factors for methane in heptane at two temperatures. In these calculations, Orye (1) followed the usual procedure; he assumed a one-fluid theory, i.e., he assumed that the equation of state of the mixture is the same as that of a pure fluid except that the characteristic constants depend on composition according to some more or less arbitrary relations known as mixing rules. Experience has repeatedly shown that at least one of these mixing rules must contain an adjustable binary constant; in this case, that constant is M_{ij} which was found by fitting to the binary data. Unfortunately, the calculated results are often highly sensitive to the mixing rules and to the value of the adjustable parameter. In this case Orye found what many others have also found, viz., that the adjustable binary parameter is more-or-less invariant with density and composition but often depends on temperature. Another example, also from Orye, is given in Figure 4 for the system methane-carbon dioxide at -65°F. The continuous line through the diamonds is not calculated but connects the experimental points of Donnelly and Katz; the calculated lines are dashed and the circles and triangles indicate particular calculations, not data. First we note that the value of M_{ij} has a strong effect, especially on the liquidus curve; a ten percent change in M_{ij} produces a large error in the bubble pressure. When M_{ij} is adjusted empirically to 1.8, much better results are achieved but note that Orye reports no calculations in the critical region. There are two good reasons for this: first, all classical analytical equations tend to be poor in the

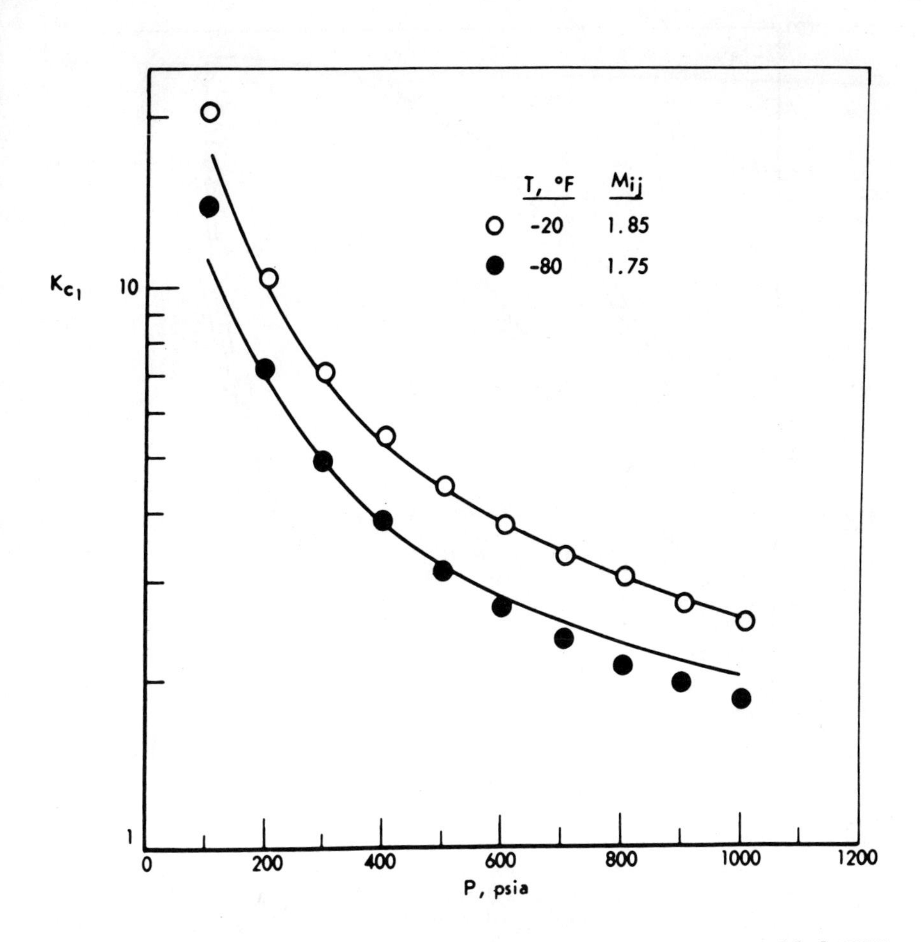

Figure 3. Methane–n-heptane (Orye, 1969) ● ○ Kohn (1961); —— Modified BWR equation

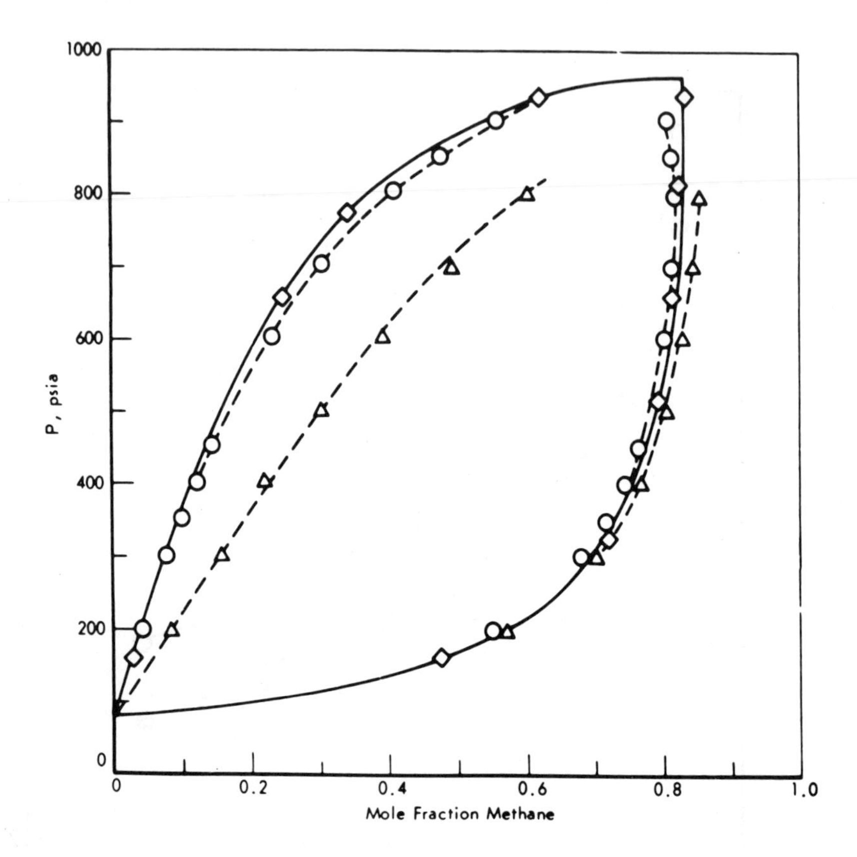

Figure 4. Methane–carbon dioxide (Orye, 1969). Temp., −65°F; ◇ Donnelly and Katz (1954); ○ modified BWR equation, $M_{11} = 1.8$; △ modified BWR equation, original mixing rule, $M_{11} = 2.0$.

critical region and second, computational problems are often severe because convergence is hard to achieve.

A similar situation is shown in Figure 5, taken from Starling and Han (2), who used on 11-constant version of the BWR equation. Again, note that an adjustable binary constant k_{ij} is required. Also, note that, contrary to usual practice, the lines represent experiment and the points represent calculations, suggesting problems in the critical region.

It is evident that the more constants in an equation of state, the more flexibility in fitting experimental data but it is also clear that to obtain more constants, one requires more experimental information. For example, a twenty-constant equation of state, essentially an extension of the BWR equation, was proposed by Bender (3) who applied it to oxygen, argon, nitrogen, and a few light hydrocarbons. For these fluids, Bender is able to obtain a highly accurate representation of experimental data over a wide density range. To illustrate one unusually fine feature of Bender's equation, Figure 6 shows the residual heat capacity of propylene for several temperatures near the critical temperature, 365°K. This is a very sensitive test and Bender's equation does a remarkable job. Bender has also applied his equation to mixtures of argon, nitrogen, and oxygen, useful for design of air-separation plants. For each binary mixture, Bender requires 3 binary parameters. With all these constants and a large computer program, Bender can calculate not only accurate vapor-liquid equilibria but also heats of mixing as shown in Figure 7. The heats of mixing here are very small and agreement between calculation and experiment is extraordinary.

However, it is clear that calculations of this sort are restricted to those few systems where the molecules are simple and small, where we have no significant polarity, hydrogen bonding or other specific "chemical forces" and, unfortunately, to those cases where we have large quantities of experimental data for both pure fluids and for binary mixtures. In the process industries we rarely meet all these necessary conditions.

If our accuracy requirements are not extremely large, we can often obtain good approximations using calculations based on a simple equation of state, similar in principle to the Van der Waals equation. The simplest successful variation of Van der Waals' equation is that by Redlich and Kwong, proposed in 1949. That equation, in turn, is to applied thermodynamics what Helen of Troy has been to literature; you recall that it was the beautiful Helen who inspired the line "...the face that launched a thousand ships." Ten years ago the Beatles turned on an entire generation of teenagers and inspired countless variations and extensions; similarly, starting about ten years ago, the Redlich-Kwong equation initiated an epoch of imitation unequalled in the history of applied thermodynamics. The number of modified RK equations is probably close to a hundred by now and, since I am amongst friends, I must confess to having constructed a few myself. A few years ago, there was an article in Chemical Engineering Science devoted exclusively to variations on

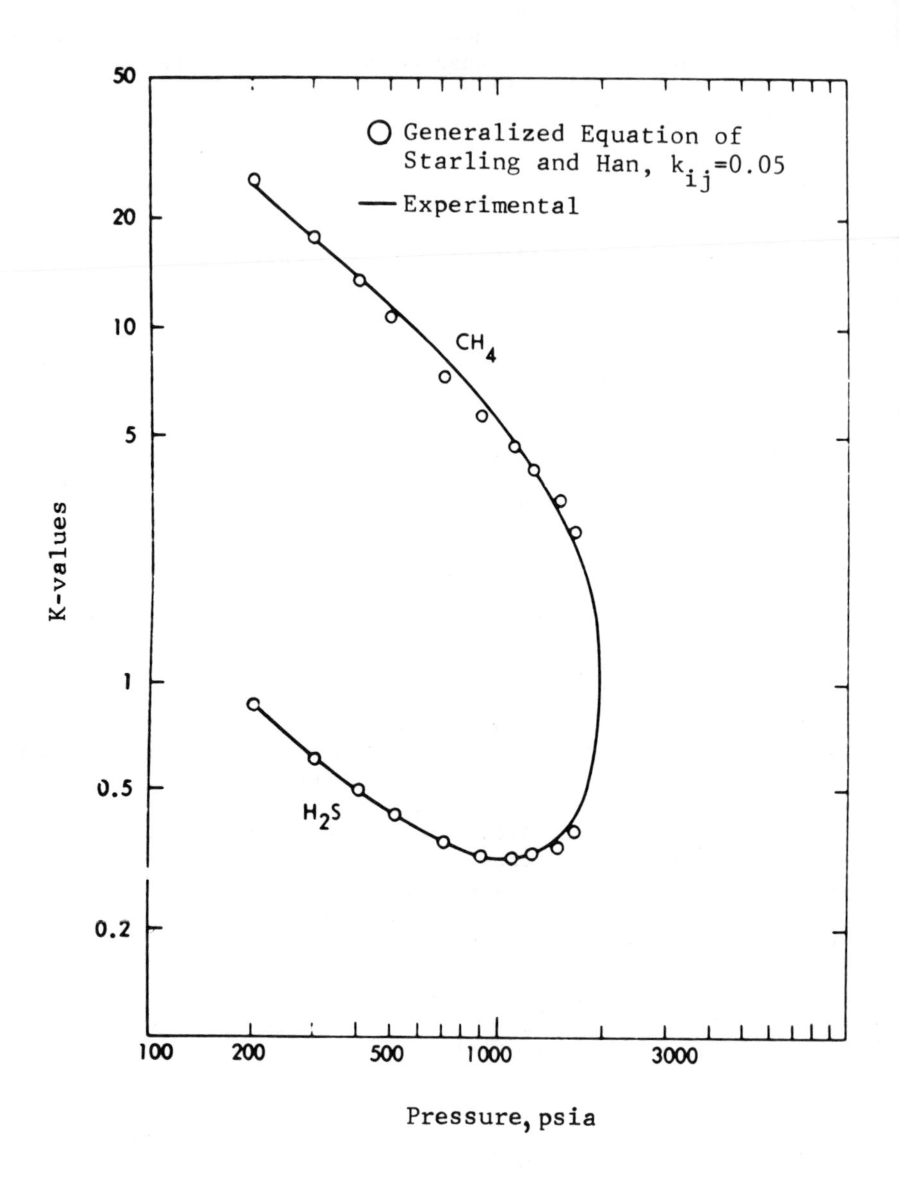

Figure 5. Predicted and experimental K-*values for the methane–hydrogen sulfide system at 40°F*

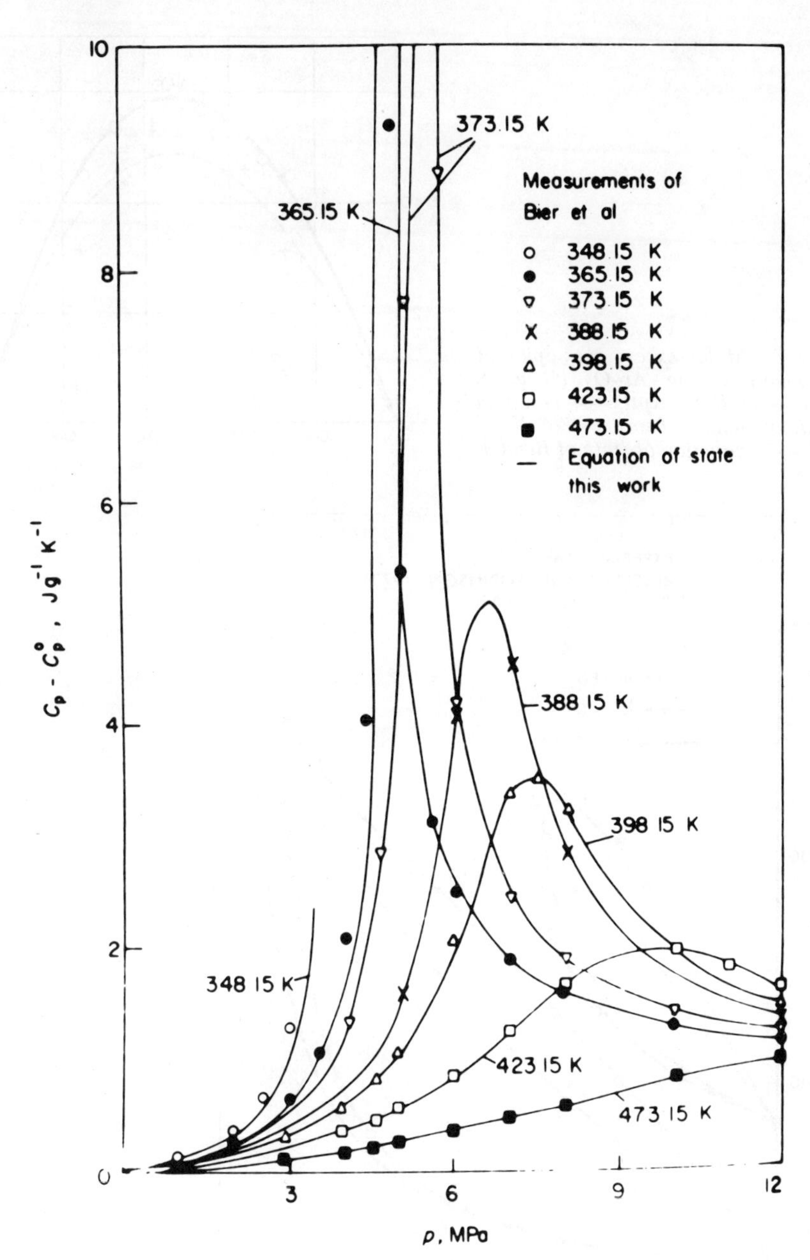

Figure 6. Comparison of the residual isobaric heat capacities of propylene of Bier et al. with those predicted by the equation of state of Bender

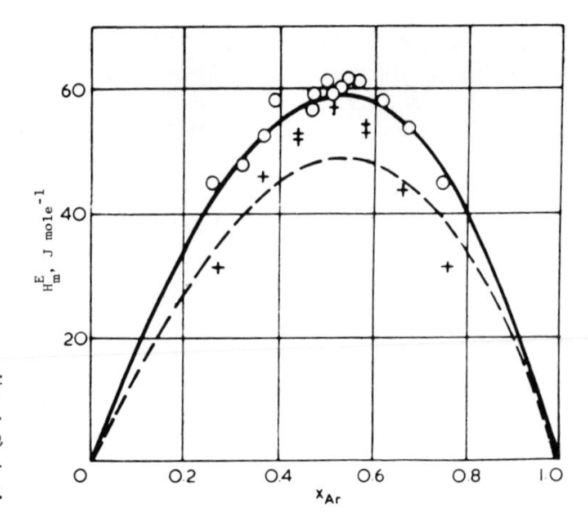

Figure 7. Molar excess enthalpies of the binary system Ar–O_2 (Bender). Temp. = 84°K: ○ exptl, —— equation of state of Bender; Temp. = 86° K: + exptl, – – – equation of state of Bender.

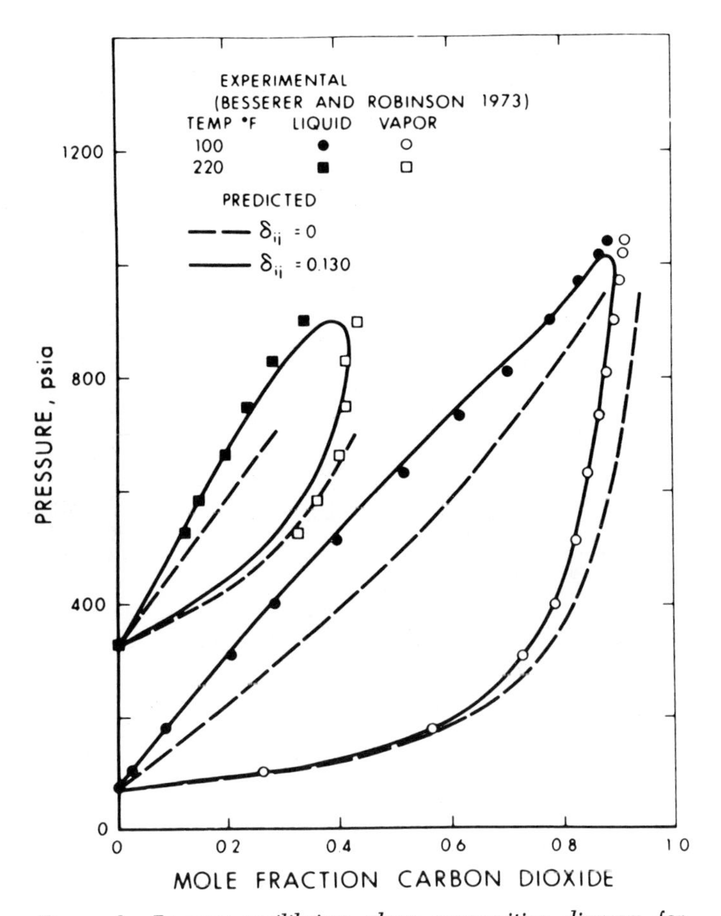

Figure 8. Pressure-equilibrium phase composition diagram for isobutane–carbon dioxide system, calculations using Peng-Robinson equation of state

the RK equation but it is now hopelessly out of date and even then, between the time the paper was written and the time it was published, seven new variations had appeared. (I get much of this information directly from Otto Redlich, who keeps a close eye on "children" of his 1949 article. Incidentally, I am happy to report that Otto, aged 80, is well, active and very pleased about the recent publication of his thermodynamics book by Elsevier. Whenever Otto has a need to feel young again, he talks with the incredible Joel Hildebrand who, at 95, is hale, hearty, in good humor and busy writing a monograph on transport properties in liquids.)

Perhaps the most successful variation on the RK equation is that proposed by Soave (4) who expresses the RK constant a by an empirical function of reduced temperature and acentric factor. This empirical function was determined from vapor-pressure data for paraffins and therefore, when Soave's equation is used with reasonable mixing rules and one adjustable binary parameter, it gives good K factors for typical light-hydrogen mixtures; however, it predicts poor liquid densities. This illustrates a point known to all workers in the equation-of-state field; it is not difficult to represent any one thermodynamic property but it is difficult, with one equation of state, to represent them all.

A comparatively recent variation on the RK equation was proposed by Peng and Robinson (5); it is similar to Soave's equation but appears to have better behavior in the critical region; an example is given in Figure 8 for the isobutane-carbon dioxide system. In this case the critical region is predicted well and the adjustable binary parameter is independent of temperature in the region 100 to 220°F.

Calculating phase equilibria from volumetric data does not necessarily require an analytical equation of state. The volumetric data can be stored in tabular or analytical form for an arbitrarily-chosen reference substance and then, using corresponding states, these data can be used to predict properties of other fluids, including mixtures. This procedure, often called the pseudo-critical method or, in a more elegant form, the theory of conformal solutions, has been applied by numerous authors. Here time permits me to call attention to only one example, a particularly useful one, initiated by Rowlinson and Mollerup and extensively developed by Mollerup in recent years (6). Using Goodwin's excellent experimental data for methane as a reference, Mollerup calculates with high accuracy thermodynamic properties of mixtures encountered in the natural-gas industry. To do so, he uses the old Van der Waals mixing rules but he pays very close attention to the all-important binary constants. Figure 9 shows excellent agreement between calculated and experimental results for the system methane-ethane from 130 to 200°K, using only one temperature-independent binary constant. Even more impressive is the excellent representation for carbon monoxide-methane shown in Figure 10 where the critical region is reproduced almost within experimental error. Finally, Figure 11 shows that the corresponding-states method also gives excellent enthalpies of

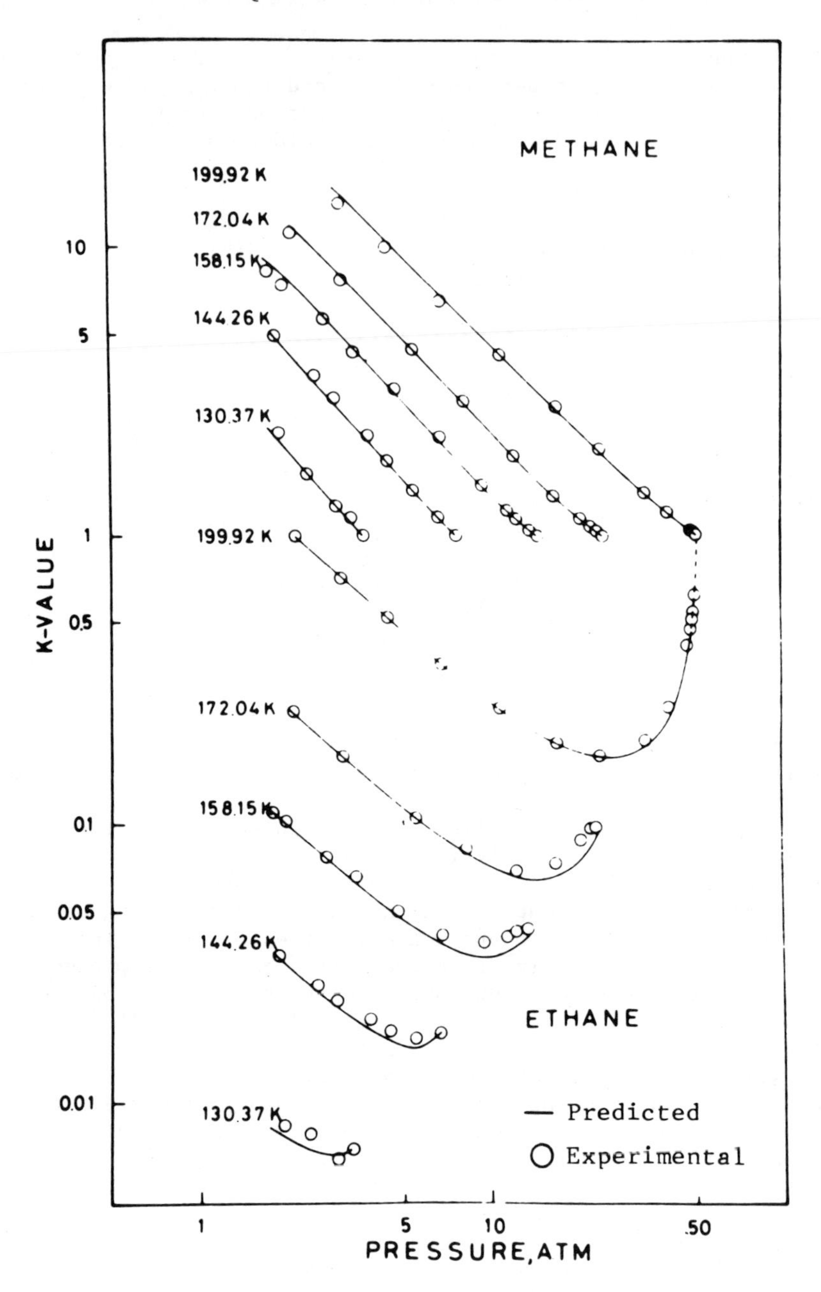

Figure 9. K-values vs. pressure for methane–ethane mixtures; corresponding-states method of Mollerup and Rowlinson

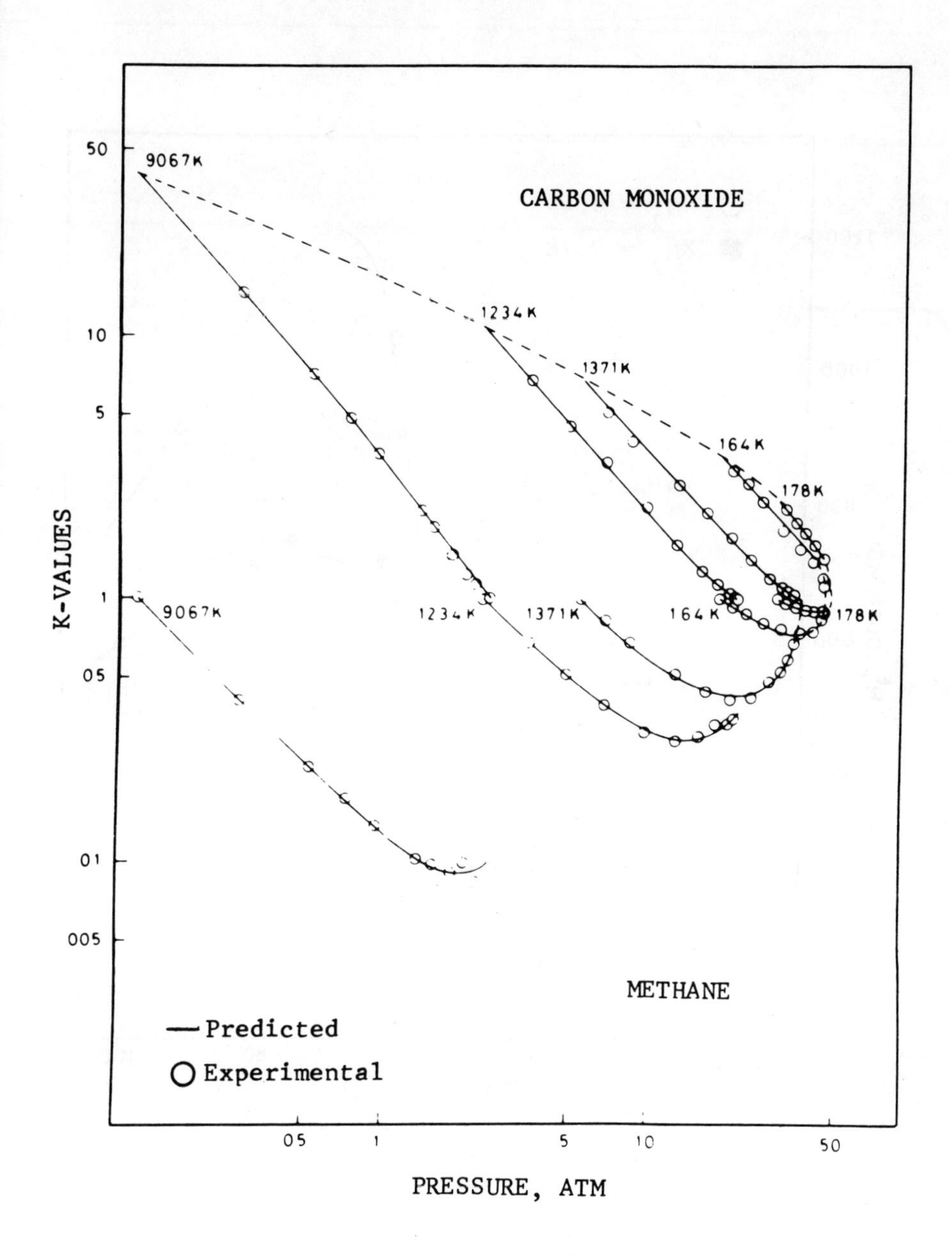

Figure 10. K-values vs. pressure for methane–carbon monoxide mixtures (Mollerup, 1975)

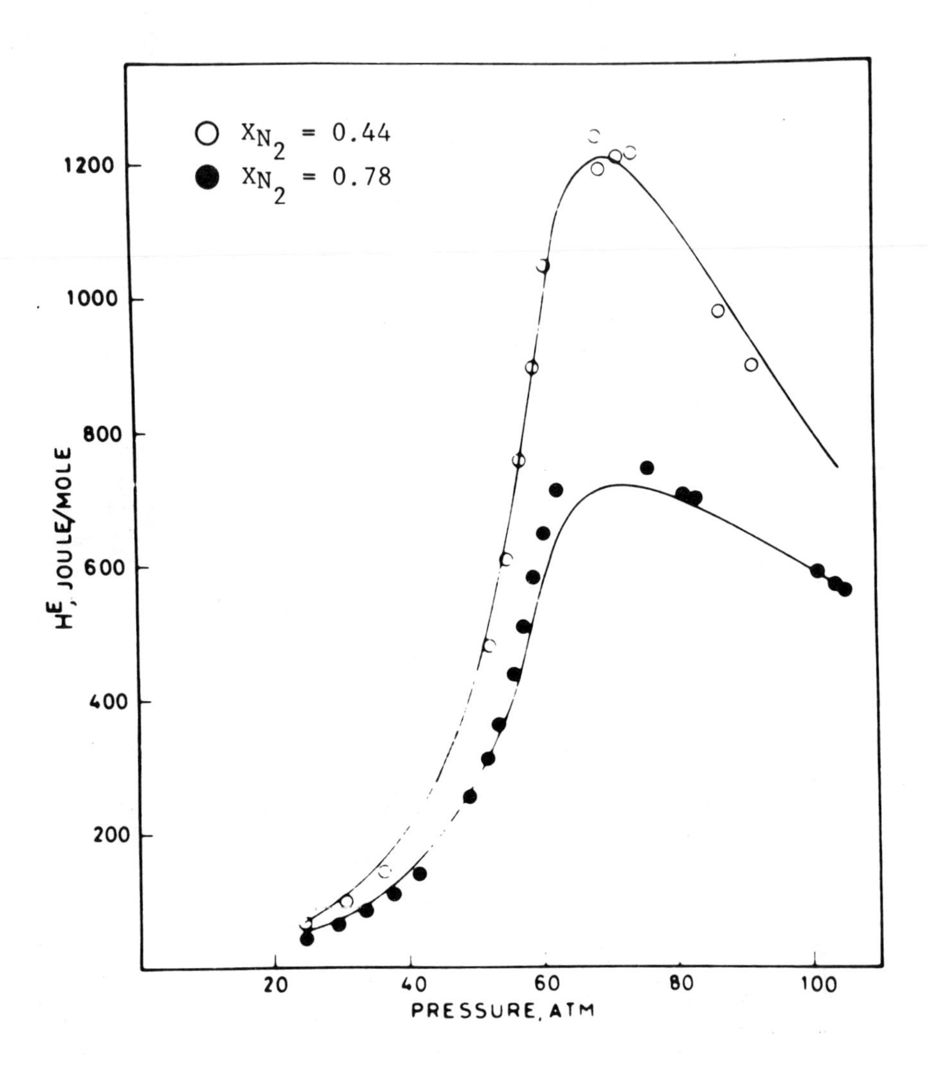

Figure 11. Excess enthalpy of methane–nitrogen mixtures at 201.2°K (Mollerup, 1975)

mixing for gaseous mixtures at high pressures, correctly reproducing the observed maxima when the excess enthalpy is plotted against pressure.

These few illustrations should be sufficient to outline our present position with respect to phase equilibrium calculations using an equation of state for both phases. I have earlier pointed out some of the advantages of this type of calculation but I want now to add one more: if we can construct an equation of state applicable to normal fluids and their binary mixtures, then we need not worry about how to calculate equilibria in ternary (or higher) mixtures. For mixtures of normal fluids, pure-component parameters and binary parameters are almost always sufficient for calculating equilibria in multicomponent mixtures. For multicomponent mixtures of normal fluids, the one-fluid theory is usually satisfactory using only pure-component and binary constants. This happy fact is of tremendous importance in chemical technology where multicomponent mixtures are much more common than binaries. Predicting multicomponent equilibria using only pure-component and binary data is perhaps one of the greatest triumphs of applied thermodynamics.

Having praised the uses of equations of state, I must also point out their contemporary limitations which follow from our inability to write sensible equations of state for molecules that are very large or very polar, or both. That is where the frontier lies. I see little point in pursuing further the obsession of modifying the Redlich-Kwong equation. We must introduce some new physics into our basic notions of how to construct an equation of state and there we must rely on suggestions supplied by theoretical physicists and chemists. Unfortunately most of these are "argon people" although, I am happy to say, in the last few years a few brave theorists have started to tackle nitrogen. Some computer-type theorists have spent a lot of time on water and on proteins but these highly complicated studies are still far removed from engineering applications. Nevertheless, there are some new theoretical ideas which could be used in formulating new equations of state suitable for those fluids that cannot now be described by the usual equations of state. Not this morning, but perhaps later in this conference, I hope to have an opportunity to say a few words about that.

Vapor-Phase Fugacity Coefficients

I now turn to what I have earlier called Method (b), that is, fugacity coefficients for the vapor phase only and activity coefficients for all condensed phases. Method (b) is used whenever we deal with mixtures containing molecules that are large or polar or hydrogen-bonded or else when all components are subcritical and the pressure is low.

At modest vapor densities, our most useful tool for vapor-phase fugacity coefficients is the virial equation of state truncated after the second term. For real fluids, much is known about second virial

$$\frac{BP_c}{RT_c} = f^{(o)}_{(T_R)} + \omega f^{(1)}_{(T_R)} + f^{(2)}_{(T_R)}$$

B = SECOND VIRIAL COEFFICIENT; $T_R = T/T_c$

P_c = CRITICAL PRESSURE; T_c = CRITICAL TEMPERATURE

ω = ACENTRIC FACTOR

$f^{(o)}$ AND $f^{(1)}$ ARE KNOWN FUNCTIONS SIMILAR TO THOSE FIRST PROPOSED BY PITZER AND CURL. TSONOPOULOS PROPOSES

$$f^{(2)} = a\,T_R^{-6} - b\,T_R^{-8}$$

FOR NONPOLAR FLUIDS $a = b = 0$.

FOR POLAR (NONHYDROGEN-BONDED) FLUIDS $a \neq 0$ BUT $b = 0$.

Figure 12. Correlation of second virial coefficients (Tsonopoulos)

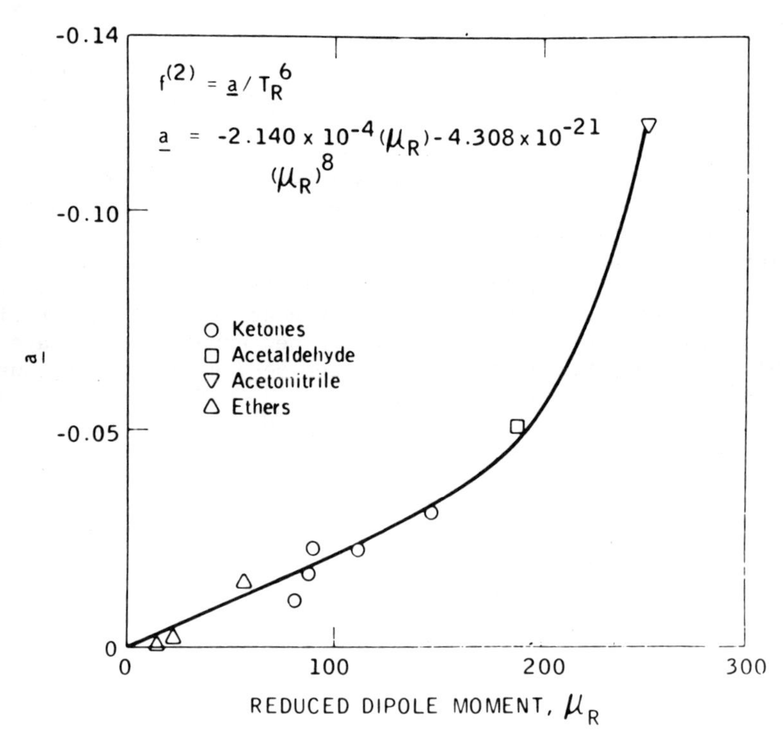

Figure 13. Dependence of a *on reduced dipole moment for nonhydrogen bonding compounds (Tsonopoulos, 1974)*

coefficients; little is known about third virial coefficients and nearly nothing is known about higher virial coefficients. Therefore, application is limited to moderate densities, typically densities up to about 1/2 the critical. There are two major advantages of the theoretically-derived virial equation: first, the virial coefficients can be quantitatively related to the intermolecular forces and second, extension to mixtures requires no additional assumptions. For engineering, the first advantage is important because it enables us to interpret, correlate and meaningfully extrapolate limited virial-coefficient data, and the second is important because we do not have to guess at arbitrary mixing rules for expressing the composition dependence of the virial coefficients.

Any standard thermodynamics text tells us how to calculate fugacity coefficients from the virial equation of state. The most important problem is to estimate the virial coefficients and here we can utilize an extended form of corresponding states, illustrated by the correlation of Tsonopoulos (7) shown in Figure 12. The first term on the right holds for simple fluids; the second term corrects for acentricity and the third term corrects for polarity and hydrogen bonding. The constants a and b cannot be completely generalized but good estimates are often possible by observing trends within chemical families. Figure 13 shows results for constant a plotted against a dimensionless dipole moment; since polarity increases attractive forces, we find, as shown, that constant a becomes more negative as the reduced dipole moment rises, giving a more negative second virial coefficient.

Figure 14 gives some results for constant b for alcohols, again plotted against reduced dipole moment. Note that the position of the OH radical has a noticeable effect. For these fluids constant a is slightly positive because the hydrogen-bonding nature expressed by constant b dominates, especially at lower temperatures.

To estimate cross virial coefficients B_{12}, one must make some assumptions about the intermolecular forces between molecules 1 and 2 and then suitably average the molecular parameters appearing in the correlation. Only for simple cases can any general rules be used; whenever we have polar components, we must look carefully at the molecular structure and use judgment which, ultimately, is based on experience.

The virial equation is useful for many cases but, when there is strong association in the vapor phase, the theoretical basis of the virial equation is not valid and we must resort to what is commonly called a "chemical treatment", utilizing a chemical equilibrium constant for dimerization. Dimerization in the vapor phase is especially important for organic acids and even at low pressures, the vapor-phase fugacity coefficients of mixtures containing one (or more) organic acid are significantly removed from unity.

Figure 15 shows some results based on the correlation of Hayden and O'Connell (8) calculated by Tom Anderson for the system propionic acid-methyl isobutyl ketone at 1 atm along the vapor-liquid saturation curve. When the mole fraction of acid is very low, the fugacity coefficients ar

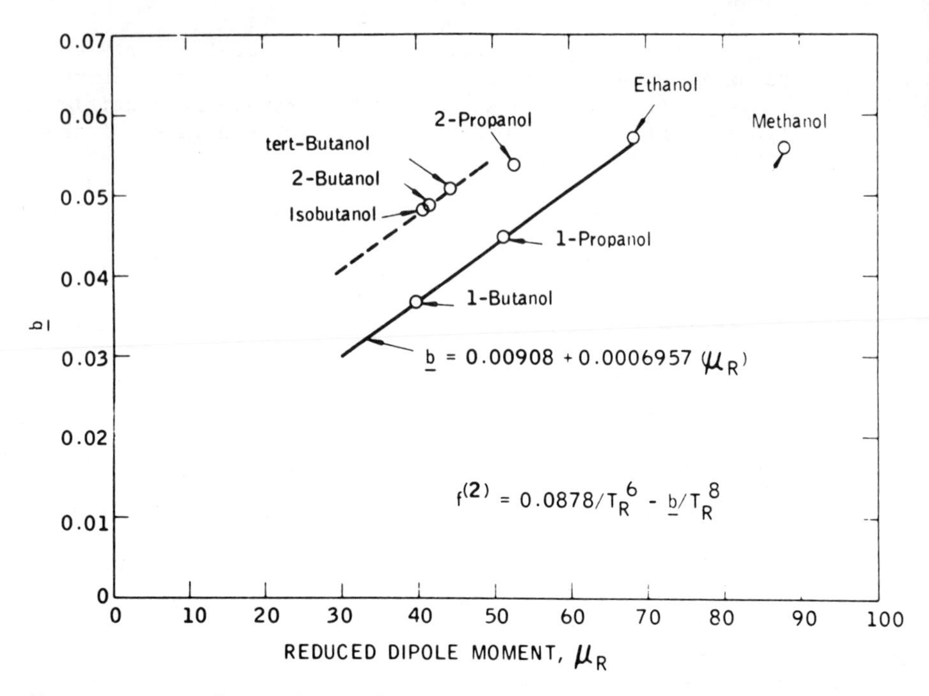

Figure 14. Dependence of b *on reduced dipole moment for alcohols (Tsonopoulos, 1974)*

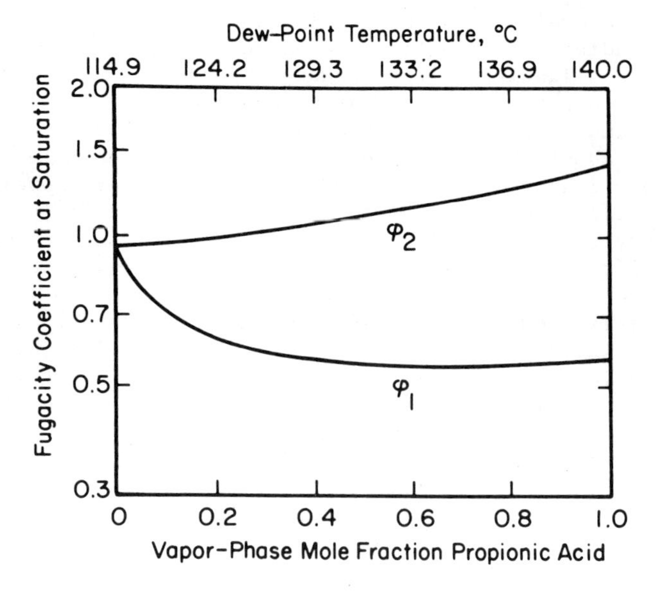

Figure 15. Fugacity coefficients for saturated propionic acid (1)—methyl isobutyl ketone (2) at 1 atm

near unity because dimerization between acid molecules is negligible. As the mole fraction of acid rises, dimerization becomes increasingly likely and therefore, on the right side of the diagram, the fugacity coefficients of both components are well removed from unity even though the temperature is reasonably high (140C) and the pressure is only 1 atm.

At high pressures, where the virial equation is no longer useful, empirical equations must be used to calculate fugacity coefficients. However, contrary to Method (a), the equation of state now need not hold for <u>both</u> the vapor phase and the liquid phase; validity in the vapor phase is sufficient.

To illustrate, I now show some results using an equation developed by de Santis and Breedveld (9) for gases at high pressures containing water as one of the components. As indicated in Figure 16, the equation is like that of Redlich-Kwong except that the attractive-force constant a is divided into a polar and a nonpolar component. For mixtures of water with nonpolar components, the cross-coefficient a_{12} is found from the geometric-mean assumption, but in this assumption only the nonpolar part of constant a for water is used because the nonaqueous component is nonpolar. Figure 17 shows that the modified RK equation gives good fugacity coefficients for aqueous water but this is hardly surprising since the constants a and b were determined from steam-table data. More gratifying are the results shown in Figure 18 which show that calculated volumetric properties at high pressures are in excellent agreement with experiment for gaseous mixtures of water and argon.

The equation of de Santis and Breedveld has recently been applied by Heidemann to the problem of wet-air oxidation. When vapor-phase fugacity coefficients are calculated from this equation of state, and liquid-phase fugacities are calculated from the properties of pure water corrected for solubility of gases in the water, it is possible to calculate the saturated water content and other equilibrium properties of combustion gases. Figure 19 shows the saturated water content in nitrogen and Figure 20 shows how that water content is enhanced when CO_2 is present in the gas phase; results are shown for two molar compositions (dry basis): 20% CO_2, 80% N_2 and 13% CO_2, 87% N_2. Especially at moderate temperatures, the pressure of CO_2 substantially raises the saturated water content. Figure 21 shows enthalpy calculations, again based on the equation of state of de Santis and Breedvelt, useful for designing a wet-air oxidation process.

In vapor-liquid equilibria according to Method (b), fugacity coefficients constitute only half the story, usually (but not always!) the less important half, while in liquid-liquid equilibria fugacity coefficients play no role at all. We now must turn our attention to the last and in some respects the most difficult topic, viz., the activity coefficient.

$$P = \frac{RT}{(v-b)} - \frac{a(T)}{T^{1/2}v(v+b)}$$

$$b_{(WATER)} = 14.6 \text{ cm}^3/\text{mole}$$

$$a(T) = \underset{(NONPOLAR)}{a^{(o)}} + \underset{(POLAR)}{a^{(1)}(T)}$$ (TABULATED VALUES OBTAINED FROM STEAM TABLES.)

FOR BINARY MIXTURES CONTAINING WATER(1) AND NONPOLAR GAS(2)

$$b = y_1b_1 + y_2b_2$$

$$a = y_1^2a_1 + y_2^2a_2 + 2y_1y_2a_{12}$$

$$a_{12} = (a_1^{(o)} a_2)^{1/2}$$

TO FIND $a_1^{(o)}$, USE B_{12} (SECOND VIRIAL COEFFICIENT) DATA FOR MIXTURES OF WATER WITH N_2, Ar, CH_4, ETC.

Figure 16. Vapor-phase equation of state for mixtures containing water (de Santis and Breedveld)

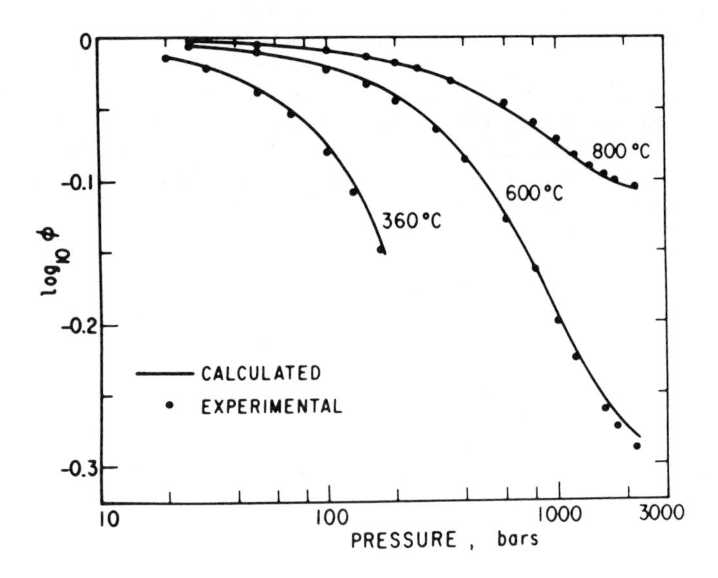

Figure 17. Fugacity coefficients for gaseous water. Calculations using equation of state proposed by de Santis et al.

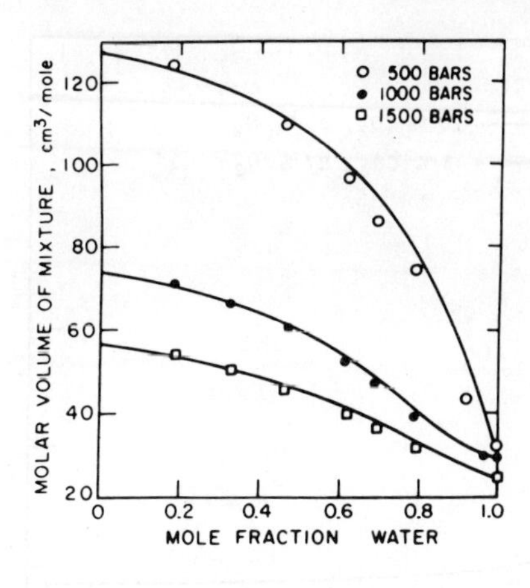

Figure 18. Molar volumes for mixtures of water and argon at 400°C; lines calculated with equation of de Santis et al.; data of Lentz and Franck

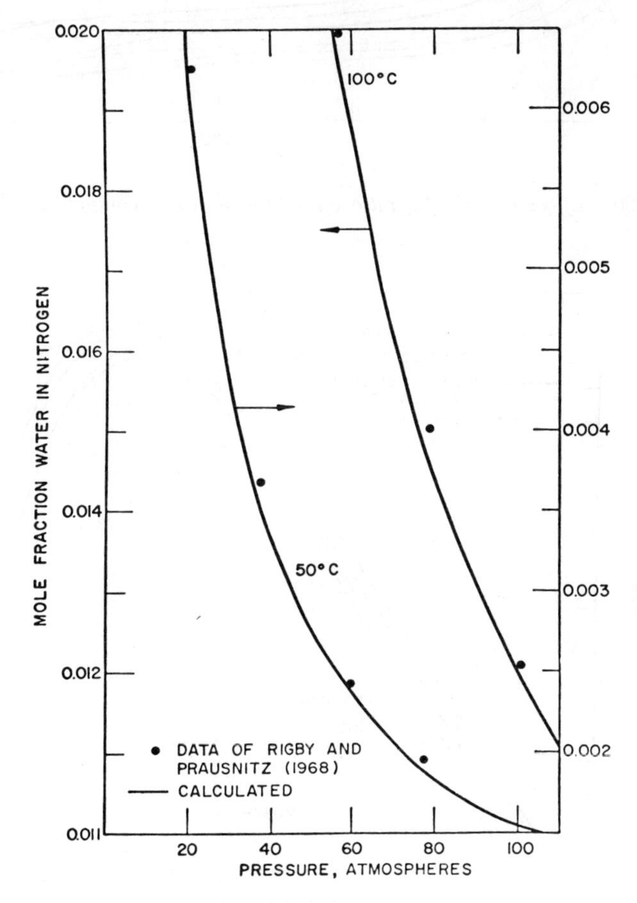

Figure 19. Water content of nitrogen; comparison with experiment (Heidemann)

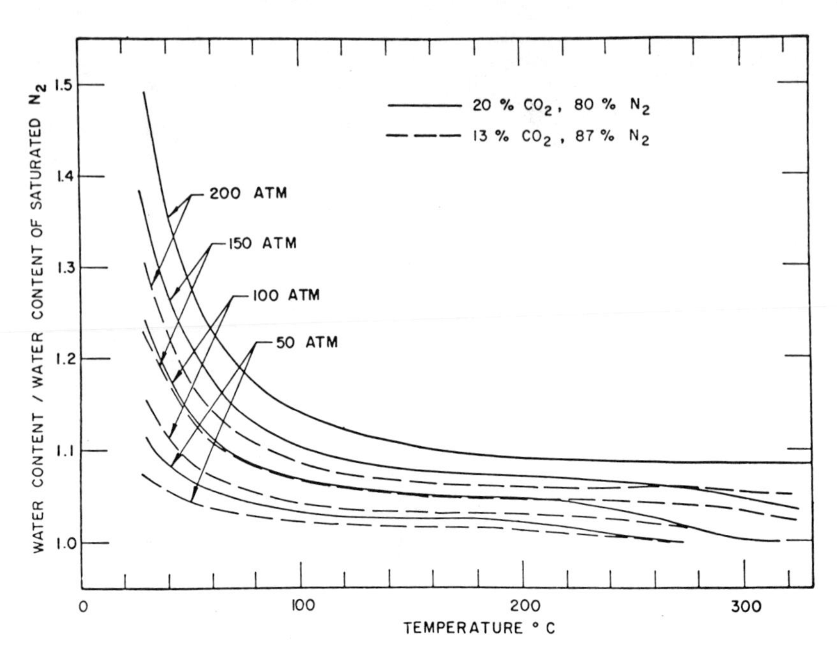

Figure 20. Effect of CO_2/N_2 ratio on saturated water content (Heidemann)

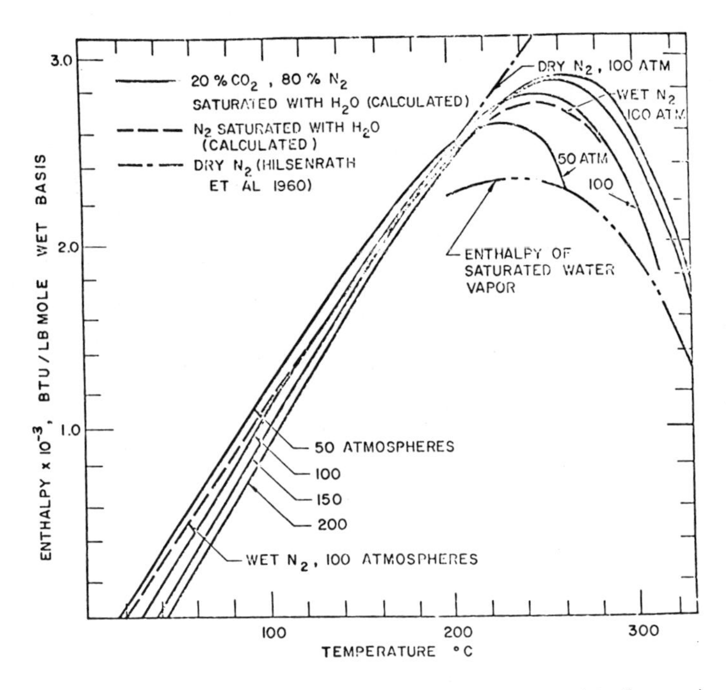

Figure 21. Enthalpy of saturated combustion gases (Heidemann)

Liquid-Phase Activity Coefficients

In principle, everybody knows that an activity coefficient has no significance unless there is a clear definition of the standard state to which it refers. In practice, however, there is all too often a tendency to neglect precise specification of the standard state and in some cases failure to give this exact specification can lead to serious difficulties. This problem is especially important when we consider supercritical components or electrolytes in liquid mixtures and, a little later, I shall have a few comments on that situation. But for now, let us consider mixtures of typical nonelectrolyte liquids at a temperature where every component can exist as a pure liquid. In that event, the standard-state fugacity is the fugacity of the pure liquid at system temperature and pressure and that fugacity is determined primarily by the pure liquid's vapor pressure.

Figure 22 reviews some well-known relations between activity coefficients and excess functions. All this is strictly classical thermodynamics and the entire aim here is that of classical thermodynamics; viz., to organize our knowledge of equilibrium properties in an efficient way so that, by relating various quantities to one another, we can minimize the amount of experimental effort required for engineering design. There are three noteworthy features in Figure 22:

1. The excess functions used here are in excess of those which apply to a particular kind of ideal solution viz. that (essentially) given by Raoult's law. This choice of ideality is arbitrary and for some situations a different definition of ideal solution may be more suitable. Further, choosing (essentially) Raoult's law as our definition of an ideal solution, we are naturally led to the use of mole fraction x as our choice of composition variable. That is not necessarily the best choice and there are several cases (notably, polymer solutions and solutions of electrolytes) where other measures of composition are much more convenient.

2. Equation (2) can be derived from Equation (1) only if we use the Gibbs-Duhem equation. Therefore, if we organize our experimental information according to the scheme suggested by Figure 22, we assure that the final results obey at least a certain degree of thermodynamic consistency.

3. The excess Gibbs energy g^E is a combination of two terms as shown in Equation (3). When we try to construct models for g^E, we often do so directly but, if we want our model to have physical significance we should instead make models for h^E and s^E because these are the physically significant quantities that can be related to molecular behavior; g^E is only an operational combination of them. The excess enthalpy is concerned primarily with energetic interactions between molecules while excess entropy is concerned primarily with the structure of the solution, i.e., the spatial arrangements of the molecules which leads us to concepts like randomness and segregation. Unfortunately, excess enthalpy and excess entropy are not

(1) $g^E = RT \sum_i x_i \ln \gamma_i$ (USES RAOULT'S LAW FOR IDEALITY)

γ_i = ACTIVITY COEFFICIENT; x = MOLE FRACTION

(2) $RT \ln \gamma_i = \left(\frac{\partial n_T g^E}{\partial n_i} \right)_{T,P,n_j}$ (BASED ON GIBBS-DUHEM EQUATION)

n_T = TOTAL NO MOLES; n_i = NO MOLES OF COMPONENT i

(3) $g^E = h^E - Ts^E$

DETERMINED PRIMARILY BY INTERMOLECULAR FORCES

DETERMINED PRIMARILY BY MOLECULAR STRUCTURE (SIZE, SHAPE, POSITION, DEGREES OF FREEDOM)

Figure 22. Excess functions and activity coefficients

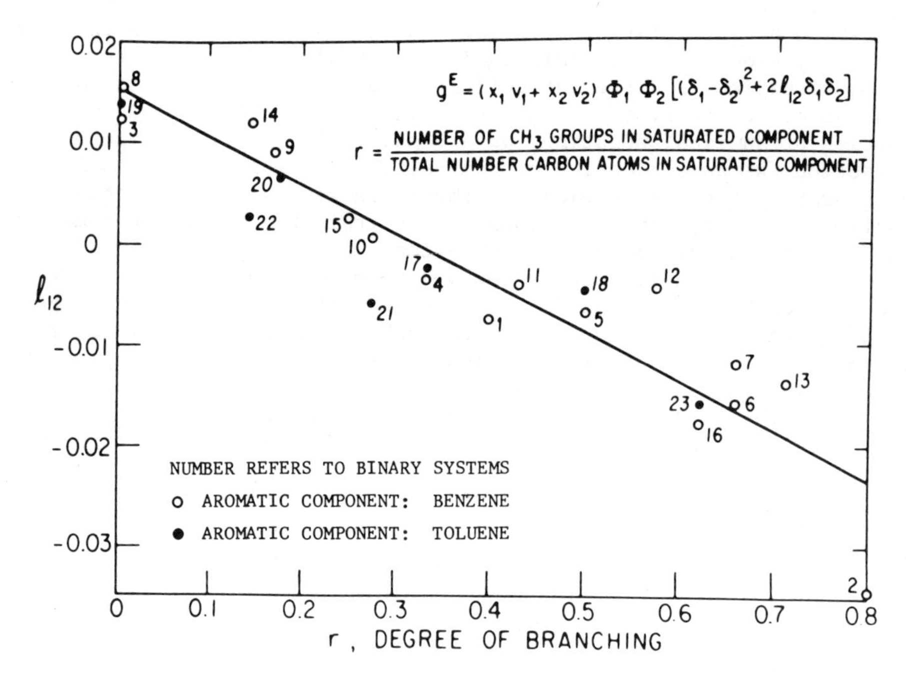

Figure 23. Binary parameters for aromatic–saturated hydrocarbon systems (Funk)

independent of one another (consider the Gibbs-Helmholtz equation) and therefore, if one builds separate models for these two quantities one must give attention to their mutual interdependence. But fundamentally h^E and s^E refer to rather different physical phenomena and historically what has happened is that those who have proposed equations for g^E have tended, often subconsciously, to give primary attention either to h^E or to s^E. For example, the well-known Wohl expansion (10) which dominated phase equilibrium thermodynamics for 20 years, is physically meaningful only for h^E (rather than g^E), and the equation of Wilson which has been modified by dozens of authors, was derived as a modification of an equation applicable to s^E. Construction of a rigorous theoretical equation for g^E which does proper justice to both h^E and s^E, is probably beyond our present theoretical capabilities. However, much effort has been expended toward establishing an expression which, while primarily directed at practical needs, is also at least grounded on some loose theoretical basis. We must keep reminding ourselves that any presently available theory for liquid mixtures is severely restricted and, when applied to real mixtures, it should not be taken too seriously. For all but the simplest molecules, molecular theory of liquid solutions provides a valuable guide, useful for interpolation and extrapolation, but it is not likely in the near future to do much more than that.

Equations for g^E which emphasize h^E and give little attention to s^E are common and one that is particularly popular is the regular-solution equation of Hildebrand and Scatchard using solubility parameters. With a little empirical modification, the regular solution equation can be extremely useful for certain kinds of mixtures. An example, shown in Figure 23, gives a correlation established by Funk (11) for mixtures containing aromatic hydrocarbons (in particular benzene or toluene) with saturated hydrocarbons. By introducing the binary parameter ℓ_{12}, agreement between calculated and experimental results is much improved and, as shown here for a limited class of mixtures, it is possible to correlate ℓ_{12} with molecular structure. The parameter ℓ_{12} is small in absolute value but it has a pronounced effect, as indicated in Figure 24.

The basic idea of solubility parameter was first described by Hildebrand about 50 years ago. It was barely known to chemical engineers until about 20 years ago but since then it has been both used and abused extensively for a variety of purposes, both legitimate and otherwise, far beyond Joel Hildebrand's wildest dreams. It shows up in the paint and varnish industry, in metallurgy, physiology, colloid chemistry and pharmacology and recently I have seen it mutilated in a magazine article on "scientific" astrology.

Before leaving the solubility parameter, I want to point out one use which, while not new, has perhaps not received as much attention as it deserves. I refer to the use of the solubility parameter for describing the solvent power of a dense gas, with particular reference to high-pressure gas extraction, certainly not a novel process but one which is receiving renewed attention in coal liquefaction and in food processing.

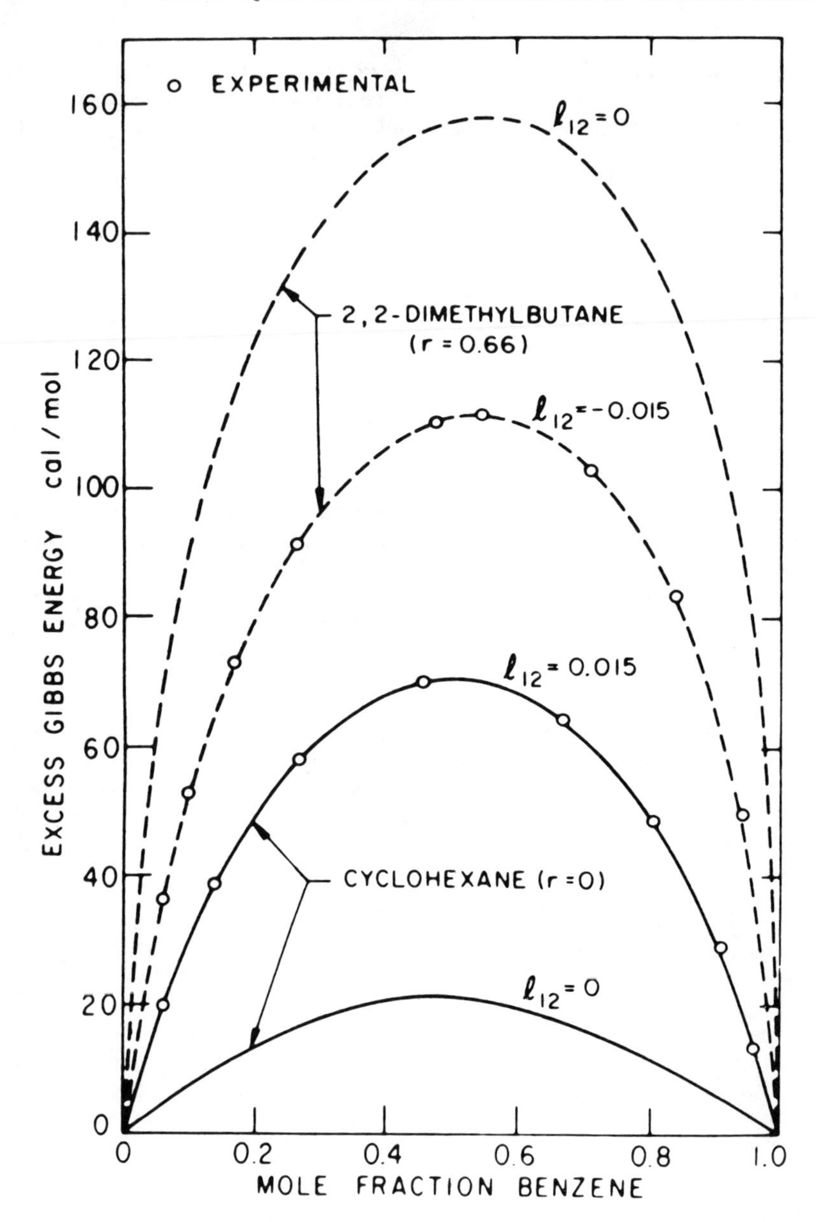

Figure 24. Experimental and calculated excess Gibbs energies for two binaries containing benzene at 50°C (Funk)

Using the well-known tables by Pitzer, it is easily possible to construct a generalized solubility parameter diagram as shown in Figure 25. In this particular case the acentric factor 0.075 is close to that of ethylene because when this chart was prepared over ten years ago, application was directed at interpreting solubility data for naphthalene in ethylene at high pressure. Just to orient ourselves, when the temperature is around 20°C and the pressure is about 400 atm, the solubility parameter is in the region 5 or 6 $(cal/cm^3)^{1/2}$, only about one or two units lower than that of a liquid paraffin like hexane. When solubility data in compressed ethylene are used with the Scatchard-Hildebrand equation to back-out a solubility parameter for liquid naphthalene, we find results shown in Figure 26. The remarkable features of this figure are first, that the solubility parameter obtained is in good agreement with what one would obtain by extrapolating the known solubility parameter of liquid naphthalene to temperatures below the melting point and second, that the backed-out solubility parameter is nearly constant with pressure. This indicates that the Hildebrand regular solution equation is useful for mixtures of nonpolar fluids regardless of whether these are liquids or gases, provided only that the density is sufficiently large.

Finally, a useful feature of the solubility parameter is shown in Figure 27 for nitrogen. Similar diagrams can be constructed for any fluid; nitrogen is here shown only as an example. Note that the solubility parameter is strongly sensitive to both pressure and temperature in the critical region. Highly selective extraction can therefore be carried out by small changes in temperature and pressure. Further, such small changes can be exploited for efficient solvent regeneration in continuous separation processes.

Local Composition to Describe Nonrandomness

For mixtures containing polar and hydrogen-bonded liquids, equations for g^E which emphasize h^E (rather than s^E) tend to be unsatisfactory because in their basic formulation such equations give little attention to the difficult problem of nonrandomness. In a regular solution, the molecules are "color-blind" which means that they arrange themselves in a manner dictated only by the relative amounts of the different molecules that are present. It is easily possible to add a correction which takes into account the effect of molecular size, as given by the Flory-Huggins expression. (Size corrections are essential for polymer solutions.) Variations on that expression [e.g., Staverman or Tompa (12)] can also account, in part, for differences in molecular shape. But, for strongly interacting molecules, regardless of size and shape, there are large deviations from random mixing; such molecules are far from "color-blind" because their choice of neighbors is heavily influenced by differences in intermolecular forces. An intuitive idea toward describing this influence was introduced by Wilson with his notion of local composition, shown schematically in Figure 28 (13). Viewed microscopically,

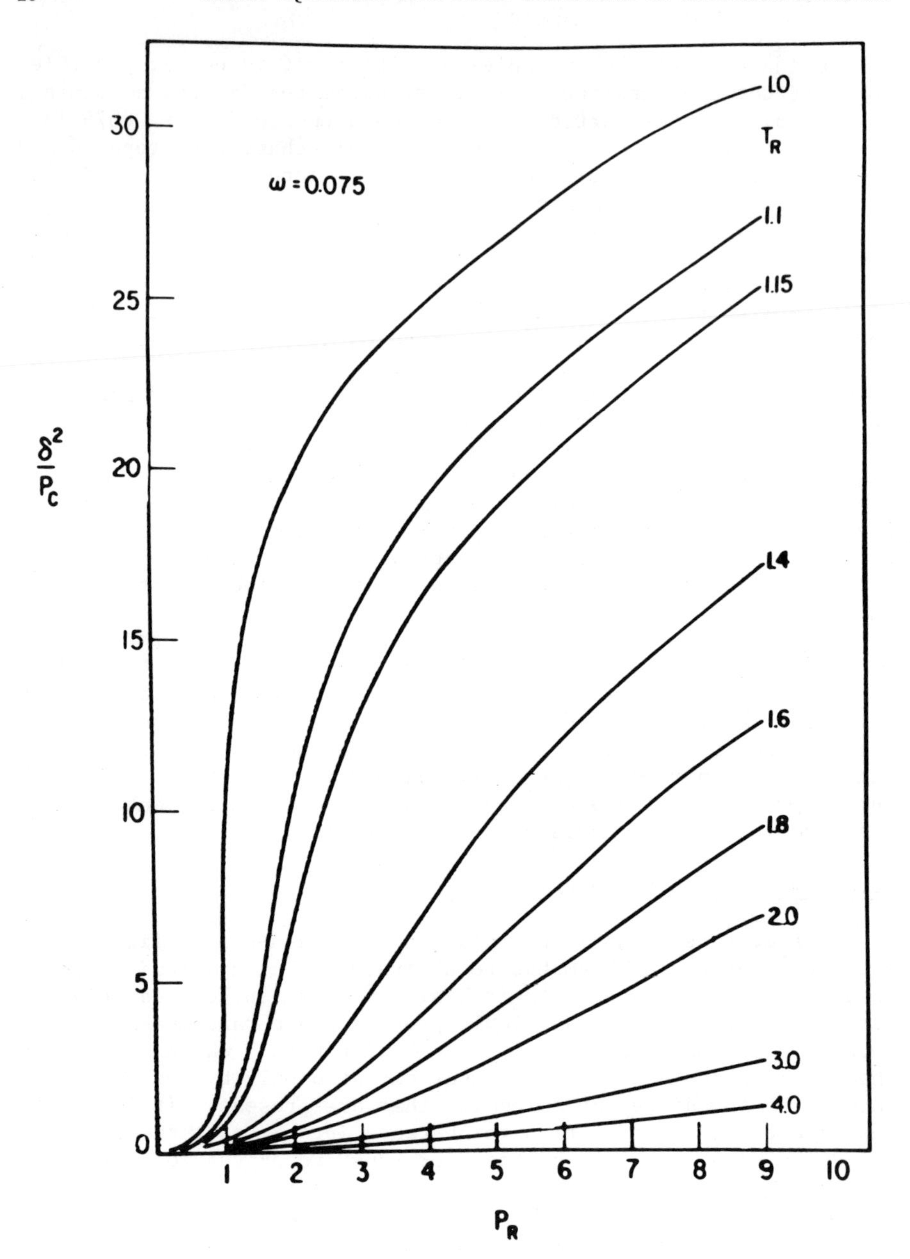

Figure 25. Solubility parameters for dense gases with an acentric factor of 0.075

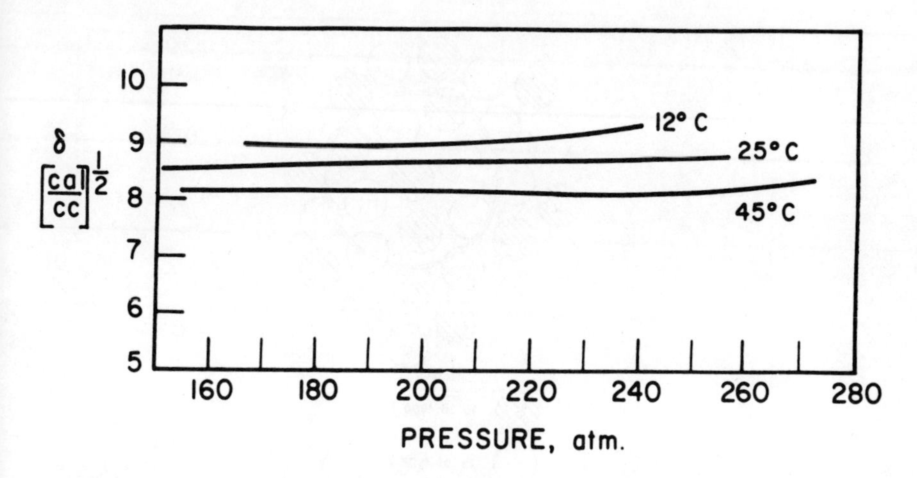

Figure 26. Solubility parameter of naphthalene calculated from solubility data in gaseous ethylene

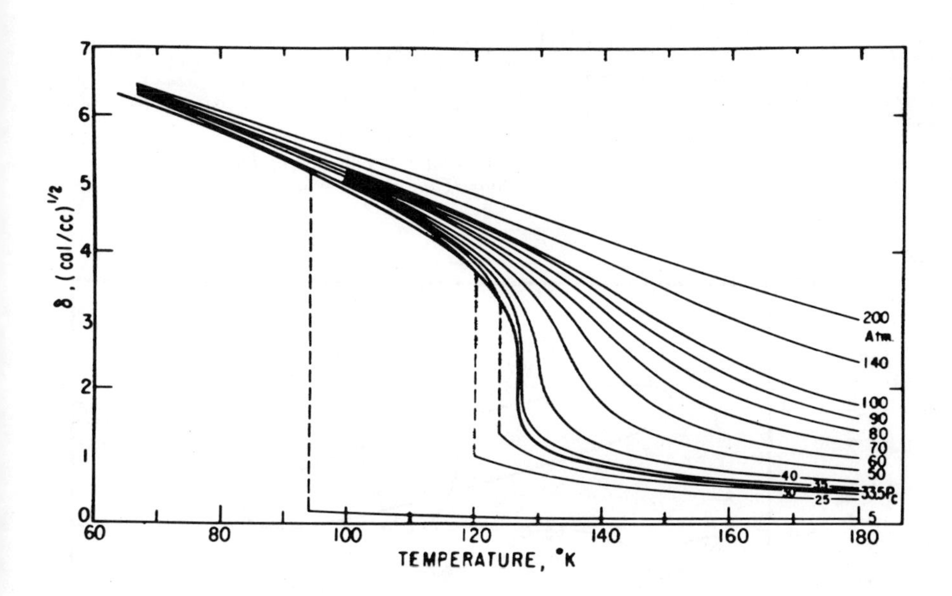

Figure 27. Solubility parameter for nitrogen

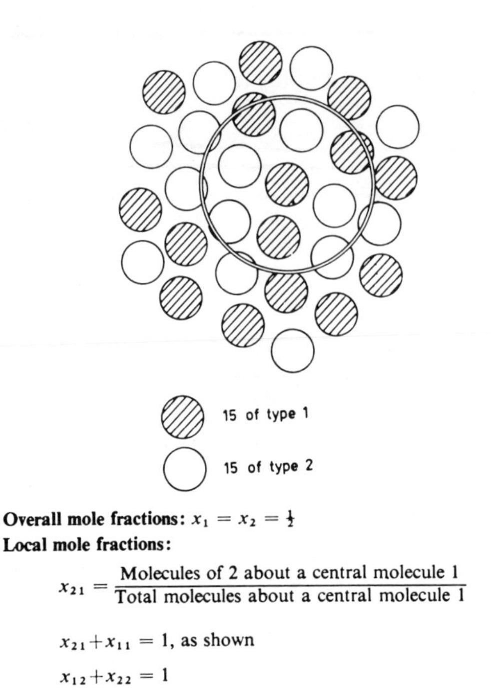

Overall mole fractions: $x_1 = x_2 = \frac{1}{2}$

Local mole fractions:

$$x_{21} = \frac{\text{Molecules of 2 about a central molecule 1}}{\text{Total molecules about a central molecule 1}}$$

$x_{21}+x_{11} = 1$, as shown

$x_{12}+x_{22} = 1$

$x_{11} \sim \frac{3}{8}$

$x_{21} \sim \frac{5}{8}$

Figure 28. Local compositions and the concept of local mole fractions (Cukor)

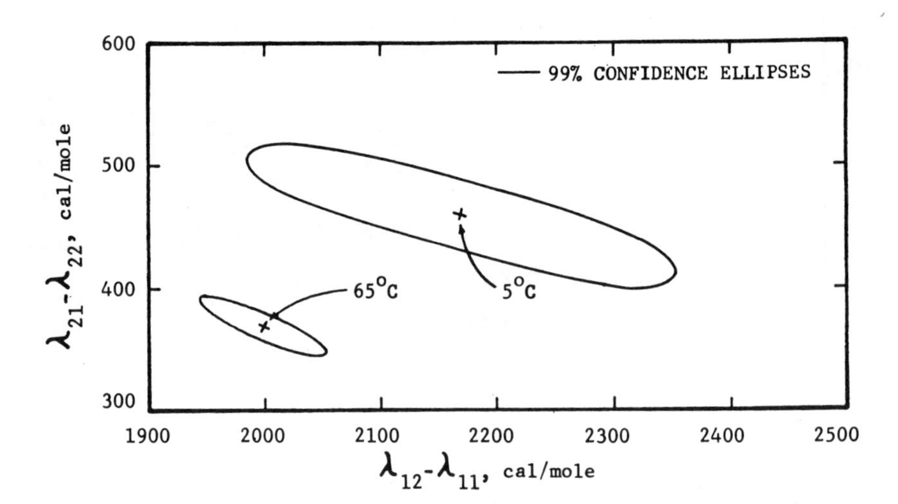

Figure 29. Wilson equation parameters for ethanol(1)–cyclohexane(2)

a solution is not homogeneous because molecules have definite preferences in choosing their immediate environment, leading to very small regions, sometimes called domains, which differ in composition. There is no obvious way to relate local composition to overall (stoichiometric) composition but the use of Boltzmann factors provides us with one reasonable method for doing so. In the last ten years, Wilson's equation for g^E has enjoyed much popularity and, for certain mixtures, notably alcohol-hydrocarbon solutions, it is remarkably good. Unfortunately, it has one major flaw: it is not applicable to partially miscible mixtures and, as a result, there has been a flurry of activity to extend and modify Wilson's equation; new modifications appear almost monthly. Time does not permit me to discuss any of these modifications but I want to point out an important development which only recently has become increasingly evident. In many cases the choice of model for g^E is not as important as the method chosen for data reduction; that is, the procedure used to obtain model parameters from limited experimental data is often more important than details within the model.

When reducing experimental data to obtain model parameters, attention must be given to the effect of experimental errors. Not all experimental measurements are equally valuable and therefore, a proper strategy for weighting individual experimental points is needed to obtain "best" parameters (14, 15). Any realistic strategy shows at once that for any given set of binary data, there is no unique set of model parameters. In a typical case, there are many sets of parameters which are equally good, as illustrated in Figure 29 prepared by Tom Anderson for the system ethanol-cyclohexane. In this particular case, the model used was Wilson's but that is not important here. The important message is that any point in the ellipses shown can represent the experimental data equally well; there is no statistical significance in preferring one point over another. The elliptical results shown in Figure 29 are typical; the parameters have a tendency to be correlated but, without additional information, it is not possible to say which point within the ellipse is the best.

Since the two ellipses do not overlap, we are justified in assigning a temperature dependence to the parameters. When we do so, we obtain the pressure-composition diagrams shown in Figure 30. In this case we assumed that the temperature dependence of one parameter is parallel to that of the other (i.e., we used three, not four, adjustable parameters since b has no subscripts) but that procedure is not always successful; frequently four parameters are required. On the other hand, if the two ellipses in Figure 29 had a region of overlap, there would be no good reason to use temperature-dependent parameters; two temperature-independent parameters would be sufficient.

Another example prepared by Tom Anderson is shown in Figure 31 for the system butanol-water; in this case the UNIQUAC model was used rather than Wilson's because we are concerned with vapor-liquid and liquid-liquid equilibria. The left side shows that when vapor-liquid and liquid-liquid equilibria are reduced separately, we obtain

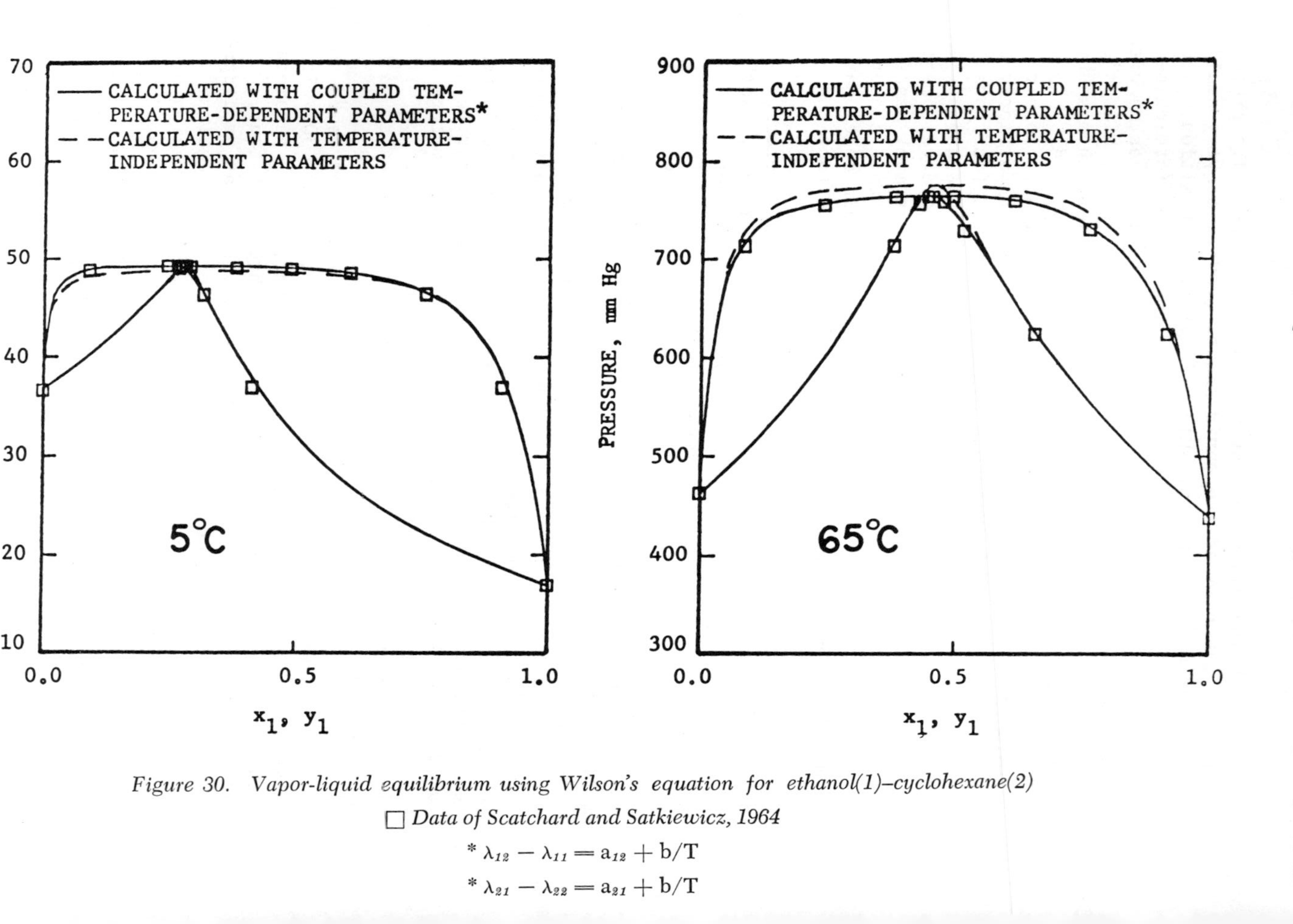

Figure 30. Vapor-liquid equilibrium using Wilson's equation for ethanol(1)–cyclohexane(2)

□ *Data of Scatchard and Satkiewicz, 1964*

$* \lambda_{12} - \lambda_{11} = a_{12} + b/T$

$* \lambda_{21} - \lambda_{22} = a_{21} + b/T$

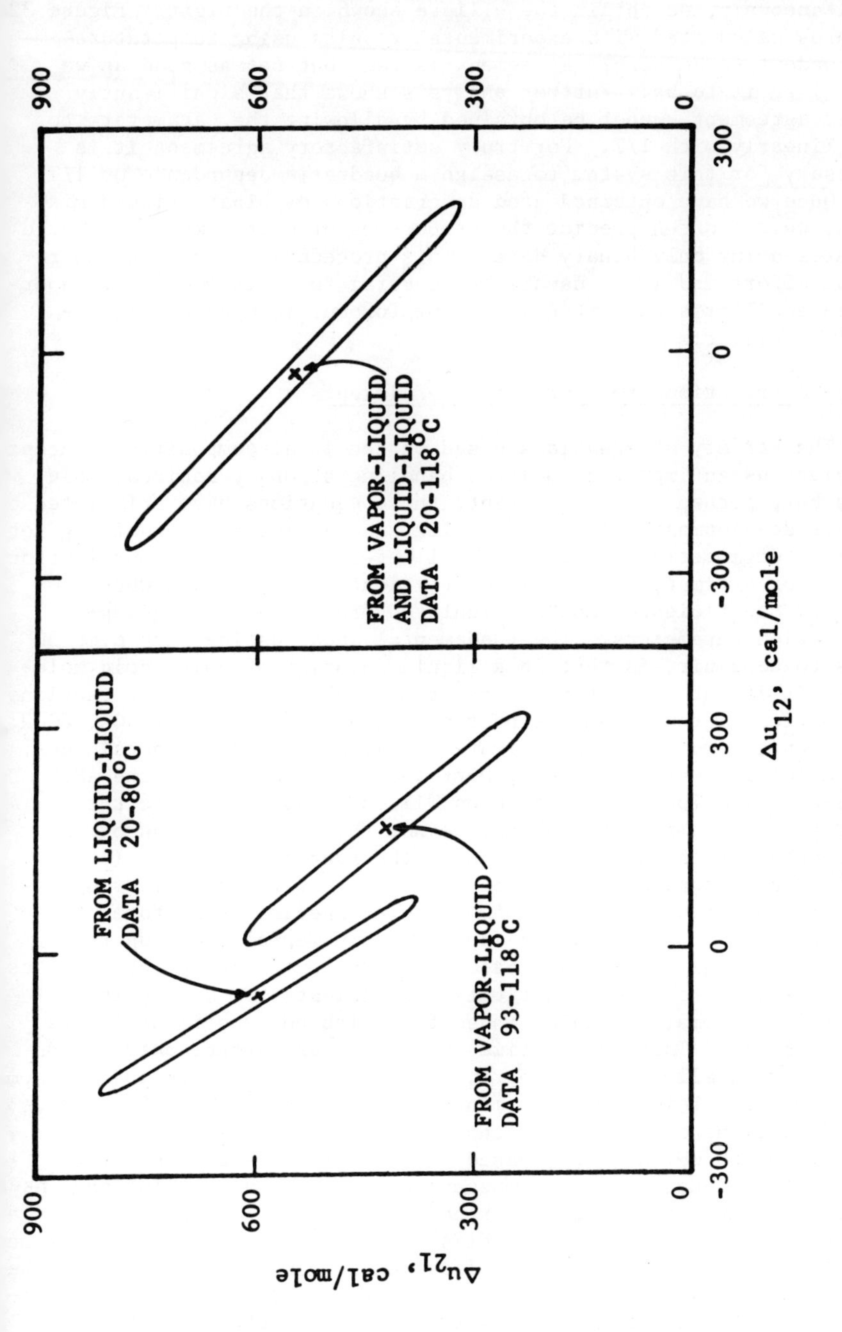

Figure 31. UNIQUAC parameters for butanol(1)–water(2)
(—99% Confidence ellipses)

two ellipses with no overlap. When both sets of data are reduced simultaneously, we obtain the ellipse shown on the right. Figure 32 compares calculated with experimental results using temperature-independent parameters. Agreement is fair but not as good as we would like it to be. Further analysis shows that significantly better agreement cannot be obtained by allowing the parameters to vary linearly with 1/T. For truly satisfactory agreement it is necessary for this system to assign a quadratic dependence on 1/T.

Once we have obtained good descriptions of binary liquid mixtures, we can often predict the properties of multicomponent liquid mixtures using only binary data. This procedure saves much experimental effort and it is usually successful for multicomponent vapor-liquid equilibria but often it is not for multicomponent liquid-liquid equilibria.

Group-Contributions for Activity Coefficients

The variety of equations based on the local composition concept has given us an improved tool for handling strongly nonideal solutions but, perhaps more important, these equations have stimulated another development which, in my view, is particularly promising for chemical engineering application. I refer to the group-contribution method for estimating activity coefficients, a technique where activity coefficients can be calculated from a table of group-interaction parameters. The fundamental idea, dating back over 50 years to Langmuir, is that in a liquid solution of polyatomic molecules, it is not the interactions of molecules, but the interactions of functional groups comprising the molecules (e.g., CH_3, NO_2, COOH, etc.) which are important; Figure 33 illustrates the general idea.

About 15 years ago, Deal, Derr and Wilson developed the ASOG group-contribution method based on Wilson's equation where the important composition variables are not the mole fractions of the components but the mole fractions of the functional groups (16, 17). In chemical technology the number of different functional groups is much smaller than the number of molecular species; therefore, the group-contribution method provides a very powerful scale-up tool. With a relatively small data base to characterize group interactions it is possible to predict activity coefficients for a very large number of systems, including those for which no experimental data are available. There is no time now to discuss group-contribution methods; we shall have an opportunity later this week to go into som details. Here I just want to mention that some of the difficulties and limitations of the ASOG method have been overcome by a similar method, called UNIFAC (18), based on the UNIQUAC equation. Very recently, Fredenslund and Rasmussen in Denmark and Gmehling and Onken in Germany have significantly extended the earlier UNIFAC work; in a publication now in press, the UNIFAC data base has been much enlarge and, therefore, the range of application is now much increased. The late Professor Ratcliff at McGill has also developed a group-contribution method and, during the last few years, Professors Chao and

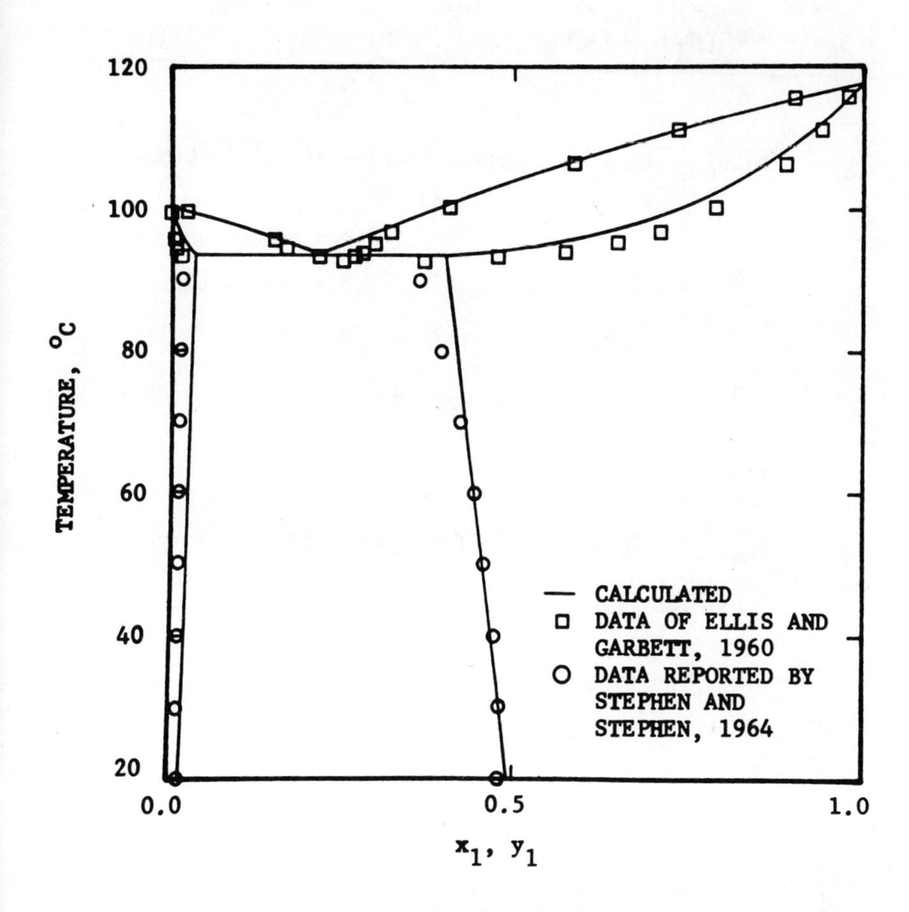

Figure 32. Temperature-equilibrium phase composition diagram for butanol(1)–water(2) system. Calculations are based on UNIQUAC equation with temperature-independent parameters.

E.G. (1) ACETONE - (2) TOLUENE

(CH3)-(C-CH3 with =O)

(CH)-(CH), (CH), (C-CH3), (CH)=(CH)

$\ln \gamma_i = \ln \gamma_i^C$ (COMBINATORIAL) $+ \ln \gamma_i^R$ (RESIDUAL)

$\gamma_i^C = F^C(X, Q, R)$

R = GROUP VOLUME
Q = GROUP AREA
X = MOLE FRACTION

$\gamma_i^R = F^R(X_K, Q, R, a_{MN}, T)$

X_K = MOLE FRACTION OF GROUP K;

a_{MN} = GROUP INTERACTION PARAMETER

Figure 33. Group contributions to activity coefficients γ_1 and γ_2

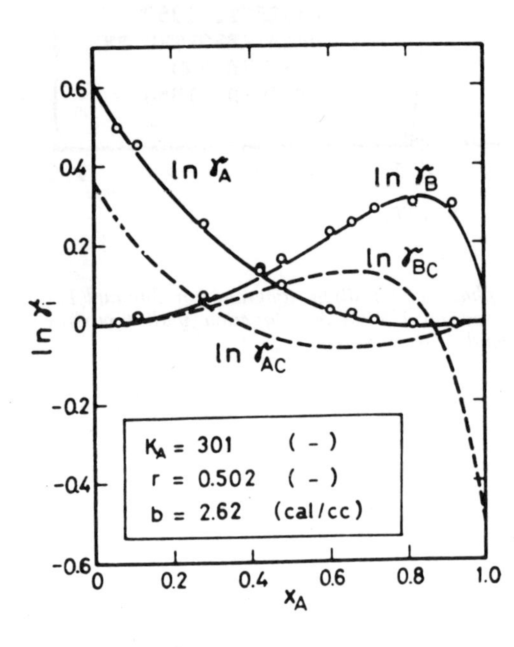

Figure 34. Activity coefficients for ethanol (A)–triethylamine (B) system at 34.85°C (Nitta and Katayama, 1973)

Greenkorn at Purdue have started to work in this area. The group-contribution method necessarily provides only an approximation but for many applications that is sufficient. For practical-minded chemical engineers this new research in applied thermodynamics represents perhaps the most exciting development since Pitzer's acentric factor.

Chemical Theory for Activity Coefficients

While the local composition concept has been highly useful for strongly nonideal mixtures, it is also possible to represent data for such solutions by assuming that molecules associate or solvate to form new molecules. It follows from this viewpoint that a binary mixture of A and B is really not a binary mixture, but instead, a multicomponent mixture containing, in addition to A and B monomers, also polymers A_2, A_3 ... and B_2, B_3 ... as well as copolymers containing both A and B in various possible stoichiometric proportions. Deviations from ideal behavior are then explained quantitatively by assigning equilibrium constants to each of the postulated chemical equilibria. This is a Pandora's box because, if we assume a sufficient number of equilibria, adjust the stoichiometry of the polymers and copolymers and also adjust the equilibrium constants, we can obviously fit anything. Nevertheless, the chemical method makes sense provided we have independent chemical information (e.g., spectroscopic data) which allows us to make sensible à priori statements concerning what chemical species are present. For example, we know that acetic acid forms dimers, that alcohols polymerize to dimers, trimers, etc. and that chloroform and acetone are linked through a hydrogen bond. Thus an "enlightened" chemical theory can often be used to represent experimental data with only a few parameters where a strictly empirical equation requires many more parameters to give the same fit. The literature is rich in examples of this sort; a recent one by Nitta and Katayama (19) is given in Figure 34 which shows activity coefficients for the system ethanol-triethylamine. Here A stands for alcohol and B for amine. Subscript C denotes chemical contribution; in addition to the chemical effects, there are physical forces between the "true" molecules and these are taken into account through the parameter b which has units of energy density. Equilibrium constant K_A, for continuous polymerization of ethanol, is obtained independently from alcohol-hydrocarbon mixture data. Parameter r is the ratio of the equilibrium constant for $A + B \rightleftharpoons AB$ to K_A. The excellent fit is, therefore, obtained with two adjustable parameters, b and r.

While chemical theories are often useful for describing strongly nonideal liquid mixtures, they are necessarily specific, limited to a particular type of solution. It is difficult to construct a general theory, applicable to a wide variety of components, without introducing complicated algebra and, what is worse, a prohibitively large number of parameters. This difficulty also makes chemical theory impractical for multicomponent mixtures and indeed, while the literature is rich with application of chemical theory to binaries, there are few articles which apply chemical theory to ternary (or higher) mixtures. For practical chemical engineering, therefore, the chemical theory of liquid solutions has limited utility.

Supercritical Components in the Liquid Phase

I have previously stressed the difficulty of standard states when we deal with supercritical components. For these components, e.g., methane or nitrogen, at ordinary temperatures, it has been common practice to ignore the problem simply by extrapolating pure-liquid fugacities to temperatures above the critical. This is convenient but ultimately unsatisfactory because there is no unambiguous way to perform an extrapolation for a hypothetical quantity. The most common method is to assume that a semi-logarithmic plot of the fugacity versus reciprocal temperature is a straight line. Experience has shown that at temperatures far above the critical, this is a bad assumption but regardless of what shape the plot is assumed to be, on semilog paper, the thickness of the pencil can already make a significant difference.

The only possible satisfactory procedure for proper use of activity coefficients of supercritical components is to use Henry's constants as the standard-state fugacity. Henry's constants are not hypothetical but are experimentally accessible; also, at least in principle, they can be calculated from an equation of state. Remarkably little attention has been given to the formal thermodynamics of liquid mixtures containing supercritical components.

Using Henry's constants introduces a variety of problems but they are by no means insurmountable. Yet, chemical engineers have stubbornly resisted using Henry's constants for standard-state fugacities; whenever I have tried to interest my industrial colleagues in this possibility I felt like a gun-control enthusiast talking to the National Rifle Association.

As long as the solution is dilute, Henry's constant is sufficient but as the concentration of solute rises, unsymmetrically normalized activity coefficients must be introduced and at present we have little experience with these. While binary mixtures can be handled with relative ease, major formal difficulties arise when we go to multicomponent mixtures because, unfortunately, Henry's constant depends on both solute and solvent and, therefore, when we have several solvents present, we must be very careful to define our standard states and corresponding activity coefficients in a thermodynamically consistent way.

About ten years ago the late Ping Chueh and I wrote a monograph on the use of unsymmetric activity coefficients for calculating K factors in hydrocarbon and natural-gas mixtures, but it never caught on. About five years ago I was window shopping in the Time Square section of New York and to my amazement I saw a copy of our monograph on a table in a used book store, completely surrounded by books on pornography. It appeared that my colleagues were trying to tell me something.

However, times change and now that pornography is accepted with little opposition, maybe Henry's constants for standard-state fugacities can be accepted too. John O'Connell at Florida has been working on this and in Figure 35 we see some results for excess Henry's constants for ethylene, carbon dioxide and carbon monoxide in binary solvent mixtures (20). To a first approximation, the logarithm of Henry's constant for a gas in a mixed solvent is given by a simple mole-fraction average; Figure 35 shows deviations from that first approximation.

Figure 36 presents another example, given by Nitta and Katayama (21); it shows Henry's constant for nitrogen in mixtures of n-propanol and iso-octane. Two plots are shown, one against mole fraction and the other against volume fraction of the solvent mixture. Since iso-octane is a much larger molecule than propanol, it is not surprising that the volume-fraction plot is more nearly linear than the mole-fraction plot, but, nevertheless, there is noticeable departure from straight-line behavior.

Katayama applies his chemical model for explaining the deviation with results shown in Figure 37. One contribution of the Flory-Huggins type corrects for size differences, another (chemical) contribution corrects for association of alcohol molecules and finally, a physical contribution corrects for differences in intermolecular forces. The sum of the corrections gives good agreement with experiment. Since the corrections for size and association were calculated from other data, only one adjustable parameter was used in preparing the final plot.

Aqueous Solutions of Weak Volatile Electrolytes

I have indicated earlier that the chemical theory of liquid mixtures presents some difficulties and that the use of Henry's constants also gives us headaches. However, when we come to solutions of volatile electrolytes we are really in a bad way because now we must use not only the awkward chemical theory but in addition, those unpleasant Henry's constants. We have no real choice here because in dilute aqueous solution, weak volatile electrolytes (e.g., ammonia, hydrogen sulfide, sulfur dioxide) dissociate into ions and thus there is real chemistry going on which we cannot ignore. Further, since ions are nonvolatile, we must use unsymmetrically normalized activity coefficients; the fugacity of a pure volatile electrolyte liquid which is not ionized doesn't tell us anything that would be useful for a dilute aqueous solution where the solute is, at least in part, in ionic form.

The situation we must describe is shown schematically in Figure 38. The horizontal equilibrium is chemical, characterized essentially by a dissociation constant, and the vertical equilibrium is physical, characterized essentially by Henry's constant. Detailed development toward quantitative results also requires unsymmetrically normalized activity coefficients, i.e., those activity coefficients which go to unity not as the composition approaches the pure solvent,

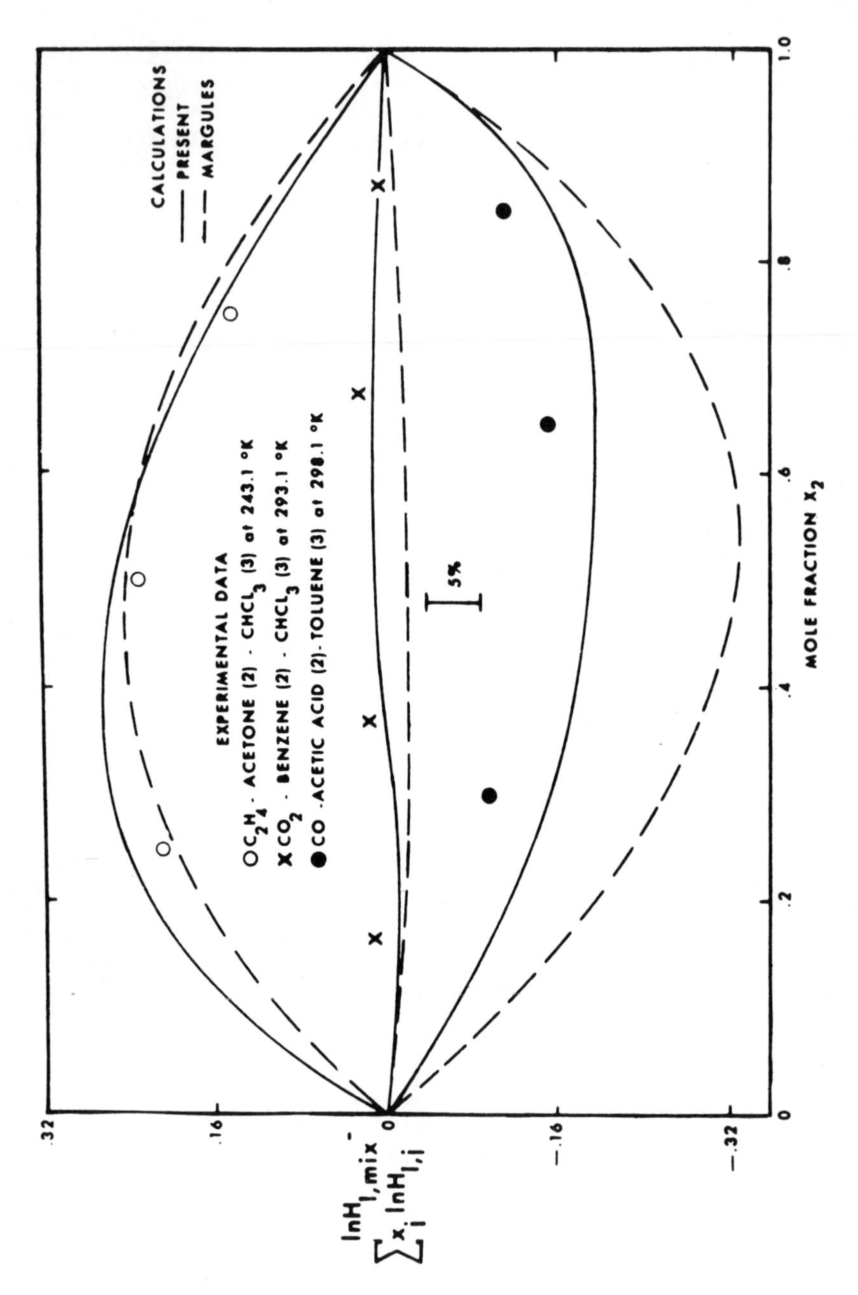

Figure 35. Deviation of Henry's constant from that in an ideal solution (O'Connell)

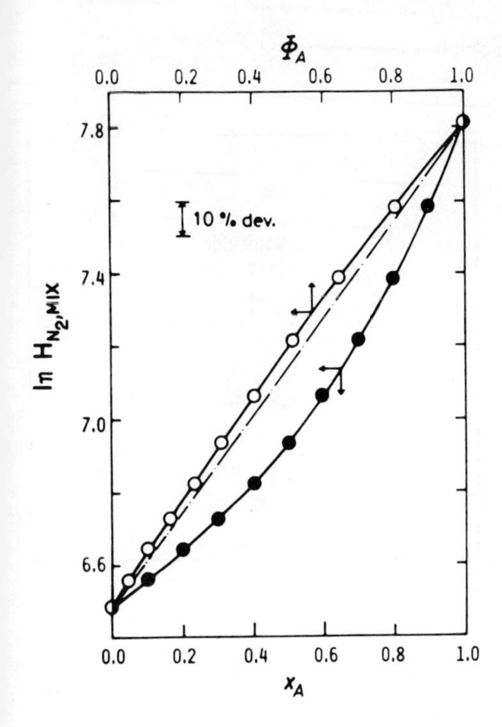

Figure 36. Henry's constants for nitrogen in n*-propanol (A)–isooctane (B) mixture vs. mole fraction and volume fraction (Katayama et al., 1973)*

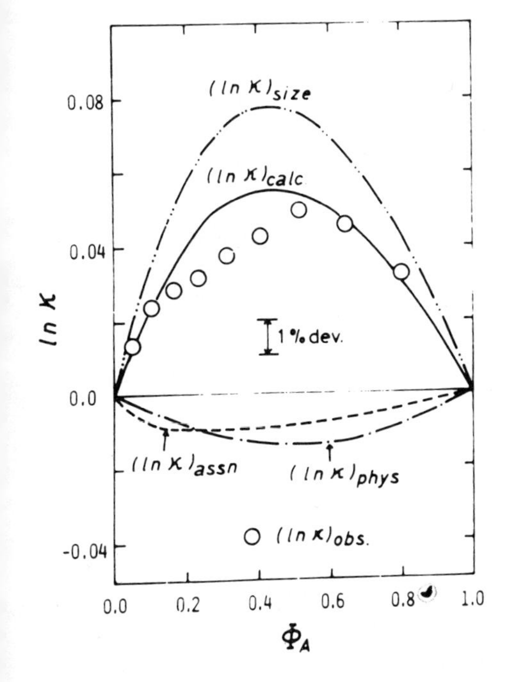

Figure 37. Experimental and calculated ln κ-values vs. volume fraction for n*-propanol (A)–isooctane (B) mixtures (Katayama et al., 1973)*

$$\ln \kappa = \ln H_{N_2,m} - \sum_j \Phi_j \ln H_{N_2,j}$$

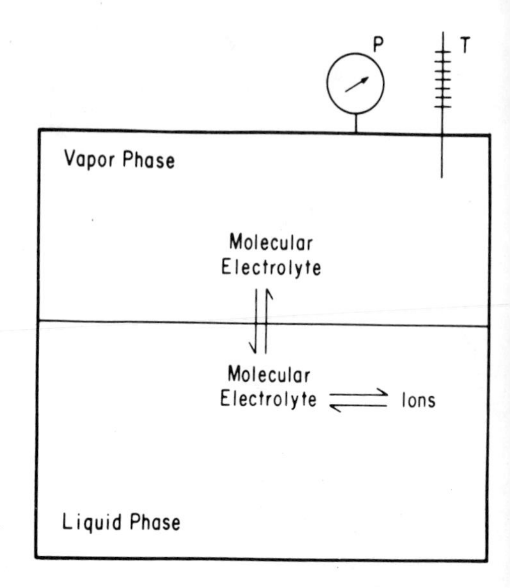

Figure 38. Vapor-liquid equilibrium in a single-solute system

(1) MASS BALANCE

$$m_A = m_a + \frac{1}{2}(m_+ + m_-)$$

m = MOLALITY

SUBSCRIPT A = STOICHIOMETRIC

SUBSCRIPT a = MOLECULAR

(2) DISSOCIATION EQUILIBRIUM

$$K = \frac{a_+ a_-}{a_a}$$

$a_i = m_i \gamma_i^* =$ ACTIVITY

$\gamma_i^* \rightarrow 1$ AS $m_i \rightarrow 0$

(3) ELECTRONEUTRALITY

$$m_+ = m_-$$

(4) VAPOR-LIQUID EQUILIBRIA

$$y_a \varphi_a P = m_a \gamma_a^* H(PC)$$

H = HENRY'S CONSTANT

PC = POYNTING CORRECTION

φ = VAPOR-PHASE FUGACITY COEFFICIENT

Figure 39. Aqueous solutions of weak volatile electrolytes

but instead, as the aqueous solution becomes infinitely dilute.

Figure 39 indicates the four conditions that must be satisfied. For most solutes of interest, chemical equilibrium constant K is known as a function of temperature; the major difficulty lies in calculating H and γ^*. Figure 40 shows two equations for γ^*. Both start out with the Debye-Hückel term which depends primarily on ionic strength but that term alone is applicable only to very dilute solutions. Guggenheim adds an essentially empirical first-order correction and this is sufficient for ionic strengths to about 1 or 2 molar. For more concentrated solutions, Pitzer has proposed a semi-theoretical equation which, however, has many parameters and all of these depend on temperature (22).

Time does not permit a detailed discussion but, in view of the importance of these solutions in chemical engineering, let me quickly show a few results. Figure 41, from Edwards et al. (23), shows how experimental data in the dilute region for aqueous ammonia can be reduced to yield Henry's constants (the intercept) and Guggenheim's coefficient β (the slope). Note that the abscissa gives the molecular molality of ammonia, not the total molality; therefore, Figure 41 implicitly includes the effect of ionization as determined by the independently-measured chemical dissociation constant. A similar analysis was made for solutions of CO_2 in water.

Figure 42 gives partial pressures for the ternary system ammonia-carbon dioxide when the total ammonia molality is 0.128; these results were predicted using only binary data; no ternary data were used. In this example the solution is dilute and Guggenheim's equation is adequate; for higher concentrations, Pitzer's equation is required as shown in Figure 43 based on very recent (and as yet unpublished) work by Renon and coworkers. The line on the right is the same as the one shown in the previous figure; the molality is low. The line on the left is at higher ammonia concentration and, as we proceed to higher ratios of carbon dioxide to ammonia, the total molality goes well above 2. We see that Edwards' line, based on Guggenheim's equation, is satisfactory at first but shows increasing deviations as the total molality rises. The results shown here are again based on binary data alone. At 20°C, experimental data are relatively plentiful and it was possible to evaluate all the parameters in Pitzer's equation but at higher temperatures, where good data are scarce, it is not easy to use Pitzer's equation until some reliable method can be found to estimate how temperature affects the parameters. Finally, I should mention that the calculations shown here are based on simultaneous solution of 14 equations. A good computer program is an absolute necessity.

Conclusion

I have tried this morning to present a survey of the present status of applied phase-equilibrium thermodynamics. In one sense, the survey is much too long because I am sure your patience has been pushed well beyond its elastic limit. In another sense, it is much

GUGGENHEIM

$$\ln \gamma_i^* = -\frac{A z_i^2 \sqrt{I}}{1+\sqrt{I}} + 2\sum_j \beta_{ij} m_j$$

WHERE: z = CHARGE

A = KNOWN CONSTANT

I = IONIC STRENGTH $= \frac{1}{2}\sum_j z_j^2 m_j$

PITZER

$$\ln \gamma_i^* = -\frac{A z_i^2}{3}\left[\frac{\sqrt{I}}{1+b\sqrt{I}} + \frac{2}{b}\ln(1+b\sqrt{I})\right]$$

$$+\sum_j m_j \left\{2\beta_{ij}^{(o)} + \frac{2\beta_{ij}^{(1)}}{\alpha^2 I}\left[1 - e^{-\alpha\sqrt{I}}\left(1 + \alpha\sqrt{I} - \frac{\alpha^2 I}{2}\right)\right]\right\}$$

$$+\sum_j \sum_k C_{ijk} m_j m_k$$

WHERE: $\alpha=2$ AND $b=1.2$

IF i AND j ARE MOLECULAR SPECIES, $\beta_{ij}^{(1)} = 0$.

Figure 40. Activity coefficients in electrolyte solutions

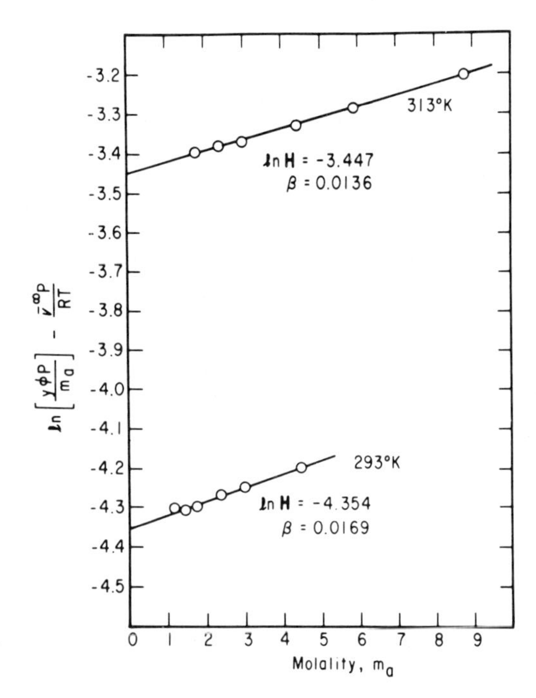

Figure 41. Data reduction for ammonia–water

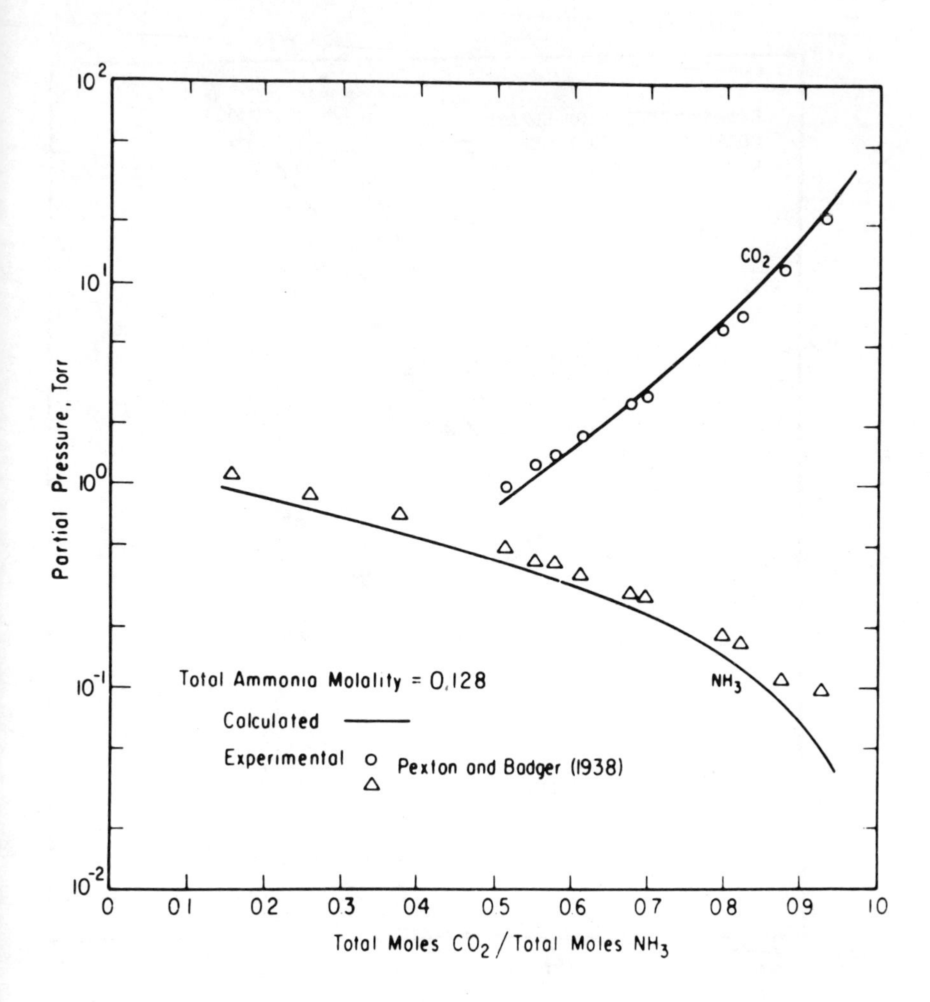

Figure 42. Vapor-liquid equilibria at 20°C for ammonia–carbon dioxide–water containing excess ammonia

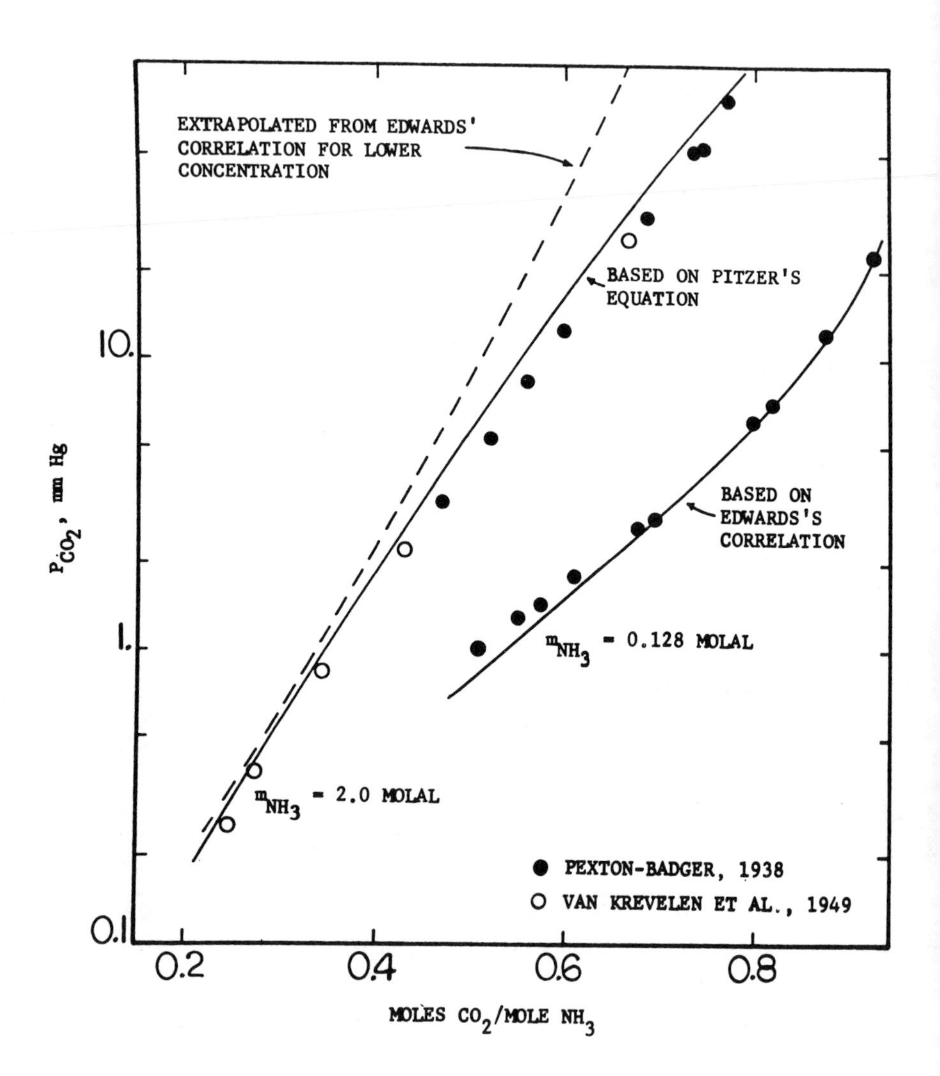

Figure 43. Calculated and experimental partial pressures of CO_2 at 20°C for the NH_3–CO_2–H_2O system: Effect of total concentration (Renon et al.)

too short because I have had to omit many worthwhile contributions. I already fear the hurt and insulted looks that I am likely to receive from some of you for the rest of the week, if not longer!

Let me quickly summarize what to me are the main implications of this frankly personalized survey.

First, those of us who are in the universities must get over our argon-fixation and start thinking boldly about molecules that are large, non-spherical, polar and hydrogen bonded. In other words, let us pay more attention to the real world. For chemical engineers it is better, I think, to solve approximately new and real problems than to improve marginally solutions to old problems. I am hopeful that this conference will contribute toward that end.

Second, we must stop the game of composing variations on old themes. The Redlich-Kwong equation, the BWR equation, the Wilson equation, all these represent fine moments in our history. However, we should not honor them by minor imitation. Rather, we should regard them as great monuments, inspiring us toward tackling new frontiers.

Where, then, are these new frontiers which demand our attention? I can here mention only six that I find particularly challenging intellectually and industrially important:

1. Construction of approximate, but physically sensible, equations of state applicable to complex molecules in both gaseous and liquid phases.

2. Vapor-phase experimental work (PVT and gas-saturation measurements) to provide fundamental information on intermolecular forces in asymmetric binary mixtures, i.e., those mixtures where the two components are strongly different, either in molecular size, or polarity, or both.

3. Vapor-liquid equilibrium experiments on mixtures of complex molecules, including polynuclear aromatics, polymers and highly polar solvents such as glycols, phenolics and other "nasty" liquids. The systems water-ethanol and benzene-cyclohexane have each been studied about 50 times. Enough of that. Let's measure equilibria in systems where we cannot now estimate the results within even an order of magnitude.

4. More attention must be given to data reduction methods. Some data are clearly more valuable than others and we must incorporate that distinction into our experimental plans.

5. We can usually do a pretty good job calculating vapor-liquid equilibria for multicomponent mixtures of typical nonelectrolytes. However, for multicomponent liquid-liquid equilibria the situation is much less favorable and we should give more attention to those equilibria.

6. Finally, let us learn to use more the powerful methods of statistical mechanics; let us overcome our fear of partition functions and let us not hesitate to introduce some enlightened empiricism into their construction.

This assembly of over 100 scientists and engineers represents a wide variety of knowledge, interests and experiences. Our meeting here provides us with a unique, unprecedented opportunity to exchange views, stimulating us all to new achievements. I have presented this rapid overview of previous accomplishments with the conviction that these accomplishments must serve us not as ground for self-congratulation but as a firmament on which to build toward a brighter future. My feeling--and I hope it is yours, too--must be that of the French philosopher who said, "From the altars of the past let us carry the fire, not the ashes."

Acknowledgment

For financial support extending over many years, the author is grateful to the National Science Foundation, the Gas Processors Association, the American Petroleum Institute, the Donors of the Petroleum Research Fund (administered by the American Chemical Society), Union Carbide Corporation and Gulf Oil Chemicals Company. Special thanks are due to Mr. Thomas F. Anderson for extensive assistance in preparing this report.

Literature Cited

1. Orye, R. V., Ind. Eng. Chem. Process Des. Dev., (1969) 8, 579.
2. Starling, K. E., and Han, M. S. Hydrocarbon Processing, (1972), 51, 107.
3. Bender, E., Cryogenics, (1973) 13, 11; (1975) 15, 667.
4. Soave, G., Chem. Eng. Sci., (1972) 27, 1197.
5. Peng, D., and Robinson, D. B., Ind. Eng. Chem. Fundam., (1976) 15, 59.
6. Mollerup, J., and Rowlinson, J. S., Chem. Eng. Sci., (1974) 29, 1373; Mollerup, J., Advan. Cryog. Eng., (1975) 20, 172.
7. Tsonopoulos, C., A.I.Ch.E. Journal, (1974) 20, 263.
8. Hayden, J. G., and O'Connell, J. P., Ind. Eng. Chem. Process Des. Dev., (1975) 14, 209.
9. de Santis, R., Breedveld, G. J. F., and Prausnitz, J. M., Ind. Eng. Chem. Process Des. Dev., (1974) 13, 374.
10. Wohl, K., Trans. A.I.Ch.E., (1946) 42, 215.
11. Funk, E. W., and Prausnitz, J. M., Ind. Eng. Chem., (1970) 62, 8.
12. Tompa, H., Trans. Faraday Soc., (1952) 48, 363; Staverman, A. J. Rec. Trav. Chim. Pays-bas, (1950) 69, 163; Donohue, M. D., and Prausnitz, J. M., Can. J. Chemistry, (1975) 53, 1586.
13. Cukor, P. M., and Prausnitz, J. M., Intl. Chem. Eng. Symp. Ser. No. 32 (Instn. Chem. Engrs., London) 3:88 (1969).
14. Anderson, T. F., Abrams, D. S., Grens, E. A., and Prausnitz, J. M., paper presented at the 69th Annual A.I.Ch.E. Meeting, Chicago, Illinois, 1976; submitted to A.I.Ch.E. Journal.
15. Fabries, J., and Renon, H., A.I.Ch.E. Journal, (1975) 21, 735.
16. Derr, E. L., and Deal, C. H., Intl. Chem. Eng. Symp. Ser. No. 32 (Instn. Chem. Engrs., London) 3:40 (1969).

17. Wilson, G. M., and Deal, C. H., Ind. Eng. Chem. Fundam., (1962) 1, 20.
18. Fredenslund, Aa., Jones, R. L., and Prausnitz, J. M., A.I.Ch.E. Journal, (1975) 21, 1086; Fredenslund, Aa., Michelsen, M. L., and Prausnitz, J. M., Chem. Eng. Progr., (1976) 72, 67; Fredenslund, Aa., Gmehling, J., Michelsen, M. L., Rasmussen, P. and Prausnitz, J. M., Ind. Eng. Chem. Process Des. Dev. (in press).
19. Nitta, T., and Katayama, T., J. Chem. Eng. Japan, (1973) 6, 1.
20. Orye, R. V., Ind. Eng. Chem. Process Des. Dev., (1969) 8, 579.
21. Nitta, T., and Katayama, T., J. Chem. Eng. Japan (1973) 6, 1.
22. Pitzer, K. S., and Kim, J. J., J. Amer. Chem. Soc. (1974) 96, 5701.
23. Edwards, T. J., Newman, J., and Prausnitz, J. M., A.I.Ch.E. Journal (1975) 21, 248.

3

Industrial View of the State-of-the-Art in Phase Equilibria

T. S. KROLIKOWSKI

Union Carbide Corp., Chemicals and Plastics Div., S. Charleston, W.Va. 25303

Twenty-five years ago, in 1952, there was a series of articles in Chemical Engineering Progress (1) entitled: "Industrial Viewpoints on Separation Processes". In the section on phase equilibrium data, it was noted that "The complete representation of such data for mixtures containing more than three components becomes impractically complex". Simplified calculations for multicomponent systems were recommended, and if the predicted values did not agree with experimental data, a system of minor correction factors should be devised.

In the same year, one of the annual review articles in Industrial and Engineering Chemistry (2) mentioned that the BWR equation of state seemed to provide the most accurate method thus far developed for estimating K-factors for hydrocarbon systems. Use of the equation was deemed tedious, and a procedure for using the equation in a simplified form suitable for rapid equilibrium calculations was to be presented. Charts based on the procedure were available from the M. W. Kellogg Co., New York.

Another review article (3) observed that automatic computers have entered the field of ditillation calculations. The author remarks: "The difficulty is that the machines cannot evaluate the errors in the assumptions set up by the operator, and therefore the value of the numbers produced by the machine gives a false impression of accuracy". That statement is as valid today as it was twenty-five years ago. On the other hand, the industrial approach to phase equilibria has changed over the years. In this state-of-the-art report, I will describe our present practices and concerns.

This presentation will be subdivided according to the sequence presented in Figure 1 - HOW, WHAT and WHY, WHERE. HOW are phase equilibria problems treated in an industrial situation? WHAT methods and correlations are used, and WHY are these techniques used? WHERE should future development work be directed?

Of necessity, the conditions described here are based

HOW?

WHAT
AND
WHY?

WHERE?

Figure 1. Sequence of presentation

principally on my own work environment. They may not be universally true, but they are certainly representative of the current state of affairs in industry.

HOW Are Phase Equilibria Problems Treated?

The technical population in industry can be divided into the computer people and the non-computer people. The non-computer people tend to use simple correlations, generalized models and graphs, and estimates based on their experience or intuition. The computer people, who are in the majority, have the capability of developing very complex models. Several years ago at Union Carbide, an effort was initiated to improve the effectiveness of the computer system used by these individuals for design purposes. I would like to spend some time describing the system as it now exists.

Subprogram Library. The basis of the system is the Engineering Subprogram Library. The library consists of computer subroutines which have been written to perform engineering process and design calculations and to solve various types of mathematical problems. The subroutines are written in accordance with a standard format, and they use consistent technology. They are fully documented in a manual so that they can be easily used by other programmers.

There are three kinds of thermodynamic and physical property subroutines: monitor subroutines, method subroutines, and initialization subroutines. Their inter-relationship is illustrated in Figure 2. There is a monitor subroutine for every property; vapor molar volume, vapor fugacity coefficients, liquid activity coefficients, etc. Suppose a main program needs the value of property 1. It will call the monitor subroutine for property 1 and supply a computation method code. The monitor subroutine checks the code and calls the appropriate method subroutine to perform the calculations of property 1. If the method subroutine requires the value of property 2, it calls the monitor subroutine for property 2 and so forth. The monitor subroutines allow us to write very general main programs with a variety of options for calculating thermodynamic and physical properties. It is also very easy to add a new method to the system; one simply includes a new call statement in the monitor subroutine. The main program does not have to be reprogrammed when adding a new method. The third type of subroutine is the initialization subroutine which calculates the constant parameters associated with a method subroutine; for instance, the Redlich-Kwong equation of state constants. The main program calls the required initialization subroutines once for any given set of components.

Code Structure. As mentioned earlier, a monitor subroutine checks a method code to determine the calculational method. A flexible scheme has been developed for specifying method codes. The codes required for vapor-liquid equilibrium calculations are shown in Figure 3. Each column represents a set of five codes transmitted to a monitor subroutine for the property designated

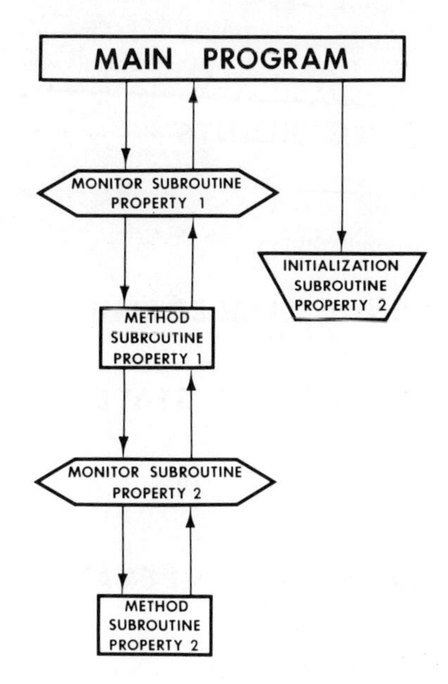

Figure 2. Structure of the subroutine system

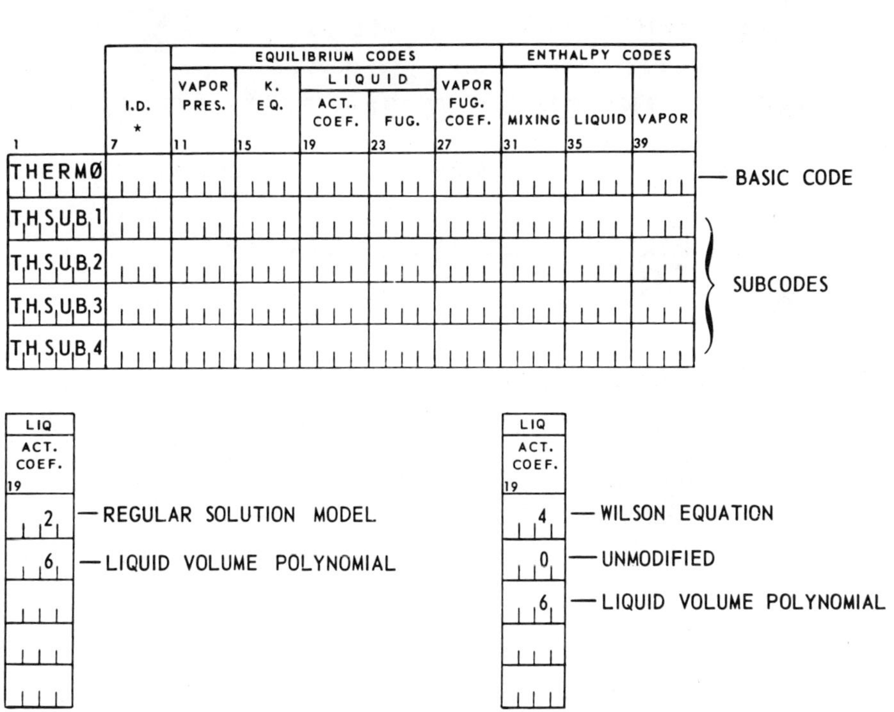

Figure 3. Method code structure

MOLECULAR WEIGHT

NORMAL BOILING & FREEZING POINTS

CRITICAL CONSTANTS

LIQUID DENSITY AT REFERENCE TEMPERATURE

HEATS OF FORMATION FOR VARIOUS STATES

PITZER'S ACENTRIC FACTOR

REFRACTIVE INDEX AT REFERENCE TEMPERATURE

SOLUBILITY IN VARIOUS SOLVENTS AT REFERENCE TEMPERATURE

SURFACE TENSION AT REFERENCE TEMPERATURE

Figure 4. Examples of single-valued properties in the date bank

PERFECT GAS HEAT CAPACITY

$$C_p^* = A + BT + CT^2 + DT^3 + ET^4$$

VAPOR PRESSURE

$$\ln P = A - B/(T + C) + D \ln T + ET^N$$

LIQUID VISCOSITY

$$\ln \mu = A + B/T + C \ln T$$

LIQUID THERMAL CONDUCTIVITY

$$\ln k = A + BT + CT^2$$

Figure 5. Examples of correlations whose constant parameters are in data bank

above it. The first code is the basic method code; each method is assigned a basic method code. The other four codes are method subcodes; the definition of the subcodes depends on the basic method code.

Let's look at the liquid activity coefficient codes as an example. A basic method code equal to 2 designates the Regular Solution model. Then, the first subcode is the basic method code for the liquid volume correlations required in this model, and the next code is the liquid volume method subcode, if one is required. On the other hand, suppose the basic method code for liquid activity coefficient is equal to 4, which denotes the Wilson Equation. Since several modifications of the Wilson Equation have been programmed in our system, the first subcode indicates which variation should be used. Now, the second subcode is the basic method code for the liquid volume calculations required in this model, followed by any requisite volume method subcodes.

The methods specified by subcodes are restricted to that particular application. Thus, one liquid volume method can be used in the activity coefficient calculations, another for the Poynting correction, and a third for pipe sizing calculations. If a subcode is missing, the basic method code for that property will be substituted. In the liquid activity coefficient examples, if the subcode for the liquid volume method is missing, the basic method code specified for liquid volume calculations will be used. If the basic method codes are missing, default values are assigned. The default method code for a property is the simplest model requiring the least input data, e.g., ideal-gas model for vapor fugacity coefficients.

Data Bank. Another part of the computer system is a Data Bank which serves as a repository for pure component data. The Data Bank contains single-valued properties of the type shown in Figure 4. It also contains the constant parameters and applicable ranges for property correlations of the sort illustrated in Figure 5. Only properties required for engineering calculations are stored in the Data Bank. All of the values in the Data Bank are internally consistent; the vapor pressure correlations do predict the normal boiling points and the critical points, the enthalpy and entropy values are based on the same reference state, etc. Every value in the Data Bank has a reference number associated with it. These relate to a list of references stored in a separate computer file. The references are comprehensive - every Data Bank value can be reproduced from the information given in the reference. The references include all of the data sources and the data-reduction methods used in obtaining the Data Bank values.

Main Programs. There are a variety of programs which call upon the Engineering Subprogram Library and the Data Bank. Several of these programs will be briefly described here. VLEFIT is a fitting program which uses vapor-liquid equilibrium (VLE) data to determine the best values for the adjustable parameters

INPUT DATA OPTIONS:

P-T-X
P-T-X-Y
P-T-Y
P-T-X-ΔH_{MIX}
T-X-γ
P-T-X-USER VARIABLE

ALLOWABLE OBJECTIVE FUNCTIONS:

P
Y
X
γ

Figure 6. Vapor-liquid equilibrium data fitting program (VLEFIT)

ABSORPTION

EXTRACTIVE DISTILLATION

AZEOTROPIC DISTILLATION

FRACTIONATION

RECTIFICATION

STRIPPING

LIQUID - LIQUID EXTRACTION

Figure 7. Scope of the MMSP (Multicomponent Multistage Separation Processes) programs

in VLE models. It will handle up to a quaternary system. The input data options and the allowable objective functions are shown in Figure 6. Any combination of input data and objective functions can be used in a given run, and weighting factors can be attached to both the input data and the objective functions.

The MMSP programs were developed for the simulation, design, and control of Multicomponent Multistage Separation Processes. These include the operations listed in Figure 7. These processes may be carried out in conventional columns (one feed and two products), in complex columns (multiple feeds, multiple products, and multiple interstage coolers or heaters), or in a series of conventional or complex columns.

A useful diagnostic option for checking VLE models is also available in the MMSP programs. Most VLE models are based on binary parameters. The data used in obtaining the binary parameters may be very limited and/or at conditions removed from those of interest. Therefore, it is wise to examine the predictions of the model for the binary pairs before attempting a multicomponent simulation. This option in MMSP will use the VLE model to produce y-x, T-x-y, P-x-y, α-x, and K-x diagrams for the binary systems. Two of these plots for the acetone-water system are shown in Figure 8.

Another program IPES (Integrated Process Engineering System) is a process simulator. It has the ability of modeling processes of any size, from a single distillation column to a complex plant. IPES is based on a building block concept. The process illustrated in the flowsheet of Figure 9-A would be modeled in IPES using the scheme in Figure 9-B. Each block corresponds to a unit or operation in the actual process. The blocks in IPES include mixer-splitters, reactors, flach units, distillation units, extractors, heat exchangers, compressors, control units, and economic units for sizing and costing process equipment. The method codes described earlier can be specified on an overall basis or individually for each block.

WHAT Models Are Used And WHY?

An efficient computing system is capable of producing meaningful results only if appropriate models are used. Therefore, at this point, I would like to describe the techniques we use for VLE, for liquid-liquid equilibrium (LLE), and for other separation processes. This will include examples of typical systems, the problems they pose, and how we attempt to cope with them.

Vapor-Liquid Equilibrium Models. I will begin by describing the models available in our system for VLE calculations - the field in which we have had the most experience, and in which the most effort has been expended. The calculations are based on VLE ratios - K values. The K value of a component in a mixture is related to its fugacity in the liquid phase and its fugacity in the vapor phase. The equations for determining the fugacities are given in Figure 10. The vapor fugacity is expressed in the usual fashion in terms of a fugacity coefficient, $\bar{\phi}^G$. There are

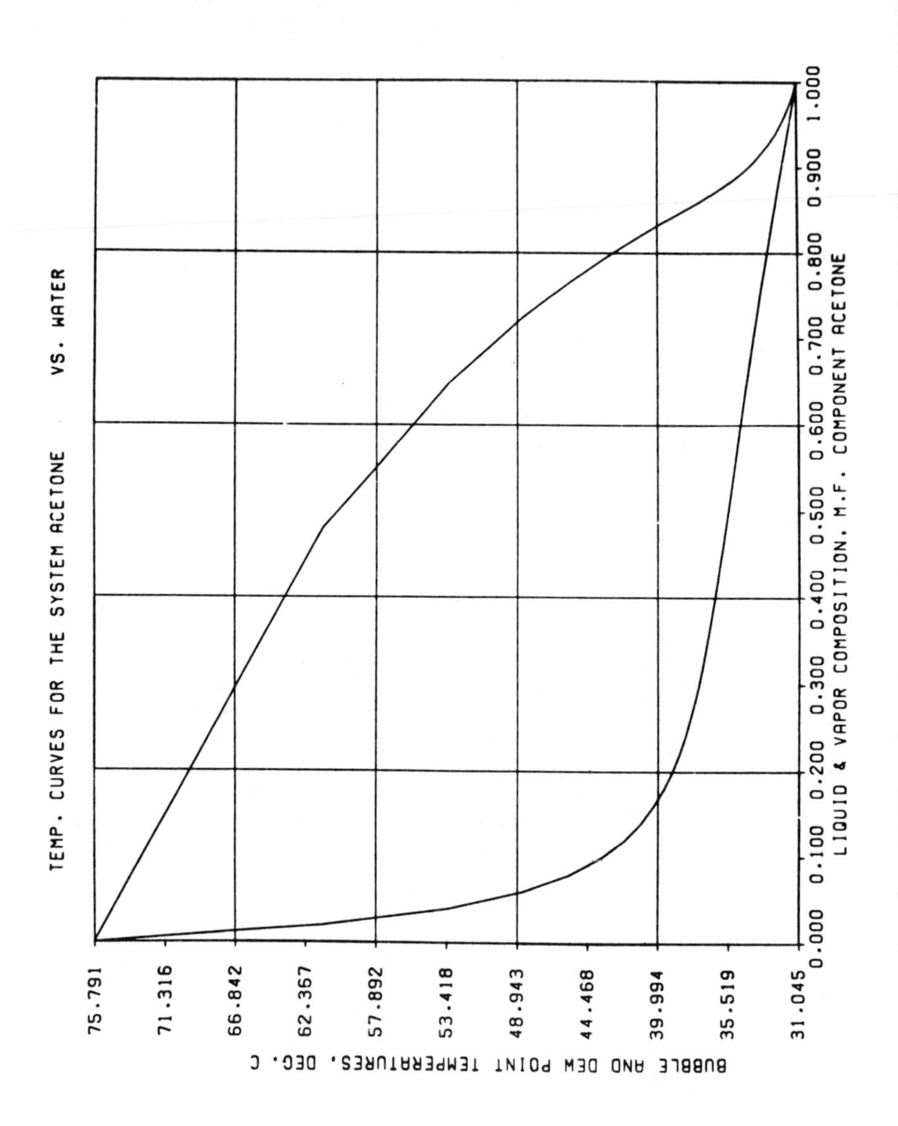

Figure 8. Diagnostic options available in MMSP

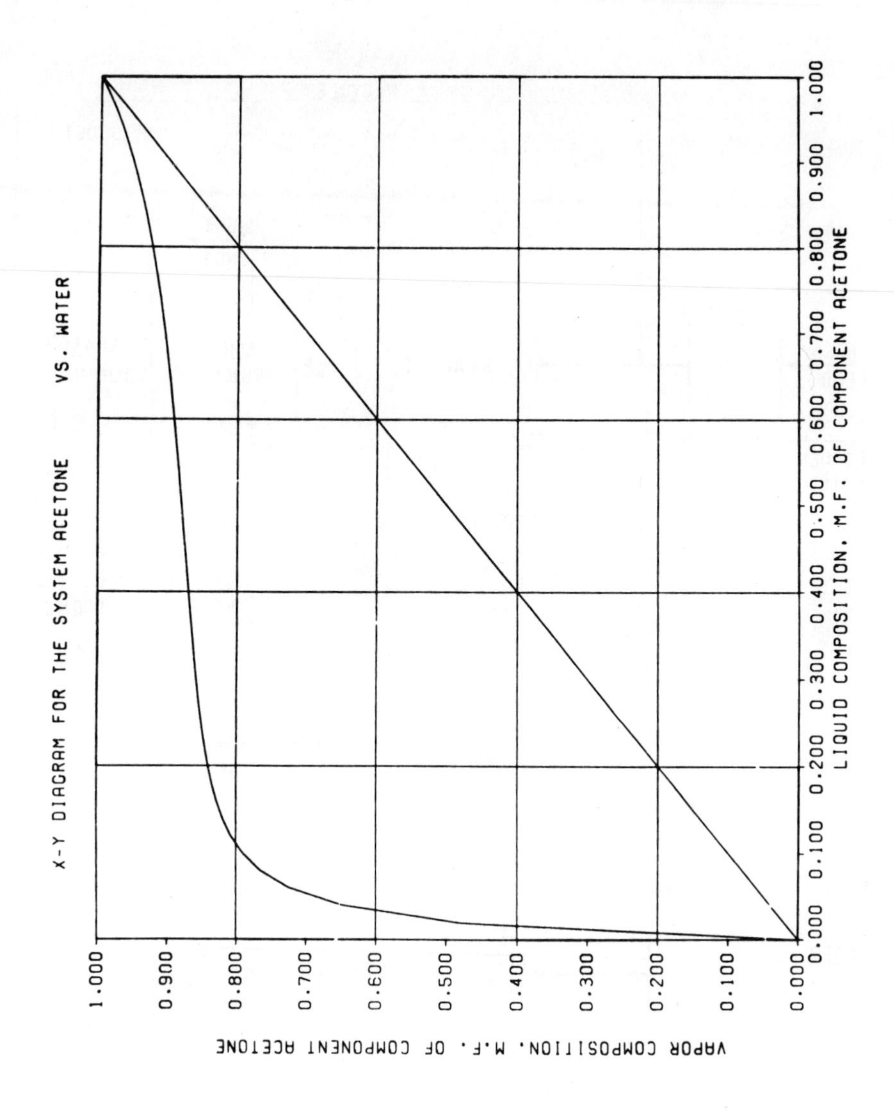
X-Y DIAGRAM FOR THE SYSTEM ACETONE VS. WATER
LIQUID COMPOSITION, M.F. OF COMPONENT ACETONE
VAPOR COMPOSITION, M.F. OF COMPONENT ACETONE
0.000
0.100
0.200
0.300
0.400
0.500
0.600
0.700
0.800
0.900
1.000

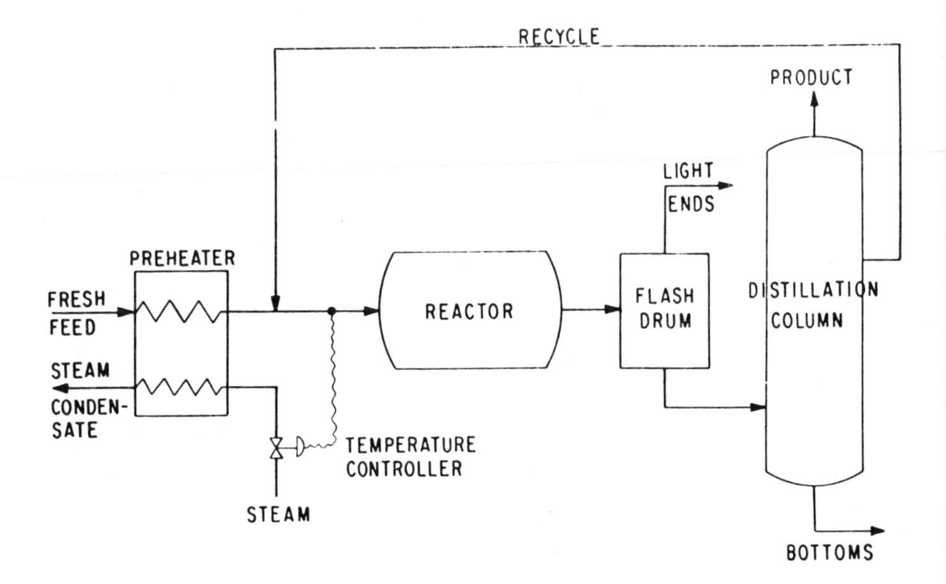

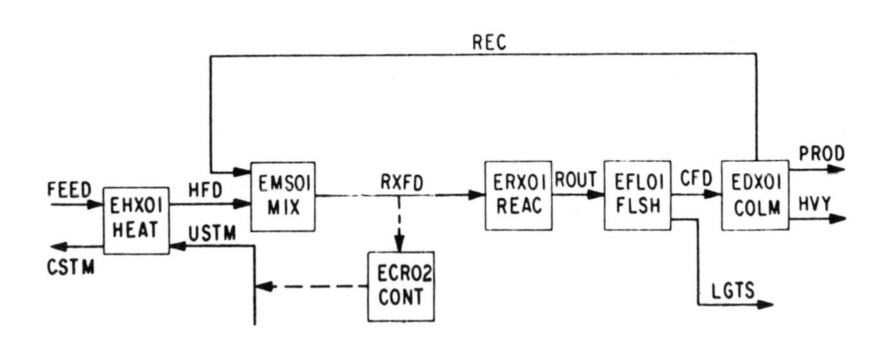

Figure 9. Process simulation program

three expressions for the liquid fugacity. The first is the traditional approach most often found in textbooks based on a standard state equal to the total system pressure; this is the option most frequently elected. In the second approach, the standard state is a fixed reference pressure. In the third equation, the liquid fugacity is expressed in terms of a fugacity coefficient, ϕ^L. Alternately, the K value can be calculated using an empirical correlation depending only on pressure and temperature.

The fugacity coefficients are determined from equations of state. The equations available for the vapor phase fugacity coefficients are listed in Figure 11. The two B-W-R equations and the Soave equation are also used for liquid phase fugacity coefficients. The Soave equation and the Hayden-O'Connell virial equation are very recent additions to the system. Therefore, if they are eliminated from consideration, the Prausnitz-Chueh modification of the Redlich-Kwong equation and the BWR equations have received the most usage.

The liquid reference fugacity and the liquid activity coefficient models are listed in Figures 12 and 13. At the present time, Henry's Law for supercritical components can be used only with the UNIQUAC equation; the unsymmetric convention is not included in the other liquid activity coefficient models. Vapor pressure with a Poynting correction is usually used for the liquid reference fugacity. The Wilson equation and the NTRL equations are the most commonly utilized liquid activity coefficient models. UNIQUAC and UNIFAC have just been added to the system, and if they fulfill our expectations, they will become the most commonly used models.

Enthalpy models are also necessary in VLE design calculations. The procedures followed for vapor and liquid enthalpy calculations are specified in Figure 14. The equation of state approach 1-(i) is the most popular for vapor enthalpy. For liquid enthalpy, the equation of state approach 1-(i), the Yen-Alexander correlation 1-(ii), and method 2 of mole fraction averaging the pure component enthalpies are used with about equal frequency. Method 3 for liquid enthalpy includes a heat of mixing term, ΔH_{mix}; it is not used very often because we have discovered that heats of mixing predicted by the liquid activity coefficient models are unreliable unless such data were included in the fitting procedure.

All of the property monitor subroutines have a group of codes which allow individuals to supply their own user-subroutin they find the available methods inadequate. The user-subroutines have fixed names and argument lists which are described in the documentation for the monitor subroutines.

Why do these lists contain so many models - some mediocre models, in fact. New models are always being added to the system, but old models are never discarded. Once a process has been successfully designed using a certain model, that model must always be available for future process calculations, expansion of

I. $K_i = Y_i/X_i = (\hat{f}_i^L/X_i)\,/\,(\hat{f}_i^G/Y_i)$

VAPOR: $\hat{f}_i^G = \bar{\Phi}_i^G\, Y_i\, \pi$

LIQUID: 1. $\hat{f}_i^L = \gamma_{i(\pi)}^L\, X_i\, f_{i(Pref)}^{OL}\, e^{\frac{1}{RT}\int_{Pref}^{\pi} V_i^L\, dP}$

2. $\hat{f}_i^L = \gamma_{i(Pref)}^L\, X_i\, f_{i(Pref)}^{OL}\, e^{\frac{1}{RT}\int_{Pref}^{\pi} \bar{V}_i^L\, dP}$

3. $\hat{f}_i^L = \bar{\Phi}_i^L\, X_i\, \pi$

π = SYSTEM PRESSURE
Pref = REFERENCE PRESSURE

II. $\ln(K\pi) = A + B/T + C/T^2 + D/T^3$

Figure 10. Vapor-liquid equilibrium equations

IDEAL ($\bar{\Phi} = 1$)

REDLICH – KWONG

VIRIAL (WOHL CORRELATION)

B-W-R

PRAUSNITZ–CHUEH MOD. REDLICH–KWONG

BARNER – ADLER

STARLING MOD. B-W-R

PERSOHN MOD. REDLICH-KWONG

SOAVE MOD. REDLICH-KWONG

HAYDEN-O'CONNELL VIRIAL

Figure 11. Equations of state

f_i^{OL} = VAPOR PRESSURE

$$f_i^{OL} = P_i * \phi_i^G(P_i) * e^{V_i^L/RT\,(Pref - Pi)}$$

CHAO - SEADER

CHAO - ROBINSON

PERSOHN

B - W - R

STARLING MOD. B - W - R

HENRY'S LAW FOR SUPERCRITICAL COMPONENTS

Figure 12. Liquid reference fugacity models

IDEAL ($\gamma = 1$)

REDLICH - KISTER

HILDEBRAND REGULAR SOLUTION (SOLUBILITY PARAMETER)

NRTL

WILSON

MARGULES

UNIQUAC – SYMMETRIC AND UNSYMMETRIC FORMS.

UNIFAC

Figure 13. Activity coefficient models

VAPOR AND LIQUID:

1. $H = H^{*} + (H - H^{*})$

H^{*} — IDEAL GAS ENTHALPY

$(H - H^{*})$ — (i) EQUATION OF STATE

(ii) YEN-ALEXANDER

2. $H = \sum_i \begin{pmatrix}\text{MOLE}\\ \text{FRACTION}\end{pmatrix}_i \begin{pmatrix}\text{TEMPERATURE}\\ \text{POLYNOMIAL}\end{pmatrix}_i$

LIQUID:

3. $H = H_{\text{IDEAL SOLN.}} + \Delta H_{\text{MIX}}$

$H_{\text{IDEAL SOLN.}}$ — (i) LIQUID REFERENCE FUGACITY MODELS

(ii) $\sum X_i \begin{pmatrix}\text{TEMPERATURE}\\ \text{POLYNOMIAL}\end{pmatrix}_i$

ΔH_{MIX} — LIQUID ACTIVITY COEFFICIENT MODELS

Figure 14. Enthalpy models

HYDROGEN	ACETALDEHYDE
METHANE	CROTONALDEHYDE
ETHYLENE	ETHYL ETHER
ETHANE	ETHANOL
1 – BUTENE	TERT-BUTANOL
n – BUTANE	WATER
2 – METHYLPENTANE	ETHYLENE GLYCOL
n – OCTANE	

T: -10 °C ⟶ 350 °C

P: 5 ⟶ 1100 psia

Figure 15. Typical VLE problem, Example 1

the unit, or the design of a new unit. The effort required to redo an acceptable old model cannot be justified.

Vapor-Liquid Equilibrium-Sample Systems. In preparing this presentation, I took an "attitude survey" among my colleagues who are using this VLE calculational system. The individuals involved in the simulation of hydrocarbon units were generally satisfied with the current state of affairs. They have years of experience in making these calculations, the correlations are appropriate for the compounds of interest to them, and they have accumulated corroborative plant data. However, they did admit that the following still present a challenge: high pressure critical phenomena, cryogenic separations, very heavy hydrocarbons - tars, and the presence of hydrogen in the system.

The individuals involved in the design of 'chemical' units had more grievances. Let's look at a few examples in Figures 15-17. In the first example, there are 15 compounds - not an unusual number for most models. The compounds vary markedly from each other. The person working on this project was until recently teaching and doing research at a university. She claims that what she misses most from the academic life is the 'nice compounds' worked with there. The model is expected to work over a broad range of temperatures and pressures; therefore, a number of the compounds will be supercritical sometimes - but not always. Pure component data are scarce for a number of these compounds; e.g. crotonaldehyde. What are the critical properties of ethylene glycol--it decomposes before reaching a critical point. VLE data are not available for all of the binary pairs; this is not surprising since there are 105 binary pairs. Some of the VLE data are contradictory; e.g., acetaldehyde--ethanol. What equation of state is applicable for hydrocarbons, aldehydes, and alcohols? How should the liquid enthalpy be treated with so many light compounds?

Pure component data which cannot be easily measured are estimated--including the critical properties of ethylene glycol. In many cases, the estimation techniques are crude, or they have not been developed for the type of compound under consideration. If VLE data are missing for any of the important binary pairs, they are measured.

For the VLE model in this example, the Prausnitz-Chueh Redlich-Kwong equation of state was selected for the vapor phase fugacity coefficients and the Wilson equation for the liquid phase activity coefficients. Vapor pressure with a Poynting correction and a saturated vapor fugacity coefficient correction was used for the liquid reference fugacity; it was suitably extrapolated into the hypothetical liquid region for supercritical components. The Wilson equation parameters were determined from the available data with the fitting program VLEFIT. The hope was that in the fitting process, the inadequacies of the model would be absorbed in the activity coefficient parameters. The Wilson equation parameters for the binary pairs with no VLE data were estimated by examining the parameters for similar pairs. It is hoped that UNIFAC will

ARGON	ACETIC ACID
NITROGEN	WATER
OXYGEN	ACETALDEHYDE
ETHYLENE	ETHYL ACETATE
ETHANE	VINYL ACETATE
CARBON DIOXIDE	GLYCOL DIACETATE
ACROLEIN	VINYL PROPIONATE

T: 0 °C – 200 °C
P: ~1atm.

Figure 16. Typical VLE problem, Example 2

HYDROGEN CHLORIDE
METHYL CHLORIDE
DIMETHYL ETHER
METHANOL
WATER
SODIUM CHLORIDE
SODIUM HYDROXIDE

T: 10 °C – 130 °C
P: 35 – 135 psia

Figure 17. Typical VLE problem, Example 3

provide us with a more systematic approach for this estimation procedure. Finally, the equation of state was also used for vapor enthalpy calculations. The liquid enthalpy was calculated as the mole fraction average of the pure component enthalpies; once again, the hypothetical liquid state was used for supercritical components.

The system in the second example is plagued with all of the usual problems and, in addition, it contains acetic acid. The person who developed a model for this system decided that it was necessary to account for the vapor phase association of acetic acid. Thus, a user equation of state subroutine was written, wherein the association was rigorously treated; then appropriate correction factors were determined for the vapor fugacity coefficient and vapor enthalpy of the apparent species. The flexibility in the computing system made this possible.

This is a highly nonideal system. After an adequate model was developed, the subsequent column calculations were very difficult to converge. This brings up the question of what is the best way to initialize such calculations--the optimum method would require no input from the user and would be quick and failsafe.

If this system had contained another associating compound like propionic acid, or if the system pressure had been higher, the analysis would have been complicated by several orders of magnitude, and the development of such a custom model might prove to be too time consuming. The "chemical theory" of Nothnagel has been included in the new Hayden-O'Connel virial equation subroutine to provide a standardized method for dealing with such systems.

The third example highlights another problem. How does one cope with several salts in an already nonideal solution? In this case, the salt effects were accounted for by modifying the vapor pressure of the solvents; a vapor pressure depression factor was calculated as a function of salt concentration. We recognize that a better technique than this should be available in the computing system. What should it be? Expensive and/or rare salts are being used as catalysts in reaction systems. Predicting their effects on the subsequent separations systems and the design of catalyst recovery systems require good models.

Vapor-Liquid Equilibrium Plus Chemical Reaction. To conclude this discussion of VLE calculations, I'd like to mention that we have initiated a program to allow VLE plus chemical reaction in column calculations. A modification of Frank's dynamic simulation method (4) is being used. The initial trials with complex VLE and kinetic models have been moderately successful, but further testing is required. Other simulation techniques are also being investigated.

Liquid-Liquid Equilibrium. The next phase equilibrium situation I would like to discuss is LLE. In our organization, this field has received less attention than it deserves. The available models are shown in Figure 18. Most of the activity coefficient models mentioned previously are applicable in the first option with the notable exception of the Wilson equation. The new UNIQUAC model may prove to have real value in this application; the developers of the model state that it gives a good representation of LLE for a wide variety of mixtures. Until recently, our fitting program did not have the facility to deal with LLE data. We still need guidance in deciding what constitutes a "good" fit, what the objective function for fitting should be, and what is the proper data to use. The second option in Figure 18 is based on a fit of multicomponent experimental data in the range of interest.

In our limited experience, we have learned the following valuable lessons. One cannot expect to model a complex system by collecting large amounts of data and attempting to fit these data by brute force. Developing the proper empirical equations for systems with several chemically dissimilar, interactive compounds is very difficult. Statistically designed experiments for collecting data proved not to be helpful in developing such models. The liquid activity coefficient approach for predicting distribution coefficients is viable, although it is still difficult. In order to develop an accurate model, both VLE and LLE data must be used in determining the parameters for the liquid activity coefficient correlation.

An example of a system which has been successfully modeled using the NRTL liquid activity coefficient correlation is shown in Figure 19. First, a set of NRTL parameters was generated by fitting binary VLE and LLE data. Then, these parameters were optimized using multicomponent LLE data. Calculations using this model show excellent agreement with plant data.

Vapor-Liquid-Liquid Equilibrium. We have had limited experiience in rigorous three phase equilibrium calculations, vapor-liquid-liquid, primarily in single stage flash units. The implementation of such a three-phase equilibrium model in column calculation is scheduled in the future. Presently, a method also exists wherein complete immiscibility in the liquid phase can be specified between one component and all of the other components in the system; e.g., between water and a set of hydrocarbons. The VLE ratios are normalized on an overall liquid basis so that the results can be used in conventional two-phase liquid-vapor equilibrium calculations.

Other Separation Processes. The final type of phase equilibrium I'd like to mention could be titled unusual separation processes--adsorption, ion exchange, membranes. The individuals involved in these areas are definitely non-computer people. There are no models in the present computer system which satisfy their particular needs, but more importantly, they can't recommend any that should be added to the system. They firmly believe that empirical correlations based on experimental data are best. For new systems, they use a similarity approach based on previous experience. They volunteered the opinion that an analysis with a

1. DISTRIBUTION COEFFICIENTS FROM ACTIVITY COEFFICIENTS

2. DISTRIBUTION COEFFICIENTS FROM EMPIRICAL CORRELATIONS OF TEMPERATURE AND COMPOSITION

Figure 18. Liquid-liquid equilibrium models

BENZENE	OCTANE
TOLUENE	CYCLOPENTANE
p – XYLENE	CYCLOHEXANE
CUMENE	METHYLCYCLOPENTANE
n – PENTANE	METHYLCYCLOHEXANE
n – HEXANE	WATER
HEPTANE	TETRAETHYLENE GLYCOL

T: 100 °C – 160 °C

P: 1 – 15 atm.

Figure 19. Typical LLE problem

theoretical basis might be nice, but they didn't really expect that it was possible.

Model Selection. As a point of interest, I'd like to remark at this time that the models in a design system are decided upon by a few individuals. In a large company, there is usually a technology section responsible for such matters; in smaller organizations, the decisions rest with one or two persons. For both, it is impossible to review and evaluate everything being published nowadays. Expert consultants are employed to give us guidance in these matters. We tend to concentrate on the areas in which our deficiencies are limiting the accuracy of design calculations.

Our interest will not be aroused by another modification of the Redlich-Kwong equation, an involved explanation of what the Wilson parameters really mean, a correlation valid for the noble gases, or a correlation limited to pure components. We look for methods that are applicable to a wide variety of compounds, especially nonhydrocarbons, and that can be used over a broad range of temperatures and pressures. The correlations must be reasonably accurate. We favor correlations which are simple in form and which require easily accessible input data. To optimize the efficiency of iterative computer calculations, we prefer correlations which can be solved directly or which converge rapidly.

To aid us in the evaluation and possible implementation of a method, we would appreciate details and sample calculations, which could be published or available as supplementary material. After a method is added to the system, we have no control over how it will be used. Therefore, we build checks into the subroutines versus applicable temperatures, pressures, acentric factors, etc., --if we've been told what they are. We have also had the unfortunate experience of checking correlations within the published applicable range and discovering erratic behavior: don't promise more than you can deliver!

How do we know that we have selected the proper models to implement in our system? We receive feedback on the effectiveness and on the inadequacies of the models from the process engineers who are using them. The shortcomings of the system are commented upon vociferously; the virtues are accepted matter-of-factly. Laboratory and plant tests are often made to confirm the validity of a process model or to indicate areas where discrepancies exist. In the latter case, the model is reviewed and appropriately adjusted to eliminate the differences.

WHERE Should Future Development Work Be Directed?

This question produces a variety of responses depending upon whom you ask, when you ask, and what is being worked on at the time. The two main categories are shown in Figure 20.

Highly Nonideal Systems. Highly nonideal systems are of principal concern; the nonideality may be caused by the components present and/or by the operating conditions. This category includes subjects like critical phenomena, extractive/azeotropic distillation, solvent selection guidelines, salt effects, high molecular

HIGHLY NONIDEAL SYSTEMS

EXPERIMENTAL TECHNIQUES AND DATA

Figure 20. Important areas for future development

weight polymer systems, and liquid mixtures at high pressures with supercritical components. Dealing with supercritical components is one of the major problems today; a good general method does not exist. The Henry's Law approach with the unsymmetric convention for activity coefficients is not well understood. Using an equation of state to represent both phases would be satisfactory, but currently this technique is only applicable to light hydrocarbon systems. No one expects to eliminate the need for experimental data. What we do require are correlating models that are appropriate for highly nonideal systems. The acid-gas system is a case in point; there is a mass of data, but no model with a firm theoretical basis. We would like to rely on the models to allow us to minimize the necessary experimental data. Estimation techniques like UNIFAC are useful because they permit us to treat the less important components of a system in a reasonable manner.

Experimental Techniques and Data. Lack of data and the dubious quality of some data also pose severe problems; this is the second area of concern. Reliable experimental techniques for measuring 'useful' equilibrium data quickly would be invaluable. Recently, there have been many data published on the excess properties of mixtures; but the usefulness of such data has yet to be defined. Many design projects have close time schedules, and so the possible data collection is severely limited because the experimental methods are so time-consuming. There also seems to be a decline in the amount of high quality, new data being published; the same systems are discussed over and over again. In approaching new systems, our literature surveys tend to lead to foreign publications more and more frequently. Has quality data collection been abandoned at a published level?

Conclusions

In conclusion, I think it is fair to say that significant progress has been achieved in the industrial treatment of phase equilibria in the past twenty-five years. We are trying to use a theoretically sound basis for our calculations, and we have the capability of using sophisticated complex models. We do keep track of new developments in the field and implement them, albeit in a cautious and conservative fashion.

We are still faced with many unsolved problems. Some of the techniques used are inadequate or inappropriate. In order to eliminate these deficiencies, we must be committed to a program of continually upgrading our skills. Hopefully, through meetings like this, we can define future development goals that will be of mutual interest and benefit to the industrial and academic communities.

Nomenclature

A, B, C, D, E - constant parameters

C_p^* - ideal gas molar heat capacity
$\hat{f}^G$ - fugacity in the vapor phase mixture
$\hat{f}^L$ - fugacity in the liquid phase mixture
$\hat{f}^{oL}$ - liquid reference fugacity
H - molar enthalpy
H* - ideal gas molar enthalpy
(H-H*) - molar enthalpy departure frm the ideal gas state
ΔH_{mix} - molar liquid heat of solution
k - liquid thermal conductivity
K - vapor-liquid equilibrium ratio
P - absolute pressure
P_{ref} - reference pressure
R - gas constant
T - absolute temperature
V^L - liquid molar volume
$\overline{V}^L$ - liquid partial molar volume
x - mole fraction in liquid phase
y - mole fraction in vapor phase
α - relative volatility
γ^L - liquid activity coefficient
μ - liquid viscosity
$\overline{\phi}^G$ - vapor phase fugacity coefficient
$\overline{\phi}^L$ - liquid phase fugacity coefficient
π - system pressure

Subscripts

i - property of component i
P_{ref} - property evaluated at reference pressure
π - property evaluated at system pressure

References

A. General

1. Hachmuth, K. H., Chem. Eng. Prog., (1952) 48, 523, 570, 617.
2. Pigford, R. L., Ind. Eng. Chem., (1952) 44, 25.
3. Walsh, T. J., Ind. Eng. Chem., (1952) 44, 45.
4. Franks, R. G. E., "Modeling and Simulation in Chemical Engineering,", Wiley-Interscience, New York: 1972.

B. Equations of State

5. Redlich-Kwong: Redlich, O. and Kwong, J. N. S., Chem. Rev., (1949) 44, 233.
6. Virial (Wohl Correlation): Wohl, A., Z. Phys. Chem., (1929) B2, 77.
7. B-W-R: Benedict, M., Webb, G. B., and Rubin, L. C., Chem. Engr. Prog., (1951); 47, 419;J. Chem. Physics. (1940) 8, 334;ibid., (1942) 10, 747.
8. Prausnitz-Chueh modified Redlich-Kwong: Chueh, P. L. and Prausnitz, J. M., Ind. Eng. Chem. Fund. (1967) 6, 492.

9. Barner-Adler: Barner, H. B. and Adler, S. B., Ind. Eng. Chem. Fund., (1970) 9, 521.
10. Starling modified B-W-R: Starling, K. E., Hydro. Proc., (1971) 50, (3), 101.
11. Persohn modified Redlich-Kwong: Persohn, T. F., Unpublished method used at Union Carbide Corporation.
12. Soave modified Redlich-Kwong: Soave, G., Chem. Engr. Sci., (1972), 27, 1197.
13. Hayden-O'Connell Virial: Hayden, J. G. and O'Connell, J. P., Ind. Eng. Chem. Proc. Des. Dev., (1975) 14, 209. Nothnagel, K. H., Abrams, D. S., and Prausnitz, J. M., Ind. Eng. Chem. Proc. Des. Dev., (1973) 12, 25.

C. Liquid Reference Fugacity

14. Chao-Seader: Chao, K. C. and Seader, J. D., AIChE Journal, (1961) 1, 598.
15. Chao-Robinson: Robinson, R. L. and Chao, K. C., Ind. Eng. Chem. Proc. Des. Dev., (1971) 10, 221.
16. Persohn: Persohn, T. F., Unpublished generalized method used at Union Carbide Corporation.

D. Activity Coefficient

17. Redlich-Kister: Smith, B. D., "Design of Equilibrium Stage Processes", Chapter 2, McGraw-Hill, New York, (1963).
18. Hildebrand: Prausnitz, J. M., "Molecular Thermodynamics of Fluid Phase Equilibria", Chapter 7. Prentice-Hall, Englewood Cliffs, New Jersey; 1969.
19. NRTL: Renon, H. and Prausnitz, J. M., AIChE Journal, (1968) 14, 135.
20. Wilson: Wilson, G. M., J. Am. Chem. Soc., (1964) 86, 135.
21. Margules: Wohl, K., Trans. Am. Inst. Chem. Engrs., (1946) 42, 215. Brown, G. M. and Smiley, H. M., AIChE Journal (1966) 12, 609.
22. UNIQUAC: Abrams, D. S. and Prausnitz, J. M., AIChE Journal, (1975) 21, 116. O'Connell, "Application of the UNIQUAC Equation for Excess Gibbs Energy to Systems Containing Supercritical Hydrogen, Nitrogen and/or Methane", presented at Sixty-Eighth Annual Meeting of the American Institute of Chemical Engineers, November 17, 1975, Los Angeles, California.
23. UNIFAC: Fredenslund, A., Jones, R. L. and Prausnitz, J. M., AIChE Journal, (1975), 21, 1086.

E. Enthalpy

24. Yen-Alexander: Yen, L. C. and Alexander, R. E., AIChE Journal, (1965) 11, 334.

4

Measurement of Vapor-Liquid Equilibrium

MICHAEL M. ABBOTT

Rensselaer Polytechnic Institute, Troy, N.Y. 12181

The importance accorded the measurement of vapor-liquid equilibrium (VLE) data needs little elaboration. A glance at the annual indices for the Journal of Chemical and Engineering Data makes the point nicely. For the five years 1971-1975, at least 50 papers present new VLE data, and many of them contain descriptions of new experimental techniques. Perusal of recent volumes of less specialized journals, such as the AIChE Journal or I&EC Fundamentals, reveals a similar proliferation of VLE studies.

The necessity for reliable VLE data is apparent. Many separations processes involve the transfer of chemical species between contiguous liquid and vapor phases. Rational design and simulation of these processes requires knowledge of the equilibrium compositions of the phases. Raoult's and Henry's "laws" rarely suffice as quantitative tools for the prediction of equilibrium compositions; precise work demands the availability of either the equilibrium data themselves, or of thermodynamic correlations derived from such data.

Specialists in the field tend to concentrate on one of two broad areas: high-pressure VLE, or low-pressure VLE. My major interests are in the latter area, and hence the thrust of my talk will be in this direction: the measurement and reduction of low-pressure VLE data.

Low-pressure VLE experimentation differs from high-pressure experimentation on two major counts. First, the problems of equipment design and operation are less formidable than for high-pressure work. Secondly, more effective use can be made of the thermodynamic equilibrium equations, both in the data reduction process and in the design of the experiments themselves. It is this second feature--the strong interplay of theory with experiment--which to me most distinguishes low-pressure from high-pressure VLE work.

The usual product of a low-pressure VLE study is an expression for the composition dependence of the excess Gibbs function G^E for the liquid phase. If experiments are done at several temperatures, then the temperature dependence of parameters in the equation for G^E may also be determined. At low pressures, the pressure dependence of G^E is weak, and may usually be ignored.

Certain advantages derive from this kind of representation, for the description of VLE is compact and yet has thermodynamic significance. Back-calculation of equilibrium curves requires values for the pure-component vapor pressures, an expression for G^E in terms of x and T, and equations of state (usually of simple form) for the vapor and liquid phases. Approximations, when they must be made, are well-defined and often subject to independent verification.

Thermodynamic Considerations

At equilibrium, the fugacities of each component i must be the same in the liquid and vapor phases:

$$\hat{f}_i^{\ell} = \hat{f}_i^{v} \tag{1}$$

The vapor-phase fugacity is related to the vapor-phase fugacity coefficient $\hat{\phi}_i$:

$$\hat{f}_i^{v} = y_i \hat{\phi}_i P \tag{2}$$

Fugacity coefficient $\hat{\phi}_i$ is calculable from an equation of state for the vapor phase; mixing rules must be available for the equation-of-state parameters.

The liquid-phase fugacity is related to the liquid-phase activity coefficient γ_i:

$$\hat{f}_i^{\ell} = x_i \gamma_i f_i^{o} \tag{3}$$

Here, f_i^o is the standard-state fugacity for species i. If the equilibrium temperature is lower than the critical temperature of pure i (the usual case for low-pressure VLE), the standard state is taken as pure liquid i at the system T and P. Thus Eq. (3) becomes

$$\hat{f}_i^{\ell} = x_i \gamma_i f_i \tag{4}$$

Determination of f_i requires the availability of equations of state for pure vapor and liquid i, and a value for the vapor pressure P_i^{sat} of pure i. Often, f_i is approximately equal to P_i^{sat}.

Combination of Eqs. (1), (2), and (4) yields the form of the equilibrium equation most commonly used in low-pressure work

$$x_i \gamma_i f_i = y_i \hat{\phi}_i P \tag{5}$$

It is through Eq. (5) that low-pressure VLE measurements provide the experimental input required for a quantitative thermodynamic description of liquid-phase nonidealities.

The excess Gibbs function G^E plays a central role in experimental solution thermodynamics, for its canonical variables (T, P, and x) are those most susceptible to measurement and control. Activity coefficients are related to mole-number derivatives of the dimensionless excess Gibbs function $g \equiv G^E/RT$:

$$\ell n\gamma_i = [\partial(ng)/\partial n_i]_{T,P,n_j} \tag{6}$$

Thus $\ell n\gamma_i$ is a partial molar property with respect to g. Accordingly, we have the additional useful relationship

$$g = \Sigma x_i \ell n\gamma_i \tag{7}$$

Classical thermodynamics provides the following expression for the total differential of g:

$$dg = -\frac{H^E}{RT^2}\,dT + \frac{V^E}{RT}\,dP + \Sigma \ell n\gamma_i dx_i \tag{8}$$

Here, H^E is the excess enthalpy (heat of mixing) and V^E is the excess volume (volume change of mixing). Both H^E and V^E are subject to direct experimental determination. Taking the total differential of Eq. (7), and comparing the result with Eq. (8), we obtain the Gibbs-Duhem equation:

$$\Sigma x_i d\ell n\gamma_i = -\frac{H^E}{RT^2}\,dT + \frac{V^E}{RT}\,dP \tag{9}$$

Equations (1) through (9), or variations upon them, constitute the usual thermodynamic basis for the reduction and interpretation of low-pressure VLE data.

Isothermal vs. Isobaric Data

Precise determination of G^E through low-pressure VLE measurements generally requires the availability of data spanning the entire range of liquid compositions. The experimentalist still has the option, however, of collecting his data either at isobaric or at isothermal conditions. The question then arises whether one type of data is more useful than the other.

Let us assume the availability of two complete sets of error-free VLE measurements. The first set has been taken at constant pressure, and consists of values of T, x, and y. The second is isothermal, consisting of values for P, x, and y. We wish to reduce both sets of data, so as to obtain values for G^E.

For each data set, we may calculate point values of γ_i from Eq. (5):

$$\gamma_i = y_i\hat{\phi}_i P/x_i f_i \tag{10}$$

The corresponding values for g are then computed from Eq. (7):

$$g = \Sigma x_i \ln \gamma_i$$

It is desirable for purposes of correlation and interpretation that experimental values of g be at a <u>single</u> T and P, with x the only variable. Clearly, this is not the case for either of our experiments. Corrections must be made in both cases to reduce the values of g to a common basis. According to Eq. (8), this requires the use of independently determined values for H^E or V^E. Thus, for isobaric data,

$$g(T',P,x) = g(T,P,x) - \int_T^{T'} \frac{H^E}{RT^2} dT$$

Similarly, for isothermal data,

$$g(T,P',x) = g(T,P,x) + \int_P^{P'} \frac{V^E}{RT} dP$$

Here, T' is a reference temperature for the isobaric data, and P' a reference pressure for the isothermal data.

Even if the required values of H^E or V^E are available--and often they are not--the above corrections complicate the data analysis. One naturally asks whether one or the other of the corrections might more safely be ignored. Calculations for real systems show that the temperature correction is often substantial, whereas the pressure correction is frequently negligible; at low pressures, g may depend strongly on T, but rarely upon P.

Thus low-pressure isothermal VLE data are more easily reduced to useful form than are isobaric data; they are to be preferred for this reason. Much of the older low-pressure work is isobaric; the best modern studies are isothermal.

The argument is sometimes advanced that isobaric VLE data are more useful for process design, because separation processes are more nearly isobaric than isothermal. This argument ignores the fact that pressure drops in distillation columns can be substantial, and that they are accounted for in modern design procedures. The best thermodynamic tool for low-pressure distillation column design is an expression for G^E in terms of T and x, with values for the parameters determined from carefully executed isothermal VLE experiments.

<u>Direct vs. Indirect Determination of Vapor Compositions</u>

We consider the determination of G^E via the reduction of low-pressure isothermal P-x-y data. The procedure is simple and direct Values of γ_i are computed from Eq. (10), and the corresponding values of g are calculated from Eq. (7); the values of g are then smoothed with respect to liquid composition. The smoothing procedure may be graphical or numerical, but the eventual product is an analytical expression for g as a function of x.

In the above procedure, no use is made of the Gibbs-Duhem equation, Eq. (9). Because the pressure-dependence of G^E is weak at low pressures, this equation reduces for isothermal conditions to

$$\Sigma x_i d\ell n\gamma_i = 0 \tag{11}$$

Equation (11) in fact imposes a thermodynamic constraint upon the liquid-phase activity coefficients, a constraint not necessarily satisfied by values of γ_i computed via Eq. (10) from real (and therefore possibly imperfect) data. However, values of γ_i generated from a smoothing equation for g via Eq. (6) do satisfy Eq. (11). Comparison of generated with experimental values of γ_i constitutes an example of a popular exercise known as a thermodynamic consistency test. Many consistency tests have been proposed, both for low- and high-pressure VLE data. Van Ness et al (1) and Christiansen and Fredenslund (2) present readable discussions of such tests.

Instead of serving as a basis for the testing of redundant data, the Gibbs-Duhem equation may be used in quite a different manner, one which aids the experimenter in the design of a VLE experiment of minimal complexity. For purposes of discussion, we assume ideal-gas behavior for the vapor phase, and pressure-independence of liquid-phase properties. In this case, Eq. (10) reduces to

$$\gamma_i = y_i P / x_i P_i^{sat} \tag{12}$$

and the Gibbs-Duhem equation becomes, for a binary system,

$$x_\ell d\ell n\gamma_1 + x_2 d\ell n\gamma_2 = 0 \tag{13}$$

Equations (12) and (13) yield, on combination and simplification,

$$\frac{dy_1}{dP} = \frac{y_1(1-y_1)}{P(y_1-x_1)} \tag{14}$$

Equation (14) is a restricted form of the binary coexistence equation. The important content of the equation is as much conceptual as mathematical: it illustrates that simultaneous measurement of P, x, and y is unnecessary, that vapor compositions can in principle be computed from measurements of just P and x. Once the y are determined by integration, values of γ_i follow directly from Eq. (12). Van Ness (3) presents a detailed discussion of the characteristics and application of Eq. (14).

Reduction of VLE data via the coexistence equation is "indirect", in that it makes no direct use of experimentally-determined vapor compositions. Other indirect approaches are possible. One of the most popular is that first proposed by Barker (4). Barker's method is based upon the equation

$$P = \sum_i (x_i \gamma_i P_i^{sat} / \Phi_i) \tag{15}$$

which follows from Eq. (5). Quantity Φ_i is defined by

$$\Phi_i \equiv \hat{\phi}_i (P_i^{sat} / f_i)$$

Activity coefficients in Eq. (15) are replaced by the expressions

$$\gamma_i = \exp \left[g - \sum_{k \neq i} x_k \left(\frac{\partial g}{\partial x_k}\right)_{T,P,x_{\ell \neq k,i}} \right] \tag{16}$$

The composition dependence of g is expressed by an equation of suitable form, and values of undetermined parameters in this equation are found by a regression procedure that yields a best fit of the P vs. x data. The Φ_i depend upon the y_i, which are not initially known; thus an iterative procedure is required. The major requirement for successful application of Barker's method is the availability of an analytical expression for g that is capable of producing a fit to the P-x data to within the limits of experimental uncertainty. Abbott and Van Ness (5) and Abbott (6) discuss the development and selection of such equations.

There has been much discussion of the relative merits of direct measurement, as opposed to indirect calculation, of vapor compositions. Although the difficulty of precise measurement of y's is generally conceded, there remains a strong body of opinion that the redundant information provided by experimental values of y (however much in error they might be) somehow enhances the reliability of the eventual correlation for g. I do not subscribe to this view. The difficulty of carrying out the necessary calculations by hand certainly at one time constituted a reasonable argument against indirect determination of y's, but the electronic computer has removed this computational barrier. The experimenter's time is best spent in achieving accuracy in the measurement of the minimum number of variables required to characterize the system, not in devising ingenious ways to collect redundant data.

Extension of Isothermal VLE Data with Temperature

According to Eq. (8), the excess enthalpy is proportional to the temperature derivative of g:

$$H^E = -RT^2 \left(\frac{\partial g}{\partial T}\right)_{P,x} \tag{17}$$

Equation (17) suggests an alternative procedure to the direct determination of isothermal VLE at several temperatures, namely, the use of heat-of-mixing data for extrapolation or interpolation of g with T. Suppose for example that isothermal data are available at a single temperature T_1, and that H^E data are available at two temperatures near T_1. Suppose further that H^E/RT is known to vary approximately linearly with T:

$$H^E/RT = a + bT$$

where a and b depend upon composition only. Values of g at some other temperature T_2 can then be found by integration of Eq. (17):

$$g(T_2) = g(T_1) - a \ln (T_2/T_1) - b(T_2-T_1)$$

Obvious extensions to this simple example are possible. Thus, the method can be used for computation of isobaric VLE data. More comprehensive expressions for the temperature dependence of H^E can be employed, which allow incorporation of several sets of isothermal VLE data and H^E data. Finally, Cp^E data can be incorporated into the procedure through use of the thermodynamic equation

$$C_P^E = \left(\frac{\partial H^E}{\partial T}\right)_{P,x} \qquad (18)$$

An elegant example of the simultaneous use of isothermal VLE and H^E data to provide a complete and internally consistent set of excess functions over a wide temperature range is the recent work of Larkin and Pemberton (7) on the ethanol-water system.

Measurement of Low-Pressure VLE Data

Hála et al (8) classify VLE measurement techniques in five major groups: distillation methods, circulation methods, static methods, dew-point/bubble-point methods, and flow methods. One could add to this list some rather specialized techniques, but in the main most modern low-pressure VLE work is done on two major types of equipment: circulation stills and static equilibrium cells.

In low-pressure vapor-circulation stills, a liquid mixture is charged to a distilling flask, and brought to a boil. The evolved vapors are condensed externally into a receiver; excess condensate returns through an overflow tube back into the distilling flask, where it mixes with the boiling liquid. The compositions of the boiling liquid and the condensate change continuously with time, until a steady-state condition is reached. The steady-state compositions of the boiling liquid and the vapor condensate are taken to be the equilibrium liquid and vapor compositions.

Hála et al (8) discuss at great length the design and principles of operation of circulation stills. The better stills are complicated devices, which may involve circulation of boiling liquid as well as of vapor condensate. Operation is often isobaric, with pressure regulation provided by a manostat. Special

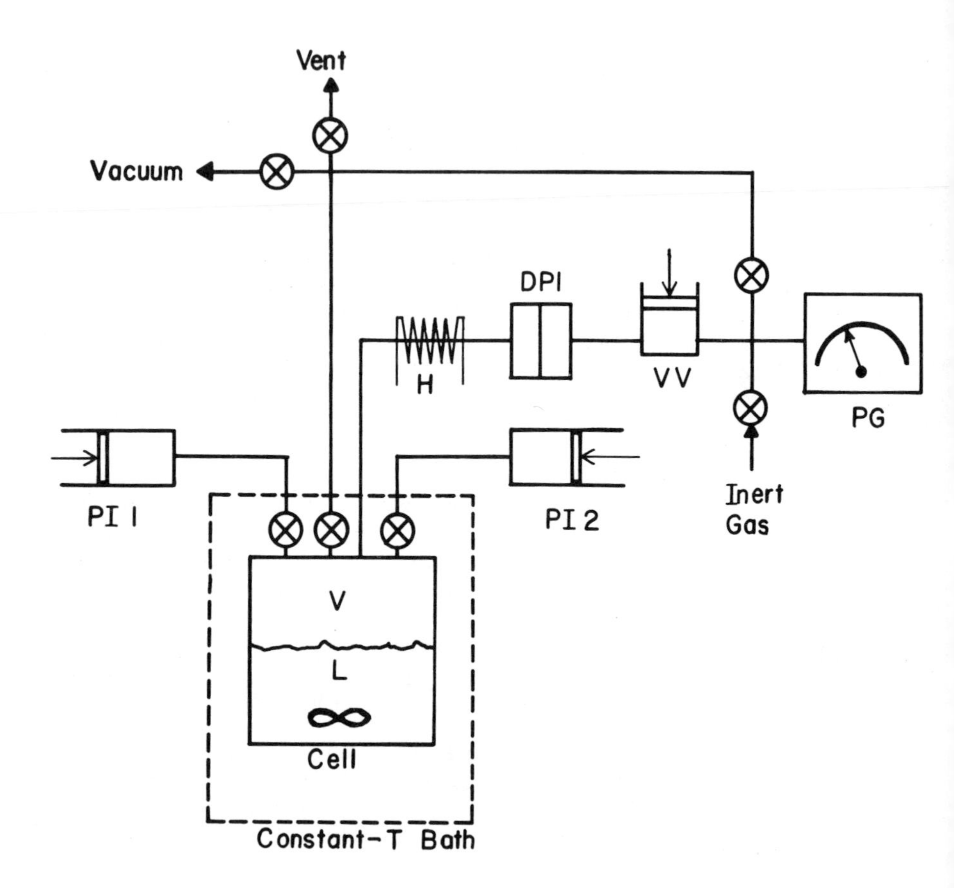

Figure 1. Schematic diagram of static VLE apparatus

care must be taken when withdrawing samples for analysis, particularly if the temperature level is high and especially for the vapor condensate, for loss by vaporization of the lighter components can be significant.

In the static method, a liquid mixture is charged to an evacuated equilibrium cell, immersed in a constant-temperature bath. Equilibration of the phases is brought about by vigorous stirring of the liquid phase or, in some designs, by rocking the cell. When the system has come to equilibrium, the pressure is read from a gauge.

Measurement of the composition of the vapor phase is the most difficult part of the static method, and probably accounts for its infrequent use until relatively recently. At low pressures, the total mass of vapor in the cell is small, so the vapor sample must be extremely small in order that equilibrium not be markedly disturbed on withdrawal of the sample. This difficulty can be avoided entirely if one establishes vapor compositions indirectly, by calculation. Liquid compositions can be determined gravimetrically, and the P-x data reduced by Barker's method, or by integration of the coexistence equation.

Figure 1 is a schematic drawing of a modern low-pressure static apparatus, designed by Gibbs and Van Ness (9). Pure liquids 1 and 2 are confined in calibrated piston-injectors PI1 and PI2. The equilibrium cell is immersed in a constant-temperature bath, and pressures are read from a Texas Instruments fuzed-quartz pressure gauge PG. The vapors are isolated from the gauge by a differential pressure indicator DPI; fine adjustment of the inert gas pressure actually measured by the gauge is provided by a sensitive variable-volume device VV. Heater H prevents condensation of vapors in the line connecting the cell with the DPI.

The cell is evacuated before the beginning of a run. A measured amount of liquid 2 is then injected into the cell, the magnetic stirrer is activated, and after a short period of time a vapor-pressure reading is taken. This pressure is the vapor pressure P_2^{sat} of pure 2. Small amounts of liquid 1 are successively added to the cell, and pressure readings are taken after each injection. The process is continued until the liquid-phase mole fraction of 1 is about 0.6. The cell is then emptied and evacuated, component 1 is added to the cell, and small amounts of 2 injected until x_2 is about 0.6; pressure readings are taken after each injection. The complete set of pressure-composition measurements for both experiments constitutes the experimental P-x curve for the binary system. The method is easily extended to ternary systems [Abbott *et al* (10)]; one merely adds another piston-injector to accommodate the third component.

The low-pressure static method has been developed to a high degree of sophistication by the NPL Division of Chemical Standards group [Pemberton (11); Larkin and Pemberton (7)]. In their work on the ethanol-water system between 30(°C) and 90(°C), they claim an accuracy in measured vapor pressures of ±0.02% or ±2.6(Pa), whichever is greater. The relative simplicity of the method and the ease of operation of the equipment seem likely to make the static technique a standard procedure for determination of low-pressure VLE.

A critical requirement for successful use of the static technique is that the liquids charged to the cell be thoroughly degassed. The cell is a closed system, so even small amounts of dissolved gases can result in erroneous pressure readings. A common--and time-consuming--degassing technique involves the repeated freezing and off-gassing of the pure liquids. We are now experimenting with degassing by distillation; our preliminary results confirm the suitability of the method.

Conclusions and Remarks

(1) The most useful low-pressure VLE data are those taken at constant temperature. I would discourage the routine collection of low-pressure isobaric VLE data. Their reduction and interpretation is difficult, and there is nothing to be obtained from them that can't be found from a couple of isothermal experiments, or, preferably, a single isothermal experiment and some heat-of-mixing data.

(2) Measurement of redundant data is often difficult and frequently unnecessary. For example, rigorous thermodynamic methods permit the determination of vapor compositions from low-pressure isothermal P-x data. If vapor compositions are not the most difficult quantities to measure, then alternative procedures may be devised in which for example P is calculated from isothermal x-y measurements, or liquid compositions from isothermal P-y measurements. The point to be made is this: thermodynamics can be used either for consistency testing, or to lessen the burden on the experimentalist by reducing the number of variables he must measure.

(3) The pure-component vapor pressures play a key role in the reduction of low-pressure VLE data. Since they appear in the equilibrium equations as normalizing factors for the activity coefficients, they directly affect the numerical values of G^E through every data point. However, vapor pressures are highly sensitive to temperature and to sample purity. Thus, to guarantee internal consistency with the rest of the data set, they should be measured with the same equipment and on the same lots of materials as are the mixture vapor pressures. This unfortunately is not always done, and one sometimes has to analyze an otherwise satisfactory set of data with foreign vapor pressures, computed for example from an Antoine equation extracted from the literature. This can be risky. I would encourage all workers in this field to measure and report pure-component vapor pressures along with their mixture data.

(4) The vapor-phase fugacity correction, although not always large, is often the weakest link in the data-reduction process.

The source of uncertainty is usually the mixed second virial coefficient B_{ij}. Experimental data are rarely available for this quantity, and one must resort to correlations. Good correlations are available, but their mixture data base is small and needs to be extended. There is a special need for vapor-phase volumetric data on mixtures containing polar and hydrogen-bonding species.

(5) It would be useful if VLE experimentalists could define by mutual agreement half a dozen or so referee test systems, as the calorimetry people have done. These would be binary systems for which pure-component vapor pressures were well known, and which had been carefully studied by several investigators on different types of equipment of proven reliability. The systems would be used for testing new equipment designs and new VLE data-reduction methods.

(6) There is a need for experimental VLE work on systems comprised of non-off-the-shelf chemicals. Industrial colleagues have confided to me the suspicion that many academic studies are based primarily on the availability of certain reagents in chromato-quality grade. This is not entirely an unfair assessment. Whatever the basis for current neglect of such systems, the need is there, and must eventually be met.

(7) We require more multicomponent experimentation. The hope remains that theory will eventually produce methods for reasonable estimation of multicomponent phase equilibria from data on the constituent binaries. A solid data base is needed for the development and testing of such theories.

(8) It is appropriate to acknowledge the role of the computer in modern VLE experimentation. Data reduction is a key--but potentially tedious--step in VLE work. The availability of high-speed computational techniques has I think put this crucial step back into perspective: no longer a major hurdle between experimentation and theory, but a valuable adjunct to both.

Acknowledgment

I appreciate financial support provided by National Science Foundation Grant No. ENG75-17367.

Notation

C_P^E = excess constant-pressure heat capacity
f_i = fugacity of pure species i
f_i^o = standard-state fugacity of species i
$\hat{f}_i$ = fugacity of species i in solution
g = G^E/RT
G^E = excess Gibbs free energy
H^E = excess enthalpy
ℓ = as superscript, designates liquid phase
n = total number of moles
n_i = number of moles of species i

P = pressure
P_i^{sat} = vapor pressure of pure i
R = universal gas constant
T = absolute temperature
v = as superscript, designates vapor phase
V^E = excess volume
x_i = mole fraction of species i in liquid phase
y_i = mole fraction of species i in vapor phase
γ_i = activity coefficient of species i
$\hat{\phi}_i$ = fugacity coefficient of species i in solution
Φ_i = factor in VLE total-pressure expression, Eq. (15)

Literature Cited

1. Van Ness, H. C., Byer, S. M. and Gibbs, R. E., AIChE J. (1973) 19, 238.
2. Christiansen, L. J. and Fredenslund, Aa., AIChE J. (1975), 21, 49.
3. Van Ness, H. C., AIChE. J. (1970), 16, 18.
4. Barker, J. A., Austral. J. Chem. (1953), 6, 207.
5. Abbott, M. M. and Van Ness, H. C., AIChE J. (1975), 21, 62.
6. Abbott, M. M., "Selection of Correlating Expressions for G^E", International Symposium on Reduction of Vapor-Liquid Equilibrium Data, U.E.R. Scientifique de Luminy Printing Office, France (1976).
7. Larkin, J. A. and Pemberton, R. C., "Thermodynamic Properties of Mixtures of Water + Ethanol Between 298.15 and 383.15K", NPL Report Chem. 43, National Physical Laboratory, Division of Chemical Standards, Teddington, England, (1976).
8. Hála, E., Pick, J., Fried, V. and Vilím, O., "Vapour-Liquid Equilibrium, Second Edition" Pergamon Press, Oxford, 1967.
9. Gibbs, R. E. and Van Ness, H. C., Ind. Eng. Chem. Fundam. (1972), 11, 410.
10. Abbott, M. M., Floess, J. K., Walsh, J. E., Jr., and Van Ness, H. C., AIChE. J. (1975), 21, 72.
11. Pemberton, R. C. "Thermodynamic Properties of Aqueous Non-electrolyte Mixtures. Vapor Pressures for the System Water + Ethanol at 303.15-363.15K Determined by an Accurate Static Method," Communications of 4th International Conference on Chemical Thermodynamics, VI, 137 (1975).

5

Equilibria in Aqueous Electrolyte Systems at High Temperatures and Pressures

E. U. FRANCK
University of Karlsruhe, Karlsruhe, Germany

Aqueous electrolyte solutions are not restricted to moderate temperatures and low pressures. They can exist and have been investigated to and far beyond the critical temperature of pure water and to pressures of several kilobars. Research on equilibria in electrolyte solutions in such a wide range gives valuable scientific information on intermolecular and interionic interaction and kinetic phenomena in dense fluids. Practical applications of the unusual combinations of properties of dense supercritical fluids can also be anticipated.

Water will always be the predominant electrolytic solvent. The temperature and density determine its properties. The relation of these functions with pressure can be derived from static experimental pvT-determinations which have in recent years been extended to almost 1000 C and 10 kbar. At present, the internationally recognized steam tables cover the region to 800 C and 1 kbar (1). It can be expected, however, that critically compiled tables for a wider range will be available to the industrial user in the near future. Fig. 1 gives a temperature-density diagram (2,3,4,5,6) for water to 1000 C and a density of 1.6 g/cm^3. The critical point, CP, (374C, 221 bar), and triple point, TP, are indicated on the gas-liquid coexistence curve. The points on the broken line extending from TP to the right denote the transitions between different high pressure modifications of ice. At pressures above 25 kbar, water densities have been derived from shock wave measurements (7). At 500 C and 1000 C, pressures of about 8 and 20 kbar are needed to produce the triple point density of liquid water. The region to about 500 C and 5 kbar is at present of particular interest for laboratory experiments, possible industrial application and geochemical phenomena.

The discussion of phase equilibria and ionic dissociation of salts in water for a wide range of temperatures and densities requires a knowledge of phase diagrams. Only a few salt-water phase diagrams have been extensively investigated experimentally.

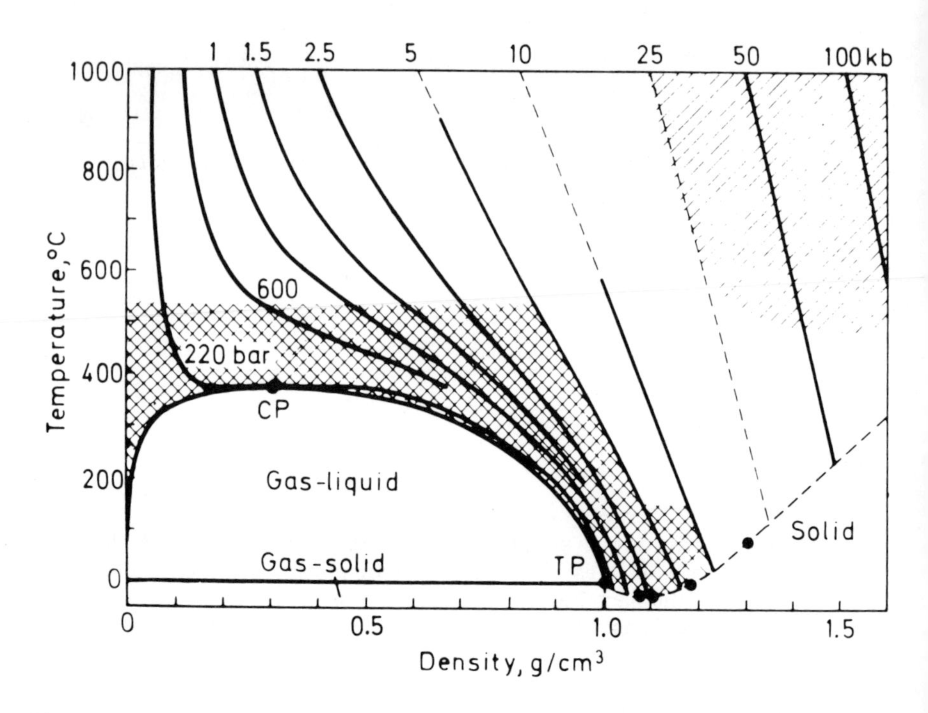

Figure 1. Temperature-density diagram of water. Cross-hatched area is where most of the hydrothermal experiments were made.

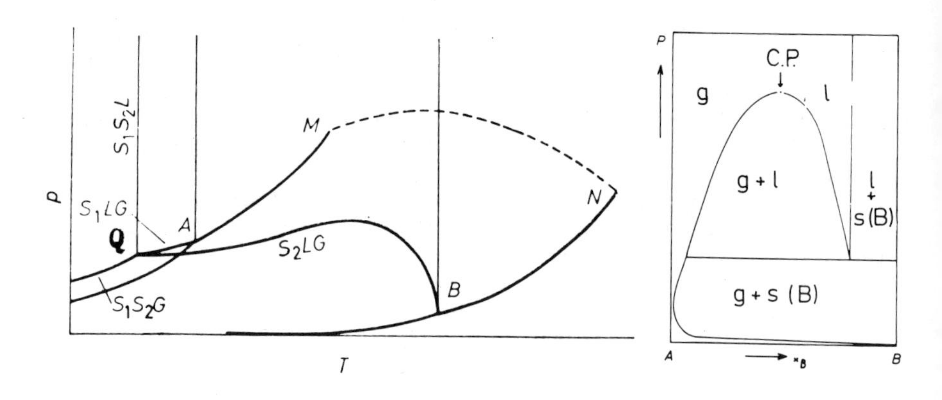

Figure 2. Schematic P-T- *and* P-x-*diagrams of a simple two-component system*

This applies to the sodium chloride-water system which is an exemplary case. Fig. 2 gives a schematic p-T-diagram of a two component system with very different critical temperatures. One observes a critical curve which is a projection from the three-dimensional pressure-temperature-composition diagram on the p-T-plane. The curve extends uninterrupted between the critical points of the two pure components. The projection of a three-phase-surface, S_2LG, between a quadruple point and the triple point T_2 can also be seen. It does not intersect the critical curve in this example. In the right part of Fig. 2 we have a P-x-section taken between M and B in the previous diagram. At CP the critical curve penetrates the P-x-plane.

In Fig. 3, a selection of experimental isotherms corresponding to the g-ℓ-curves of the right part of Fig. 2 is shown for NaCl-H_2O (8). A portion of the critical curve is shown for this system and it is reasonable to assume, that it is of the uninterrupted type and will eventually extend to the critical point of pure NaCl, which has not yet determined but will very probably be above 2000 C. It follows from Fig. 3, that even above the critical temperature of pure water and at pressures readily available, homogeneous fluid phases with much more than 10 weight percent NaCl can exist over a wide range of conditions. This is undoubtedly true for other alkali salts. Solutions of this type exist as hydrothermal fluids within the earth's crust and can possibly be used for practical purposes too. Many other aqueous systems, however, exhibit interrupted critical curves and different behavior. An example is the SiO_2-H_2O-system (9).

Critical Curves and Miscibility

The electrolytic properties of water can be varied by expansion. Similar variations can be achieved by admixing an inert volatile component to supercritical water at high pressures. Such mixtures can have interesting properties without adding salt. In recent years various binary aqueous systems have been investigated by several groups and the two-phase surfaces determined. If the nonaqueous component is nonpolar or only slightly polar, the critical curves are mostly of the interrupted type and the upper branch, which begins at the critical point of the pure water, proceeds to rather high pressures. Fig. 4 shows a number of examples (6). These critical curves are a sort of envelope of the two-phase region. At the high temperature side of the curves, the components are completely miscible at all total densities. Above 400 C the rare gases, nitrogen and light alkanes all appear to be miscible with water. Two of the curves have a different character than the others. Benzene-water has a critical curve with a minimum below 300 C at about 200 bar. Thus, dense homogeneous mixtures of water with benzene and other similar aromatic compounds can be obtained relatively easily (10). The curve for H_2O-CO_2 also proceeds to

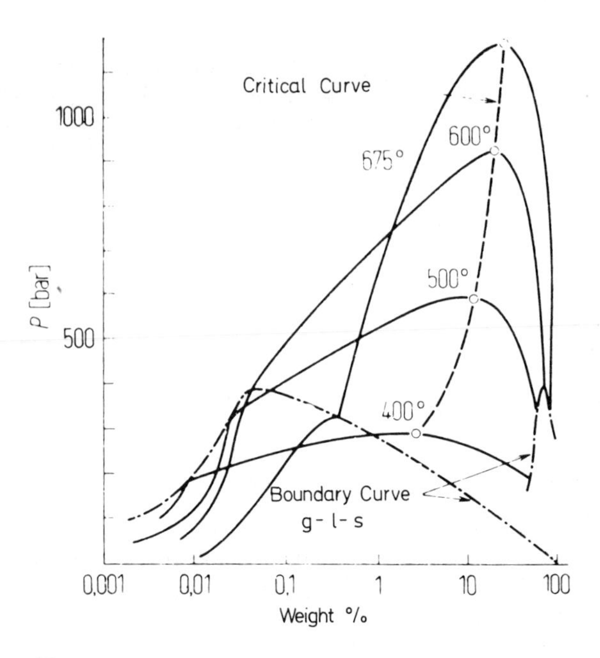

Figure 3. Experimental isotherms for the NaCl–H_2O system in a pressure-salt concentration-diagram

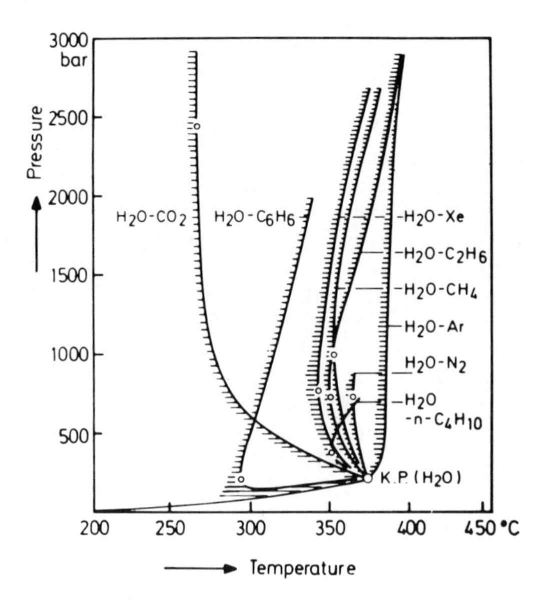

Figure 4. Critical curves of binary aqueous systems

rather low temperatures, indicating a certain degree of attractive interaction between the two compounds. This does not mean stochiometric association, however, as shown by recent high temperature-high pressure Raman measurements (11).

The two-phase boundary surface and the pvT-behavior in the supercritical region has been extensively investigated (12,13,14). In analogy to "salting out" effects at low temperature, one should expect that a dissolved salt as a third component also affects the miscibility of CO_2 and H_2O in the critical region. This is indeed the case as was shown some years ago (15). Recently extended measurements with the H_2O-CO_2-NaCl system has been performed at several concentrations and the results are shown in Fig. 5. The curves are for constant water-salt concentration ratios. The CO_2 content varies. This shows that the immiscibility range and probably also the critical curve is shifted to higher temperatures by 50C or more (16). Beginning with the critical point of pure water, a part of the critical curve of the binary system H_2O-NaCl is also shown. It appears that by making use of the critical data for the applied water-salt ratio, a certain unified, reduced representation of the various curves can be obtained. A more rigid theoretical treatment of a ternary system of this kind does not seem to be possible at present. The experimental work is being continued. This system is certainly one of the most important in geochemistry.

Irrespective of the incomplete quantitative knowledge of the interactions for a ternary system like H_2O-CO_2-NaCl one can attempt to make estimates. Fig. 6 gives a tentative triangular diagram for a constant pressures of 1 kbar to the melting temperature of pure sodium chloride. The immiscibility curves of the binary systems H_2O-CO_2 and H_2O-NaCl are relatively well determined by experiments. The lower right curve in the H_2O-NaCl plane of the prism is based on the solubility curve of NaCl in water at equilibrium pressures (8) and can be only qualitatively correct at 1 kbar. Fig. 6 suggests at least, that there is a rather extended range of homogeneous fluids in the H_2O-CO_2-NaCl ternary system at high pressure between 400 and 600 C. It is hoped that current work will permit the construction of more reliable diagrams for different pressures in the future.

Values of the thermodynamic functions based on experiments for the two binary systems H_2O-NaCl and H_2O-CO_2 in the fluid one-phase region at high temperatures and pressures have not yet been sufficiently determined. New work in this field is being done at present. As one example, Fig. 7 shows partial mola volumes of NaCl in H_2O which were calculated recently (17) from a critical compilation of existing data. Fig. 8 gives excess Gibbs free energy values of H_2O-CO_2 mixtures for two supercritical temperatures.

These are part of the results from calculations now being done at Karlsruhe using both older and new experimental data and a Redlich-Kister type equation of state.

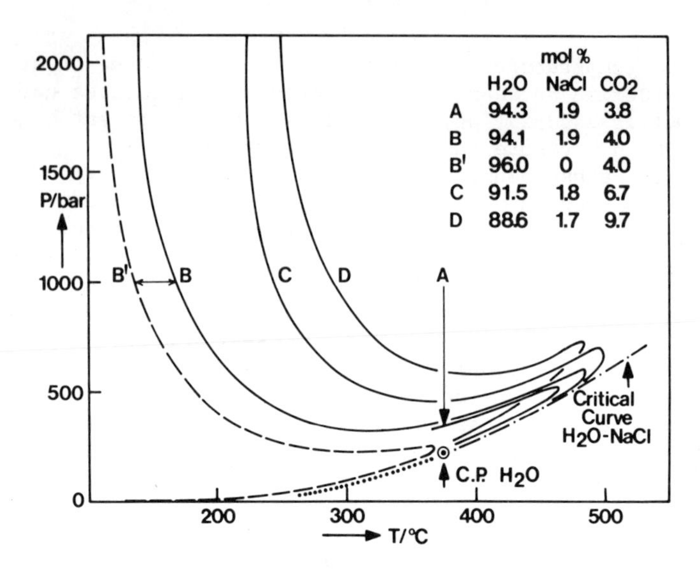

Figure 5. Phase boundary curves for constant concentrations (isopleths) for the ternary system H_2O–CO_2–NaCl

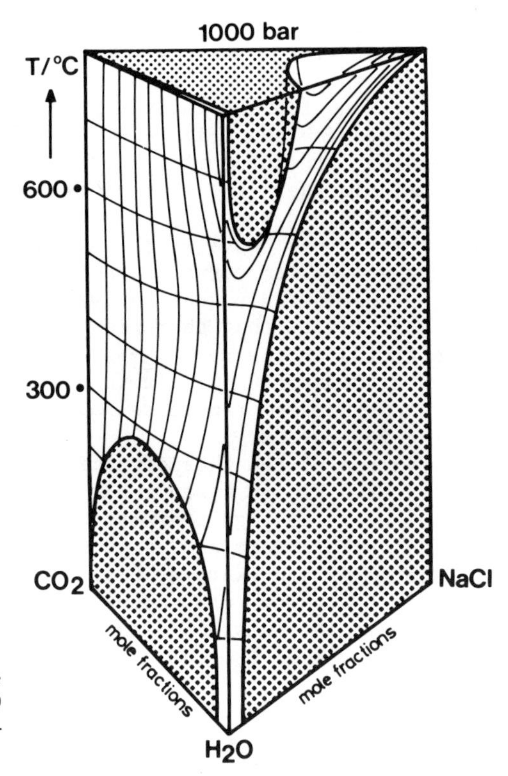

Figure 6. Tentative boundary surface for one-phase fluid conditions at 1000 bar for the ternary system H_2O–CO_2–NaCl

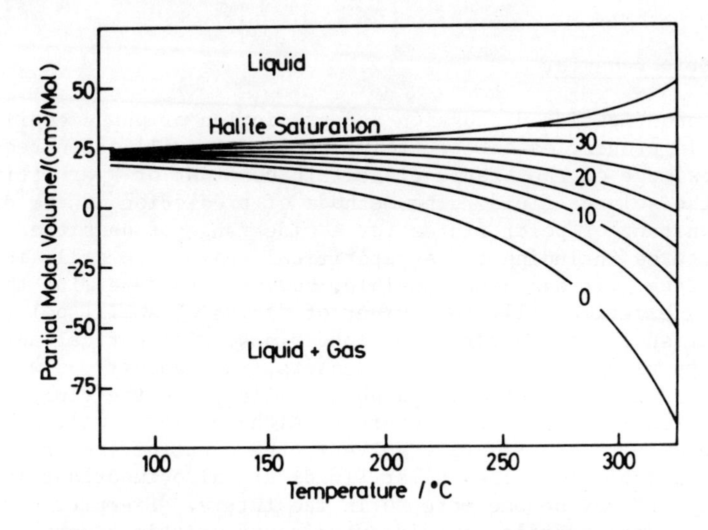

Figure 7. Partial molal volume of NaCl in the liquid phase for vapor-saturated NaCl solutions from 0 wt % to halite saturation between 80 and 325 C. (calculated by J. L. Haas, Jr., U.S. Geological Survey, Reston, Va., 1975)

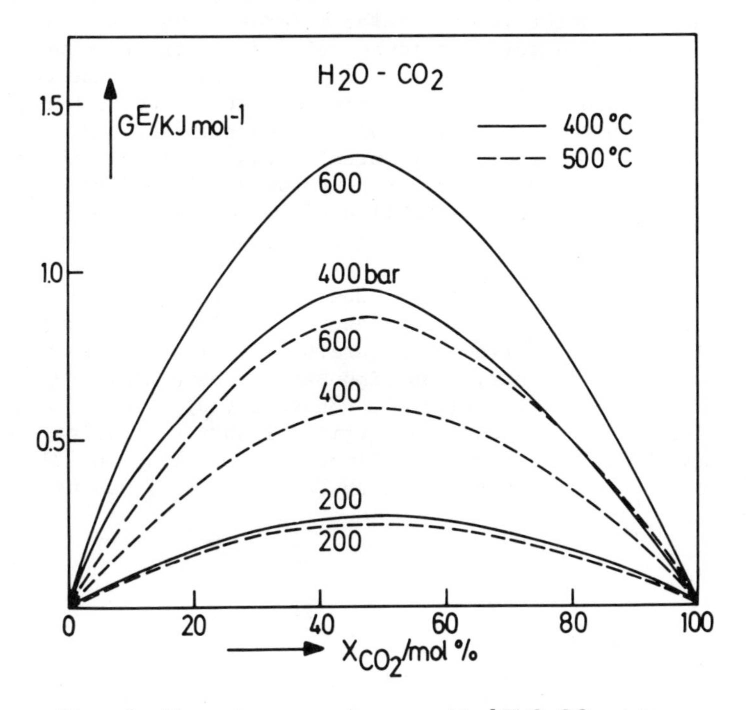

Figure 8. Excess free energy for supercritical H_2O–CO_2 mixtures

Dielectric Behavior

A quantitative discussion and prediction of phase equilibria and fluid properties of electrolyte solutions will often require the knowledge of the static dielectric constant or "permittivity", ε, of the solvent fluid. The methods of prediction of the dielectric constant of polar fluids for a wide range of densities and temperatures including the supercritical region are still in an early stage. It has been possible, however, to determine this quantity experimentally for number of fluids of small, polar molecules at sub- and supercritical conditions. Cylindrical autoclaves have been used with built-in condensers, the capacity of which could be determined and varied while filled with the fluid samples at high temperatures and pressures. Although water will always be the predominant electrolytic solvent, the electrolytic properties of a number of other less polar fluids are also important in industry and may become more so in the future. Examples are methanol, acetonitrile, liquid ammonia and certain Freons. Fig. 9 gives experimental results for ε of methylchloride along the isobars (18). As expected, increasing the temperature and decreasing the pressure lowers the dielectric constant. Since the dipole moment of the isolated CH_3F molecule of 1.85 debye units is almost equal to that of an isolated water molecule (1.84 debye units) a comparison with water is particularly interesting. The higher ε-values of dense water have to be related to its structure caused by hydrogen bonds. The temperature and pressure dependence of the dielectric constant of methylfluoride appears to be rather similar to those of several of the more polar Freons and of ammonia. It may be possible to describe and predict quantitatively the dielectric properties of such polar, non-hydrogen bonded fluids on the basis of the Debye-Onsager-Fröhlich-Kirkwood theory by devising suitable expressions for the "Kirkwood Correlation Factor."

For comparison in Fig. 10 an ε-T-diagram for HCl is given (19). Even in the high density supercritical region ε is much lower than in CH_3F. This mainly due to the lower dipole moment of HCl (1.05). Fig. 11 gives a temperature-density diagram of water with a number of isobars. Superimposed on these are curves of constant values of the dielectric constant (20). These latter ε-curves are based on an extensive critical survey of 47 papers with experimental data which were published between 1920 and 1975. As a result of this survey an equation for $\varepsilon = f(t,\rho)$ was derived of the form:

$$\begin{aligned}\varepsilon = 1 &+ a_1\tau^{-1}\rho \\ &+ (a_2\tau^{-1} + a_3 + a_4\tau)\rho^2 \\ &+ (a_5\tau^{-1} + a_6\tau + a_7\tau^2)\rho^3 \\ &+ (a_8\tau^{-2} + a_8\tau^{-1} + a_{10})\rho^4\end{aligned}$$

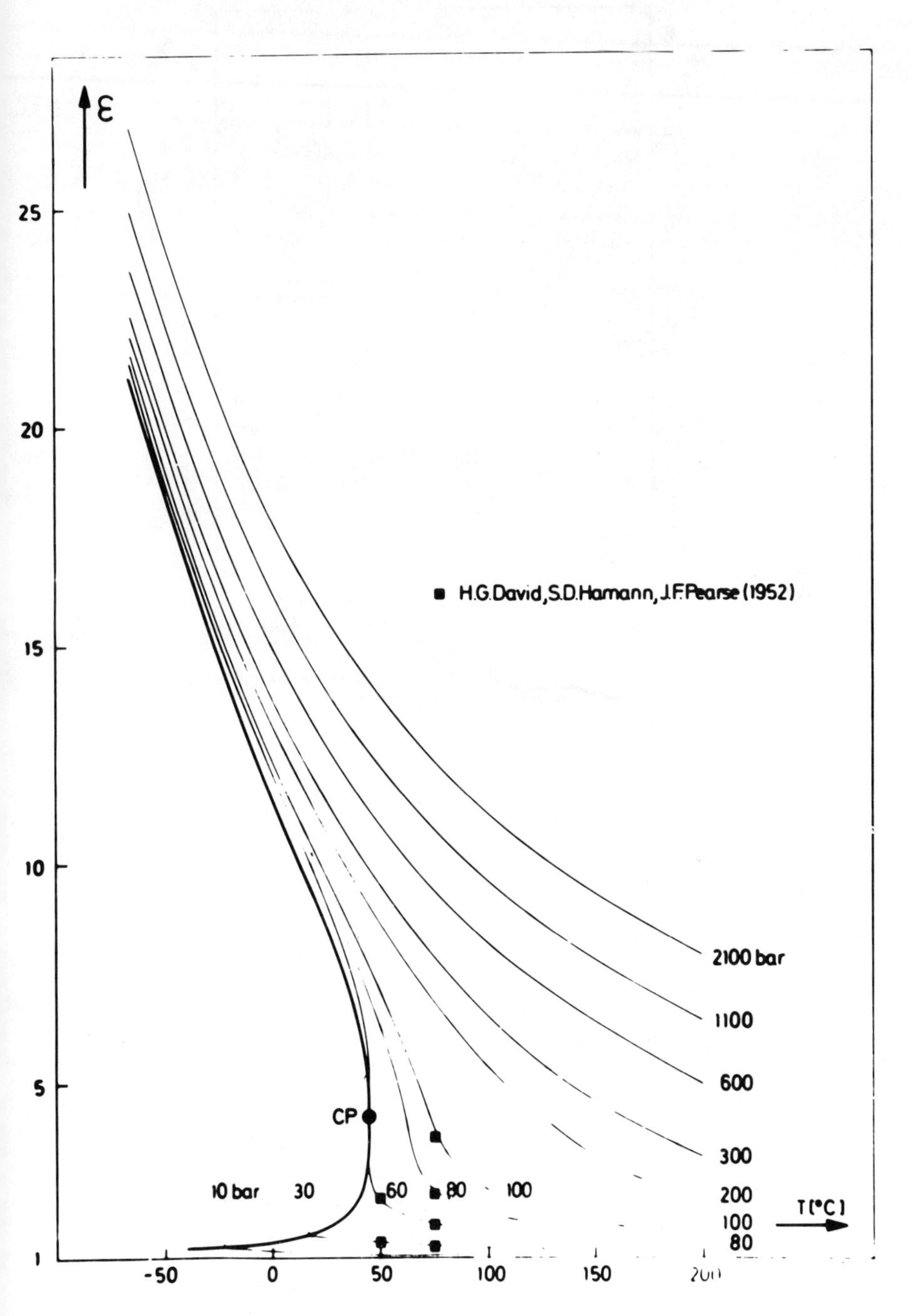

Figure 9. Isobars of the static dielectric constant for CH_3F as a function of temperature

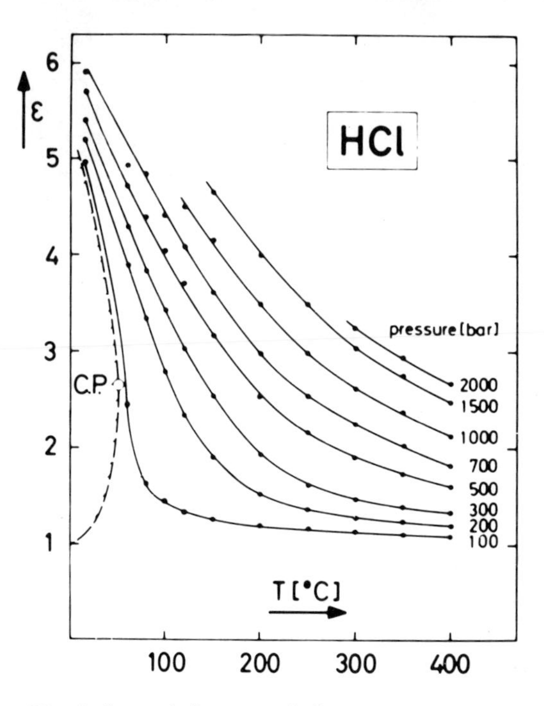

Figure 10. Isobars of the static dielectric constant ε of HCl as a function of temperature, T

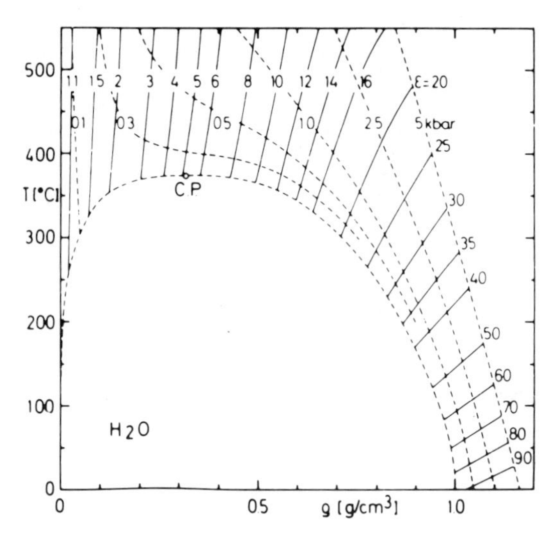

Figure 11. Isodielectric constant curves on a T-g *diagram*

(τ = T/100, T in K and ρ in g/cm^3). This equation represents the existing data.

It is obvious that even in water very high dielectric constants occur only at relatively low temperatures and high densities. There exists, however, a wide range of supercritical conditions, where the dielectric behavior is comparable to that of the more polar organic solvents which to some extent can act as electrolytic solvents. Finally, Fig. 12 shows a reduced temperature-density diagram of HCl, H_2O and CH_3CN with experimentally determined ε-values of these three compounds, the dipole moments which nearly increase as 1 : 2 : 4 from HCl to CH_3CN. Such reduced diagram may eventually permit first estimates of the dielectric properties of other polar fluids.

Ionization and Complex Formation

Some polar fluids exhibit a small degree of ion formation even in the liquid state at room temperature and at ordinary pressure. This self ionization can be of considerable influence on the dissociation equilibria of dissolved electrolytes. The phenomena of solvolysis or hydrolysis in aqueous systems depend on the

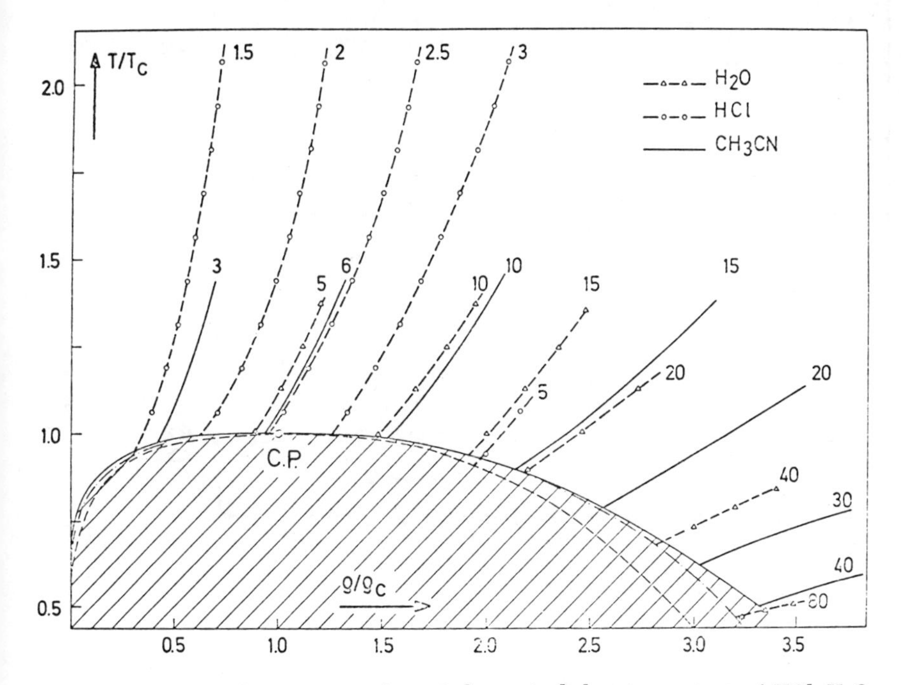

Figure 12. Curves for constant values of the static dielectric constants of HCl, H_2O and CH_3CN on a reduced temperature–density diagram

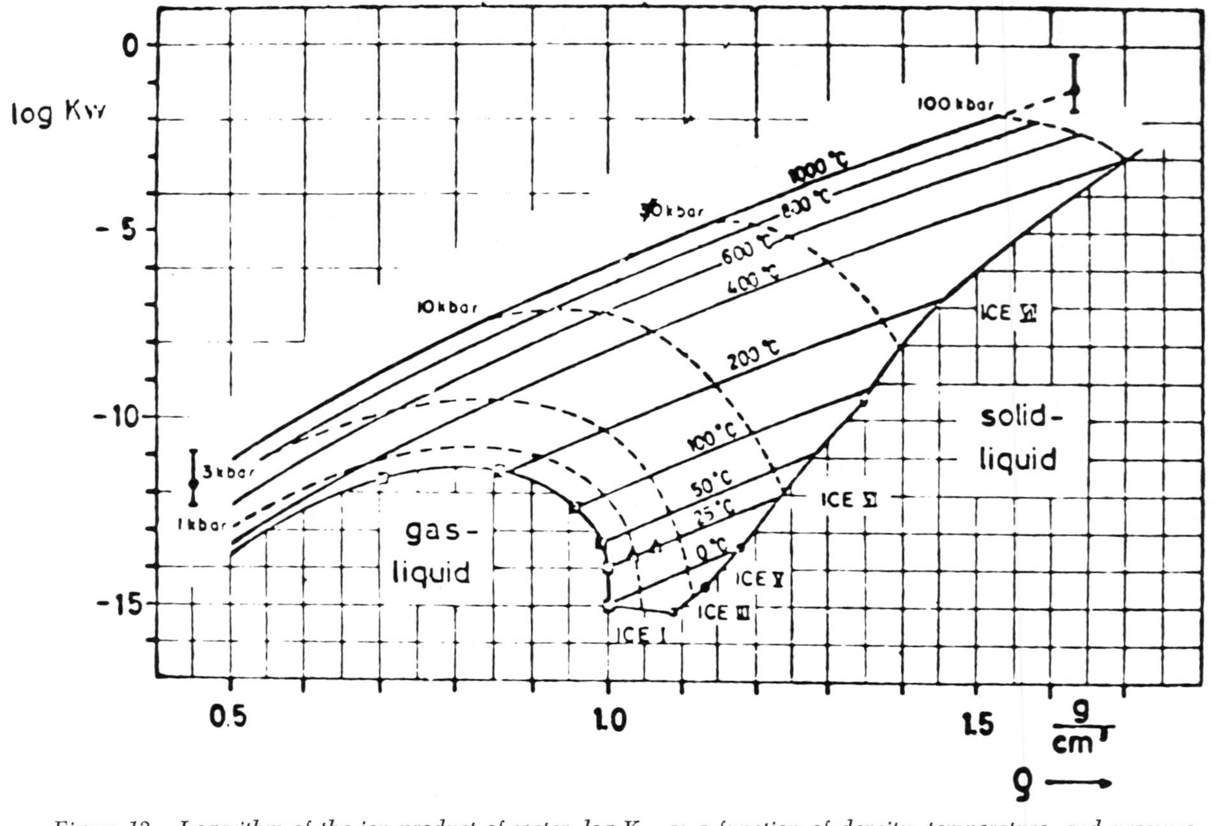

Figure 13. Logarithm of the ion product of water, log K_w*, as a function of density, temperature, and pressure*

ion dissociation of the solvent and may thus be of influence on certain phase equilibria also. This is particularly true for high temperature aqueous solutions. Examples are the "hydrothermal" fluids of the earth's crust, which are important as natural transport media for minerals. Hot aqueous fluids of this kind can also occur during exploration procedures for oil or for geothermal energy.

The ion product of water, K_w, that is the product of the activities or concentrations of hydroxyl ions and hydrogen ions in mol l^{-1} is close to 10^{-14} mol^2 l^{-2} at room temperature and normal pressure. It can be expected that this quantity will increase at elevated temperatures as well as elevated pressures. This has been confirmed in recent years by many different experiments, some of them extending to 100 kbar and almost 1000 C. Fig. 13 shows the results in a diagram with log K_w as a function of the density in the form of isotherms and isobars (6). At 400 C and 1 kbar, log K_w has increased to -11 and at 500 C and at the triple point density of 1 g/cm^3, log K_w is -8. This means that one hundredth of a percent of the water is dissociated into ions at these conditions. At such conditions even alkali salts of the strong mineral acids can be hydrolyzed to an extent comparable to that of acetates in aqueous solutions at ordinary conditions.

If one considers the decrease of the dielectric constant of water at higher temperatures, one could expect a lowering of the ionic dissociation of dissolved electrolytes. This is indeed the case shown by electrolytic conductance measurements. One example is shown in Fig. 14 (21). It gives isobars of the specific conductance of 0.01 molal KCl solutions to 800 C. Up to 300 C the conductance rises at all pressures because of decreasing water viscosity and increasing ion mobility. Beyond 300 C, however, the declining isobars indicate a reduction of charge carrier concentration caused by ion association and decreasing dissociation constants. Fig. 14 gives results only for a very dilute salt solution. However, the hot natural brines which are found in certain places often have very high salt concentrations. The conductance and dissociation equilibria in such solutions approach the behavior of fused salts. This is demonstrated in Fig. 15. It shows experimental values of the molar conductance of NaCl in aqueous solutions at the supercritical temperatures of 400 and 500 C as a function of the mole fraction of the salt (22). Measurements could not be made beyond x = 0.1. The molar conductance of the fused salt has been determined, however. It is not very temperature dependent, so that an extrapolation of the conductance to an assumed supercooled liquid at 400 C or 500 C seems to be possible. These values are shown in Fig. 15. It is clear that solutions with salt mole fractions of x = 0.2 and higher have properties closer to the pure fused salt than to the dilute solutions as far as the conductance is concerned. A large fraction of the ions is already in some way associated which must influence the thermodynamic functions accordingly.

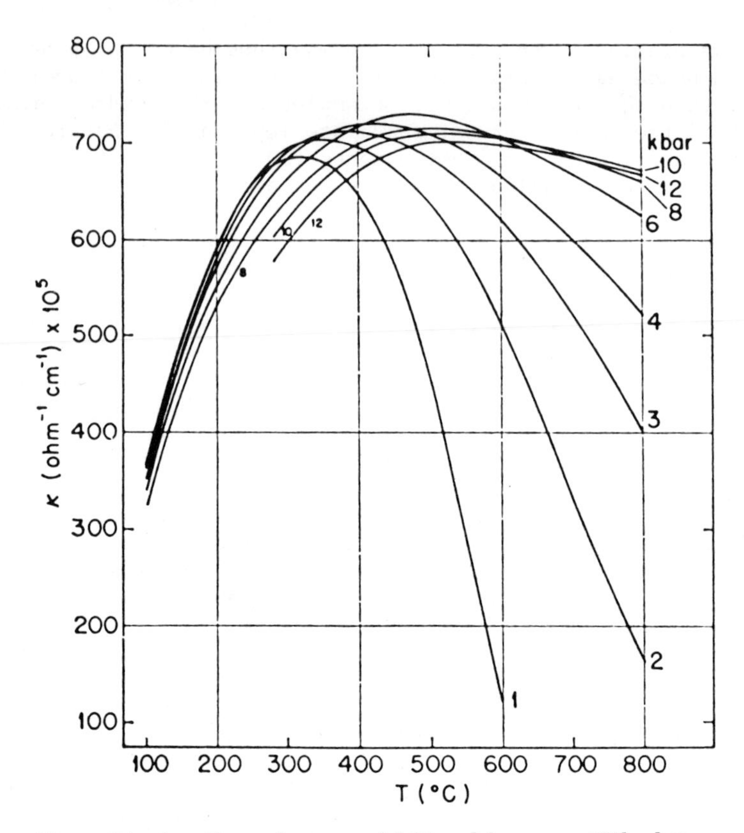

Figure 14. Specific conductance of 0.01 molal aqueous KCl solutions as a function of temperature and pressure

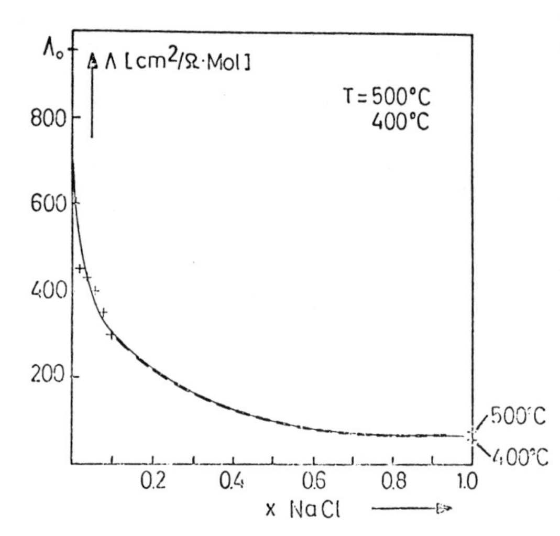

Figure 15. Molar conductance of aqueous NaCl solutions to mole fractions of NaCl of 0.1 and estimated molar conductances of the supercooled fused salt

The observation that supercritical dense water can ionize dissolved salts suggests that one attempt electrolytic decomposition at such conditions. It is well known, that an increase in temperature reduces overvoltages and it has been shown very early that at room temperature with water, high pressures act in the same direction. Both effects are shown by a number of simple current-voltage curves for dilute, aqueous NaOH solutions in Fig. 16. It is interesting to observe, that the curves increase quickly at very low voltages if the temperature is high and that beyond 400 C the decomposition potential is no longer clearly visible. A complete analysis of this behavior has not yet been made. It is certain, however, that it is caused by the combined influence of high ion mobilities, high diffusion constants of neutral molecules, complete miscibility of oxygen and hydrogen in the aqueous solution and reduced adsorption at electrode surfaces at high temperatures. As a consequence the electrodes appear to be almost nonpolarizable above 400 C. Very high current densities have been obtained at such conditions. In order to consider the possibilities of technical application, a knowledge of phase boundary surfaces and critical curves for water-oxygen and water-hydrogen systems would be desirable.

In the fields of power plant operation and high temperature corrosion, in metallurgy and in geochemistry, a knowledge of stability regions and other equilibrium data for heavy metal complexes in high temperature and supercritical aqueous solutions are of interest. At present the relevant information is still limited. Most of the results come from geochemical investigations (23,24). Spectroscopy in the visible and near ultraviolet complex equilibria of halides of cobalt (25), nickel (26), copper (27), zinc (28), lead and other heavy metals (23). Fig. 17 gives one example with adsorption spectra for cobalt chloride at 300 C and 500 C.

The extinction coefficient at 20 C and normal pressures is shown which corresponds to the pink colored solutions of bivalent hexaquo cobalt complexes. At 300 C and the relatively modest pressure of 350 bar, the adsorption is considerably increased and shifted to greater wavelengths. The solutions are blue, tetrahedral four-ligand cobalt complexes prevail. This tendency is even more pronounced at 500 C. In equilibrium, such compressed supercritical solutions contain the bivalent metal ions mainly as complexes with the lower coordination number of four. Analogous behavior has been found for nickel and copper. Stability constants have been determined, which may be useful in the discussion of corrosion products in supercritical steam and of the composition of hydrothermal fluids.

Conclusion

As a concluding remark it may be pointed out that properties of the kind discussed above may also be investigated with other classes of fluids. Fluid salts and fluid metals can be mentioned as examples. Fig. 18 shows the specific conductance of mercury

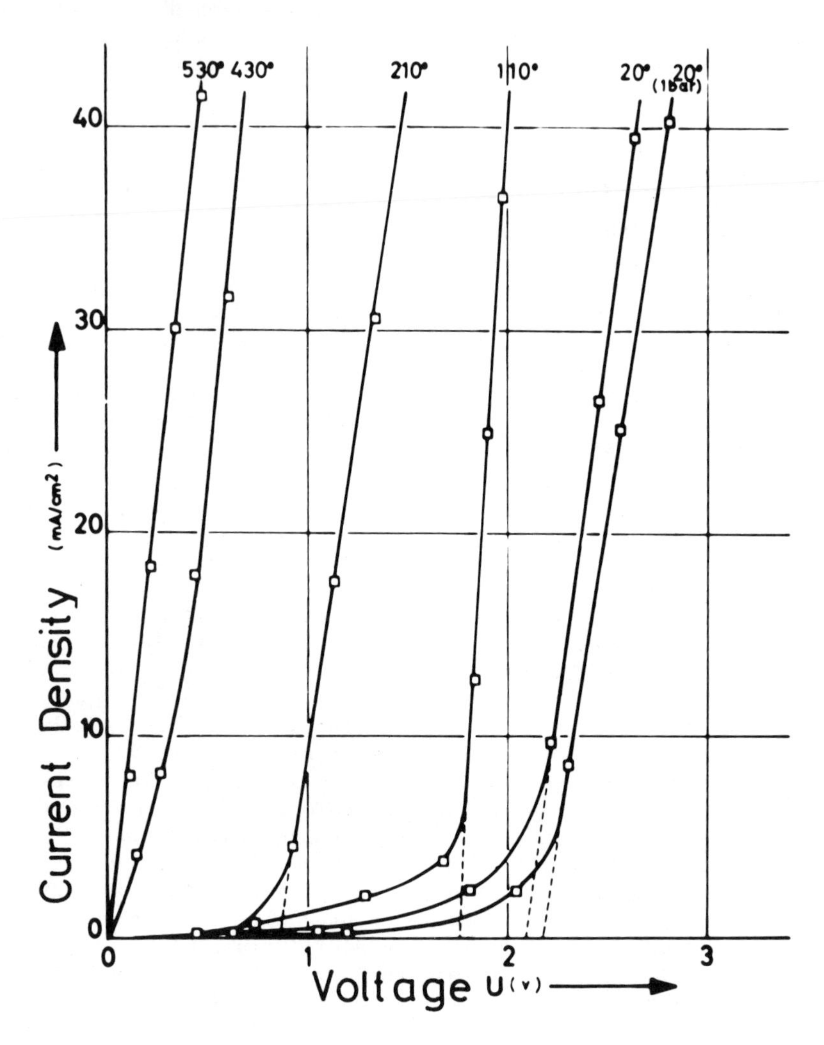

Figure 16. Current–voltage curves obtained for 0.01 molal aqueous NaOH solutions at high temperatures and pressures

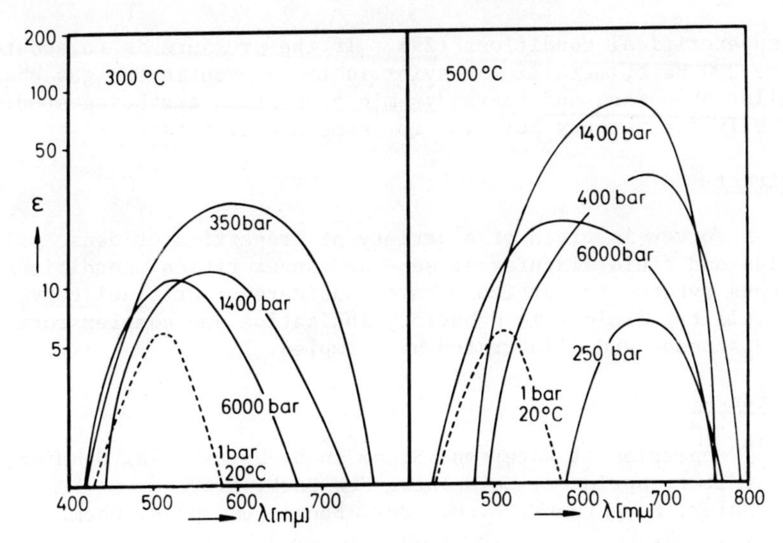

Figure 17. Absorption spectrum of $CoCl_2$ (Molality = 0.01) in water at 300° and 500°C between 250–6000 bar

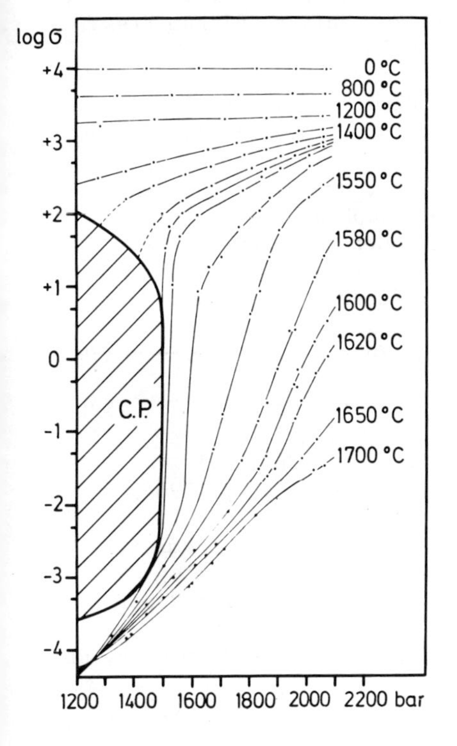

Figure 18. Specific conductance, σ, of mercury as a function of temperature and pressure at supercritical conditions

at supercritical conditions (29). If the pressure is raised to about 200 bars, metallic behavior is found even in the gas phase. Similar phenomena and thermodynamic properties are being studied not only for mercury but also for some alkali metals.

Abstract

A survey is given of a variety of properties of dense polar fluids and fluid mixtures at sub- and supercritical conditions. Aqueous systems in particular are considered. Critical curves and miscibility, dielectric behavior, ionization and complex formation are discussed and illustrated by examples.

References

1. "Properties of Water and Steam in SI-Units". E. Schmidt, editor, Springer, Heidelberg-New York, 1969.
2. Maier, S., Franck, E. U. Ber. Bunsenges. phys. Chem. (1966) 70, 639.
3. Köster, H., Franck, E. U. Ber. Bunsenges. phys. Chem. (1969) 73, 716.
4. Burnham, C. W.; Holloway, J. R.; Davis, N. F. Am. J. Sci., (1969) 267 A, 70.
5. "Water at High Temperatures and Pressures", K. Tödheide in "Water" Vol. I, p. 463, F. Franks, editor, Plenum Press, N. Y. London 1972.
6. Franck, E. U. Pure and Applied Chemistry (1974) 38, 449.
7. Rice, M. H., Walsh, J. M. J. Chem. Physics (1957) 26, 824.
8. Sourirajan, S., Kennedy, G. C. Amer. J. Sci. (1962) 260, 115.
9. Kennedy, G. C.; Wasserburg, G. J.; Heard, H. C.; Newton, R. C. Publ. No. 150, Inst. of Geophysics, UCLA, 1960.
10. Schneider, G. M. in "Topics in Current Chemistry" Vol. 13, p. 559, Springer, Heidelberg, New York, 1970.
11. Kruse, R., Franck, E. U. Ber. Bunsenges, physik. Chem. (1976) 80, 1236.
12. Franck, E. U., Tödheide, K. Z. Physik. Chem. N. F., (1959) 22, 232.
13. Tödheide, K., Franck, E. U. Z. Physik. Chem. N. F., (1963) 37, 387.
14. Greenwood, H. J. Amer. J. Sci. (1969) 267, 191.
15. Takenouchi, S., Kennedy, G. C. Amer. J. Sci. (1964) 262, 1055.
16. Gehrig, M. Thesis, 1975, Inst. for Physical Chemistry, University of Karlsruhe.
17. Haas, J. L., Jr. U. S. Geol. Survey, Open File Report 75-675, (October 1975).
18. Reuter, K. Thesis 1974, Inst. for Physical Chemistry, University of Karlsruhe.

19. Harder, W. Thesis 1972, Inst. for Physical Chemistry, University of Karlsruhe.
20. Uematsu, M.; Harder, W.; Franch, E. U. "The Static Dielectric Constant of Water to 550 C and 5 kbar". Report of the "International Association on the Properties of Steam" (IAPS) Sept. 1976.
21. Quist, A. S.; Marshall, W. L.; Franck, E. U.; v. Osten, W. J. Physical Chemistry, (1970) 74, 2241.
22. Klostermeier, W. Thesis 1973, Inst. for Physical Chemistry, University of Karlsruhe.
23. Helgeson, H. C.; Kirkham, D. H. Amer. J. Science, (1976) 276, 97.
24. Franck, E. U., J. Solution Chemistry, (1973) 2, 339.
25. Lüdemann, H. D., Franck, E. U. Ber. Bunsenges. Physik. Chem. (1967) 71, 455.
26. Lüdemann, H. D., Franck, E. U. Ber. Bunsenges. Physik. Chem. (1968) 72, 514.
27. Scholz, B.; Lüdemann, H. D.; Franck, E. U. Ber. Bunsenges. Physik. Chem. (1972) 76, 406.
28. Schulz, K. Thesis 1974, Inst. for Physical Chemistry, University of Karlsruhe.
29. Hensel, F., Franck, E. U. "Experimental Thermodynamics" Vol. II, ed.: B. Le Neindre, B. Vodar, p. 975, Butterworths, 1975 (London).
30. Hensel, F. Angewandte Chemie, (1974) 86, 459.

6

Polymer Equilibria

A. BONDI

Shell Development Co., Houston, Tex. 77001

The purpose of the present paper is to provide a generalized view of those phase equilibrium characteristics of polymers or polymer dominated systems which are unique to the "polymer world". This uniqueness may flow from the high molecular weight of technically important polymers or from the associated high viscosity or from the glassy conditions which characterize the useful solid state of many members of that class.

Another purpose of this paper is to assess the technical and the economic significance of ignorance in this area of research, because the purpose of this conference is served more by exhibiting what we do not know than by the proud display of a seemingly impregnable, coherent body of validated theoretical understanding.

The scope of this presentation is a sweeping survey with a few in-depth excursions into valleys of significant ignorance.

General Principles

The dominant problem of phase equilibria involving a polymer phase is the measurement problem: When has (or can) equilibrium be called "established". The usual means for rapid equilibration, forced convection, is not only difficult to implement, it may even be impossible to do, when the required energy input would actually break chemical bonds along the polymer molecule's backbone chain.

Another problem is that of data correlation, or of data generalization by means of well founded theory. At the rough approximation level we are (thanks to the work by Flory, Huggins, Prigogine, Prausnitz, and others) in good shape, and heretofore that has been quite adequate. But we shall see that recent studies of concentrated systems exhibit vapor/liquid equilibria which are not even described qualitatively by existing theory, assuming that the measurements are reliable.

In this discussion we shall take for granted that the reader is familiar with the body of theory, according to which polymers are characterized by their molecular weight, their molecular weight distribution, their cohesive energy density (or other measure of pair potential between unbonded neighbors) and the flexibility of chains, measured empirically or characterized by barriers to internal rotation and energies of rotational isomerization of main chain bonds or bond groupings, the crosslink density, if any, and electrostatic charge density, especially in the case of polyelectrolytes. The latter will not be dealt with in this survey.

Single Component, Two Phase Systems Solid/Solid Equilibria in Crystalline Polymers

Polymorphism is far less common among crystalline polymers than among crystals composed of small molecules. The reason for this paucity is, of course, the constraint imposed by their one-dimensional infinity upon rotational disorder, the primary cause of polymorphism among organic solids. The few known cases have been assembled on Table 1.

The situation is quite different, if we admit association polymers as legitimate polymers. An outstanding example is the class of the alkali and earth alkali metals soaps of long chain fatty acids. Their polymorphic behavior is well established (1), and is ascribed to the increasing mobility of the alkane groups with increasing temperature. Typical examples are shown on Table 2.

Polymeric crystalline long chain esters of vinyl compounds are, of course, the C-C backbone chain equivalent of those association-polymers.

Solid/Liquid Equilibria. Few polymer equilibria have been studied as thoroughly as those at the melting point of crystalline polymers. In large measure this interest may be caused by the desire to understand the peculiar chain folding propensity of many crystalline polymers. The currently accepted theory (2) associates chain folding with an entropy phenomenon, such that chain folding should disappear with increasing temperature; and, indeed, if one raises the melting temperature high enough by conducting the solidification from the melt at high enough hydrostatic pressure, chain folding can be avoided, and completely straight chain crystals are formed without chain folds (3). Thus the liquid/solid, pressure/temperature phase diagram of polymeric crystalline solids, illustrated on Figure 1, is not quite comparable with those of simpler compounds, because the two ends of each curve represent materials in two quite different crystal morphologies. As this change is gradual, and related to the molecular weight distribution, the T_m vs. pressure curves on Figure 1 are without discontinuity.

A unique aspect of high polymer solid/liquid equilibria is the orientation-induced crystallization at temperatures above T_m (4). The best known instances of crystallization of $T > T_m$ under

TABLE 1

EXAMPLES OF POLYMORPHISM AMONG CRYSTALLINE POLYMERS[a)]

Polymer	Crystal Morphology	t_m, ^oC
poly-1-butene	rho	135/6
	ter	122/4
	ortho	106
poly-1-pentene	mono	130
	port	80
poly-4-methylpentene-1	tet	235/50
	2	125
	2	75
poly 1,3 butadiene, trans I	phex	100
	mono	96
II	phex	141
	hex	148
poly cis isoprene	mono	23
	ortho	14
poly-cyclo-octene, trans	tri	67/73
	mono	62/77
vinyl cyclo pentene	tri	292
	tet	270
poly-p-xylylene, α	mono	375
	ortho	420
poly arcylonitrile, synd.	hex	317
	ortho	341
poly vinylchloride	ortho	273
	mono	310
poly vinylfluoride	hex	200
	ortho	230
poly 8-amino caprylic acid, α	mono	185
	phex	220

[a)] From Polymer Handbook, Brandrup, J. and Immergut, E. H., John Wiley, New York, 1975.

TABLE 2

POLYMORPHISM OF ASSOCIATION POLYMERS

EXAMPLE: THE ALKALI SALTS OF LONG CHAIN FATTY ACIDS[a]

Solid/Solid Transition Temperatures and Final Melting Points, °C									
	Carbon Number of Fatty Acid:								
Cation	C_6	C_7	C_8	C_9	C_{10}	C_{12}	C_{14}	C_{16}	C_{18}
Li m.p.						237	233	223	226
						227	208	215	215
								197	191
Na m.p.	361	363	360	355	348	329	311	302	283
	235	242	243	243	245	246	246	251	255
	-	-	-	-	218	218	215	212	208
	210	198	189	185	181	179	171	168	165
					140	141	138	136	135
							113	117	116
K m.p.	>400	>400	>400	>400	>400	395	375	362	348
	305	301	291	282	277	273	271	269	267
							-	195	170
Rb m.p.						∿400	380	375	358
						300	291	284	281
C_S m.p.						385	370	358	345
						295	290	279	273

[a] From Baum, E.; Demus, D.; and Sackmann, H. Wiss. Z. Univ. Halle (1970), 19, (5) 37.

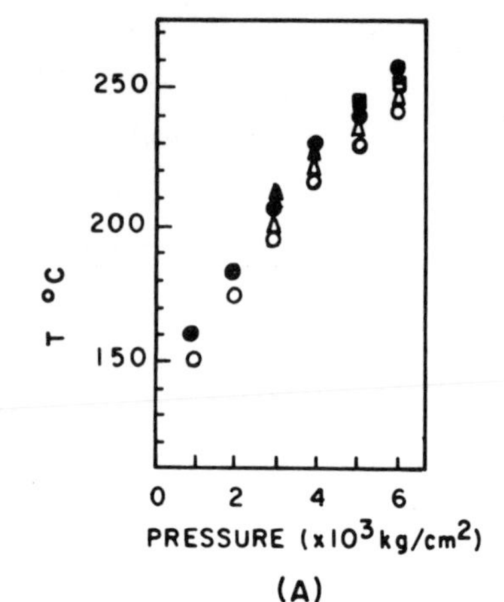

Journal of Macromolecular Science

Figure 1. Phase diagrams of crystalline polymers (6)

Figure 1A. Phase diagrams of polyethylene. Melting (solid points) and crystallization (open points) temperatures of: (● ○) fcc, (▲ △) ecc, (■ □) unknown structure

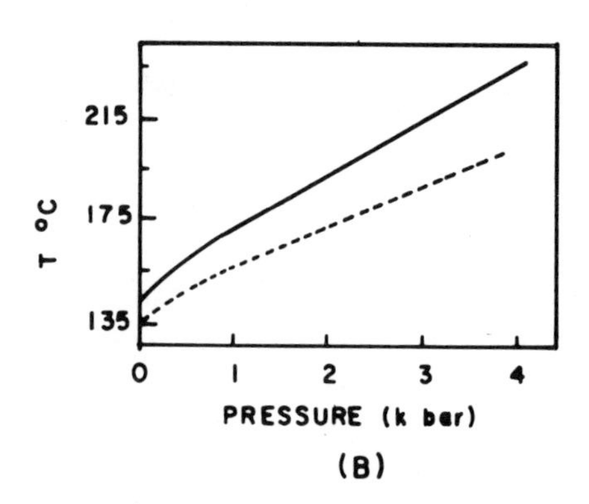

Figure 1B. Pressure dependence of the melting temperature, T_m, of polyethylene-chain crystals (——) and folded-chain crystals (– – –)

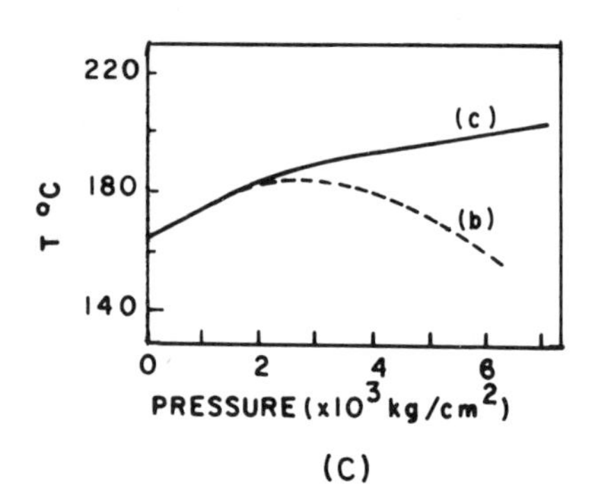

Figure 1C. Phase diagram of Teflon

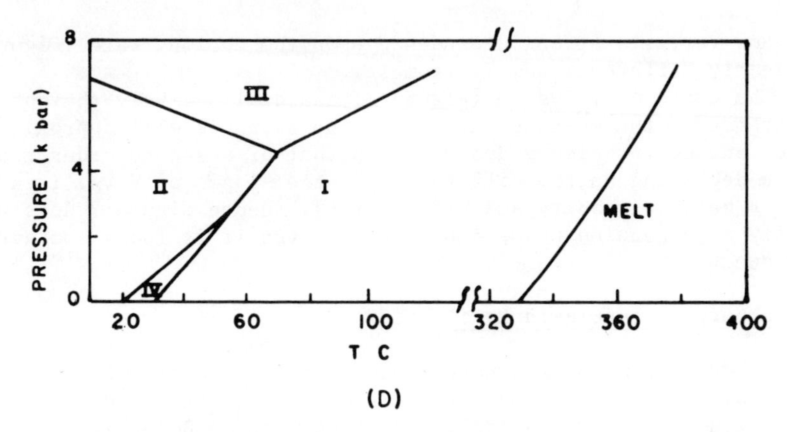

Figure 1D. Pressure dependence of the melting temperature, T_m, *of Penton for samples (a) crystallized at atmospheric pressure, and (b) crystallized at the measurement pressure*

conditions of tensile elongation are those of natural rubber and, occasionally, of melt spinning. Similarly, orientation-induced crystallization has been observed in melt flow, such as in injection molding, at high shear rates. A very high degree of stereoregularity or "tacticity" is necessary, so that long, uninterrupted parallel aligned stretches of molecule chains can form stable crystal crystal nuclei that will grow rapidly into macroscopic crystals. A theoretical treatment of strain induced crystallization with quantitative predictive power is still in its infancy.

The relation of T_m or better of ΔS_m and ΔH_m (and ΔV_m) to molecular structure of crystalline polymers has been thoroughly discussed elsewhere (5) and its display would go beyond the scope of this review. Suffice if to say here, that T_m at atmospheric pressure is measured so easily that its estimation by predictive methods seems an uneconomical thing to do. The estimation of the slope of the T_m vs. P line, (6) the so-called melting curve of crystalline polymers seems no more difficult than of other substances. However, given the high viscosity of the polymer melt, especially at high pressures, nucleation may be so retarded, to make experimental melting point determination quite uncertain.

The Glass/Rubbery State "Equilibrium"

Glassy polymers are of such technical (and economic) importance that it would seem pedantic to ignore the very obvious physical change at the glass transition temperature. Morphologically, both phases are liquid like, i.e. highly disordered on an atomic scale. Furthermore the dramatic change in viscosity at the glass transition temperature (T_g) is not much more gradual than it is at

T_m. However, its strong dependence upon the cooling rate (Figure 2) clearly differentiates T_g from T_m.

The curve of T_g vs. P (Figure 3) has qualitatively the same appearance as the melting curve, but the slope is well approximated, but not precisely described as that of a second order transition depending on the differences in the slopes of V vs. T and V vs. P between rubbery and glassy state. Hence Figure 3 does not qualify as a genuine phase equilibrium, even if it looks and acts like one.

The Rubbery State/Liquid Equilibrium

There is now increasing evidence for the reality of the blips in the differential thermal analysis at $T > T_g$, at a temperature identified as T_{11} or liquid/liquid transition by Boyer (7). The physics of this phenomenon have yet to be studied both phenomonologically as well as in terms of molecular motions. According to some conjectures (8) T_{11} is the temperature (range) characterizing the change from entanglement of neighboring chains to a state of freer intermolecular movement. The parallel change in T_{11} and in viscosity with moelcular weight (shown on Figure 4) is the source of this conjecture. Existence of a relation between the viscosity

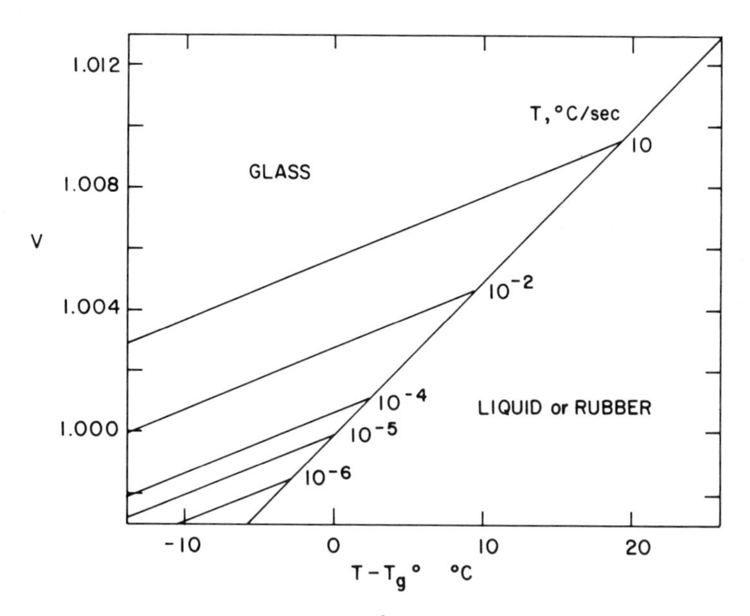

Figure 2. Effect of cooling rate $\dot{T}$ on the observed glass-transition temperature and specific volume according to the theory of Saito et al. The rate 10^{-5} C/sec was chosen for the standard for which $T_g = T°_g$ and $V_g = 1.000$; dV_g/dT was assumed independent of cooling rate. (A. Bondi, "Physical Properties of Molecular Crystals, Liquids and Glasses," Wiley, New York, N.Y., 1968)

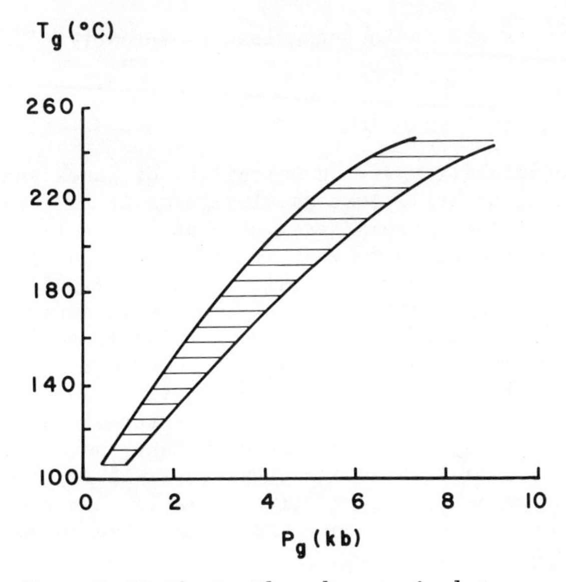

Figure 3. Vitrification Phase diagram of polystyrene. Note that vitrification takes place over a pressure range.

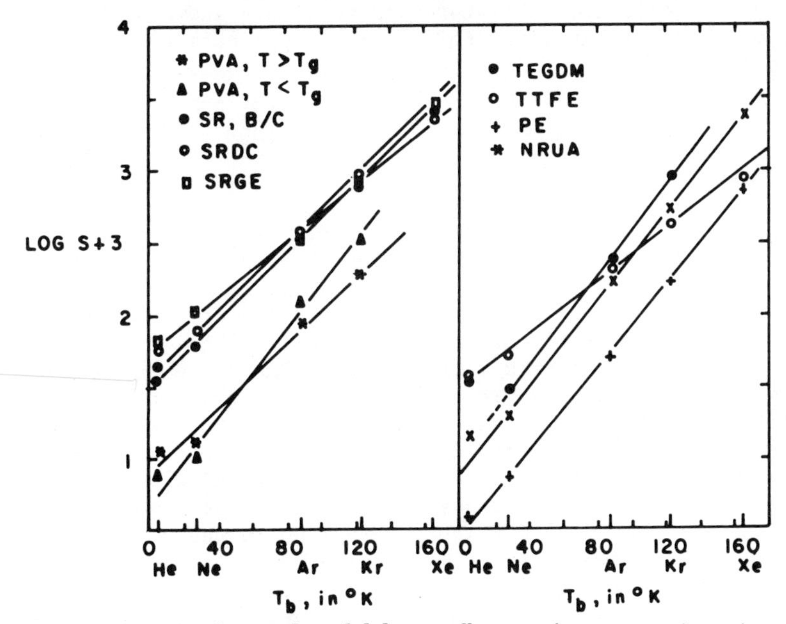

Figure 4. Logarithm of the solubility coefficients of rare gases in various homopolymers vs. the boiling point, T_b, of the gas. [Lundstrom, J. E., Bearman, R. J., J. Polym. Sci., Polym. Phys. (1974) 12, 97] PVA = poly(vinyl acetate), SR = silicone rubber, SRDC = silicone rubber, SRGE = 5% phenyl silicone rubber, TREGEM = poly(tetraethyleneglycol dimethacrylate), PE = polyethylene, NRUA = natural rubber.

of polymer melts and chain entanglement is fairly widely accepted (9).

Two Component, Two-Phase System Melt/Gas

The predictability of the solubility of gases and vapors in polymer melting is still just as elusive as it is for simple liquids. Some hoary generalizations about the relative effects of the molecular force constants (ε and r_o) of gases or of their critical constants, and the solubility parameter of the polymer melt can be used for interpolation purposes in a narrow group of systems. But a good theory, or even a reliable engineering correlation for a wide range of such systems have yet to be developed.

The solubility of polymer melts in highly compressed gases (at $T > T_c$ and $P >> P_c$) is of potential interest as a fractionation medium for polymers by molecular weight, because under some conditions pressure is a more convenient variable than solvent composition, and a less destructive variable than temperature. However so far only exploratory experiments have been published in this field (10).

Polymer Melt/Monomer Liquid(s)

The solubility of polymer melts in single component and in binary monomeric liquid mixtures is so widely known and frequently reviewed both with respect to practical applications, especially for polymer fractionation, and with respect to theory that little need be said here. The properties of monomer liquids and of polymer melt which determine their mutual miscibility have been assembled on Table 3 and are seen to be quite similar to those which determine the interaction between gases and polymer melts on Figure 4.

Four cases are of special interest, the precipitation of a dissolved polymer melt by addition of a non-solvent, the separation of chemically different polymers by mixtures of several solvents (below their CST), (11) the dissolution of a polymer melt by mixture of two non-solvents, and the swelling of a crosslinked polymer melt by monomeric solvents. The incidence of the first three cases is described in Figures 5 and 6, respectively, for non-specific interactions between polymer and solvent(s). Specific interactions are qualitatively just as predictable as in the case of specific interactions between monomeric liquids, say molecular compound formation or acid/base interaction. But quantitative predictions are even more difficult here than in the monomer case.

Although polymers are commonly mixtures with a molecular weight distribution of finite width we treat them as single components, or better as quasi-single components. If we do so, all the consequences of the phase rule are found to be valid. Moreover,

TABLE 3. PREDICTABILITY OF TWO VERSUS MULTICOMPONENT - TWO PHASE SYSTEMS

	Phase Diagrams and Properties of			
	Glass/Glass Blends	Glass Plasticizer Systems	Mixtures of Gas, Solvents with Block Copolymers	Adsorption Equilibria
Constraints	Time and Proximity of Tg	Time and Proximity of Tg	Size of Individual Blocks; Proximity to Block: Tg	Polymer Concentration Geometry of Adsorption Surface; Time
Required Data	δ (1), δ (2), $M_{1,2}$ D/A characteristics	$\delta_1;\delta_2$, $T_g(1);T_g(2)$ $\eta_2=f(T)$		
Available from Experiment	Nothing, most properties cannot be measured in the glassy state.	$T_g(1)$, $T_g(2)$, δ_2 $\eta_2 = f(T)$	The problem is too new for meaningful formulation of its specific character.	
Estimated Data	$\delta_1;\delta_2 \pm .03$ F^E of Functional Group Combination	δ_1; $\delta_2 \pm .03$ $\log \eta_2 = f(T) \pm .2$		
Prediction from Experimental Data	Impossible	Phase Boundaries - No		For dilute solutions of non-polar polymers-fair. For highly polar interactions-irreversible. For concentrated solution absorption is often dominated by dynamic phenomena. Predictions not achievable now.
Prediction from Estimated Data	Too few data so far on glas/glass phase boundaries	Single phase properties only very approximately.		

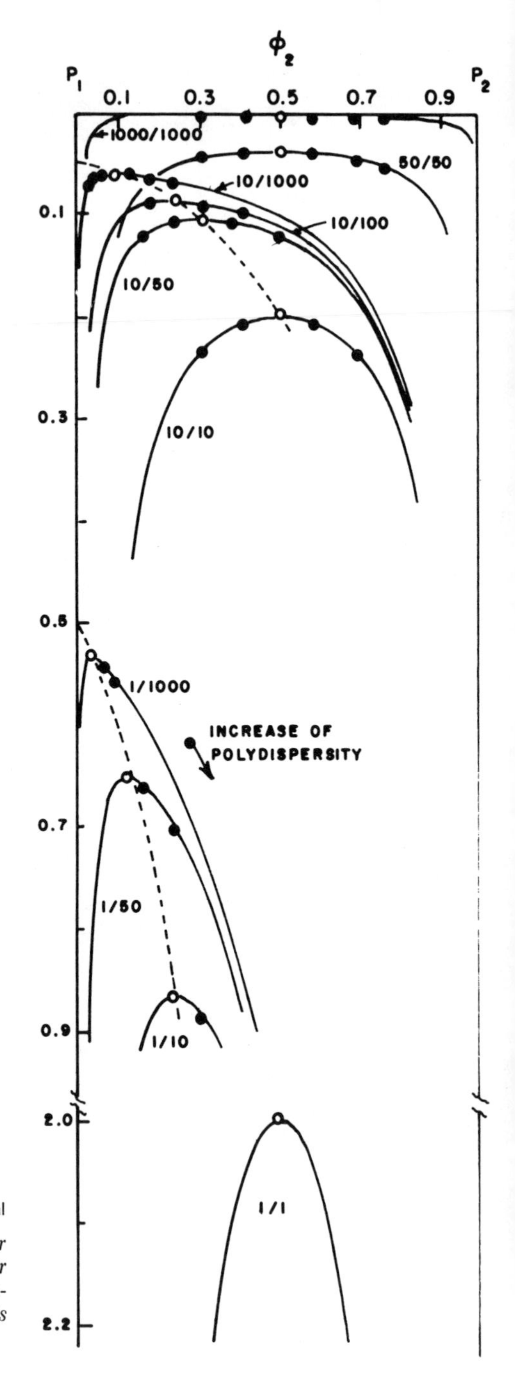

British Polymer Journal

Figure 5. Calculated spinodals for various macromolecular systems for indicated ratios of weight average molecular weights of mixture components (12)

the incidence and shape of miscibility gaps is predicted qualitatively correctly by molecular theory. But no existing theory can predict quantitatively the interesting miscibility gap geometries shown on Figures 6 through 8 (12).

The swelling of crosslinked polymer melts is just a limited dissolution. The only difference is that the effective molecular-weight for interaction with the solvent is (M_c) that of the chains between crosslinks. Hence minimization of swelling is achieved by the choice of polymer and solvent chemistry that minimizes mutual compatibility.

There is increasing evidence that the activity of the solvent such as plasticisers in highly concentrated systems is not adequately represented by existing formulations, such as the Flory-Huggins relations and their various higher approximations (13).

In view of the increasingly convincing evidence for the aggregation of polymer molecules into "superaggregates" in concentrated solutions (14,15) these deviations are not surprising. "Concentrated" means concentrations in excess of $[\eta]^{-1}$. Far more work needs to be done on the systematics of the aggregation constants as function of concentration, solvent-polymers interaction, and temperature before theories of the equilibria from concentrated polymer solutions can even be formulated.

Melt/Melt Interaction. The simplest formulation of polymer solubility in terms of regular solution theory on Table 3 or Figure 4 makes it very clear that high molecular weight (or volume) of the solvent must be compensated for by small differences in their solubility parameter, if mutual compatibility is to be maintained. When both components of the mixture are high polymers, compatibility is commonly achieved only when their solubility parameters differ by less than .05 $(cal/cm^3)^{1/2}$. In other words, most polymer melts are incompatible with each other. Even dilute solutions of two different polymers in a single solvent will separate into dilute phases. The rare exceptions of compatible yet different polymers are shown on Table 4.

Hence most so-called "polyblends" are highly dispersed two phase systems which are prevented from separating into large scale phases by potential energy barriers impeding flow (16) or by the skillful introduction of crosslinks (19).

Glass/Gas (or Vapor). Polymers in the glassy state are plasticized by most monomeric substances with which they are compatible. Hence we are concerned here with those solutions of substances in the glass for which T_g > T (observation). Measurements of the partial pressure of the volatile solute over such glassy solutions are generally better represented by a combination of a Langmuir isotherm with Henry's law. The common explanation is that the bulk of the volatile solute is adsorbed on the surfaces of minute voids in the glass, rather than being dissolved in the glass (18,19). These

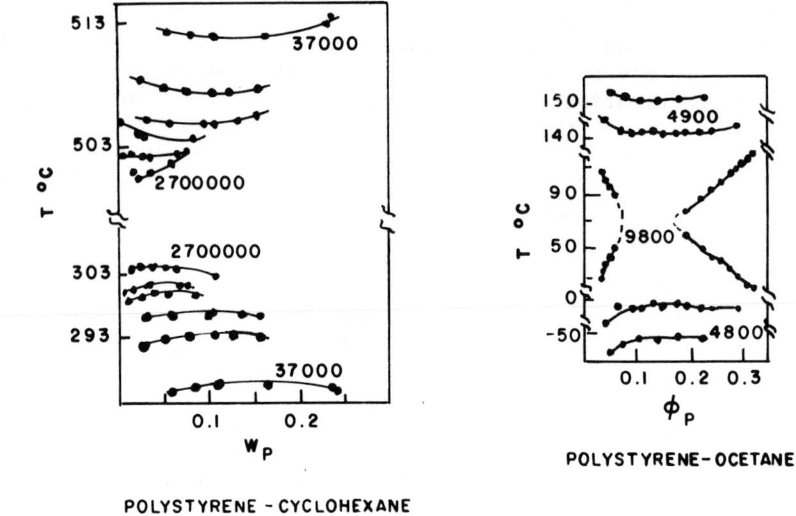

British Polymer Journal

Figure 6. Experimental examples of the various types of miscibility gaps in polymer solutions (12)

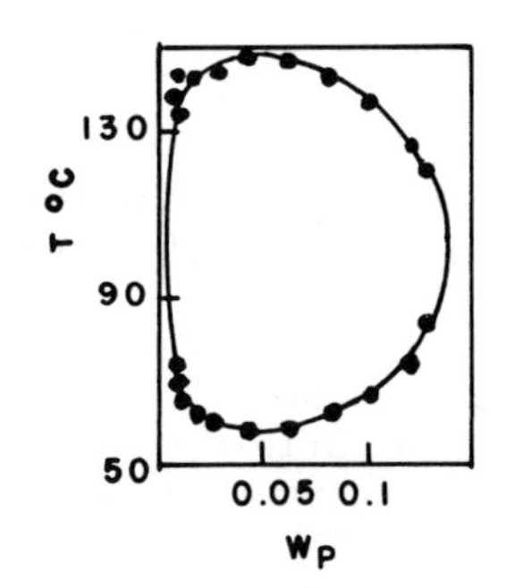

British Polymer Journal

Figure 7. Experimental example of the miscibility gap for polyvinyl alcohol–water, a hydrogen bonding system (12)

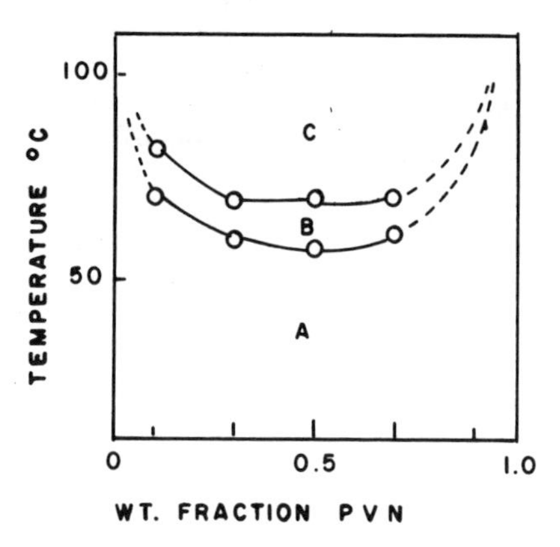

Figure 8. Lower critical solution temperature observed in glassy state by mechanical relaxation spectroscopy [Akiyama, S., et al., Kob. Roab. *(1976)* 5, *337.]*

TABLE 4

COMPATIBLE POLYMER PAIRS[a)]

Polymer 1	Polymer 2
Polystyrene	poly-2-methylstyrene
"	poly-methylvinyl ether
"	poly-2,5 dialkyl-phenylene oxide
"	benzyl-cellulose
poly caprolactam	polyvinyl chloride
"	various poly ethers
"	nitrocellulose
polyvinylidene fluoride	polymethyl methacrylate[b)]
"	" ethyl "
polyvinyl chloride	butadiene/acrylonitrile copolymer
"	ethylene/vinyl acetate copolymer
Polymethyl methacrylate (iso)	same syndiotactic
"	nitrocellulose

[a)] From data by S. Davidson and by D. R. Paul.

[b)] Confirmed by Brillouin-Scattering, Patterson, G. D. et al., Macromol. (1976), 9, 603.

voids are said to be the result of the inability of the glass to acquire its equilibrium density in the face of viscous resistance to bulk flow.

The practical importance of this observation and of the increasing evidence for the orientation dependence of gas or vapor solubility in glassy systems derives from the increasingly widespread use of glassy polymers as gas and vapor barrier films. Their permeability for the gas is, of course, the product of gas solubility and diffusion coefficient.

Given this practical importance, it is obviously awkward that the equilibrium constants of the dual mode sorption equations cannot be correlated with the properties of the polymer and of the permeant. Such correlations are reasonably successful in the rubbery state. Owing to the effects of mechanical and thermal histor of the glass on the incidence and effective surface area of the voids, a reliable generalized correlation for prior estimations of the three constants for gas sorption in glassy polymers may never be possible.

Yet more complicated is the sorption equilibrium of more soluble substances in glassy polymers. Since they plasticize the glass far more than the less soluble gases do, sorption causes drastic changes in the polymer, including the building up of stresses along the sorption front. Sorption equilibration under such circumstances is with a glass only at very low solvent activities, while at higher solvent activities it would be with a rubbery substance. The "vitrification concentration" of the solvent which separates these two regimes is, of course, well known from plasticization experiments, but has rarely been systematized for other solvents.

Crystal/Solvent. Given the difficulty of finding suitable solvents for the technically important high melting crystalline polymers, it is perhaps not surprising that only little work has been done on the discovery of eutectics between crystalline polymers and solvents of nearly similar melting points. These solvent then must be completely miscible with the polymer in their respective melt states, and completely immiscible as solids. Such systems have acquired practical significance because of the unique fibrillar morphology of crystalline polymer which precipitates at or near the eutectic point. In the current context it is importar to note that the experimentally observed phase diagram, Figure 9, differs substantially from that predicted by the Flory-Huggins theory. At present it is not clear whether that difference is real or whether it is due to retarded crystallization of the polymer (12,20).

Glass/Plasticizer. Basically this system is just a variant of the liquid/liquid system. However, the effect of phase separation on the glass transition temperature is not only time dependent, as expected. but can also be quite unique. This is best

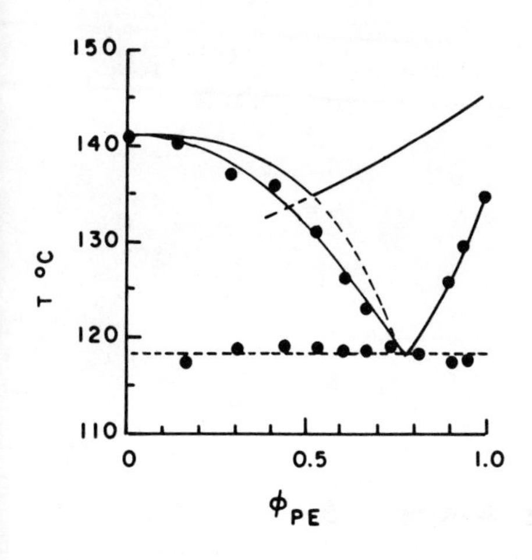

Figure 9. Eutectics in quasi-binary systems: crystalline polymer plus crystalline, high-melting-point solvents

British Polymer Journal

Figure 9A. Phase diagram of the system linear polyethylene–1,2,4,5-tetrachlorobenzene. (——) Flory-Huggins theory calculations, (●) melting temperatures of mixtures quenched at 87 C. (20)

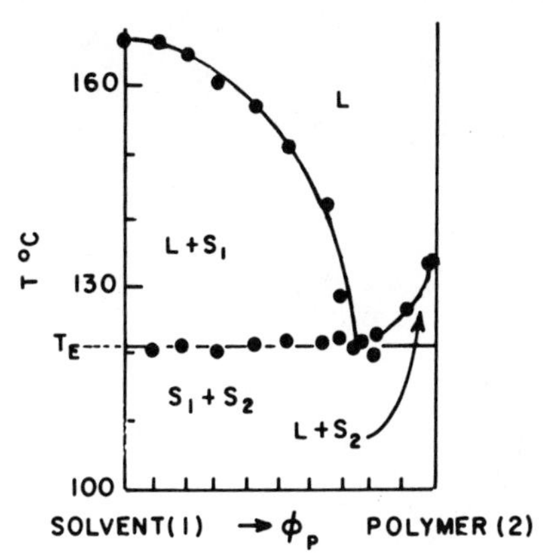

British Polymer Journal

Figure 9B. Phase diagram of the quasibinary system hexamethylbenzene–polyethylene (12)

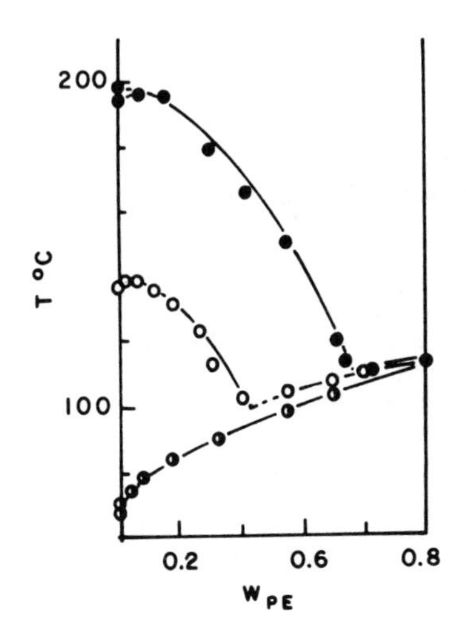

British Polymer Journal

Figure 9C. Phase diagram for polyethylene with various solvents: W_{pe} = wt fraction of polymer; ●, nitrobenzene; ○ amyl acetate; ○ xylene (12)

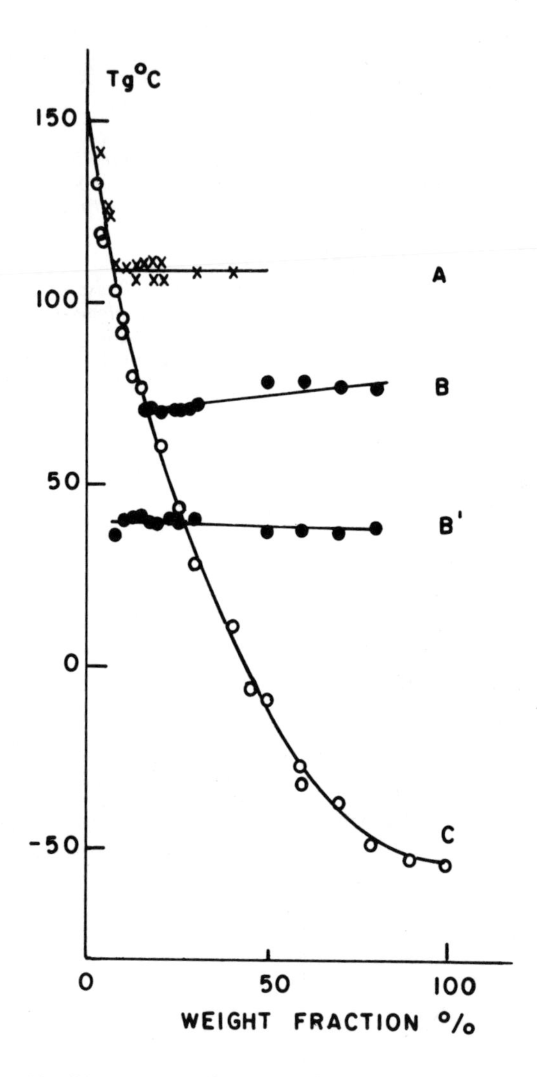

Figure 10. Variation of the DSC glass-transition temperature of bisphenol-A polycarbonate with concentration of plasticizer. Curve A: Pentaerythriol tetra nonanoate; Curve B: Trimellitic acid tridecyloctyl ester T_g *(1) and* T_g *(2) of two polymer solutions in equilibrium with each other after storage for one day at room temperature; Curve C: Tritolyl phosphate.*

illustrated by the recent thorough studies of polycarbonate (21). Typically, in the case of partial miscibility, there is just a separation into pure plasticizer and a polymer phase with fixed plasticizer concentration and constant, correspondingly depressed T_g, regardless of the initial mixture composition. In one case, with tri(decyloctyl) trimellitate as plasticizer, the polymer-plasticizer phase separates after some time into two polymer phases with widely differing plasticizer content and correspondingly two constant T_g's across the entire phase diagram, as shown on Figure 10. That figure also shows the curve for the normal depression of T_g with increasing plasticizer concentration. None of the otherwise successful theories of polymer solubility behavior predict this curious phenomenon.

Multi-Component/Multi-Phase Systems

Block Copolymers. The chemically stabilized permanent two phase systems which are obtained by block-copolymerization of incompatible chains exhibit the rather unexpected gas solubility characteristics shown on Figure 11. Tentatively this peculiar behavior has been ascribed to the special properties of the "interfacial" region between the two polymer phases (19). If the reality of the special properties of such an interfacial region is validated by independent experiments, we can speak of a component phase. Given the early development stage of this field, a definitive theory must be based on far more experimental work than has been produced so far.

Co-Extruded Films. The limited compatibility of polymers makes it possible to coextrude an arbitrary number of different polymer films or fibers, and maintain their separate identity indefinitely, even though they are in contact on a molecular scale. Even large differences in density are not likely to lead to observable Taylor instabilities over practically significant time intervals at temperatures at which the films are effectively solid. Viscosity reduction through composition changes or elevation of the temperature would, of course, allow the development of such buoyancy driven instabilities.

One economic driving force for making such multilayer films is the need for films of low permeability for several different gases, which can rarely be achieved by a single polymer. The solubility of different gases in the different films is independent of each other.

Adsorption Equilibria

The adsorption of polymers out of their solutions at liquid/solid, liquid/liquid, and liquid/gas interfaces is at present a very frustrating subject for broad generalizations. This frustration is caused by the appearance of ever more independent variables

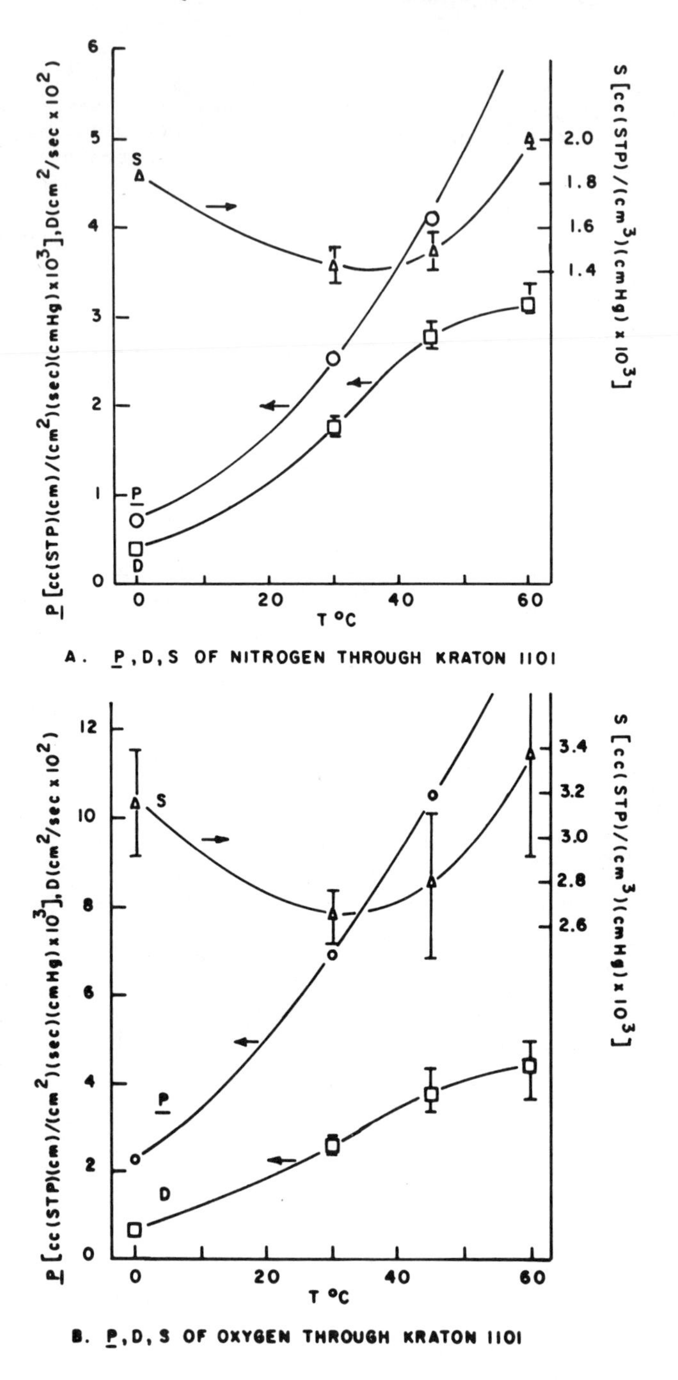

Figure 11. Solubility (S) of N_2 and O_2 in Kraton Block Copolymer, determined in unsteady state experiments; note the minimum near 30 C (22)

which complicate the situation. Adsorption of non-polar or slightly polar polymers out of dilute athermal or near athermal solution on planar surfaces can be described reasonably quantitatively by the established techniques of statistical mechanics and thermodynamics of polymer solutions (23). The outcome is that the polymer coil attaches to the surface at comparatively few spots. Consequently the thickness of the adsorbed layer is of the same order as the diameter of the random coil in the dilute solution from which it has been adsorbed.

Conversely, very polar polymers, strongly adsorbed out of thermal solutions are found to be strongly attached in most uncoiled configuration, i.e., nearly flat on the planar surface, especially if their molecular weight is not very high (24). In such cases a substantial fraction of the adsorbed polymer can be so strongly adsorbed to be virtually unelutable. Polymers in helical configuration in solutions are also adsorbed as cylindrical helices lying flat in the solid/liquid interface (25).

Adsorption of polymers approaches equilibrium very slowly because in general the lowest molecular weight components are adsorbed first, even if least strongly. They are then slowly displaced by the higher molecular weight components which dominate the adsorption equilibrium, especially in athermal solutions. If the adsorbent is porous, the higher molecular weight components may be excluded on geometrical grounds, and the steady state does not represent the thermodynamic equilibrium of the corresponding planar surface.

Adsorption equilibria for polymers out of concentrated solutions as function of concentration frequently exhibit very pronounced maxima (Fig. 12). These unusual curves can be accounted for if one assumes that the adsorbed species are in aggregation equilibrium in the solution, depending upon the amount of surface area per unit volume of solution. Hence one expects that the adsorption equilibrium out of concentrated polymer solution may not only be approached with "infinite" slowness but is also a function of the system characteristics, and the definition of reproducible conditions contains many more variables than one is used to from the more common work with dilute solution. This complexity is particularly awkward when one deals with the important case of competitive adsorption of polymers out of concentrated multicomponent solutions, a common phenomenon in many industrial processes, such as paint adhesion, corrosion prevention, lubrication, especially wear prevention, etc.

Conclusions

This quick survey of non-electrolyte polymer equilibria demonstrates that phenomenologically, of course, systems containing polymers do not differ from monomeric systems. An important complication is the everpresent polymolecularity of individual

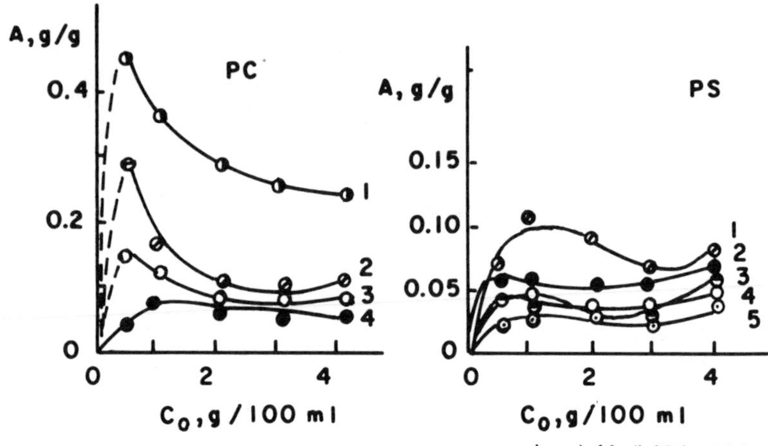

Figure 12. Adsorption equilibria for polycarbonate (PC) and polystyrene (PS) out of concentrated solution in CCl_4 at 25°C onto different amounts of Aerosil Silica:

	(1)	(2)	(3)	(4)	(5)
g/l.	5	10	20	30	40

A = g *(Adsorbate)*/g *(Adsorbant)* (15)

polymer species, which systematically distorts all phase diagrams of polymeric systems. But exact relations between the shape of phase diagrams and the width of molecular weight distributions have yet to be developed.

Similarly, molecular theory successfully describes one aspect of polymeric systems, namely their dilute solution behavior, but on occasion does not even predict qualitatively the phase equilibria of concentrated polymer systems, and rarely does so quantitatively. The high viscosity of such systems often prevents or certainly slows the approach to equilibrium, so that the measurement problem can be substantial.

The glassy "phase" owes its separate existence to the very high viscosity. Moreover, the practical importance of polymers in the glassy state has led to far more studies of "phase" equilibria involving this pseudo solid, than has ever been devoted to monomeric glasses. Hence the present survey includes glassy systems as well. Needless to say that a definitive theory of glass transition pseudo phase boundaries has yet to be developed, many attempts in that direction not withstanding.

Finally, what would be the economic impact of improved theoretical understanding of polymer equilibria, given that polymer processing and applications technology have grown faster than those of any novel material, except perhaps solid state electronics. In the absence of multistage separations processes involving polymers, one of the most important economic driving forces of industrial thermodynamic studies is missing from this field. Where, on the other hand, improved theoretical insight leads to qualitative changes in the understanding of polymer physics, new vistas would be opened for imaginative inventors of new compositions of matter or of novel processes. Those developments could have a far greater impact than improving the precision of our estimates of the location of phase boundaries from molecular parameters. One area promising qualitatively novel conclusions might be a good theory of superaggregate formation in concentrated polymer solutions.

Abstract

The present survey examines the current state of ignorance wherever polymers form part of a phase equilibrium in non-electrolyte systems, whether in or with solids, liquids, gases or surfaces (by adsorption). Overall we note the absence of a reliable means to estimate the effect of molecular-weight distribution on the shape of phase equilibria. An exception may be fractional precipitation by solvent/non-solvent combinations.

Among the more significant ignorance islands we find: tensile strain and strain rate induced crystallization; equilibria in concentrated polymer solutions, the detailed shape of miscibility gaps; solubility of polymers in high pressure gases; vapor-liquid equilibria in block copolymer systems; adsorption of polymers out of their concentrated solutions. Ignorance in these areas is noted by the engineer insofar as he cannot even make crude a priori estimates of these equilibria, and after asking a chemist for a few experimental data points, is hard pressed to put those few data into a correlational framework for expansion into the data network that he commonly needs for process design calculations, or for materials properties estimation over the usual range of the independent variables.

The absence of large scale multistage separations processes involving polymers removes the economic incentive which drives phase equilibrium studies in the world of simple chemical substances. Moreover, given the large number of composition variables available to and exploited by the polymer chemist as well as the multidimensionality of the measurement problem, we can expect the elimination of ignorance islands by the private sector only where continued ignorance is too costly to bear as a source of design and performance uncertainties. We can only hope that important new insights from investigators in academia will encompass one or more ignorance islands incidental to the resolution of more basic problems in polymer physics.

Literature Cited

1. Ekwall, Per, Adv. in Liquid Crystals 1, 1 (1975), Academic Press.
2. Fischer, E. W., Koll. Z. & Polymers (1966), 213, 113, 93; (1969), 231, 472; (1971), 247, 858.
3. Wunderlich, B., et al., J. Polymer Sci. (1964), A2, 3694.

4. Mandelkern, L., "Crystallization of Polymers", McGraw-Hill, New York, 1964.
5. Bondi, A., Chem. Rev. (1965), 67, 565.
6. Bhateja, S. K. and Pae, K. D., J. Macromal. Sci-Revs. Macromol. Chem. (1975), 13, (1).
7. Boyer, R. F., J. Polym. Sci. (1966), C14, 267.
8. Boyer, R. F., Stadhicki, S. J. and Gilham, J. K., J. Appl. Polymer Sci. (1976), 20, 1245.
9. Porter, R. S. and Johnson, J. F., Chem. Rev. 66, (1966) 1 (see also Ref. 8).
10. Ehrlich, P. and Woodbrey, J. C., J. Appl. Polymer Sci. (1969), 13, 117.
11. Kuhn, R., Makromol. Chem. (1976), 177, 1525.
12. Koningsveld, R., Br. Polym. J. (1975), 7, 435.
13. Bonner, D. C., J. Macromol. Sci. Revs. Macromol. Chem. (1975), C13(2), 263.
14. Lipatov, Y. S. and Sergeeva, L. M., Adv. Coll. and Intf. Sci. (1976), 6, 1.
15. Su, C. S., Patterson, D., and Schreiber, H.P., J. Appl. Polym. Sci. (1976), 20, 1025.
16. Y. Mori and H. Tanzawa, J. Appl. Polym. Sci. (1976), 20, 1775.
17. Sperling, L. H., et al., Polym. Eng. & Sci. (1972), 12, 101.
18. Vieth, W. R., Frangoulis, C. S. and Rionda, J. A., J. Coll. Intf. Sci. (1966), 22, 454.
19. Koros, W. J., Paul, D. R., and Rocha, A. A., J. Polym. Sci-Polym. Phys. (1976), 14, 687.
20. Pennings, A. J. and Smith, P., Br. Polym. J. (1976), 7, 460.
21. Onu, A., Legras, R., Mercier, J. P., J. Polym. Sci-Polym. Phys. (1976), 14, 1187.
22. M. I. Ostler, "Structure and Properties of Ordered Block Copolymer Membranes", PhD Thesis, Case-Western Reserve U., 1975.
23. Silberberg, A., Polym. Sci. (1970), Part C30, 393.
24. Fowkes, F. M., Schick, J., Bondi, A., J. Coll. Sci. (1960), 15, 531.
25. Eirich, F. R., et al., Preprints 48th Nat'l Colloid Symp. U. Texas, June 1974, p. 165/7.

Discussion

J. P. O'CONNELL

The papers presented in these two sessions describe well both the generalities and specifics of the state-of-the-art in this area which is so essential in the design and modeling of chemical processes. It would be redundant for me to repeat what is said in them, but it may be appropriate for me to summarize the impressions that I received and note some of the more relevant discussions which followed.

First, I think it is correct to say "you've come a long way, baby!" in the sense that the thermodynamic and physical basis has been laid for nearly all of the important situations encountered, there are many correlations which are at least adequate for most practical purposes, and there are many data available on systems of importance which are continually drawn upon for screening purposes and design calculations.

For example, there are many equations of state (about which an excellent report is given elsewhere in this volume) and corresponding states' correlations which are used daily with considerable satisfaction for "normal" fluids including mixtures (nonpolar and weakly polar substances). There are numerous accurate and flexible activity coefficient correlations which can describe more complex systems provided data are available to determine their binary parameters. In the absence of this, the methodology of group contributions has passed through successive generations to a fairly high degree of generality and accuracy. There are gas-liquid chromatography for infinite dilution properties and other automated apparatuses for easy measurement of highly accurate vapor-liquid equilibrium data, particularly at lower pressures. Computational sophistication, including the fitting of data and convergence of iterative calculations, has reached impressive heights. Although perhaps less well recognized, broadly applicable theories from statistical mechanics and most of the fundamentals of the critical region have been worked out so that development toward application can be started.

However, it is also true that "you ain't there yet!" because there are many systems which cannot even be characterized and others for which present descriptions are insufficiently accurate, if not actually erroneous. There have been many "failures" (defined as designs based on physical property information whose performance deviated significantly from specification, for worse, or perhaps better, in an economic sense) and possibly "disasters" for which no alternative methods exist even now. In addition, as time passes our ability to afford large safety factors will be decreased.

For example, how should hydrocarbon fractions really be characterized? What are the species which actually exist in systems in which strong or even weak chemical reactions can occur? Is there a way of putting a "chemist in the computer" to warn of such phenomena? No equation of state presently available describes mixtures of polar and nonpolar substances in the dense gas or liquid regions (though the second virial coefficient can be used for moderately dense systems). While the article of Leung and Griffiths (Phys. Revs., 1973) shows the formalisms of the critical scaling laws for mixtures, these have not been developed for use. Generalized correlations for liquids containing significant amounts of ionized species along with nonelectrolytes are nonexistent. Most aspects have not been worked out for the thermodynamic properties and equilibria of systems of macromolecules in the crystalline, glassy, or solution form. There is much uncertainty about the use of dilute solution reference states for supercritical components, particularly in multisolute, multisolvent solutions. The prediction of liquid-liquid phase boundaries and the concomitant vapor-liquid distribution coefficients is not on a par with that of less complex systems. In addition to correlation inadequacy, even for those situations where equations are available, inadequate data bases exist for establishing parameters, particularly for many group interactions. It is accepted that, except for accurate prediction of ternary and higher critical points (such as liquid-liquid plait points), only binary interactions need to be characterized, usually by a single constant for either equations of state or activity coefficients. Yet a very large number of binaries have not yet been addressed.

To improve upon this situation, the major source of ideas must come from applying molecular concepts through adopting statistical thermodynamics. The efforts described by John Prausnitz and others in the conference are only beginnings. Yet their success warrants further investigation. One aspect of their use will undoubtedly establish new characterization methods requiring data which are different from the normal kind.

Finally, then "lets get moving!" Our speakers exhorted us to concentrate our efforts on the unsolved problems. Incremental changes in present correlations (except to correct obvious deficiencies in form under extrapolation), and new measurements of systems whose properties are subject to little uncertainty (except for checking equipment) are uninteresting. The challenge is to be creative. I believe the greatest success will occur when communication and cooperation, such as at this conference, are established. In general, the limiting factor for new ideas is the bridge of development. Let us all resolve to build it whether we are researchers or practitioners. With all the possibilities available, let it not be said that a delay or a lack of a solution to the technological problems of the present and future could be laid at our feet for want of adequate descriptions of physical properties.

Comment submitted by J. S. Rowlinson:

John Prausnitz had mentioned the excellent agreement with experiment that Mollerup had obtained in the gas-liquid critical region of binary hydrocarbon mixtures. Mollerup's results were obtained with a good reference equation of state for methane (but one which is classical in form and so which does not describe accurately the known nonclassical singularities in the thermodynamic functions at the critical point), and with a one-fluid model based on a mole fraction average (or "mole fraction based mixing rules").

It is, perhaps, worth comment that no improvement would be made by using a more correct reference equation of state (i.e., one with the correct singularities) unless the mole fraction averaging is also abandoned. To do this first step, without the second, means that the singularities are lost in the mixture calculations--the mixture critical point is still classical. If, however, the mole fraction average be abandoned also, and replaced by mixing rules based on a composition variation of the absolute activity $\lambda_i = \exp(\mu_i/kT)$, where μ is the chemical potential, then the singularities are preserved on going from pure reference equation of state to the mixture. This has been shown by Leung and Griffiths (Phys. Rev.) for ^3He - ^4He mixtures. It may be that this composition variable is worth further study in the mixing rules.

Question by D. T. Binns:

In computer calculations for distillation or flowsheeting, K-values subroutines may be called thousands of times. Routines based on complex equations of state can make the computation costly. Is this a problem here? Have you, or anyone else present, any experience of fitting data, either measured or calculated by an appropriate equation of state, to something simple for use in computer programs? For instance, it should be possible to fit the two Redlich-Kwong parameters per component instead of calculating them from critical constants. It would also be possible to incorporate enthalpy data which may be available, so helping thermodynamic consistency.

Reply by T. Królikowski:

There have been some examples, but computer costs are now so low that it is not worth the trouble. Some economy can be achieved by calculating K-values only every fifth iteration. We go ahead and use the best equation of state.

Comments by A. Bondi:

Regarding the paper presented by J. Prausnitz:

1. For the advance guess or the correlation of adsorption isotherms under supercritical conditions, I can see no alternative to the estimation of the need p_o (for the driving force p/p_o) via the extrapolation of the vapor pressure curve beyond T_c, which you deprecated.
2. Your elipses, the rather striking evidence for the use of an inappropriate data reduction scheme, should be given wide exposure. Chemical engineers are now highly skilled in making the computer represent data by quite complex regression equations. But many seem to be far less careful in watching for autocorrelation among coefficients in these equations, and in ascertaining the significance of each coefficient than are our colleagues in economics and elsewhere in teh behavioral sciences. The absence of reference to autocorrelation and coefficient quality criteria from most papers on the multiconstant equations of state is rather serious evidence for this state of affairs. Maybe this needs some more emphasis in the teaching process.

Dr. Królikowski, you and your coworkers' observation difficulties in physical properties estimation for two process streams containing aldehydes, alcohol(s) and water is not surprising. Aldehydes react with water to form hemihydrates by

$$R_1-\overset{\overset{\displaystyle O}{\|}}{C}-H + H_2O \rightleftharpoons RC\begin{matrix} \diagup OH \\ \diagdown OH \end{matrix}$$

and they react with alcohols to form hemiacetals by

$$R_1-\overset{\overset{\displaystyle O}{\|}}{C}-H + R_2OH \rightleftharpoons R_1-\underset{\underset{\displaystyle OH}{|}}{C}-O-R_2$$

In both cases there is a steep increase in viscosity, as shown by the examples of my own experience, shown on Figures 1 and 2, and naturally a substantial increase in density, as shown on Figures 3 and 4. These reactions are reversible and--as shown--of less importance at high than at low temperature. Moreover, once one has recognized what chemistry is going on, the mixture properties can be estimated with fair approximation, as shown by the compari-

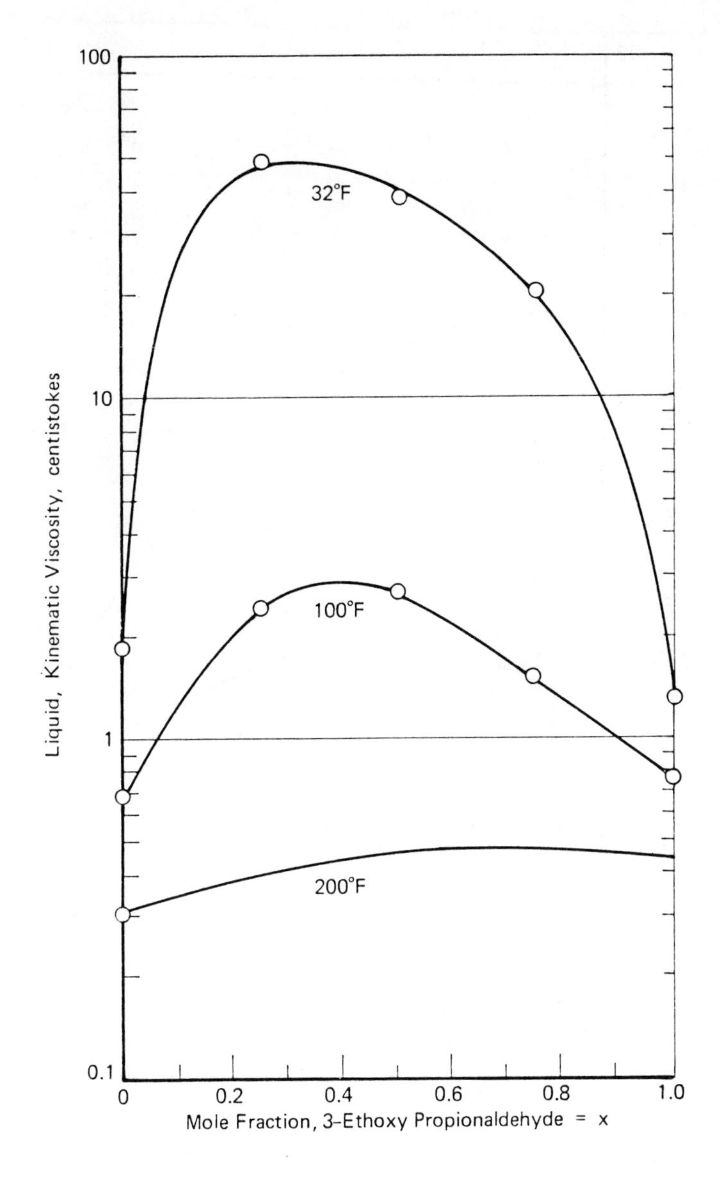

Figure 1. Viscosity of mixtures of 3-ethoxy propionaldehyde with water at various temperatures

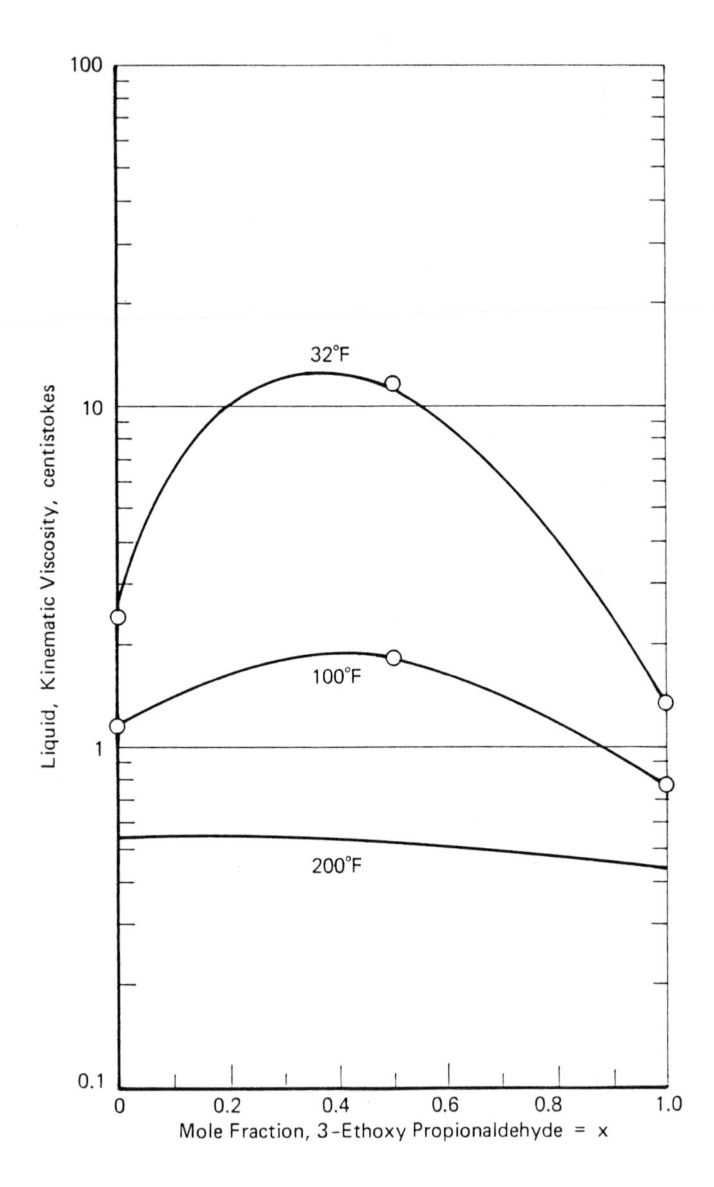

Figure 2. Viscosity of mixtures of 3-ethoxy propionaldehyde with ethanol at various temperatures

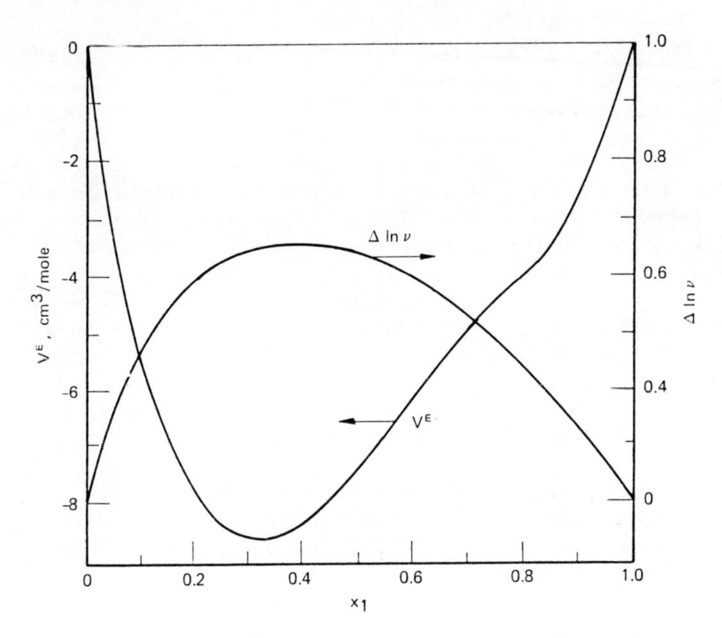

Figure 3. Volume contraction (V^E) of mixtures of 3-ethoxy propionaldehyde with water at 0°C compared with deviation of the viscosity from the ideal mixing rule

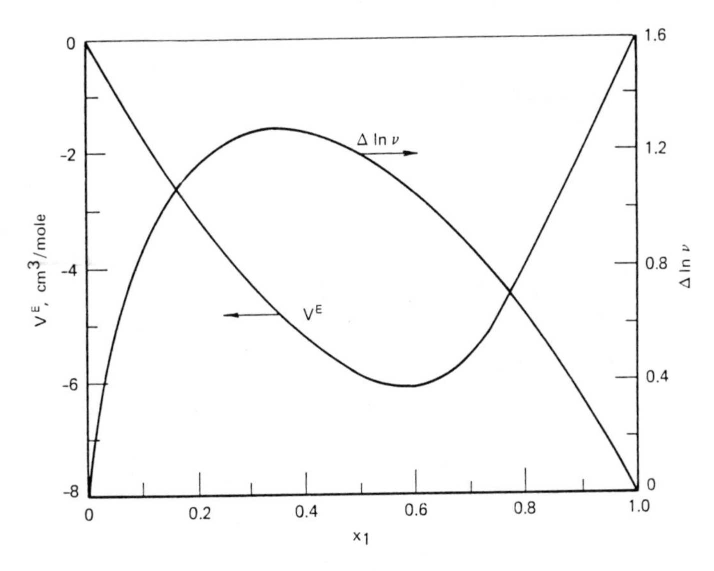

Figure 4. Volume contraction (V^E) of mixtures of 3-ethoxy propionaldehyde with ethanol at 0°C compared with deviations of the viscosity from the ideal mixing rule

sons of Table 1. A few literature data for mixtures of acetaldehyde with water show a similar behavior.

The main point of these discussion remarks is to emphasize the need for a very alert chemist on the properties estimation team. In my case it was a "gut feeling" when I asked for the mixture properties to be measured rather than slip a routine estimation into a process manual. The wise explanation of the chemistry came only as we all contemplated the unexpected experimental evidence.

The potential costliness of property estimation errors caused by the unexpected formation of (loose) covalent or molecular compounds among process streams components calls for an early remedy. One of these might be to have the editor of C.E.P. commission a well known authority on molecular compounds and related chemistry to write an article for engineers including a table with danger signals for insertion into design engineers' notebooks.

These remarkable data presented by Dr. Franck are so difficult to obtain experimentally that their estimation with paper and pencil is clearly worthwhile. Can one estimate the dielectric constants in your supercritical range from a combination of

Table 1 - Estimation of viscosity* of Hemihydrate and of Hemiacetal of 3-Ethoxy Propionaldehyde compared with observation on Figures 1 and 2.

Substance	t, °C	$\eta_{est.}$	$\eta_{obs.}$
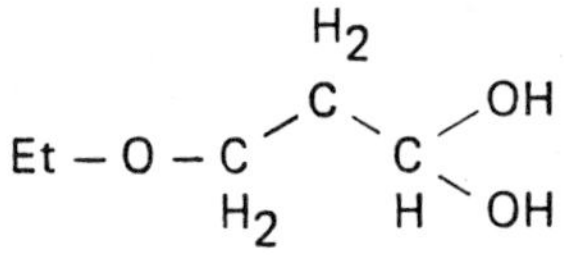 $Et-O-CH_2-CH_2-CH(OH)_2$	4	0.22	0.23
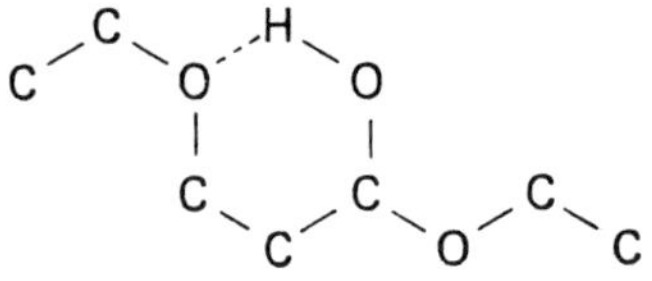 C–C–O···H–O–C(–O–C–C)–C–C–(ring to O)	0	0.18	0.12

*By method in A. Bondi "Physical Properties of Molecular Crystals, Liquids and Glasses", Wiley 1968.

dielectric polarisability, such as modern Clausius-Mosotti derivatives, with the reduced temperature (T/T_c)? Can one estimate the electrical conductivity in your supercritical range from any theory of the Walden product $\Lambda \cdot \eta$ or other semi-empirical correlation?

Dr. Franck indicated that the Calusius-Mosotti effect had been treated using the Onsager-Kirkwood corrections with agreement between experiment and theory that is only satisfactory.

Chemical reactions commonly occur between water and other molecules. Hydrocarbons start to react at 400C so the chemical equilibrium and physical equilibrium are simultaneously observed.

Solid-liquid melt lines have not been studied at the Karlsruhe laboratory. Some caution must be used at these "superpressures" because molecules deform enough that they change their chemical identity.

7

Industrial Uses of Equations of State: A State-of-the-Art Review

STANLEY B. ADLER, CALVIN F. SPENCER, HAL OZKARDESH, and CHIA-MING KUO

Pullman Kellogg Co., Houston, TX 77046

The objective of the presentations at this conference, as the authors of this paper understand them, is to document the state of the art in various areas of thermodynamics. In turn the ultimate objective is to enable academia to grasp the needs of industry, while industry becomes better acquainted with the new tools developed in the universities.

Those of us from industry on the week's program have been asked to describe, from the industrial side of the fence, the current activities, developments, and applications of particular assigned areas of phase equilibria or physical properties correlation. This paper pertains to equations of state, especially their application in phase equilibrium predictions. It is not the intention of this talk, nor would it be appropriate, to review all the equations of state that have been published. An excellent paper by Tsonopoulos and Prausnitz (1) does make such a review. Instead, this paper will present the applications and limitations of some of the principal equations of state in current use.

At the very outset--before even getting to the body of this paper--it is necessary to picture the magnitude of industrial involvement with equations of state. The two lists given in Table I summarize the basic outlines of this involvement. In one category the research aspects of industrial work in equations of state have been assembled. In a second category, the applications in which equations of state are intermixed with thermodynamic principles in day-to-day process and design problems and computations are shown. To put it another way, the first list deals with sharpening the tools; the second list with using the tools.

Characteristics of Equations of State Required by Industry

When equations of state are used in industry they must possess two essential characteristics: (1) Versatility and (2) Workability.

Versatility. To best describe the versatility possessed by equations of state, a quotation is taken from Professor Joseph Martin (2), a long time investigator in this field.

> "The second reason for developing new equations of state concerns the exceptional power and utility of an equation of state. When combined with appropriate thermodynamic relations, a well-behaved equation can predict with high precision isothermal changes in heat capacity, enthalpy, entropy and fugacity, vapor pres-

TABLE I

Research Aspects

1. Develop completely new equations.
2. Improve existing equations.
3. Test existing ones for successes and failures.
4. Study application to mixtures.
   a. new models
   b. new interaction constants
5. Extend given equations of state to other properties, modify if necessary for improvement
   e.g., Redlich-Kwong to enthalpy [Barner et al., (3)]
6. Extend to lower and lower temperatures.
7. Get constants for more and more substances.
8. Extend usefulness of equations of state to new applications.
   a. addition to minimization of free energy technique
   b. to solution of freeze-out problems
9. Develop a whole new approach--combining existing equations: one for liquid and one for the vapor, and even one for pure liquid fugacity.

Thermodynamic Applications in Process Engineering

1. Compressibility.
2. Enthalph and heats of mixing.
3. Liquid-vapor equilibria: bubble-point, dew-point, flash vaporizations; predictions in the retrograde region.
4. Liquid-liquid equilibria.
5. Liquid-vapor-solid equilibria and freeze-out problems.
6. All phases with chemical reaction equilibria simultaneously.
7. Prediction of critical points and the critical locus of a mixture.
8. Prediction and interpretation of results for thermodynamic anomalies, e.g., multiple bubble-points.
9. Incorporation of equation of state as a tool in computerized flow sheet calculations for an entire process.
10. In calculation of data charts issued company-wide in Data Books, particularly in K constant charts for liquid-vapor equilibrium.

sure, latent heat of vaporization, activity coefficients, and vapor-liquid equilibria in mixtures, not to mention the assistance it offers in transport property correlations. Unfortunately, even though the useful applications of an equation of state are so extensive and attractive, the development of a high performance equation proves to be so involved that to date no one has come close to discovering a single relation which is truly good over a wide range of density."

As Martin pointed out, the equation of state is a powerful tool. Working with it to solve problems is an exciting occupation. Nevertheless, before getting into the exciting side of the picture, it is necessary to paint the prosaic, more mundane side. The reason for doing so is that this paper was designed to give the industrial point of view. In industry we must have a tool that works, that is flexible, that can reach a solution without having a computer failure after the calculation has proceeded as far as the mid-tray in a one hundred tray distillation tower. The word flexible, mentioned in the previous sentence, covers a myriad of attributes. The equation of state, as Martin said, must be suitable for all the thermodynamic and physical properties--not just enthalpy or compressibility as many published equations of state are. As an initial requirement it should be able to generate these properties for vapors, both for pure components and mixtures. Vapor-liquid equilibrium calculations demand, as an additional requirement, application to liquids and liquid mixtures. As an added requirement it would be expedient to have it apply to petroleum fractions which are a continuum of many components that cannot individually be defined. This is truly asking a lot, yet all the requirements are integral parts of a typical process design calculation.

Some of the shortcomings of published expressions which lack such flexibility are illustrated below. These examples are based on studies performed by the Technical Data Group at Pullman Kellogg over the last fifteen years.

Some years ago Barner et al. (3) developed a modification of th Redlich-Kwong equation of state for application in enthalpy calculations. The original Redlich-Kwong equation, intended for compressibility, was being misapplied by a client for enthalpy computations. Doing so, the client disagreed with a specified duty on a flowsheet. The Redlich-Kwong equation was modified to represent enthalpy by fitting its constants to the Curl-Pitzer enthalpy tables. The end result: it helped the client, but the revised equation failed to represent fugacity adequately. Similarly, in another publication Barner and Adler (4) revised the Joffe equation to represent quite a number of vapor properties. It was, as expected, an utter failure for representing liquids, yet well applicable to the gaseous mixtures of those components for which BWR constants were not available and where it would be impractical to determine them, parti larly if the components are not often encountered.

As an illustration of an equation of state that will not be applicable to mixtures, consider that the usual mixing rules require taking square roots, cube roots, or the like. If the constants are negative for one or more of the components in the mixture, extension to mixtures is not feasible.

The most important industrial application of equations of state is, and will continue to be, in phase equilibrium predictions. All such calculations are based on the equilibrium criterion:

$$\hat{f}_V = \hat{f}_L = \hat{f}_S \tag{1}$$

where, $\hat{f}$ = component fugacity

V, L, S = vapor, liquid, and solid phases, respectively.

Equations of state are applied in several ways to compute the component fugacities. In the one-equation of state approach, the same equation is used to get the fugacity of each component in all the coexisting phases. In a second approach, one equation of state is used to get vapor fugacity, while different equation(s) are employed for the liquid or solid phases. These approaches are explained in greater detail in the next section of the paper.

Although most equilibrium calculations involve a single vapor and liquid phase only, a number of equilibrium phase combinations such as liquid-liquid-vapor, liquid-solid with or without a vapor phase, and vapor-solid phases are encountered in industry. Furthermore, any of these physical equilibrium conditions may also require computations of the chemical reaction equilibria. For example, consider the following realistic inquiry from an engineer designing a pilot plant. The chemists completed their bench-scale work for the hydrolysis of propylene to make an alcohol and an ether. They reported the optimum temperature to be 240F and the pressure to be 500 psia. The chemical engineer wanted to know how many phases the reactor effluent would have at these conditions: all vapor, a liquid phase with the alcohol product and condensed steam in equilibrium with the vapor, or perhaps three phases involving the unreacted, condensed propylene as well. This complex equilibrium problem involving both physical and chemical equilibrium with nonideal liquids and a nonideal vapor best represents what chemical process engineering is all about.

Problems like these can be solved by applying the principle of the minimization of free energy (5). In this approach the free energy is related to fugacity of each component in each phase, whereby the fugacities are calculated from equations of state. No other known approach to the solution of such problems is really satisfactory. The fact that this approach is also direct and easy to visualize as one looks at successive computer print-outs of the iterations to the final solution makes it all the more the one to be recommended. Furthermore, the other properties, enthalpy being the most important, are computed on a basis consistent with the equilibria, by means of the same equation of state. It is for this sort of problem that the versatility of the equation of state really pays off.

Workability. Even if the equation is applicable to liquids and vapors, pure components and mixtures, physical properties, thermodynamic properties, and equilibria, there is still the question of workability. In simple terms, can it accomplish what it proposes to do?

Very frequently a new thermodynamic relationship is carefully derived, tested on a few systems, and published either in the open literature or in a company engineering report. It may look promising, but until it is tested on a wide variety of systems, and a wide range of temperatures and pressures and other conditions, its real merits or shortcomings in areas such as handling close boiling components, the critical region, hydrogen systems, immiscible liquids, etc., cannot be fully understood. Any or all of these areas may be outside the applicability of the new relationship.

Some years ago a research chemist tried to develop a thermodynamic consistency test for hydrogen and helium systems where one component is above its critical temperature, and where a vapor pressure and therefore, the activity coefficient, cannot be obtained. His relationship required the use of a generalized chart to get a property--of course it was before the days of computers. When attempts were made to use the generalized chart, it was found that near the critical region an accurate property value could not be read, certainly not accurate enough to test consistency without adding as much error as would be measured in the consistency test. Unless one takes the time to sufficiently test, one does not see the problems that may be encountered.

A second consideration, which is certianly very important to industry, is workability without the intervention of man, as with a computer. Not only will the equation of state be used in an iterative fashion to converge to a solution, but also on successive stages in equipment design, to produce a whole series of results. With all these repeated uses, it must lend itself to ease of solution with scarcely a failure.

This entire presentation could be devoted to the problem of getting some equations of state to converge on the correct root. Some of the problems are listed here:

1. In applications of equations of state to vapor-liquid equilibria, avoid convergence to a trivial solution, particularly all the K values being 1.0 (where $K = y/x$ is the ratio of the mole fractions in vapor and liquid, respectively).
2. The selection of the correct root for the vapor density and then the liquid density, when the two phases are in equilibrium.
3. Treating convergence to negative roots. Some of the cookbook rules to extricate the computer from this situation

simply lead to a negative compressibility again a few trials later.

Besides providing feasible computer routes to handle the three aforementioned problems, special care must be taken to prevent the computer flash algorithm from oscillating back and forth in K constant (vapor-liquid equilibrium) calculations because the prescribed flash conditions of temperature and pressure do not fall in the two-phase region. It would be presumptuous to assume that all flash conditions are realistic; engineers do, and will continue to, submit unrealistic temperature and pressures, in the range of subcooled liquids and superheated vapors.

Proper criteria must be selected for establishing whether two phases for a given mixture really exist at the set conditions. Frequently the definition of the critical point of a pure compound is often erroneously used. Most textbooks in their coverage of equations of state define the critical state as:

$$\left[\frac{\partial P}{\partial V}\right]_{T_C} = \left[\frac{\partial^2 P}{\partial V2}\right]_{T_C} = 0 \qquad (2)$$

In simple terms, the critical isotherm has a point of inflection at the critical point. By studying the behavior of an isotherm predicted from an equation of state, it is possible to establish whether two phases exist. For example, if the isothermal second derivative changes sign over a range of pressures and volumes, a vapor and liquid are present. This test unfortunately also works for mixtures at conditions far removed from the critical region, but breaks down in the critical and retrograde region. Such a test is valid for pure components <u>only</u>. For mixtures--and K constants arise only for mixtures--the thermodynamics are far more complicated, and Eq. (2) should <u>never</u> be applied. It should be added that problems in this area are often complicated by the fact that many of the existing equations of state are not valid in the critical region and often exhibit unusual behavior in that portion of the phase envelope. Proper criteria for rapidly identifying unrealistic flash conditions are only developed after sufficient experience is gained working with the respective equation of state and properly analyzing flash results, both its successes and failures.

As pointed out in the preceding paragraph, some equations, by their very nature, complicate matters. As an additional example, consider the specific limitations of the Chao-Seader correlation and others like it, which allow only about 20% methane in the liquid. In a typical flash problem it is conventional for a first set of K constants to be furnished either by the engineer or initialized by the flash program. These K's immediately lead to vapor and liquid compositions. With these compositions the component fugacities and liquid activity coefficients are calculated, which in turn lead to a seemingly better set of K's. If the first liquid composition--admittedly far removed from the correct one--contains 80% methane in the liquid, the correlation's range of applicability has been exceeded. These K's are so far removed from the real set that the resulting second trial may be worse than the first, and convergence never obtained. Thus, even though the mixture composition answer may be well within the range of the correlation, the initial or even the intermediate trials may exceed that range, and lead to erroneous trial values from which the computer solution algorithm will never recover.

Summarizing, it takes a flexibile equation of state and the proper combination of thermodynamics, common sense, patience and programming finesse to utilize equations of state effectively in computerized vapor-liquid equilibrium and other design calculations

Equations of State Adopted by Industry

As stated earlier, two-phase liquid-vapor equilibrium is the predominant problem in industrial equilibrium calculations. To incorporate equations of state, Eq. (1) is often expanded and rewritten as:

$$x_i \cdot \gamma_{L_i} \cdot f^o_{L_i} = y_i \cdot \phi_i \cdot \pi \qquad (3)$$

where, for each component i,

x_i = mole fraction in the liquid phase

γ_{L_i} = liquid phase activity coefficient

$f^o_{L_i}$ = pure liquid fugacity

y_i = mole fraction in the vapor phase

ϕ_i = vapor phase fugacity coefficient $\hat{f}_{V_i}/y_i\pi$

π = system pressure

When the single equation of state approach is used, Eq. (3) reduces to:

$$\frac{\dfrac{\hat{f}_{L_i}}{x_i\pi}}{\dfrac{\hat{f}_{V_i}}{y_i\pi}} = K_i \qquad (4)$$

where K_i is the liquid-vapor equilibrium constant of component i. The same equation of state is then employed to get both the numerator and denominator in this expression using standard thermodynamic relationships. The work of Benedict, Webb and Rubin (6), of Starling (7), and the series of Exxon papers (Lin et al., (8); Lin and Hopke, (9))--all on various forms of the BWR, Soave, and Peng-Robinson equations of state are examples of the use of one equation of state to perform the whole calculation. It should be added that the original developments in this area treated the $\hat{f}_{L_i}/x_i$ term in the numerator as a separate entity and multiplied the final answer by $1/\pi$ for consistency. Most contemporary approaches relate both numerator and denominator to equations of state via the fugacity coefficient route, the only difference in liquid and vapor being the density and the equation constants obtained from the respective mixing rules.

In the two (multi) equation of state approach, Eq. (3) is restated as:

$$K_i = \frac{y_i}{x_i} = \frac{\gamma_{L_i} f^o_{L_i}}{\phi_i \pi} \qquad (5)$$

This so-called two equation of state method often requires three equations: one for the nonideality of the liquid, one for the nonideality of the vapor, and one for the standard state liquid fugacity to which the activity coefficient must be referred. These three relationships determine respectively, γ_L, ϕ, and f_L^o in Eq. (5). The Chao-Seader correlation and its many modifications, and the Chueh-Prausnitz correlation are examples of this approach. In the Chao-Seader correlation $f_L{}^o$ and π are combined as a pure liquid fugacity coefficient, ν, so that Eq. (5) has three distinct parts --each requiring a unique equation, as just described.

Obviously it is desirable to use a single equation of state for the whole computation. It is simpler, more consistent, and less work for both the engineer and computer programmer. However, the hard, cold fact is that no one equation of state can meet the versatility requirements stated earlier. Modifications of Starling's BWR-11 equation are excellent in the cryogenic region for predicting both VLE, and thermal and physical properties. However, the equation is not applicable in many petrochemical processing calculations because of limitations caused by the availability of constants for a number of needed components. The various modifications of the Chao-Seader correlation are useful for predicting vapor-liquid equilibrium in a wide variety of process streams; however, this correlation cannot be used to predict thermal or physical properties. The Soave and Peng-Robinson equations are steps in the right direction, but further work is needed to fill the needs of industry.

A few of the equations of state that have widespread use in industrial VLE calculations are discussed below. The comments, which deal mainly for the aforementioned category 1, "The Research

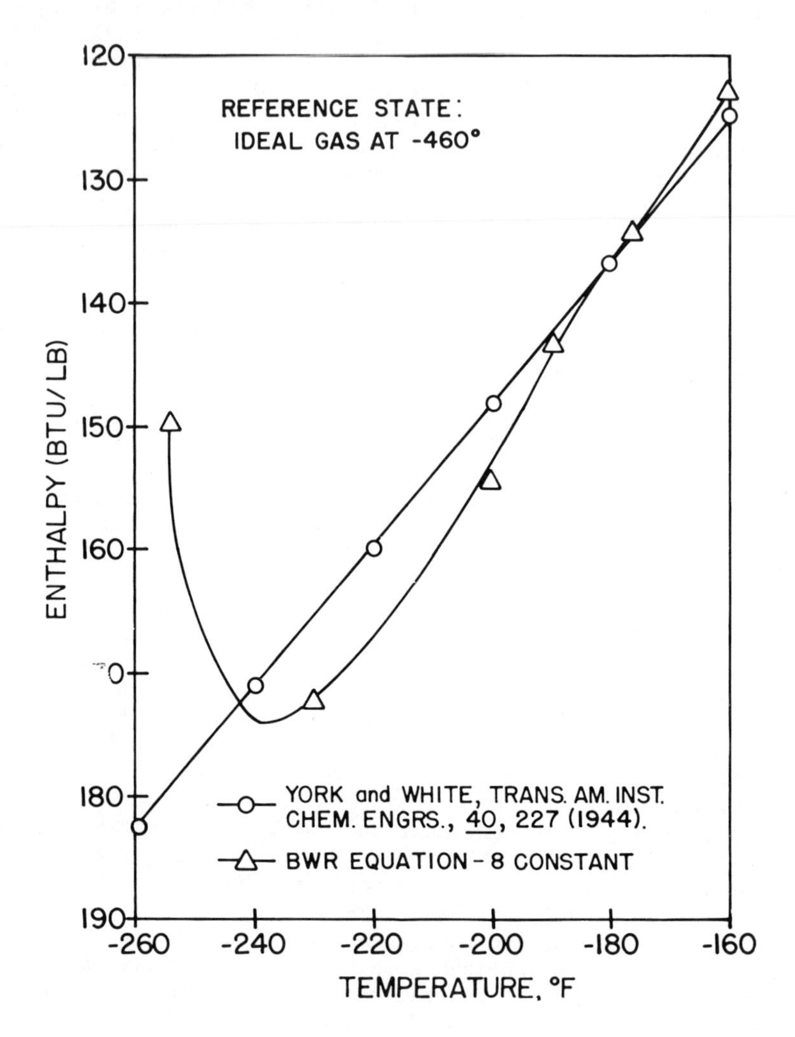

Figure 1. Ethylene — enthalpy of saturated liquid

Aspects ," in Table I, are based on personal experience and information obtained from other industrial sources. Again no attempt has been made to provide a technical treatise on each equation employed by industry.

Benedict-Webb-Rubin Equation. The original form of the BWR equation contained eight constants. It was intended primarily for compressibility, vapor-liquid equlibrium and enthalpy, although the relationships for the other thermodynamic functions were published. It was used for the phase equilibria of hydrocarbon systems in the temperature range of 26 F to 400 F and the pressure range of 65 to 2000 psia. The error in the K constants was reported as 3.4% for this range of conditions. It should be emphasized that hydrogen was not present in the systems studied, and good results cannot be obtained when it is present if the early BWR form is used.

The equation applies equally well to liquids and vapors; the first equation of state known by the authors to apply to both phases, and, almost forty years later, is still one of the most capable for doing so. Because its analytical form permitted application to mixtures, and allowed the required mathematical operations for obtaining all the thermodynamic properties in a thermodynamically consistent manner, the BWR equation attracted widespread industrial attention.

If components were either above their normal boiling points or present in small amounts, the mixture enthalpy predictions were excellent. At lower temperatures, enthalpy results were sometimes found to be so poor that a cold stream in a heat exchanger could be computed to exit at a temperature lower than it went in--"a negative specific heat"! (See Figure 1) This problem was eliminated by permitting the constant C_o to change with temperature (Barner and Adler, (11)). This in fact, added two more constants. Starling (7) added an eleventh constant and included a binary interaction coefficient in the mixing rules which extended the range of applicability of the equation.

With the greater number of constants and the elimination of low temperature negative specific heats, came new problems. Constants could no longer be obtained from pure component properties. Experimental equilibrium data, density data, and enthalpy data for mixtures had to be included in the regression procedure for obtaining the constants. To develop a more useful set of constants for a particular component, as many mixtures as possible containing that component should be included.

Experience has shown that two sets of almost identical constants can give diverse results when compared to experimental measurements of K's and vice versa. As shown in Table II, small differences in only five of the eleven BWR constants can lead to significant changes in the predicted K's. The opposite trend is observed in Table III for a third set of constants. Although four of the constants, B_o, D_o, E_o, and d, differ appreciably in magnitude

TABLE II

COMPARISON OF RESULTS OF TWO SIMILAR SETS OF BWR-11 CONSTANTS FOR ETHYLENE

Property	System		% Deviation From Experimental Data	
			Set 1	Set 2
Density	Ethylene		0.62	0.72
Vapor Pressure	Ethylene		9.24	0.76
K-Value	Methane-Ethylene	K_{C_1}	6.69	5.24
		$K_{C_2^=}$	10.70	5.98
	Ethane-Ethylene	K_{C_2}	0.90	0.88
		$K_{C_2^=}$	1.58	0.84
	Hydrogen-Ethylene	K_{H_2}	9.60	3.97
		$K_{C_2^=}$	10.65	15.26

BWR-11 Constants for C_2^- (Ethylene)

	Set 1	Set 2	Difference
C_o x10^{-10}	.180628	.183101	1.4%
γ x10^{-1}	.227978	.228344	0.2
b x10^{-1}	.257883	.265697	3.0
a x10^{-5}	.156497	.158859	1.5
α	.604935	.604567	-0.0006

:A_o, B_o, D_o, E_o, c, d are identical

TABLE III

COMPARISON OF RESULTS OF TWO DIFFERENT SETS OF BWR-11 CONSTANTS FOR ETHYLENE

Property	System		% Deviation From Experimental Data	
			Set 2	Set 3
Density	Ethylene		0.72	0.63
Vapor Pressure	Ethylene		0.76	1.70
K-Value	Methane-Ethylene	K_{C_1}	5.24	6.00
		$K_{C_2^=}$	5.98	7.76
	Ethane-Ethylene	K_{C_2}	0.88	1.24
		$K_{C_2^=}$	0.84	2.01
	Hydrogen-Ethylene	K_{H_2}	3.97	4.19
		$K_{C_2^=}$	15.26	14.96

BWR-11 Constants for C_2^-

	Set 2	Set 3	Difference
B_o	.593445	.445599	-24.9%
D_o x10^{-11}	.821161	.745192	9.3
d x10^{-6}	.845194	1.067700	26.3
E_o x10^{-11}	.263014	.404265	53.4

:A_o, C_o, a, b, c, γ, α are within 5%

the predictions are of similar accuracy. A final example is given in Table IV for the methane-propane system. The Starling results are based on the generalized form of the BWR-11 equation. The improvement gained by using Exxon's constants, which were obtained from more sophisticated regression techniques and a wider, multi-property range of input data, is evident.

It should also be mentioned that the BWR-11 equation causes some problems in the convergence of computer flash programs that might not be observed for some of the simpler equations of state. To some degree these problems have been detailed in the earlier section on workability.

Once these problems are solved, the results obtained are quite worthwhile. For one such example see Table V which gives the results for a mixture calculation in the cryogenic region, one of the areas where BWR equation is quite accurate. Note also the accuracy of Table VI for high pressure vapor-liquid equilibrium such as occur in petroleum reservoir work.

Besides Pullman Kellogg, organizations such as Exxon, Chicago Bridge and Iron Works, Northern Natural Gas Corp., Union Carbide and University of Oklahoma also use and probably prefer the BWR equation of state.

As a final comment on the BWR equation, we quote Hopke and Lin (12) on the problem of the determination of the BWR constants:

> "In our regression work, we found that attempts to fit pure component and binary mixture data alone will not yield a unique set of optimal BWR's parameters for a given component. On the contrary, many widely different parameter sets can give about the same fit to the data. Moreover, the different parameter sets will yield different results when used to predict thermodynamic properties at conditions outside of the temperature and pressure range of the data base used to determine the parameter sets. Also, extrapolation of light hydrocarbon parameters to obtain estimates of parameters for heavier hydrocarbons is impossible unless a unique set of optimal parameters is obtained."
>
> "In this work we developed a new regression procedure to obtain a unique set of optimal parameters for iso-butane, normal butane, iso-pentane, normal pentane and carbon dioxide. This new procedure, which is described in this paper, involves using multicomponent K-value data to determine one of the eleven BWR pure component parameters, and then regressing on the remaining ten parameters using density, enthalpy, vapor pressure and K-value data for pure components and binary mixtures. Including these multicomponent data has the effect of extending the temperature range of the data base to lower temperatures."

Soave Equation

The Soave equation (13) is one of the many modifications of the original Redlich-Kwong equation of state. As shown below,

TABLE IV

COMPARISON OF BWR-11 RESULTS: EXXON'S 1974 CONSTANTS
STARLING'S GENERALIZED CONSTANTS

Property		Absolute Average Deviation	
		Exxon (1974)	Starling
Pure Component:			
Density	Methane	0.60%	2.04%
	Propane	0.74%	0.88%
Enthalpy	Methane	0.78 Btu/lb	1.84 Btu/lb
	Propane	1.40 Btu/lb	1.31 Btu/lb
Vapor Pressure	Methane	0.63%	1.23%
	Propane	0.94%	2.43%
Mixture:			
Density	Methane-Propane	0.76%	1.87%
Enthalpy	Methane-Propane	1.30 Btu/lb	3.64 Btu/lb
K-Value	K_{C_1}	1.94%	9.58%
	K_{C_3}	4.41%	13.74%

TABLE V

CRYOGENIC PHASE EQUILIBRIA PREDICTED BY BWR-11
$He-N_2-C_1$ System at 2000 psia

Temp., °F	Component	K = y/x Expt'l**	K = y/x Calc'd
-306.7	He	38.7	37.5
	N_2	0.028	0.029
	C_1	*	0.001
-297.7	He	30.5	30.0
	N_2	0.044	0.045
	C_1	*	0.002
-288.7	He	24.3	24.5
	N_2	0.068	0.067
	C_1	0.007	0.004
-279.7	He	19.5	20.2
	N_2	0.10	0.097
	C_1	0.012	0.008

* Indeterminate

** Smoothed Experimental Data by Boone, DeVaney, and Stroud, Bur. of Mines, RI-6178 (1963)

TABLE VI

PHASE EQUILIBRIA AT RESERVOIR CONDITIONS BY BWR-11

	T = 120F		p = 3566 psia			
	Liquid (x)		Vapor (y)		K = y/x	
Component	Expt'l	Cal'd	Expt'l	Cal'd	Expt'l	Cal'd
C_1	0.528	0.511	0.907	0.913	1.72	1.78
C_2	0.045	0.045	0.038	0.037	0.838	0.832
C_3	0.029	0.031	0.016	0.016	0.557	0.514
C_4	0.029	0.035	0.013	0.010	0.438	0.292
C_5	0.026	0.030	0.007	0.005	0.272	0.163
C_6	0.032	0.038	0.006	0.004	0.198	0.093
C_7^+	0.312	0.311	0.014	0.016	0.044	0.052
	T = 200F		p = 4957 psia			
C_1	0.594	0.593	0.883	0.862	1.49	1.45
C_2	0.041	0.041	0.040	0.039	0.966	0.938
C_3	0.024	0.026	0.019	0.018	0.786	0.711
C_4	0.023	0.027	0.014	0.014	0.631	0.517
C_5	0.018	0.023	0.009	0.008	0.489	0.372
C_6	0.022	0.029	0.009	0.008	0.411	0.271
C_{7+}	0.278	0.262	0.026	0.051	0.092	0.195
	T = 200F		p = 6740 psia			
C_1	0.679	0.727	0.840	0.815	1.24	1.12
C_2	0.040	0.040	0.038	0.039	0.959	0.970
C_3	0.021	0.022	0.018	0.020	0.874	0.886
C_4	0.020	0.020	0.015	0.016	0.751	0.795
C_5	0.015	0.015	0.020	0.011	0.648	0.709
C_6	0.018	0.019	0.010	0.012	0.540	0.634
C_7^+	0.208	0.157	0.059	0.088	0.284	0.563

Mixture Critical: T_c = 128F p_c = 3450 psia

Expt'l Data: Roland, C., IEC, 37, 930 (1945)

Soave has introduced a temperature-dependent attractive force term into the equation:

$$P = \frac{RT}{V-b} - \frac{a(T)}{V(V+b)} \quad (6$$

a(T) is calculated from the following expression:

$$a(T) = 0.42747 \left\{\frac{R^2 T_c^{\,2}}{P_c}\right\} \alpha(T) \quad (7$$

where,

$$\alpha(T)^{0.5} = 1.0 + (0.480 + 1.574\omega - 0.176\omega^2)(1 - T_r^{\frac{1}{2}}). \quad (8$$

For mixtures,

$$a = (\Sigma\, x_i\, a_i^{\,0.5})^2 \quad (9$$

$$b = \Sigma\, x_i\, b_i. \quad (10$$

For unlike interacting pairs,

$$a_{ij} = (1 - K_{ij})(a_i a_j)^{\frac{1}{2}} \quad (11$$

where $K_{ij} = 0.0$ for hydrocarbon-hydrocarbon interactions.

The comments in the next few paragraphs are based on the work done at Pennsylvania State University, where, under the auspices o the API Subcommittee on Technical Data, the API Technical Data Boo (14) was revised.

The Penn State group presented six reports (15) on vapor-liqu equilibrium. After thoroughly checking their equilibrium data set for thermodynamic consistency, they tested a number of vapor-liqui equilibrium prediction methods with the binary hydrocarbon-hydrocarbon portion of the data set and concluded that the Soave equati and Peng-Robinson equation (16) were the most accurate and the generalized of the available methods. Additional testing with a large amount of ternary and some higher-order hydrocarbon mixture K data indicated that the use of the Soave equation gave the best results.

With respect to industrial application, the following three questions regarding the Soave equation are pertinent:

1. Is it usable in the cryogenic region?
2. Does it apply to mixtures containing inorganic gases?
3. What problems do petroleum fractions present?

With regard to point one, the Penn State group found that the original, unmodified Soave equation cannot approach the accuracy o the BWR equation in the cryogenic region. In addition, they inten to recommend the BWR for cryogenic systems in the revised vapor-liquid equilibrium chapter of the Data Book. Reference will probably be made to Starling's 11-coefficient BWR work.

The original Soave equation is not applicable to mixtures containing hydrogen. As shown in one API report on vapor-liquid equilibrium (15), this failure occurs because at high reduced temperatures $\alpha(T)$ approaches zero and the attractive term in the Soave equation vanishes. Because in practical situations hydrogen K values are required at large reduced temperatures, the Penn State group introduced a revised α-expression for hydrogen. The improvement was three-fold with regard to hydrogen K value prediction, and the revised expression was shown to be excellent for both binary and multicomponent mixtures containing hydrogen.

The Penn State group developed pair interaction terms, K_{ij}, for mixtures containing inorganic gases. These values were obtained from a large set of binary equilibrium data and some solubility data. The vapor-liquid equilibrium data were however given much more weight than the solubility data in the regression procedures. In all cases, the error in the inorganic K's were reduced by a factor of four with these modifications. These interaction coefficients were found to be relatively independent of both temperature and pressure, and could be generalized as a function of the absolute solubility parameter difference between the interacting pairs. $\Delta\delta = |\delta_{inorg.} - \delta_{HC}|$. Unique expressions are required for CO_2 - HC pairs, H_2S-HC pairs, and N_2-HC pairs. In all other cases, including H_2, $K_{ij} = 0.0$. Multicomponent system calculations based on these interaction parameters were found to agree well with the experimental data.

The Soave equation, with the modifications made by the Penn State group does very well for systems that are of interest to a wide cross-section of the chemical industry. Another point in favor of the Soave equation is that no additional adjustable parameters are required beyond T_c, p_c , ω, and δ for each compound. For companies using the Chao-Seader method, where δ is already available in computer storage, transition to Soave is certainly advantageous.

The Penn State group has not answered point three in any of their reports. However, this point was discussed at the 1976 API Sub-committee meeting in Washington, and it was felt that petroleum fractions probably would not cause any particular problems with regard to the solubility parameter correlation needed in the presence of inorganic gases. Nevertheless, it is believed that each company must draw their own conclusion in this area. Testing a few typical refinery type mixtures would suffice.

Chao-Seader Correlation. Reference was made earlier to the well known and much used Chao-Seader correlation for the prediction of vapor-liquid equilibrium for principally hydrogen-hydrocarbon systems with small amounts of CO_2, H_2S, O_2, N_2, etc. The heart of the correlation consists of several equations to represent liquid fugacity. The other two constituents, the Scatchard-Hildebrand equation for activity coefficients and the Redlich-Kwong equation for the vapor-phase nonideality, were already well established.

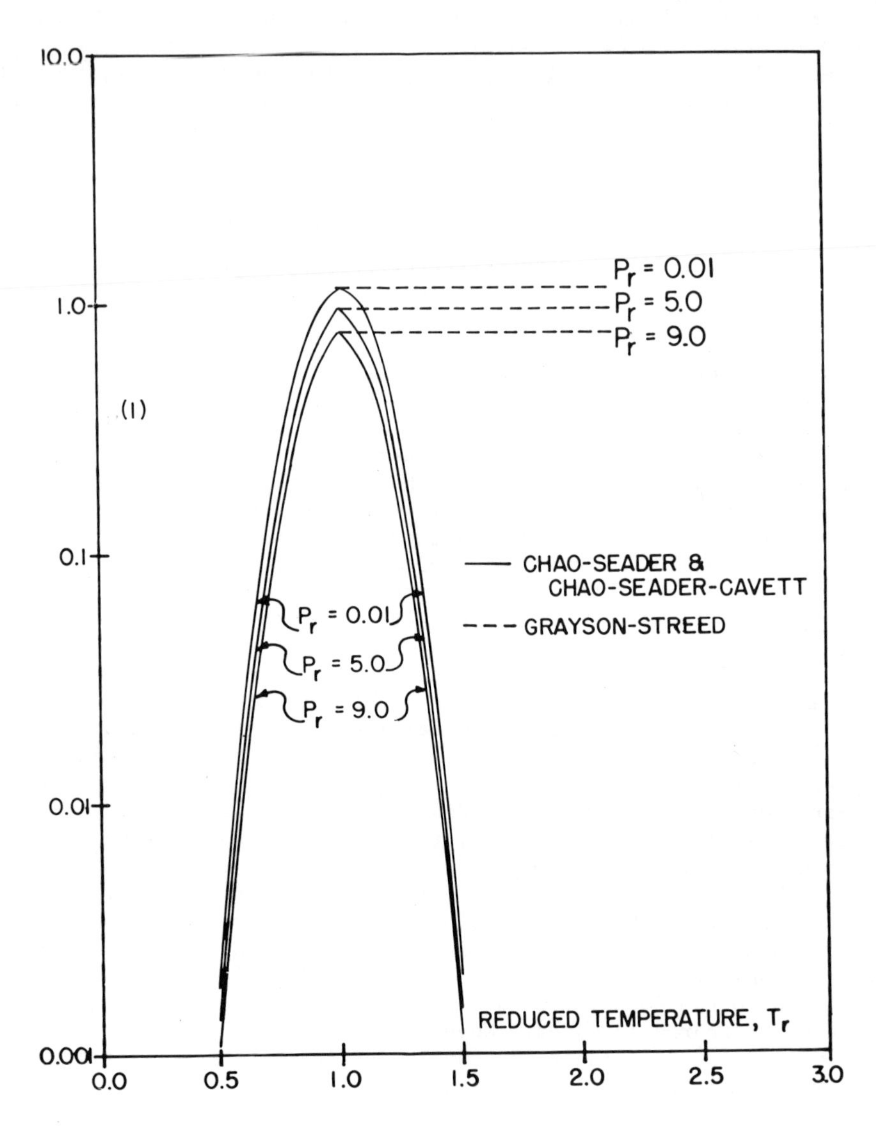

Figure 2. Comparison of Chao-Seader and Grayson-Streed liquid fugacity coefficient correction terms

Because of its simplicity and generality, the original correlation and its modifications by Cavett (17) and Grayson and Streed (18) have found wide applications in petroleum and gas processing industries. However, articles, too many to enumerate, have pointed out the weaknesses in the correlation. Some of our observations are listed below:

1. The liquid fugacity coefficient expression does not reproduce pure component vapor pressures at saturation conditions. This reason alone makes the correlation unsafe for the design of separations of close-boiling components. For such designs as the separation of isopentane from normal pentane in a gasoline isomerization unit, a 5% error in i-C_5 vapor pressure can change the number of trays required in a column by over 33%. (This is assuming an i-C_5/n-C_5 relative volatility of $\alpha = 1.15$ and applying a "rule of thumb" that the approximate number of theoretical plates required for a given separation of fixed reflux ratio is inversely propertional to $(\alpha-1)$).
2. In the original and Cavett versions of the correlation, the real-fluid correction term, $\nu^{(1)}$, in the liquid fugacity equation, $\nu = \nu^{(0)}\nu^{(1)\omega}$, leads to progressively lower hydrocarbon K's with increasing temperature at component reduced temperatures greater than one. This occurs because the numerical value of $\nu^{(1)}$ drops substantially at these conditions as shown in Figure 2. One of the Grayson and Streed modifications on the original Chao-Seader correlation was to fix $\nu^{(1)}$ at its calculated value at $T_R = 1.0$ for all temperatures above the critical temperature of a component, which increased the applicability of the correlation from 500 F to 800 F.
3. In general, with the original and the modified forms of the Chao-Seader correlation, the predicted equilibrium K values of the light components (those with K's greater than 1.0) are often too high and those of the heavy components (species with K's less than 1.0) are too low, making the relative volatility too large and the designs unsafe in certain applications. More specifically, K's for components with reduced temperatures greater than 0.9 are usually overestimated, and K's for components with temperatures less than 0.9 are usually underestimated. As shown in Figure 3 for the Grayson-Streed correlation these over- and underestimations may reach intolerable levels as the system pressure increases. A classic example of the problem of underestimation of heavy component K values can be seen in Figure 4, which is taken from a paper by Robinson and Chao (19). Referring to this figure, they comment:

   "The 0 F temperature corresponds to $T_r = 0.472$ for n-heptane; and $T_r = 0.411$ at -60 F. The comparison appears to be typical of the heavy substances at low T_r. The calculated K values are generally lower than the available

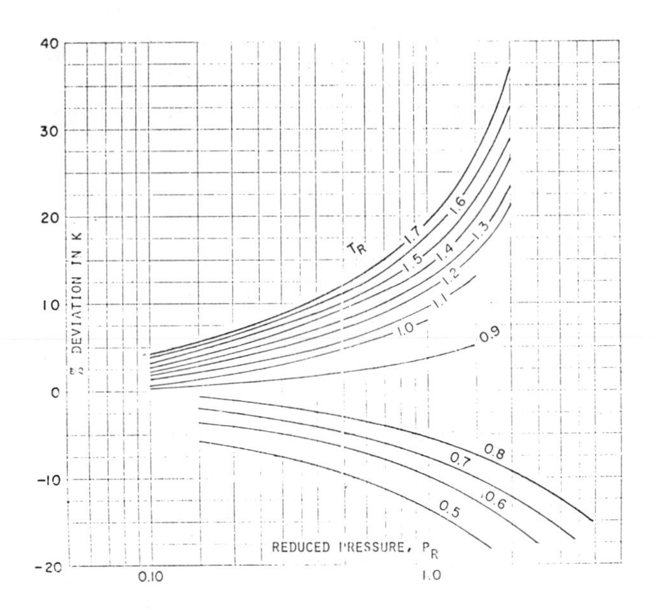

Figure 3. Percent deviation in Grayson-Streed K-value predictions vs. reduced pressure at different reduced temperatures for paraffin–paraffin binary systems

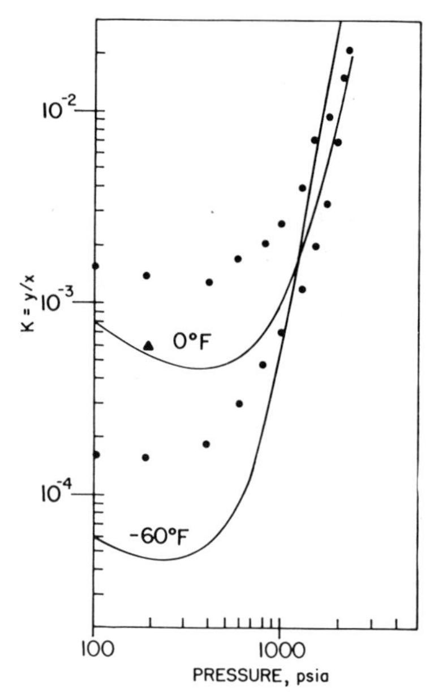

Figure 4. n-Heptane in binary mixtures with methane

experimental data, even at the lowest pressure that is reported in the experimental work, i.e. 100 psia. The cause of the deviation remains unresolved at this time."

Multiequation Approach to Vapor-Liquid Equilibria. The correlations mentioned earlier were developed specifically for hydrocarbon systems and, in general, are not applicable to systems containing polar and associating components. The vapor-liquid equilibrium correlations for systems with such components are best handled with a multi-equation of state procedure using Eq. (5). This method is also used in developing vapor-liquid equilibrium correlations for the design of separation units for close-boiling hydrocarbons.

The separation of the vapor and liquid fugacities and the activity coefficients in the fundamental equilibrium relationship allow great flexibility, and a multitude of choices, in the selection of the thermodynamic relationships or empirical equations for estimation of each of these quantities. For the vapor fugacity coefficient any of the equations of state mentioned earlier or some other, such as the virial equation, may be used. In the latter case, the virial coefficients may be determined experimentally or estimated using three- or four-parameter generalized correlations.

The standard state fugacity of the pure liquid, of course, is estimated from the exact thermodynamic relationship:

$$f_L^o = p^o \phi^o \exp \{ \frac{\bar{V} (\pi - p^o)}{RT} \} \quad (12)$$

where p^o, ϕ^o, and $\bar{V}$ are, respectively, the vapor pressure, fugacity coefficient at the standard state, and the partial molar liquid volume of the component, the latter usually taken to be equal to the saturated liquid molar volume of the pure component. When any of the components is above its critical temperature, the pure liquid fugacity may be estimated (a) by equation(s) with limited extrapolation of vapor pressure, (b) from a generalized empirical equation such as Chao-Seader (or Grayson-Streed), or (c) by back-calculation from experimental vapor-liquid equilibrium data. Alternatively, one can use Henry's constants with unsymmetric convention for activity coefficients, thereby eliminating the need for hypothetical liquid fugacities for pure components, as in the Chueh-Prausnitz correlation.

The empirical equations proposed for the correlation of the activity coefficients are certainly diverse. In addition to the classic Margules and Van Laar equations, there are local composition equations such as Wilson, nonrandom two-liquid (NRTL), Heil, UNIQUAC, and many others in both classes.

Each equation used, whether for ϕ^o, f_L^o, or γ, has its particular advantages and disadvantages, and limitations on its range of applicability; these and other factors influence the selection of the equations, or particular combination of equations used. For example, in many Kellogg design applications the four-suffix Margules form of the Wohl equation for activity coefficients is preferred because of our extensive library of constants developed through past experience, familiarity with its behavior, the flexibility offered by the addition of constants, and its applicability to immiscible systems as well as miscible ones. The multicomponent form of this equation, which may be seen in some publications occupying half of the page, can be reduced to two terms as shown below

$$\log \gamma_i = 4 \sum_{j=1}^{n} \sum_{k=1}^{n} \sum_{\ell=1}^{n} x_j x_k x_\ell \beta_{ijk\ell} - 3 \sum_{i=1}^{n} \sum_{j=1}^{n} \sum_{k=1}^{n} \sum_{\ell=1}^{n} x_i x_j x_k x_\ell \beta_{ijk\ell} \quad (1$$

where n is the number of components and $\beta_{ijk\ell}$ is related to the binary Wohl constants, A's and D's, and the ternary constant, C*, depending on the values of the integers i, j, k, and ℓ, (Adler et al., (20)).

Listed below are some systems for which the multiequation of state approach was used in correlating vapor-liquid equilibrium data needed at Pullman Kellogg either for process development studies or for the design of commercial scale plants. They have been chosen to show the wide spectrum of components. In all cases the Wohl equation was used for the activity coefficients.

High Pressure Applications:

- Isopropanol - Isopropyl Ether - Water - Propylene
- H_2 - N_2 - CH_4 - A or He
- H_2 - N_2 - CO - C_2H_6
- Ethane - Ethylene
- Propane - Propylene
- Propane - Propylene - Propadiene - Methylacetylene
- Chlorinated Hydrocarbons - HCl

Low Pressure Applications:

- Oxygenated Hydrocarbons
- HCN - Water
- Herbicide Plant Data

Table VII shows results for the first system, isopropanol - isopropyl ether - water - propylene, in which the experimental compositions in each of the three phases are compared with the values predicted by the method just described. A modified Redlic Kwong equation of state for vapor fugacity, Chao-Seader equation with adjusted parameters for liquid fugacity, and the Wohl equati for the activity coefficients were used. The predictions were based only on data for binary systems.

Practical Applications

In the previous section the use of equations of state for vapor-liquid equilibrium predictions was emphasized. This portic of the paper expands on this topic with some specific vapor-liqui equilibrium example calculations, and also deals with a number of

TABLE VII

PROPYLENE - IPA - IPE - WATER THREE-PHASE EQUILIBRIA
220 F, 600 psia

Compound	Hydrocarbon Liquid-Phase Composition (Mole Fraction)	
	Exp.	Calc.
Propylene	0.767	0.750
IPE	0.069	0.067
IPA	0.097	0.122
Water	0.068	0.061
	Aqueous Liquid-Phase Composition (Mole Fraction)	
	Exp.	Calc.
Propylene	0.002	0.003
IPE	0.0002	0.0002
IPA	0.033	0.038
Water	0.965	0.959
	Vapor-Phase Composition (Mole Fraction)	
	Exp.	Calc.
Propylene	0.922	0.910
IPE	0.021	0.021
IPA	0.032	0.035
Water	0.025	0.033

other industrial applications of equations of state. Most of thes applications fall into category two of Table I. The discussion or freeze-out problems, however, overlaps category one, The Research Aspects of Industrial Work in Equations of State.

Obviously every industrial application of equations of state can not be covered in this section of the paper. Thus, for example, calculation of Joule-Thompson coefficients, heats of mixing, and the velocity of sound, and obtaining various other parameters required in compressor calculations, which are an integral part of industrial design, will not be discussed.

Critical Properties From Equations of State. In industrial work calculating the critical locus of a mixture is frequently necessary. Because this is done at perhaps 0.05 mole fraction intervals for a binary system, the repetitive nature of the calculation demands as direct and clear cut a method of solution as possible. This precludes solving simultaneous equations, such as setting the second and third derivatives of the free energy with respect to composition equal to zero as indicated below:

$$\left\{\frac{\partial^2 G}{\partial x^2}\right\}_{T,p} = 0 \quad (1$$

$$\left\{\frac{\partial^3 G}{\partial x^3}\right\}_{T,p} = 0. \quad (1$$

This is the classical thermodynamic approach to solving this problem.

Rather, an equation of state and semiempirical correlations are used; e.g. the Chueh-Prausnitz correlation scheme (10). In this approach the critical pressure of the mixture is obtained indirectly from a modified version of the Redlich-Kwong-Chueh equa tion of state after T_{cm} and V_{cm} have been obtained directly from quadratic mixing rules which employ the respective Chueh-Prausnit interaction parameters, τ_{12} and ν_{12}. An example based on this method is given in Figure 5 for the ethane-n-heptane binary mixtu Agreement is rather good except in the immediate vicinity of the maximum critical pressure of the mixture.

A member of the Technical Data Group at Pullman Kellogg has done some promising work on the prediction of the critical properties of mixtures (Spencer (21)). In his recommended procedure the true critical pressure is obtained from the Redlich-Ngo equation of state which is restated below in a form valid at the critical point only:

$$P_{cm} = \frac{3Z_{cm}RT_{cm}}{(V_{cm}-b_m)} - \frac{a_m}{T_{cm}^{\frac{1}{2}} V_{cm}(V_{cm} + b_m)} \quad ($$

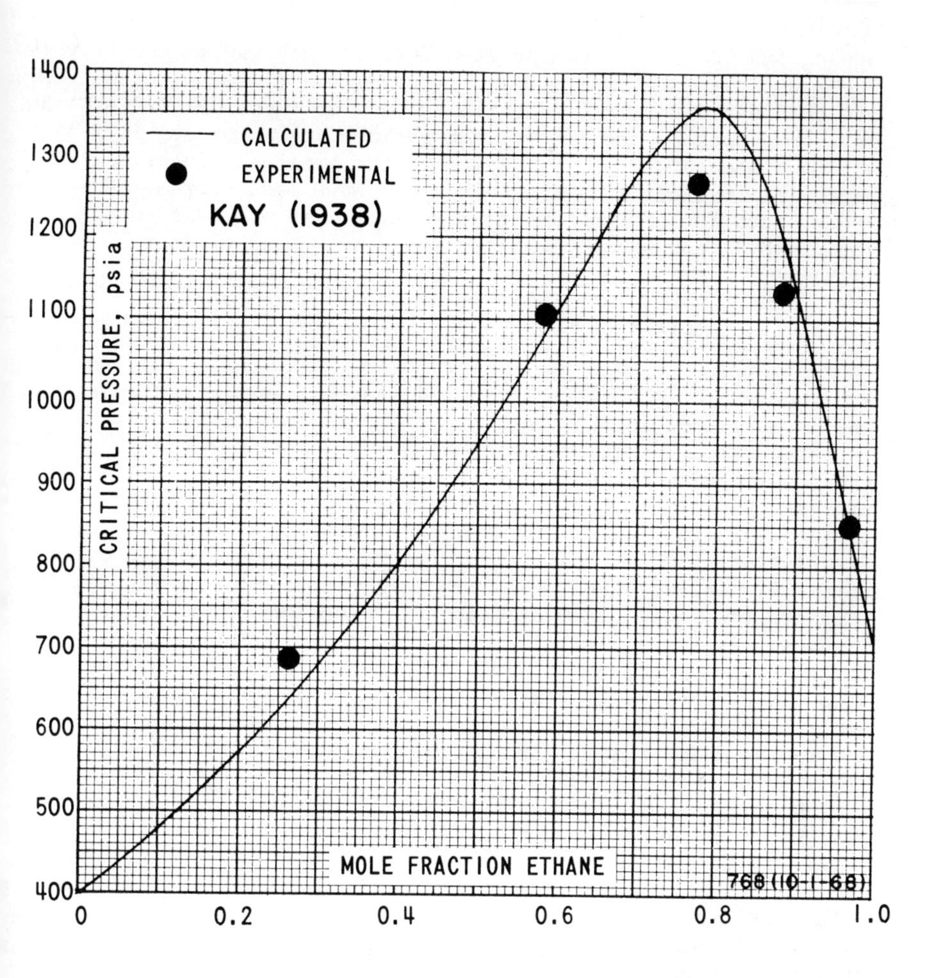

Figure 5. Comparison of calculated and experimental critical pressures for the ethane–n-heptane system

where,

$$Z_{cm} = \Sigma x_i Z_{ci} \tag{17}$$

and a_m and b_m are calculated from Eqs. (9) and (10), respectively. The true critical temperature, T_{cm}, and true critical volume, V_{cm}, are obtained directly from correlations for these properties that were developed in the study. The optimum values of a_{ij} $(i \neq j)$, which were obtained by regression of the available experimental binary mixture critical pressure data, have been correlated as a generalized function of a size parameter, the molecular weight ratio, (MWR) and shape parameter, and the moment of inertia ratio (MIR).

$$a_{12} = [A + B(MWR) + C(MIR)] \frac{a_1 + a_2}{2.0} \tag{18}$$

A unique set of coefficients is required for both binary methane and methane-free interactions.

The overall critical pressure results obtained from the modified Redlich-Ngo Procedure are given in Table VIII. The Chueh-Prausnitz entry was obtained using the original (published) version of that correlation (10). The Kreglewski equation, a semiempirical direct prediction method, is the recommended correlation for the critical pressure in the API Technical Data Book. As is clearly evident, the Redlich-Ngo procedure is superior to the other methods. Although preliminary results have indicated that this technique is excellent for multicomponent mixtures, additional study is needed, particularly for mixtures containing heavier components such as petroleum fractions, before this correlation can be incorporated in everyday industrial design calculations.

Volumetric Properties From an Equation of State. In general, most equations of state give relatively good predictions of volumetric properties at high temperature and low pressure. However, near the saturation envelope and especially in the critical region volumetric predictions based on equations of state are poor, particularly for the saturated liquid. Therefore, with the exception of vapor-liquid equilibrium calculations, where internally consistent liquid densities are needed to calculate the liquid fugacity, empirical liquid density correlations are normally used in industrial design calculations.

Nevertheless, with more sophisticated equations of state, reasonable predictions of the saturated liquid density can be obtained. Consider the results given in Table IX. A number of empirical liquid density correlations and also Pullman Kellogg's version of the BWR-11 equation of state have been evaluated with a set of LNG bubble-point density data. [The most accurate LNG data generally available are those of Klosek & McKinley (22), Shana'a (23), Miller & Rodesevitch (24), and Miller (25)].

TABLE VIII

EVALUATION OF EQUATIONS FOR PREDICTING CRITICAL PRESSURE

Method	Ave. Dev., Psia	
	Methane	Methane-Free
Modified Redlich-Ngo	90.0	11.1
Chueh-Prausnitz	278.0	28.6
Kreglewski (API)	411.3	18.8

TABLE IX

RESULTS OF EVALUATION OF EQUATIONS FOR PREDICTING THE BUBBLE POINT DENSITY OF LNG

Correlation	Percent Avg. Deviation*
BWR-11	2.81
Watson	3.33
Yu-Lu	0.29
Spencer-Danner	0.54
Chiu-Hsi-Lu # 1	0.58
Chiu-Hsi-Lu # 2	0.26

* Basis - 130 points (20 mixtures)

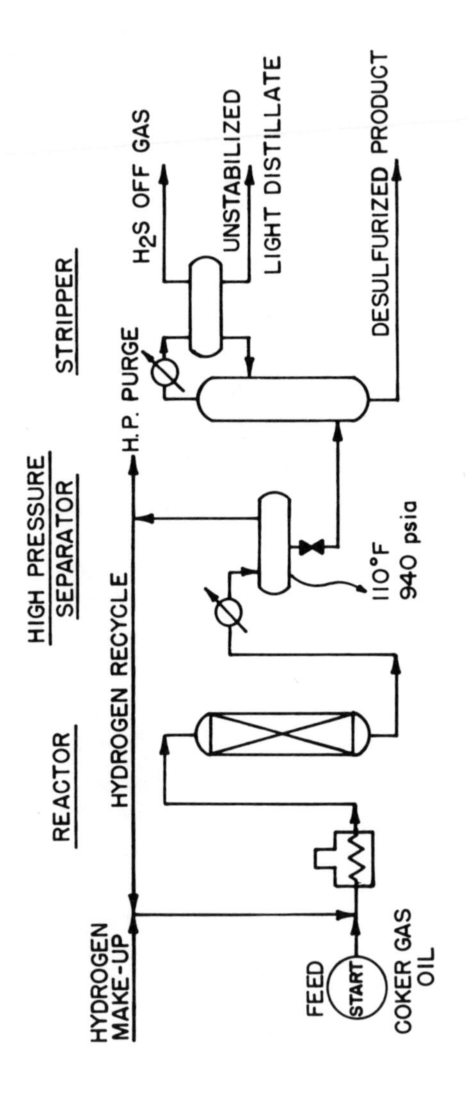

Figure 6. Distillate hydrodesulfurization

As is evident, the BWR-11 is not as accurate as most of the empirical correlations (exception Watson (26)) that have been derived specifically for predicting liquid density. Calculations involving the design of storage areas, and especially custody transfer definitely require the more accurate empirical equations.

Liquid-Vapor Equilibrium (Design Case Studies). As mentioned earlier, equations of state certainly have utility in industrial design studies requiring a knowledge of the phase behavior. This section deals with some challenging liquid-vapor equilibrium design calculations, where experience as well as a workable and flexible equation of state are required. The first example deals with a very unusual design situation that has occurred twice during the last year, multiple bubble-points.

In the first occurrence a hydrodesulfurizer unit was being designed. A simple flowsheet is given in Figure 6. The coker gas oil is preheated in a furnace and catalytically reacted with hydrogen to convert the hydrocarbon mercaptans to H_2S, which is subsequently stripped off.

A process engineer did a computer calculation involving heating of a bubble-point liquid (from the bottom of the high-pressure separator) from 110 F at a design pressure of 940 psia, to feed the H_2S stripping column, and then flashing the stream. The computer output indicated that the stream was all liquid; that is, even though it was at its bubble-point initially and was then heated to a 500 F higher temperature it was still liquid.

The composition of the bubble-point liquid is given in the following table.

HYDRODESULFURIZER HIGH PRESSURE SEPARATOR EFFLUENT

	Mole Percent		Mole Percent
H_2	3.2	300 NBP	0.36
H_2S	5.6	345 NBP	0.37
C_1	4.62	460 NBP	6.34
C_2	0.47	565 NBP	12.50
C_3	0.57	645 NBP	16.28
C_4	0.46	705 NBP	16.37
100 NBP	0.05	760 NBP	15.04
155 NBP	0.28	805 NBP	9.82
245 NBP	0.54	855 NBP	7.14

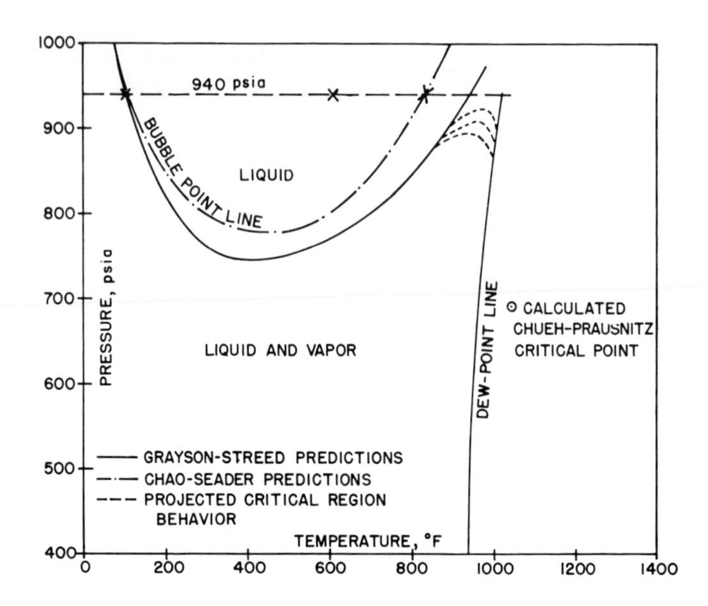

Figure 7. Pressure-temperature diagram for hydrodesulfurization effluent stream

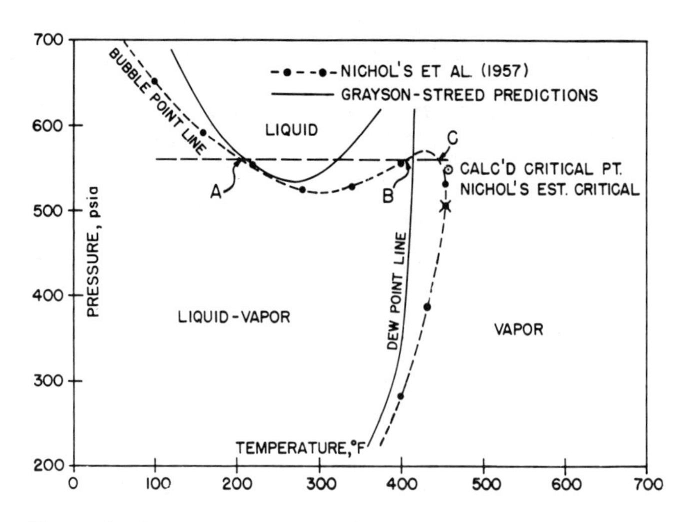

Figure 8. Pressure-temperature diagram for a 4.0 mol % H_2–96.0 mol % n-*hexane system*

Although the mixture consists of about 80% very heavy oils, the light components still influence mixture behavior. In fact, a very small amount of hydrogen can have a large effect on the phase envelope.

The P-T diagram shown in Figure 7 illustrates the situation. The solid bubble-point and dew-point lines are based on the Grayson-Streed modification of the Chao-Seader correlation and the dashed bubble-point line is predicted from the Chao-Seader correlation. Notice the very exaggerated minimum in the predicted bubble-point curve. Indeed, the engineer was performing his calculations in a single phase region, and the computer was correct; in fact, if the stream was heated 200 F higher, or to about 810 F, a second bubble-point would have been reached. This behavior perhaps can be labeled retrograde, because it occurs at a pressure higher than the predicted true critical pressure of the mixture. However, it is assuredly different from what one normally perceives to be isobaric retrograde vaporization (i.e., two bubble-points at a fixed pressure above the critical pressure separated by a two-phase region).

Because this type of behavior is seldom discussed in the literature, and also because it was our first exposure to the problem in some time, an attempt was made to locate some experimental data to confirm the predicted equilibrium behavior, and also to convince the process engineer that the situation could occur.

Data for a 4% H_2 in hexane system are given in Figure 8. The phase envelope represented by the solid lines is predicted from the Grayson-Streed correlation. The dashed line represents the experimental phase envelope. Again notice the minimum in the bubble-point curve, and that as many as three bubble-points, A, B, and C, can occur at constant pressure above the true critical pressure of the mixture. An additional set of experimental data are given in Figure 9 for varying amounts of hydrogen in a petroleum naphtha. At the lowest concentration of hydrogen the minimum in the bubble-point curve extends below the critical point of the mixture. At the medium concentration, the phenomena occurs only at pressures higher than the critical point. At the highest concentration the phenomenon was not observed. After reviewing other data, it was concluded that multiple bubble point behavior may occur when hydrogen (or more general, any slightly soluble gas) is present in the mixture at about 7 mole percent or less.

The BWR-11 equation of state, and the Chao-Seader, and Grayson-Streed computational procedures were shown to be adequate for predicting this type of behavior. However, without some external guidance, a commercial flash routine based on one of these correlations will not generate multiple bubble-points. For example, consider the general case shown in Figure 10. At the design pressure P' P", if the moles in the vapor phase to the total moles in feed is set equal to 0.0, indicating a bubble-point calculation, a computer flash program will converge to only one of the bubble-points (A, B, C). For instance, if the temperature is initialized

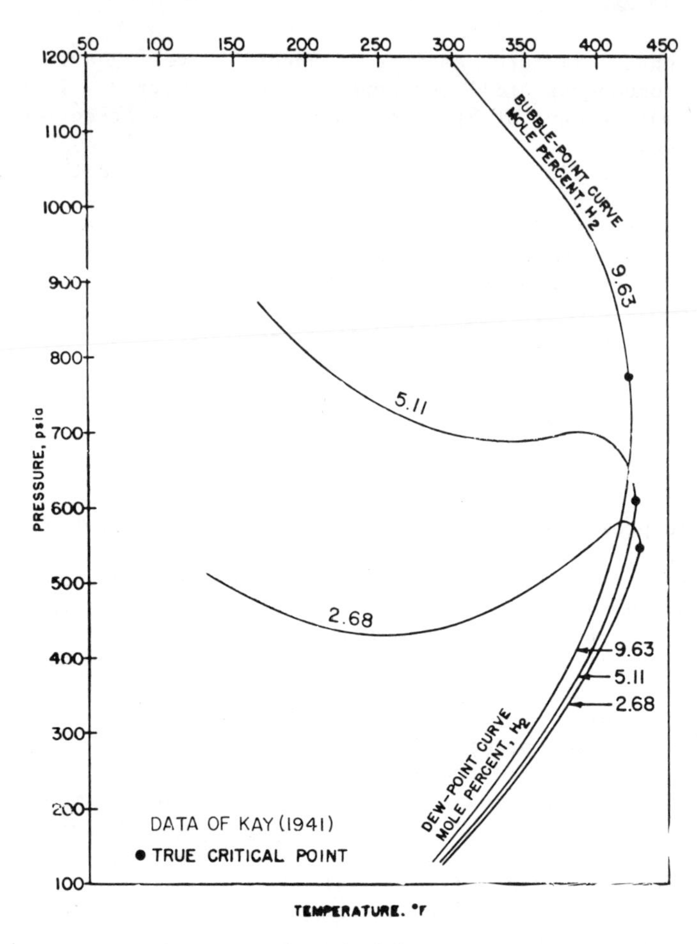

Figure 9. Hydrogen–petroleum naphtha pressure-temperature diagram

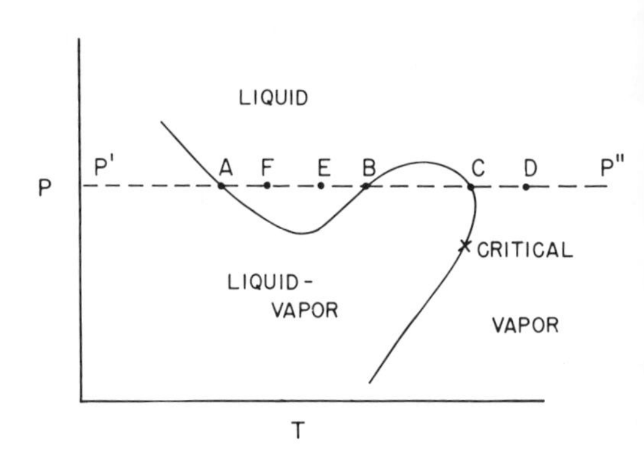

Figure 10. Pressure-temperature diagram for a mixture having a minimum in the bubble point

at F the flash would probably converge to bubble-point A. Likewise, if the initial guess was at point E, the bubble-point at B would be obtained. In general, the process engineer is not aware of this behavior and could be misled by the results. Therefore, certain checks should be built into a flash program to allow for the presence of multiple bubble-points. For instance, if the design pressure exceeds the true critical pressure, or if a small amount of hydrogen or other slightly soluble gas is present in the mixture, the flash vaporization calculation could be programmed so that after convergence is obtained with the initial temperature guess, additional bubble-point calculations are automatically attempted using a temperature guess twice the initial temperature or even using the critical temperature itself. Using such a procedure, all three, and for some systems, four bubble-points could be predicted. The Chao-Seader and Grayson-Streed correlations can never converge to point C (because of the funnel effect which will be described later) and is an inherent weakness of both methods. For some mixtures, if the BWR-11 or BWR-8 equation is used in the flash calculation, the bubble-point at C can be obtained. However, in some case studies, if the initial temperature guess was D, the bubble-point at B would be obtained rather than C. In these cases, going one step further, even if the initial temperature guess was temperature C the program would converge on bubble-point B. This could be a problem in the convergence portion of the flash routine itself or perhaps a general stability problem with the equation of state.

As mentioned earlier this phenomenon occurred twice within the past year, and the second time was certainly easier to understand. The second case is shown in Figure 11. A process engineer was trying to confirm a 10 degree difference between the predicted bubble-point temperature A and a client's value at point B for a given design pressure (solid line). The system is the bottoms of a nitrogen wash tower which contains nitrogen, methane, and a small amount of hydrogen. A series of computer flash calculations between A and B and then at temperatures higher than B were made. For these calculations the message "subcooled liquid" was printed out. By increasing the temperature in large increments a two phase region was located at point E. Because of the aforementioned work in this area, multiple bubble-point phenomena were suspected, and the phase envelope for the mixture was generated using the BWR-11 equation of state. Because (i) the BWR-11 equation is, in general, excellent for all three components in the mixture, (ii) our predicted critical point at D fell in line with the envelope, and (iii) of the experience gained from Figures 7 to 10, it was felt that a true picture of the nitrogen mixture behavior was obtained.

The second example deals with a dew-point calculation for a typical natural gas mixture based on the BWR-11 equation of state. As shown in Figure 12, at a design pressure of 620 psia the solution algorithm initially diverged but finally approached a solution in the region of the true critical point of the mixture, -100 F. However, complete convergence was not obtained until 22 F, 122°F beyond the true critical temperature of the mixture! It's certainly

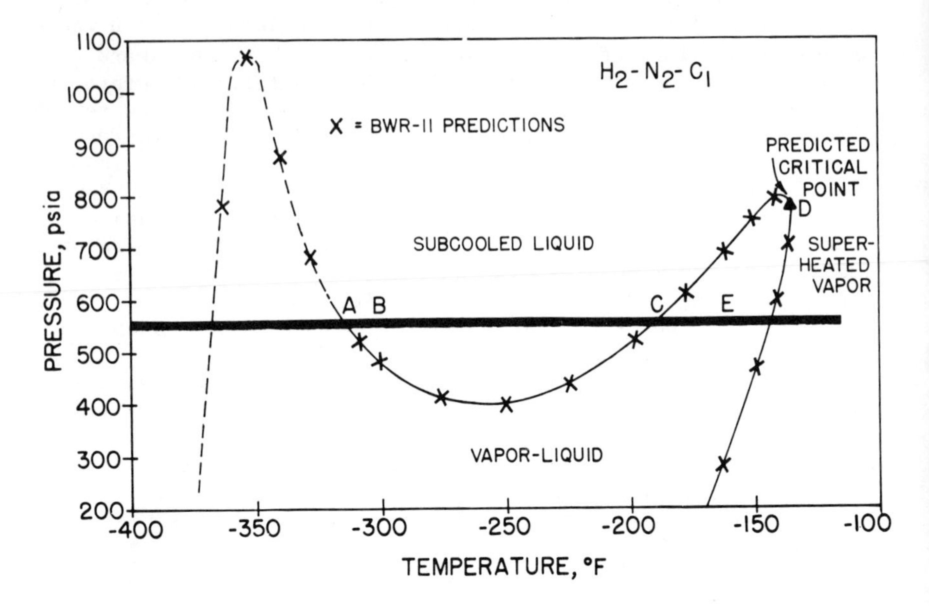

Figure 11. Nitrogen wash tower bottoms phase diagram

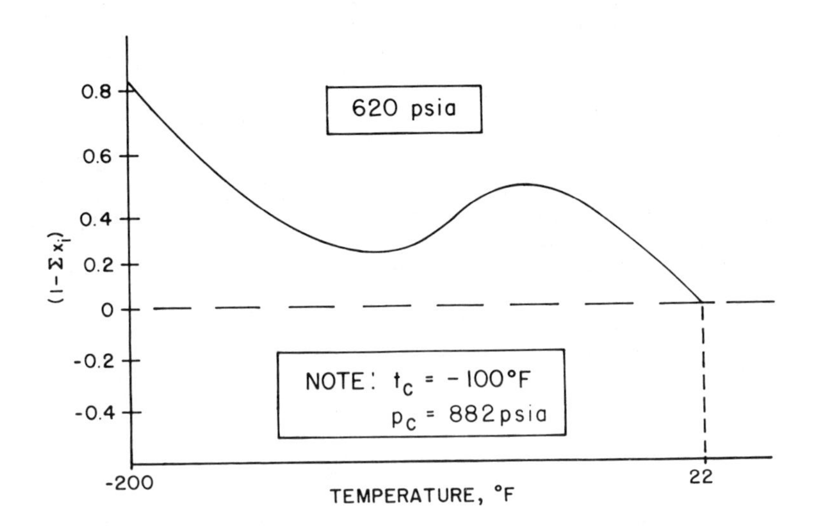

Figure 12. Dew-point calculation (trial and error)

possible to have a solution at 22 F, because the retrograde bulge in a natural gas can extend appreciably beyond the true critical point of the mixture. For a solution close to the true critical point as might be indicated by the results, the design pressure must exceed the true critical pressure of the mixture, which in this case was 882 psia.

Figure 13 shows the experiment phase behavior for this mixture. +22 F is an upper retrograde dew-point. If the design pressure was lowered a second dew-point could also be located at a temperature of +22 F. This is, of course, the lower dew-point pressure.

The dew-point of a natural gas is very sensitive to the heavy ends characterization. Consider a natural gas stream with the following composition:

Component	Moles
N_2	0.69
C_1	950.0
C_2	92.0
C_3	12.0
C_4^+	<10.0

The different characterizations of the heavy ends, which make up less than 1% of the entire mixture, were considered. In Case 1 normal butane was selected to represent the heavy ends, in Case 2 normal pentane, and in Case 3 normal hexane. The phase envelopes for the three cases were predicted with the BWR-11 equation of state. The results are given in Figure 14. The bubble-point curves almost coincide, and the true critical points are not very sensitive to composition. However, as shown, the dew-point is greatly affected by the characterization of the heavy ends composition, and there is about a 70 F difference in the maximum dew-points of the three mixtures. Therefore, when working with a natural gas type system, if the breakdown of the C+ fraction into additional compounds is available, it should be used, especially in flash calculations, to get an accurate representation of the phase behavior.

The last example deals with some troublesome flash calculations which were based on the Chao-Seader correlation.

In a particular design study a dew-point calculation and bubble-point calculation failed to converge for a C_2-C_5 system at a pressure of 1015 psia. In other words, a pressure of 1015 psia apparently exceeded the border curve of the mixture. Some ethane systems do have two-phase regions above 1000 psia. However, without considering the mixture composition or without calculating the critical properties, it is difficult to provide a reasonable explanation of these flash calculation failures.

The composition and predicted critical properties of the mixture are given below:

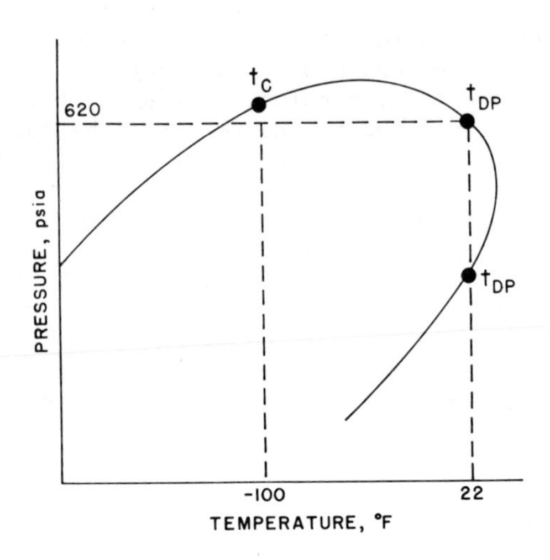

Figure 13. Phase diagram for a typical natural gas

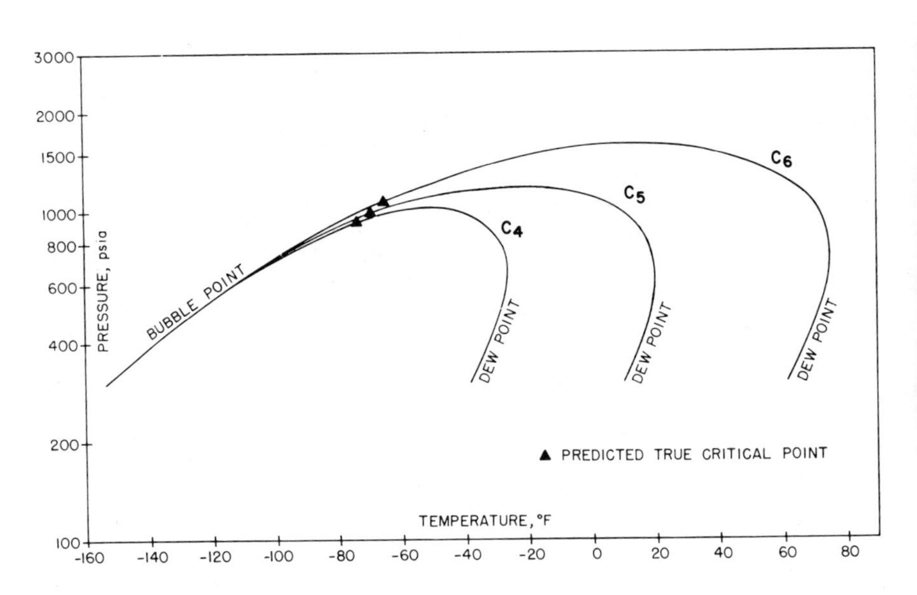

Figure 14. Pressure–temperature diagram for natural gases with a varying heavy concentration

C_2-C_5 System

	Mole Fraction
Ethane	.0004
Propylene	.1269
Propane	.0758
Isobutane	.3786
Butane	.1107
Isobutylene	.3013
Isopentane	.0063

T_{cm} = 276.05 F

P_{cm} = 584.75 psia

The mixture is essentially a C_3-C_4 mixture; there is no extreme size difference in the prominent components. Therefore, the mixture should be close-boiling and also have a critical point located between the cricondenbar and cricondentherm of the phase envelope. The predicted critical properties are based on a modified Chueh-Prausnitz technique. Based on the composition of the mixture, it is unlikely that the two-phase region will extend appreciably beyond the true critical point. Thus, considering the predicted critical pressure, a reasonable estimate of the cricondenbar would be about 675 psia. Therefore, qualitatively, it was concluded that the design pressure of 1015 psia far exceeded the two-phase region and that the process engineer was attempting these flashes in a homogenous liquid region.

To substantiate these conclusions, a bubble-point and dew-point curve for the mixture were predicted using the Chao-Seader correlation. The results are given in Figure 15. The dashed line to the left is the predicted bubble-point curve. The dashed line to the right is the predicted dew-point curve. The small triangle represents the predicted critical point of the mixture. It appears that the envelope should close in the area of the true critical point. The solid line in the upper right hand portion of the graph represents the attempted design pressure, certainly far above the two-phase region.

In this study, attempts were made at calculating dew-points and bubble-points at 50 psia intervals, from 550 psia to the design pressure of 1000 psia. At most pressures, convergence was not achieved. However, at a pressure of 900 psia solutions were obtained for both the bubble-point and dew-point. These unrealistic roots are an unfortunate result of Chao-Seader type predictions if the limits of the correlation, $T_R < 0.97$, are not observed. This is a problem with the correlation, rather than of the flash algorithm. (A more extreme situation will be discussed shortly.) The end result may have been that the engineer, realizing that the design pressure of 1015 psia was too high, lowered his conditions to about 900 psia and unknowingly accepted the results. To avoid these problems, one should remember that for this type of mixture

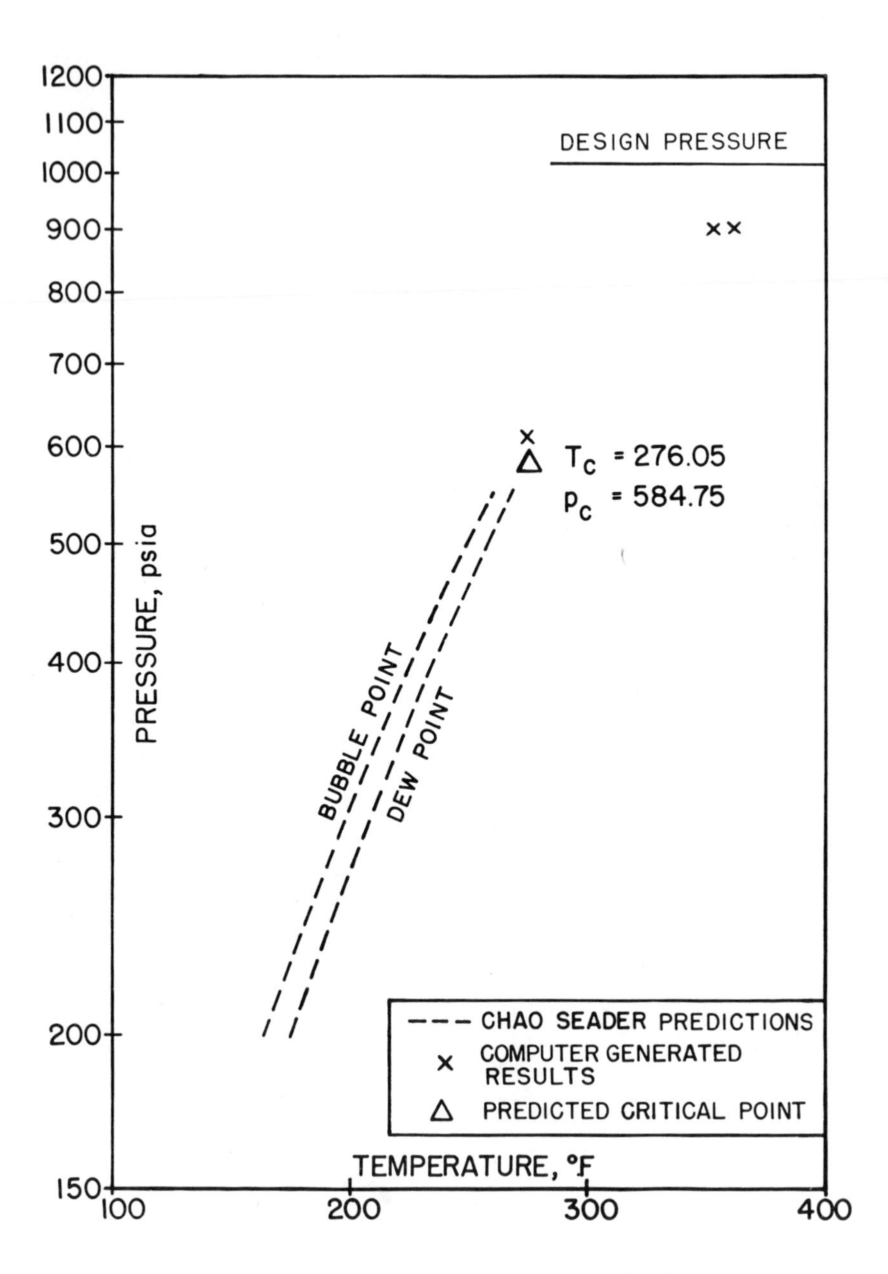

Figure 15. Pressure–temperature diagram for a C_2–C_5 system

there is no way a bubble-point can exist at a temperature greater than the true mixture critical temperature.

The observations of the previous problem were based on predicted properties, for both the phase envelope and the critical point. To justify these conclusions, a similar mixture was studied, but one for which the experimental critical region phase envelope and critical point data were available. The composition and critical properties of the mixture are given below:

C_5-C_9 System

	Mole Fraction
Pentane	.246
Hexane	.217
Heptane	.192
Octane	.178
Nonane	.162

T_{cm} = 514.6 F

P_{cm} = 448.6 psia

At about the equimolar composition of the mixture, it is unlikely that the bubble-point curve will exist at temperatures greater than the true critical temperature of the mixture.

The Chao-Seader generated phase envelope for this mixture is given in Figure 16. The dashed line represents the critical region phase envelope, that is, the experimental phase envelope. The remainder of the phase envelope coincided relatively well with the predicted values, and is not included in the diagram.

Although the critical region clearly terminates at a pressure of about 450 psia, bubble-point and dew-point predictions were obtained up to 600 psia before the flash program finally failed to converge. This behavior, which has been labeled the "funneling effect", is specific to Chao-Seader type correlations and indicates these correlations need improvement. Most of these erroneous results can be rejected by comparison with results for similar mixtures and with the predicted critical properties. The funneling effect does not occur, at least for the cases that have been considered, when a single equation of state approach is used in a flash program.

Freeze-out of a Solid from a Liquid Solution. Thermodynamic textbooks and journal articles, for the most part, give greatest coverage to the phase equilibria of fluid phases but, with few exceptions, little to solids. However, prediction of solid-fluid equilibria is an important industrial problem. This is especially true for process designs related to natural gas and cryogenic plants, where the solubility limits at the incipient solid phase separation of, for example, water, CO_2 and heavy hydrocarbons must be accurately known to prevent costly plant shutdowns with

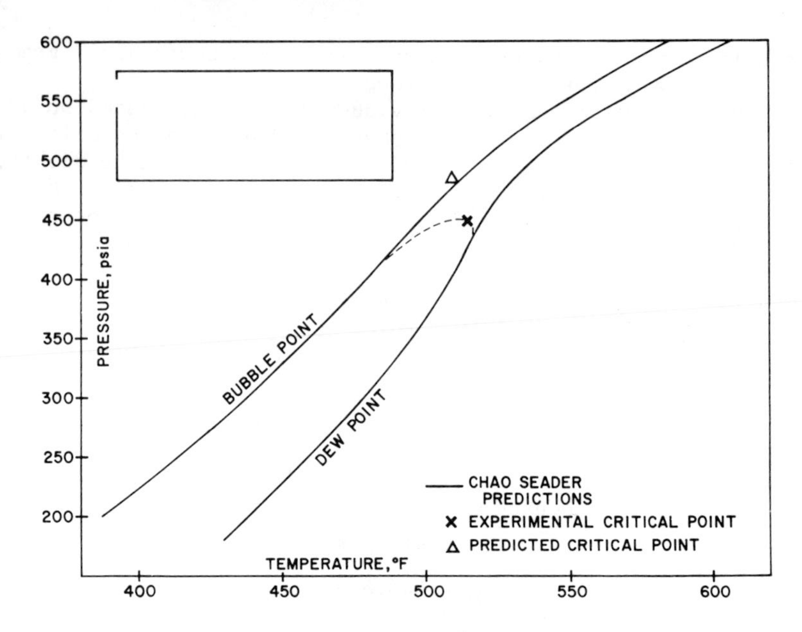

Figure 16. Pressure–temperature diagram for a C_5–C_9 system

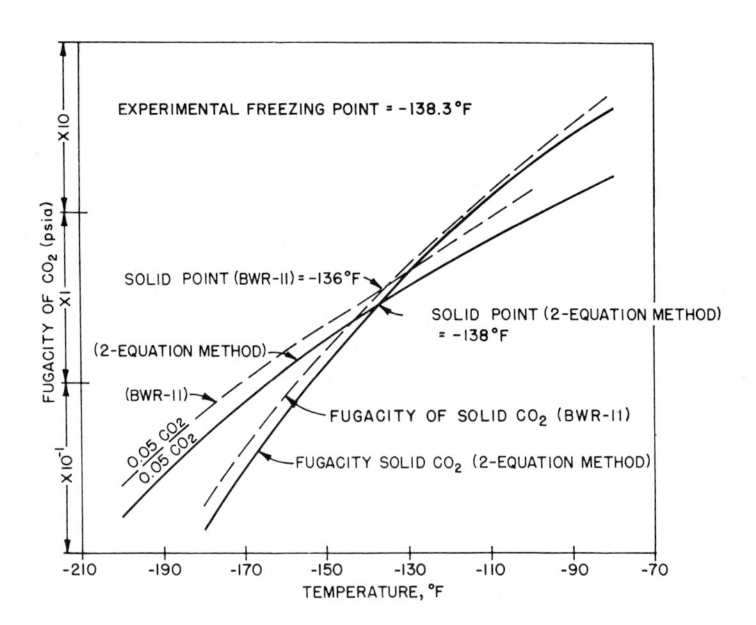

Figure 17. Estimation of freeze-out temperatures by intersection of fugacity curves

plugged pipelines or heat exchanger units due to excessive accumulation of solids formed by freeze-outs.

Equations of state can be used for determining the equilibria in solid-liquid systems. The starting point is the observation that at the point where the first minute solid particle precipitates out, the composition of the liquid is not altered from its initial composition. Therefore, the fugacity of the liquid solution can be calculated at successive temperatures until a temperature is reached where the computed fugacity of the precipitating component in the liquid equals that of the pure solid. The fugacity of the pure solid must first be computed by application of the same equation of state to the solid in similar solutions, for which experimental solid-liquid or solid-liquid-vapor data are available. When appropriate, the two-equation of state approach can, of course, be applied.

An application is illustrated in Figure 17 for estimating the temperature at which solid carbon dioxide would start to form in a condensed natural gas system containing 5% dissolved carbon dioxide at a pressure of 440 psia. In one case, as indicated by the dashed lines, the 11-constant BWR equation of state is used to solve the problem. In the other, a two-equation approach in which the liquid activity coefficients are correlated by the Wohl equation is used. For BWR application, curve CC represents the fugacity of pure solid carbon dioxide as back-calculated from experimental data compiled by Kurata (29) on the solubility of solid carbon dioxide in liquid hydrocarbons. Curve FF relates the computed fugacity of a 0.05 mole fraction CO_2 liquid solution over a temperature range where freezing might be expected. At the point of intersection, the fugacity of the component in liquid solution equals the fugacity of the pure solid; therefore, precipitation occurs. The temperature of the intersection is the solution to the problem.

In the two-equation of state approach, the Redlich-Kwong-Chueh equation of state was used for vapor-phase fugacities and the Wohl equation for liquid activity coefficients in the reduction of the experimental vapor-liquid and vapor-liquid-solid equilibrium data. The activity coefficients so obtained were subsequently correlated as a function of temperature. Curves C'C' and F'F' have the same significance as curves CC and FF. While neither of the "pure solid" fugacity curves in Figure 17 may represent the real values, each is consistent with the equations used in the data reduction and, therefore, reliable for prediction of the needed information as long as this consistency is maintained. As shown in Figure 17, the predicted freezing temperature for the problem stated earlier is -136 F by BWR-11 application, and 0138 F by the two-equation of state approach.

The graphical technique of intersecting curves is, of course, only one of the variety of ways this problem can be solved. Once the fundamental thermodynamic parameters are established, computer programs can be used to obtain the answers.

As mentioned earlier, the minimization of the free energy is the universal method of solving chemical and phase equilibrium problems. A computer program developed to handle multicomponent-multiphase systems can be designed not only to determine the freeze-out temperatures, but, for a given mixture, one can follow complete changes in the equilibrium phases as temperature and pressure conditions change.

An interesting method from a fundamental point of view is the graphical approach to free energy minimization. Since at equilibrium, i.e. when total free energy of the system is minimum, the chemical potential (partial molal free energy) of a component is the same in all phases, a common tangent line to the free energy curves drawn for each phase would establish the equilibrium concentrations. A common example for this is given in most textbooks for liquid-liquid equilibria (King, (30); Prausnitz, (31)).

An application of this principle for solid-liquid equilibria is shown in Figure 18 for the carbon dioxide-methane system at -100 F. The free energies plotted in this figure are relative values which are related to the molal free energy of each phase as follows:

Pure solid CO_2 phase; $$G_S = G^o + RT \ln \frac{f_S}{f_L} \quad (19)$$

Liquid phase; $$G_L = \Sigma\, x_i\, G_i^o + \Delta G^M \quad (20)$$

where the standard reference state is taken as the pure subcooled liquid at the system temperature. The free energy of mixing, ΔG^M, is related to the activity coefficients by

$$\Delta G^M = RT\, \Sigma\, x_i \ln x_i\, \gamma_i \quad (21)$$

From Eqs. (19), (20), and (21), the relative molal free energies may be defined as

$$\left[\frac{\Delta G}{RT}\right]_{solid\ CO_2} = \frac{[G_S - G^o]}{RT} = \ln \frac{f_S}{f_L} \quad (22)$$

$$\left[\frac{\Delta G^M}{RT}\right]_{liquid} = \frac{[G_L - \Sigma\, x_i\, G^o]}{RT} = \Sigma\, x_i \ln x_i\, \gamma_i \quad (23)$$

The relative partial molal free energy of carbon dioxide in the liquid solution for a given concentration may be obtained by drawing a tangent to the relative free energy of mixing curve at that concentration (Lewis and Randall, (32)). The intercept of this tangent on the ordinate, $x_{CO_2} = 1.0$, is the relative partial molal free energy of carbon dioxide in the liquid and should have the same numerical value for carbon dioxide in all equilibrium phases. Inversely, a tangent line drawn from the point correspon[ding] to the relative molal free energy of CO_2 in the solid phase ($x = 1.0$ for pure CO_2), would touch the liquid relative free ener[gy] curve at a concentration which is in equilibrium with the pure

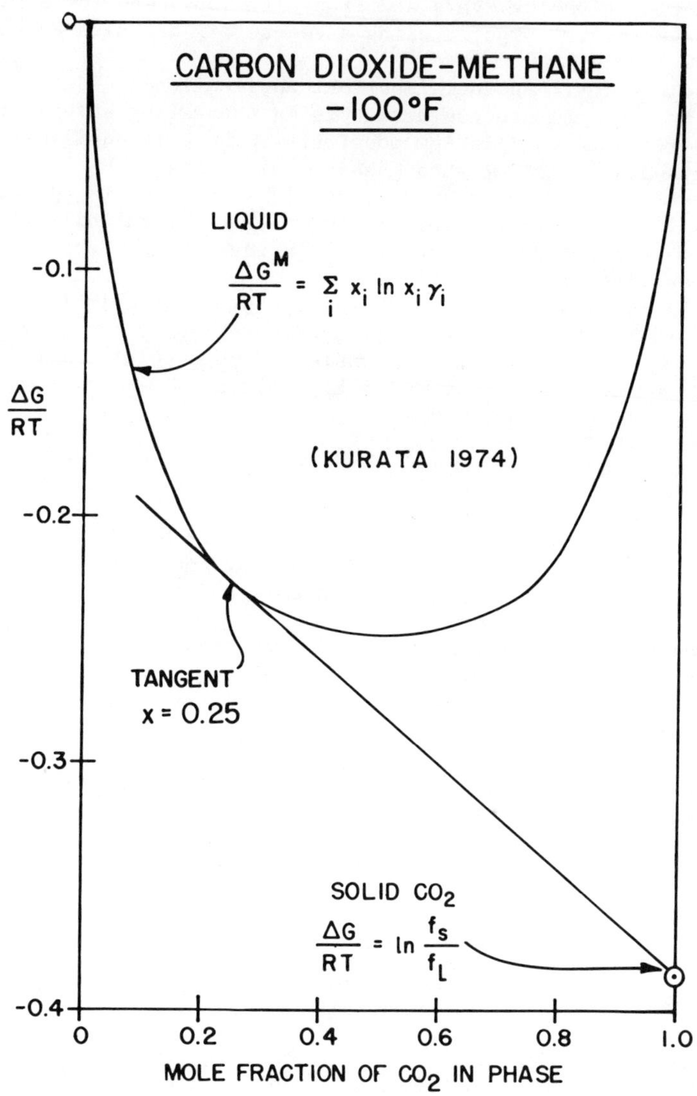

Figure 18. Estimation of the solubility of solid carbon dioxide in methane by the tangent method

solid carbon dioxide. The (f_S/f_L) value for this example is taken from a plot given by Myers and Prausnitz (33), and the activity coefficients are computed using a two-suffix Margules equation.

Solubility. Another important application of equations of state in chemical process design is in converting solubility data to vapor-liquid equilibrium constants. The only equilibrium data that are available for such gaseous components as H_2, N_2, O_2, CH_4, CO, CO_2, H_2S, the acetylenes, etc., in such solvents as water, methanol, DMF, for example, are in the form of solubility data which may be found in Seidell (35) in the form of Bunsen and Ostwald coefficients.

A simple way to deal with the low pressure solubility data, which are usually at atmospheric pressure, is to assume the vapor phase to be an ideal gas and consist of pure solute, i.e., y_i = 1.0, ϕ_i = 1.0, thereby reducing Eq. (5) to:

$$K = \frac{1.0}{x} = \frac{\gamma f_L^o}{\pi} \tag{24}$$

from which activity coefficients can be computed and correlated. When warranted, some improvement can be made by allowing for the vapor pressure of the solvent so that for the solute gas $y = \bar{p}/\pi$. Further details and refinements on this point are given by Friend and Adler (34).

Once the activity coefficients are established, the K's at higher pressures are calculated using Eq. (5) and an equation of state to calculate ϕ as a function of temperature, pressure and vapor composition, the latter being established by a trial and error procedure.

Whether or not the fugacity correction is also included in the reduction of the solubility data for more precise activity coefficients, and whether or not these activity coefficients are developed and correlated with the standard state of Henry's constants or hypothetical liquid fugacities, it is obvious that equations of state again make possible an important chemical engineering computation.

Summary

This paper has dealt with the characteristics of equations of state required by industry, has discussed a number of equations that are used in industrial vapor-liquid equilibrium calculations, and has covered a number of everyday and sometimes unusual practical applications of equations of state. In all three areas an attempt was made to analyze the shortcomings, deficiencies, and handicaps of specific equations of state as well as equations of state in general, from an industrial viewpoint. It is hoped that some of the material discussed in this paper will prove advantageous in future equation of state development work.

It is realized that the demands of industry may be steep and, in some cases, unreasonable; however, every published equation of state has some limitations. The available equations of state were developed and originally applied in relation to certain systems, and their greatest value is in the interpolation and extrapolation of those systems. In many cases without modification, application of the equation of state beyond the original parameters is invalid. Perhaps, the development of a wholly flexible and workable equation of state, as defined earlier, is an unattainable goal. However, to be more optimistic, the authors believe that this challenge can be met.

What then has to be done to make the equation of state a flexible, inclusive, successful tool? Firstly, a general purpose, three-constant, three-parameter equation is needed. Its use should be unrestricted as to system or state. The correlating parameters utilized in these developments should be easily accessible, not difficult to estimate in their own right. These parameters should be developed from multiproperty (volumetric, enthalpy, vapor-liquid equilibrium) pure component and defined mixture data. Extension to polar and associated substances is needed and eventually must include testing the correlation against all available reliable data. If interaction parameters are required in the treatment of mixtures, an attempt should be made at obtaining temperature and pressure-independent values which can be generalized as a function of a convenient size-shape or polarity parameter. Finally, the new developments should aim at eliminating the shortcomings of the existing equations of state enumerated earlier.

As a concluding thought, if a flexible and truly workable equation of state is developed, there remains a final factor which is specific to industry. For example, some twenty-five years ago--and that is before the availability of computers--the senior author of this paper was developing a set of charts for the enthalpy of petroleum fractions. In addition to the usual parameters of temperature and pressure, these charts also were functions of gravity, of the slope of their distillation curve and of a characterizing factor peculiar to petroleum fractions. With all these variables, months of hand calculations were entailed. During this work it was discouraging to hear a furnace design engineer comment that the charts would never be used, because the heat of cracking correlations used with the petroleum fraction enthalpy curves were derived from the old enthalpy charts that would be replaced. Of course, the heat of cracking correlation could also be modified. So it is with any new correlation produced today. If it is introduced, every chart, graph, correlation and analytical expression in a computer program that was derived from the proceeding correlation must be revised. This can best be seen by looking at the ingredients of the Chao-Seader correlation. If the Redlich-Kwong equation of state is replaced with its Chueh modification, the liquid-fugacity equation, which originally took up the slack in developing the overall correlation from experimental data, also must be revised. Considering that it is based on more than 3000 experimental points and consists of about 20 constants, this is a big task. And will the overall result be better? In summary, when a new piece of work is introduced in the literature no matter how good it is, it can not immediately be put into use.

Acknowledgment

Acknowledgment is made to other members of our staff who assisted in the preparation of this paper; Kathleen Cowan, Denise Hanley, Janice Lambrix, and Rosalyn Quick.

Nomenclature

a = Constant
A = Constant in Eq.(18)
A_o = Constant in BWR-11 Eq. (Table II)
b = Constant
B = Constant in Eq.(18)
B_o = Constant in BWR-11 Eq. (Table II)
c = Constant in BWR-11 Eq. (Table II)
C = Constant in Eq.(18)
C_o = Constant in BWR-11 Eq. (Table II)
d = Constant in BWR-11 Eq. (Table II)
D_o = Constant in BWR-11 Eq. (Table II)
E_o = Constnat in BWR-11 Eq. (Table II)
f = Fugacity; f, component fugacity; f^o, standard state fugacity or fugacity of pure component
G = Gibbs molar free energy; G^o, Gibbs molar free energy of pure component at standard state
K = Equilibrium constant, y/x
K_{ij} = Binary interaction coefficient in Eq.(11)
n = Number of components, used in Eq.(13)
P,p = System pressure
P_c,P_c = Critical pressure
P_{cm} = True critical pressure of mixture
$\overline{p}$ = Partial pressure of component in mixture
p^o = Vapor pressure of pure component
P_r = Reduced pressure, P/P_c
R = Gas constant
T = Temperature
T_c = Critical temperature
T_{cm} = True critical temperature of mixture
T_r = Reduced temperature T/T_c
V = Molar volume
$\overline{V}$ = Partial molar liquid volume of component
V_{cm} = True critical molar volume of mixture
x = Mole fraction of component in liquid phase
y = Mole fraction of component in vapor phase
Z_{ci} = Critical compressibility factor of component i
Z_{cm} = Critical compressibility factor of mixture

Greek

α = Constant used in BWR-11 Eq. (Table II) and parameter in Eqs. (6) and (7)
β = Coefficient in Wohl Eq.(13)
γ = Constant used in BWR-11 Eq. (Table II)
γ = Activity coefficient of a component in liquid
δ = Solubility parameter
$\nu^{(0)}$ = Simple fluid liquid fugacity coefficient
$\nu^{(1)}$ = Liquid fugacity coefficient correction term
π = System pressure
τ = Chueh-Prausnitz interaction parameter
ϕ = Vapor phase fugacity coefficient
ϕ^{o} = Fugacity coefficient at standard state
ω = Acentric factor

Subscripts

c = Critical point
i,j,k,l= Components in a mixture
L = Liquid phase
m = Mixture
S = Solid
V = Vapor

References

1. Tsonopoulos, C. and Prausnitz, J. M., Cryogenics, (1969), 9, 315.
2. Martin, J. J., Appl. Thermodyn. Symp., (1967), 59, 34.
3. Barner, H. E., Pigford, R. L., and Schreiner, W. C., Proc. Am. Petrol. Inst., Vol. 46 (III), (1966), 244.
4. Barner, H. E. and Adler, S. B., Ind. Eng. Chem. Fundamentals, (1970), 9, 521.
5. Dluzniewski, J. H. and Adler, S. B., I. Chem. E. Symp. Ser., (1972), No. 35, 4:21.
6. Benedict, M., Webb, G. B., and Rubin, L. C., Chem. Eng. Progr., (1951), 47, 449.
7. Starling, K. E., Fluid Thermodynamic Properties for Light Petroleum Systems, p. 221, Gulf Publishing Company, Houston, Texas (1973).
8. Lin, C-J., et al., Can. J. Chem. Eng., (1972), 50(10), 644.
9. Lin, C-J. and Hopke, S. W., A.I.Ch.E. Symp. Ser., (1974), 70 (140), 37.
10. Chueh, P. L. and Prausnitz, J. M., A.I.Ch.E.J., (1967), 13, 1107.
11. Barner, H. E., and Adler, S. B., Hydrocarbon Process, (1968) 37 (10), 150.
12. Hopke, S. W. and Lin, C-J., Application of the BWRS Equation to Natural Gas Systems, Presented at the 76th National A.I.Ch. E. Meeting, Tulsa, Oklahoma, March 10-13, 1974.

13. Soave, G., Chem. Eng. Sci., (1972), 27, 1197.
14. American Petroleum Institute, Technical Data Book--Petroleum Refining, 2nd Edition, Chap. 4,7, & 8, Washington, D.C., 1977.
15. American Petroleum Institute, Report Nos. API-1-74, API-4-74, API-3-75, API-4-75, API-1-76, and API-3-76, The Pennsylvania State University, University Park, Pennsylvania (Private Communications).
16. Peng, D-Y. and Robinson, D. B., Ind. Eng. Chem. Fundamentals, (1976), 15, 59.
17. Cavett, R. H., Proc. Am. Petrol. Inst., Vol. 42 (III), (1962) 351.
18. Grayson, H. G., The Influence of Differences in Phase Equilibria Data on Design, Presented at the Midyear Meeting of the American Petroleum's Division of Refining, San Francisco, California, May 14, 1962.
19. Robinson, R. L., Jr. and Chao, K. C., Ind. Eng. Chem. Process Design Develop., (1971), 10, 221.
20. Adler, S. B., Ozkardesh, H., and Schreiner, W. C., Hydrocarbon Process., (1968), 47(4), 145.
21. Spencer, C. F., A Semi-Empirical Study of the True Critical Properties of Defined Mixtures, Ph.D. Thesis, Pennsylvania State University (August, 1975).
22. Klosek, J. and McKinley, C., Proceedings, First International Conference on LNG, Chicago, Illinois (1968).
23. Shana'a, M. Y. and Canfield, F. B., Trans. Faraday Soc., (1968), 64, 2281.
24. Miller, R. C. and Rodesevitch, J. B., A.I.Ch.E.J., (1973), 18, 728.
25. Miller, R. C., Chem. Eng., (1974), 81 (23), 134.
26. Watson, K. M., Ind. Eng. Chem., (1943) 35, 398.
27. Nichols, W. B., Reamer, H. H., and Sage, B. H., A.I.Ch.E.J., (1957), 3, 262.
28. Kay, W. B., Chem. Rev., (1941), 29, 501.
29. Kurata, F., Solubility of Solid Carbon Dioxide in Pure Light Hydrocarbons and Mixtures of Light Hydrocarbons, Research Report RR-10, Gas Processors Association, Tulsa, OK (1974).
30. King, M. B., Phase Equilibrium in Mixtures, p. 65, Pergamon Press, Inc., New York, New York (1969).
31. Prausnitz, J. M., Molecular Thermodynamics of Fluid-Phase Equilibria, p. 234, Prentice-Hall, Inc., Englewood Cliffs, New Jersey (1969).
32. Lewis, G. N. and Randall, M., Thermodynamics, 2nd Edition, Revised by K. S. Pitzer and L. Brewer, p. 207, McGraw Hill Book Co., Inc., New York, New York (1961).
33. Meyers, A. L. and Prausnitz, J. M., Ind. Eng. Chem. Fundamentals, (1965) 4, 209.
34. Friend, L. and Adler, S. B., Chem. Eng. Progr., (1957) 53, 452.

35. Seidell, A., Solubilities of Inorganic and Metal Organic Compounds, 4th Edition, Vol. 1 & 2, D. Van Nostrand Company, Inc., New York, New York (1958).

8

Applications of the Peng-Robinson Equation of State

DONALD B. ROBINSON, DING-YU PENG, and HENG-JOO NG

University of Alberta, Edmonton, Alberta, Canada

The recently proposed Peng-Robinson equation of state (1) incorporates the best features of the Soave (2) treatment of the Redlich-Kwong (3) equation into a new model which has some significant advantages over earlier two-parameter equation of state models. The purpose of this contribution is to indicate how the equation performs when it is used for calculating fluid thermodynamic properties for systems of industrial interest.

The flexibility of the equation and the generality of the situations for which it can be used to give answers of acceptable reliability at a reasonable cost are illustrated through example calculations of vapor pressure, density, vapor-liquid equilibrium, critical properties, three phase L_1L_2G equilibrium for systems containing water, and HL_1G, HL_1L_2G, and HL_1L_2 equilibrium in hydrate forming systems.

The equation shows its best advantages in any situation involving liquid density calculations and in situations near the critical region, but it is usually better than other two parameter models in all regions.

In developing the new equation certain criteria were established at the outset. In order for the equation to be acceptable the following conditions had to be met.

a. The equation was to be a two-constant equation not higher than cubic in volume.
b. The model should result in significantly improved performance in the vicinity of the critical point, in particular with Z_c and liquid density calculations.
c. The constants in the equation should be expressable in terms of P_c, T_c and ω.
d. The mixing rules for evaluating mixture constants should not contain more than one fitted binary interaction parameter, and if possible this parameter should be temperature, pressure, and composition independent.

e. The equation should be sufficiently general in its applicability so that the one equation could be used to handle all fluid property calculations for systems normally encountered in the production, transportation, or processing or natural gas, condensate, or other related hydrocarbon mixtures.

The Peng-Robinson Equation

The details of the development of the Peng-Robinson (PR) equation are given in the original paper (1). The final results are summarized here for convenience.

The equation has the form:

$$P = \frac{RT}{v - b} - \frac{a(T)}{v(v + b) + b(v - b)}$$

In this equation:

$$b = 0.07780 \frac{RT_c}{P_c}$$

$$a(T) = a(T_c) \cdot \alpha(T_R, \omega)$$

$$a(T_c) = 0.45724 \frac{R^2 T_c^2}{P_c}$$

$$Z_c = 0.307$$

$$\alpha^{\frac{1}{2}}(T_R, \omega) = 1 + \kappa(1 - T_R^{\frac{1}{2}})$$

$$\kappa = 0.37464 + 1.54226\omega - 0.26992\omega^2$$

The values of b and $a(T_c)$ are obtained by equating the first and second derivative of pressure with respect to volume to zero along the critical isotherm at the critical point, solving the two equations simultaneously for a and b, and then using the equation of state to solve for Z_c.

The value of κ for each pure component was obtained by linearizing the $\alpha^{\frac{1}{2}}(T_R, \omega)$ vs $T_R^{\frac{1}{2}}$ relationship between the normal boiling point and the critical point. This is slightly different that the method of Soave (2) which assumed a linear relationship from $T_R = 1.0$ to $T_R = 0.7$.

The influence of ω and κ was obtained by a least squares fit of the data for all components of interest.

TABLE 1. Comparison of Vapor Pressure Predictions (Reprinted with Permission from Ind. Eng. Chem. Fundamentals, 15, 293 (1976a). Copyright by The American Chemical Society.)

		Relative Error, %.					
	No. of Data	AAD[a]		BIAS[b]		RMS[c]	
Component	Points	SRK	PR	SRK	PR	SRK	PR
Methane	28	1.44	0.66	0.47	0.38	1.57	0.77
Ethane	27	0.70	0.34	-0.10	-0.34	0.95	0.38
Propane	31	0.98	0.36	0.87	0.31	1.10	0.42
Isobutane	27	1.06	0.32	0.82	0.16	1.18	0.34
n-Butane	28	0.75	0.37	0.47	-0.22	0.86	0.42
Isopentane	15	0.46	0.54	0.17	-0.53	0.49	0.60
n-Pentane	30	0.92	0.58	0.50	-0.29	1.02	0.66
n-Hexane	29	1.55	0.90	1.31	0.37	1.75	1.06
n-Heptane	18	1.51	0.79	1.48	0.63	1.88	1.04
n-Octane	16	1.99	1.04	1.97	1.02	2.24	1.26
Nitrogen	17	0.56	0.31	0.00	-0.02	0.75	0.37
Carbon Dioxide	30	0.53	0.62	0.50	-0.49	0.63	0.71
Hydrogen Sulfide	30	0.66	0.96	0.34	0.42	1.00	1.48
Average, %		1.01	0.60	0.68	0.11	1.19	0.73

[a] $AAD = \frac{\sum_{i=1}^{N} |d_i|}{N}$ [b] $BIAS = \frac{\sum_{i=1}^{N} d_i}{N}$ [c] $RMS = \sqrt{\frac{\sum_{i=1}^{N} d_i^2}{N}}$

d_i = error for each point

The mixing rules which are recommended are as follows:

$$b_m = \sum_i x_i b_i$$

$$a_m = \sum_i \sum_j x_i x_j a_{ij}$$

$$a_{ij} = (1 - \delta_{ij}) a_i^{\frac{1}{2}} a_j^{\frac{1}{2}}$$

In the equation for a_{ij}, δ_{ij} is a fitted parameter preferably obtained by optimizing the prediction of binary bubble point pressures over a reasonable range of temperature and composition.

Applications

Vapor Pressure. One of the important considerations for any equation of state that is to be used for vapor-liquid equilibrium calculations is whether or not it can accurately predict the vapor pressure of pure substances. Table 1 shows a comparison between the PR equation and the Soave-Redlich-Kwong (SRK) equation for predicting the vapor pressure of ten paraffin hydrocarbons and three commonly encountered non-hydrocarbons.

It will be noted that the use of the PR equation has improved the RMS relative error by a factor of about 40 percent over the SRK predictions using the same data. Further, it is seen that the SRK predictions are biased high in every case but one and that PR predictions are evenly split between positive and negative departures with a resulting overall positive bias that is only 16 percent of the value obtained using the SRK equation.

Densities. It is well known that the SRK equation tends to predict liquid molal volumes that are too high, and that this is particularly noticeable in the vicinity of the critical point. The fact that the PR equation gives a universal critical compressibility factor of 0.307 compared to 0.333 for the SRK equation has improved the ability of the PR equation to predict liquid densities.

Figure 1 illustrates the performance of the two equations for predicting the molal volume of saturated liquids and vapors for pure n-pentane. At reduced temperatures above about 0.8, the average error in liquid density has been reduced by a factor of about 4 by using the new equation. At lower reduced temperatures, the predictions by the new equation are better by a factor of about 2. Both equations give acceptable predictions of the vapor density.

The ability of the new equation to predict the specific volume of saturated liquids in multicomponent systems is clearly illustrated in Figures 2 and 3. Figure 2 shows the calculated liquid volume percent in the retrograde region for a 6 component paraffin hydrocarbon mixture containing components from methane through n-decane. Figure 3 shows the same kind of information for a 9 component system

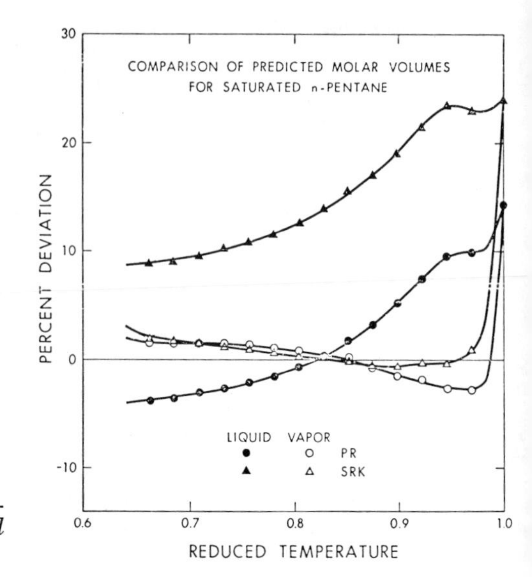

Figure 1. Comparison of molar volumes of saturated n*-pentane predicted by the PR and SRK equations*

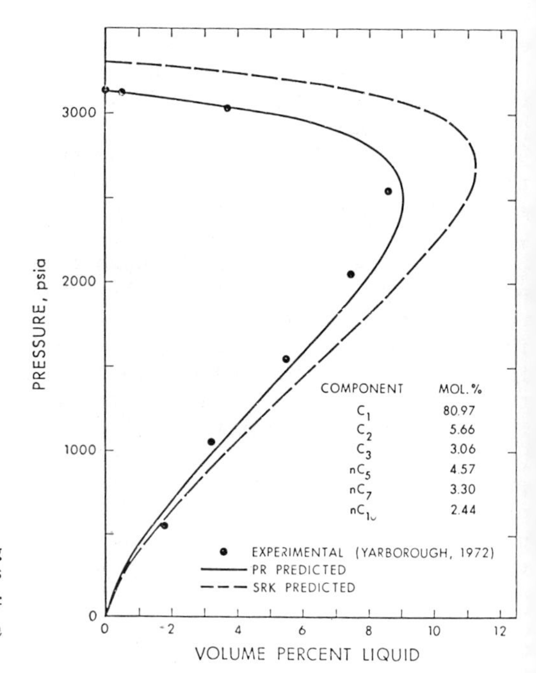

Industrial and Engineering Chemistry, Fundamentals

Figure 2. Volumetric behavior of six-component paraffin hydrocarbon mixture in retrograde region (13)

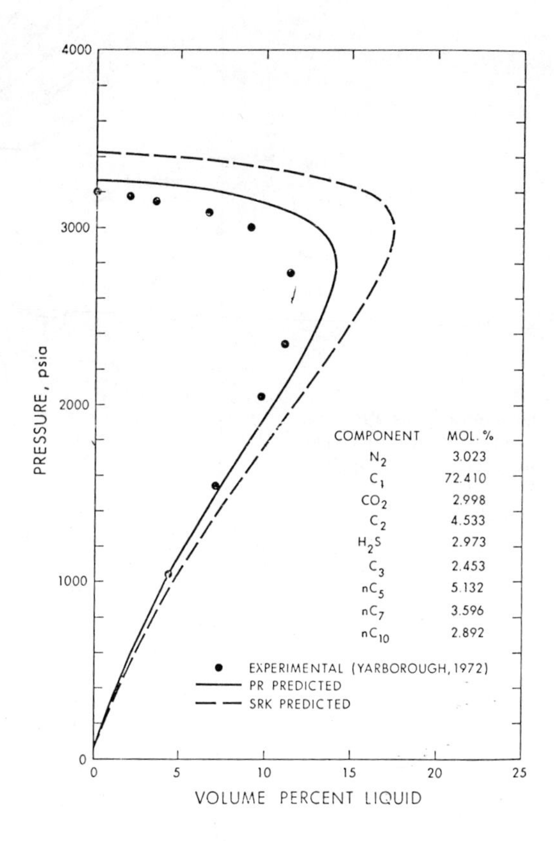

Figure 3. Volumetric behavior of nine-component condensate-type fluid containing hydrocarbons, nitrogen, carbon dioxide, and hydrogen sulfide at 250°F

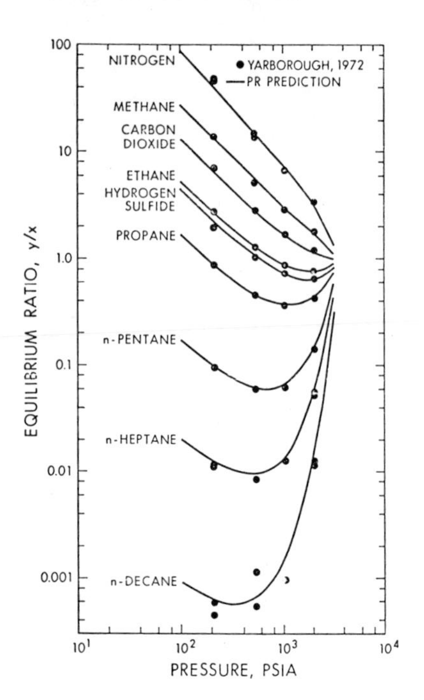

Figure 4. Experimental and PR predicted equilibrium ratios for a nine-component system containing hydrocarbons, nitrogen, carbon dioxide, and hydrogen sulfide at 100°F

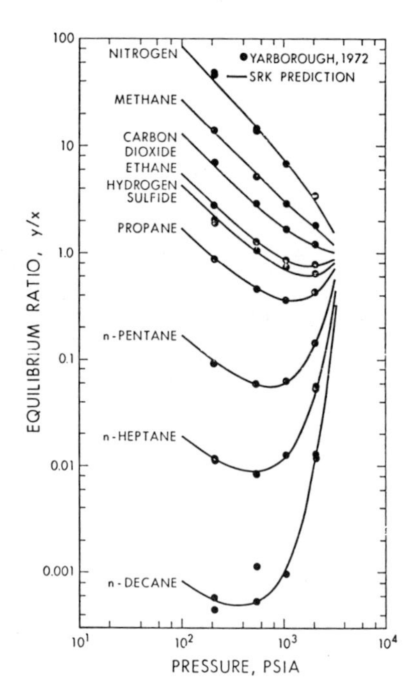

Figure 5. Experimental and SRK predicted equilibrium ratios for a nine-component system containing hydrocarbons, nitrogen, carbon dioxide, and water at 100°F

containing N_2, CO_2 and H_2S in addition to the 6 hydrocarbons.

The greatly improved performance of the PR equation is obvious, particularly in the upper retrograde region, where an improvement from about 20 to over 100 percent results. The ability of the equation to predict upper dew points for these systems is also evident. It should perhaps be emphasized that none of the experimental data on the mixtures was used in evaluating the parameters for making the predictions.

Vapor-Liquid Equilibrium. One of the advantages of using a simple two-constant cubic equation of state is the relative simplicity with which they may be used to perform vapor-liquid equilibrium calculations. The time-consuming iterative procedures required by multiconstant equations of state for calculating the coexisting phase densities can be avoided by obtaining the roots analytically from the cubic equation.

The ability of the PR equation to predict equilibrium phase compositions and phase envelopes has been evaluated for a variety of systems, including those containing both hydrocarbon and non-hydrocarbon components. Examples of some of these are illustrated in Figures 4, 5, 6, and 7.

Figures 4 and 5 show comparisons between equilibrium ratios determined experimentally and those calculated by the PR and SRK equations for a nine component system containing hydrocarbons from methane to n-decane, together with nitrogen, carbon dioxide, and hydrogen sulfide. The composition of this mixture is given in Table 2. An inspection of the figures shows that in general the agreement between the experimental data and both predictions is good. However, a more detailed analysis shows that the SRK prediction would yield a somewhat higher convergence pressure for the system than that predicted by the PR equation. Although it cannot be seen from Figures 4 and 5, the K-factors converge to unity for the PR prediction at 3400 psia and by the SRK prediction at 3700 psia.

The undesirable consequences of this may be explained as follows. Each equilibrium ratio curve for the heavier components passes through a minimum at a certain operating pressure. For these curves to converge to unity at a lower pressure means that the equilibrium ratios must increase more rapidly as pressure increases from the value at the minimum to the value at the convergence pressure. By the same reasoning, the ratios for the lighter components, in particular nitrogen and methane, must decrease faster in the same pressure interval. However, it is the equilibrium ratios for the heavier components that control the dew point pressures. Thus if the PR equation predicted the dew point pressure for a particular system in the upper retrograde region, the SRK equation would have to go to a higher pressure to find the same set of equilibrium ratios and consequently it would predict a higher dew point pressure. The errors resulting from this characteristic are clearly seen in Figures 2 and 3.

Figure 6 shows the experimental and predicted phase envelope and critical point for a four-component lean gas mixture. Although the agreement along the bubble point locus is good, the agreement

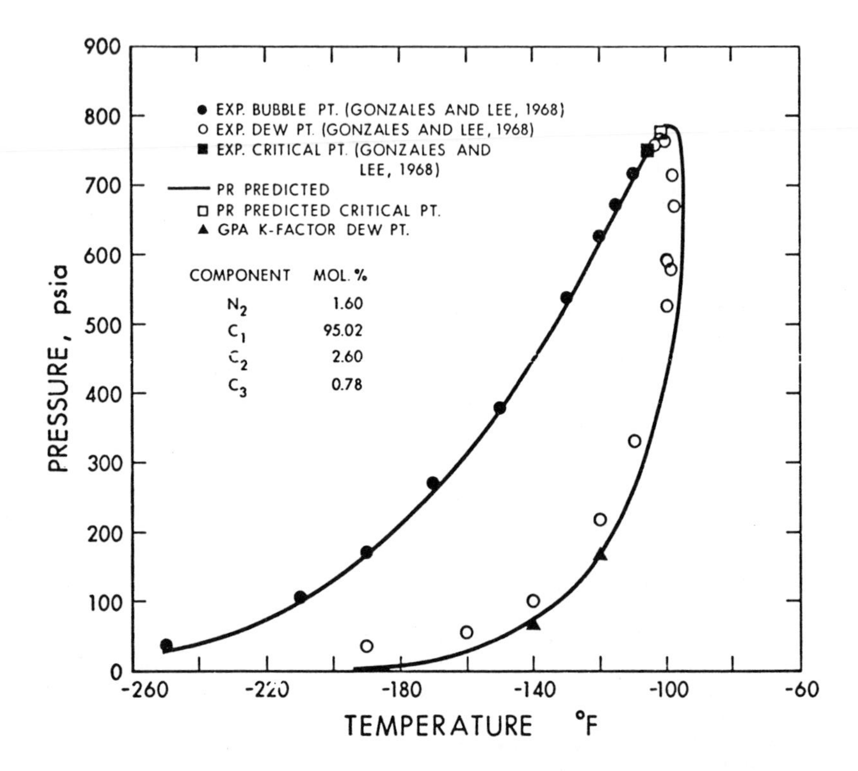

Figure 6. Experimental and predicted phase envelope and critical point for lean gas mixture (data from Ref. 16)

Table 2

Composition of Sour Gas Mixture
Used for Equilibrium Ratio Comparisons
Yarborough (15)

Component	Mole Fraction
N_2	0.0299
C_1	0.7180
CO_2	0.0302
C_2	0.0456
H_2S	0.0377
C_3	0.0251
nC_5	0.0533
nC_7	0.0377
nC_{10}	0.0300

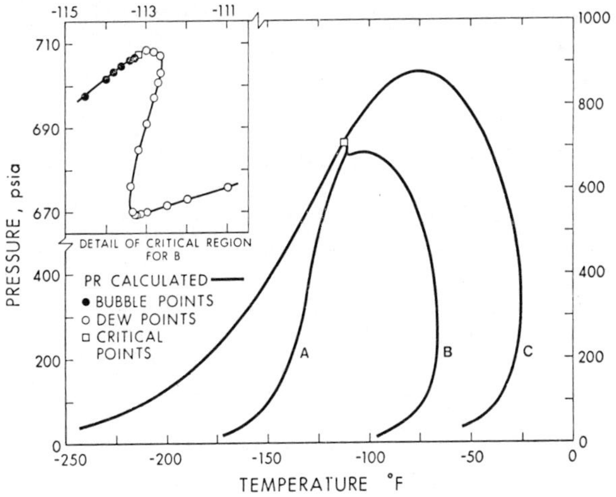

Figure 7. Predicted phase envelopes and critical point for selected lean gas mixtures containing varying butanes, pentanes, and heptane

along the dew point locus does not appear to be as good. However, measurements of this type are exceedingly difficult to make and the pressures of initial liquid formation may be open to some uncertainty. Initial liquid formation at -120° and -140°F as calculated from the GPA Engineering Data Book (Revised 1976) agrees with the predictions. As a further assessment of the discrepancy, flash calculations were carried out at each of the experimental dew points, and the results indicated that the system was more than 99.9 volume percent vapor in each case. Thus it becomes a question of whether or not one can actually observe something less than 0.1 volume percent liquid.

Figure 7 illustrates the response of the equation to slight changes in the concentration of components heavier than propane in a multi-component mixture. The three systems shown each contain a small amount of nitrogen and carbon dioxide, approximately the same amount of methane, ethane, and propane, but varying small amounts of butanes, pentanes, and heptane. The compositions of the three mixtures together with the calculated critical properties are shown in Table 3. It will be noted that the phase envelopes for mixtures A and C are typical, but that the phase envelope for mixture B exhibits a double retrograde region. A detail of this region and the critical point are shown in Figure 7. The authors are unaware of other reports of a double retrograde region in multicomponent systems although the phenomenon is well known in binary systems (Chen et al, (4, 5)).

Critical Properties. A method for calculating the critical properties of multicomponent systems using an equation of state has recently been developed by Peng and Robinson (6). The method is based on the rigorous thermodynamic criterion for the critical state enunciated by J. Willard Gibbs (7) and uses the PR equation of state to describe the fluid properties. In this study, the predicted and experimental critical temperature and pressure were compared for 32 mixtures containing from 3 to 12 components. The systems included paraffin hydrocarbons from methane to n-decane, nitrogen, carbon dioxide, and hydrogen sulfide. The average absolute error in pressure was 25 psia or 2.33% and the average arithmetic error was +5.7 psia or 0.13%. For temperature the corresponding errors were 7.7 ^{o}R or 1.31% and +7.0 ^{o}R or 1.14%. The range of critical pressures varied by a factor of 4.43 and the range of critical temperatures by a factor of 2.88.

Figure 6 illustrates a typical comparison between the predicted and experimental critical point for a four-component system and Figure 7 and Table 3 illustrate the calculated critical properties for systems containing up to 10 components.

Water-Hydrocarbon Systems. The application of the PR equation to two and three-phase equilibrium calculations for systems containing water has recently been illustrated by Peng and Robinson (8). As in the case of other hydrocarbon-non-hydrocarbon mixtures, one fitted binary interaction parameter for water with each of the hydrocarbons is required. These parameters were obtained from experimental data available in the literature on each of the water-hydrocarbon binaries

Table 3

Selected Mixtures Used for Phase Envelope and Critical Point Calculations

Component	Mixture A Mole %	Mixture B Mole %	Mixture C Mole %
Nitrogen	1.67	1.76	1.27
Carbon Dioxide	0.20	0.17	0.34
Methane	97.21	97.18	97.57
Ethane	0.805	0.78	0.78
Propane	0.090	0.079	0.02
Isobutane	0.010	0.009	--
n-Betane	0.015	0.016	--
Isopentane	--	0.002	--
n-Pentane	--	0.001	--
n-Heptane	--	0.003	0.02
Critical Pressure, psia	706.4	706.6	707.2
Critical Temperature, ^{o}F	-113.0	-112.7	-113.2

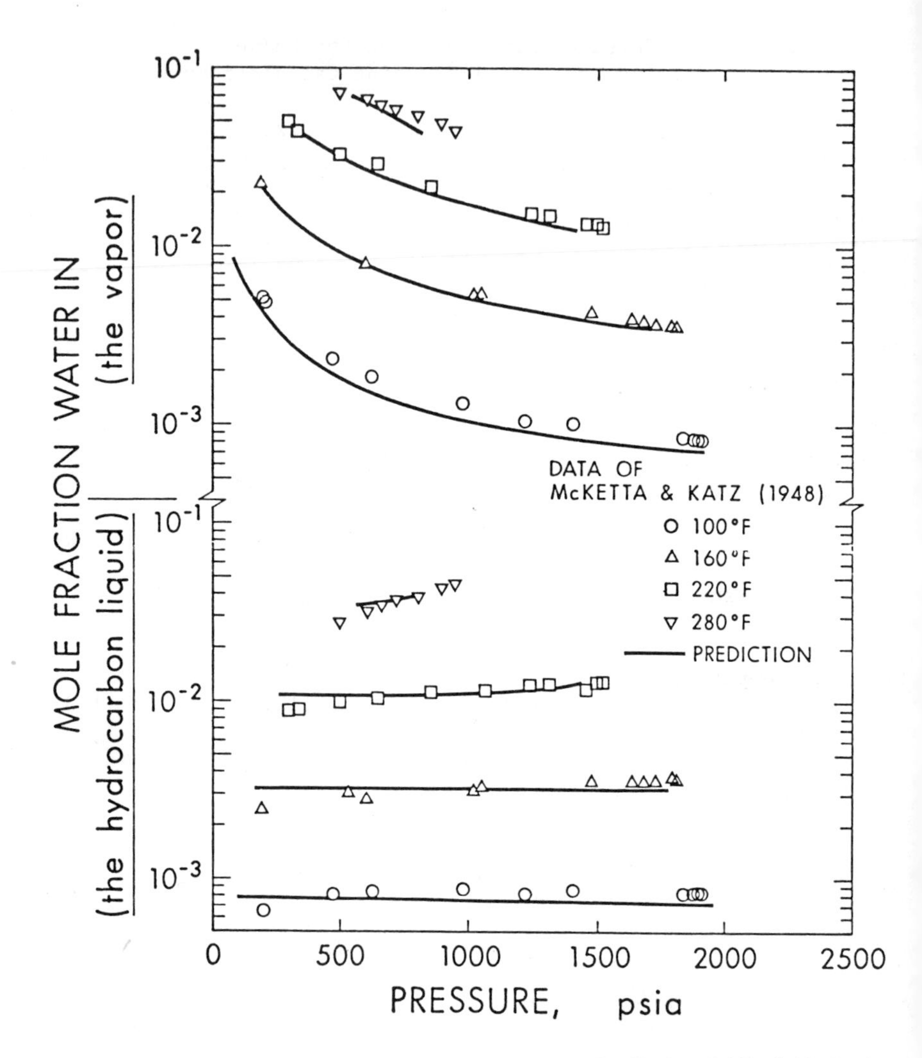

Canadian Journal of Chemical Engineering

Figure 8. Water content of hydrogen-rich liquid and vapor in a three-phase system containing water, methane, and n*-butane*

The three phase program was then evaluated by comparing predicted and experimental results for the methane-n-butane-water and n-butane-1 butene-water systems as reported by McKetta and Katz (9) and Wehe and McKetta (10).

Figures 8 and 9 illustrate the type of agreement that was obtained for the water content and the hydrocarbon distribution in both the hydrocarbon liquid and vapor phases for the methane n-butane-water system. It can be seen from Figure 8 that the predictions reproduce the water content very well at all temperatures. Figure 9 shows that the agreement between experimental and predicted hydrocarbon concentrations in the vapor and liquid phases is good at 100°F, although the agreements does not seem to be as good at 2220°F. The experimental data for 220°F may be open to question since the critical pressure for methane-n-butane mixtures at 220°F is reported to be about 1350 psia by Roberts et al (11). The experimental data of these authors on the methane-n-butane system at 220°F are included for comparison. It seems doubtful that the presence of water in this system would increase the critical pressure to about 1550 psia as indicated by the three component data. In view of this, the predicted results are thought to be just as good at 220°F as at 100°F.

The programs used to make the above prediction have been successfully used for three phase flash, bubble point, and hydrocarbon and/or water dew point calculations. Generally the calculations converge rapidly and normally require only about 20 iterations. The programs predict hydrocarbon concentrations in the aqueous liquid phase which are several orders of magnitude lower than the reported experimental data. In order to obtain a good prediction of these very dilute concentrations a different and possibly temperature dependent δ_{ij} may be required.

An additional example of the ability of the PR equation to predict the water content of gases saturated with water is shown in Figure 10. The comparisons shown here are for the PR predictions and the GPA Engineering Data Book (Revised 1976) values for the water content of natural gases. It will be noted that the GPA data and the predictions are practically coincident over the entire range of pressures and temperatures.

Hydrates. Programs have been developed for predicting hydrate forming conditions for the HL_1G, HL_1L_2G, and HL_1L_2 equilibia. These use the approach developed by Parrish and Prausnitz (12) but with the PR equation used throughout for all fluid property calculations. Details of the method have been reported by Ng and Robinson (13).

Typical results for the HL_1G equilibrium are shown for the carbon dioxide-propane-water system and the methane-isobutane water system in Figure 11. The predicted and experimental pressures are compared at the experimentally determined hydrate temperature. The mixtures of carbon dioxide and propane included concentrations from 6 to 84 mole percent propane on a water-free basis in the gas phase and the mixtures of methane and isobutane included concentrations from 0.4 to 63.6 mole percent isobutane on the same basis. It will be seen that the predicted and experimental results compare favorably

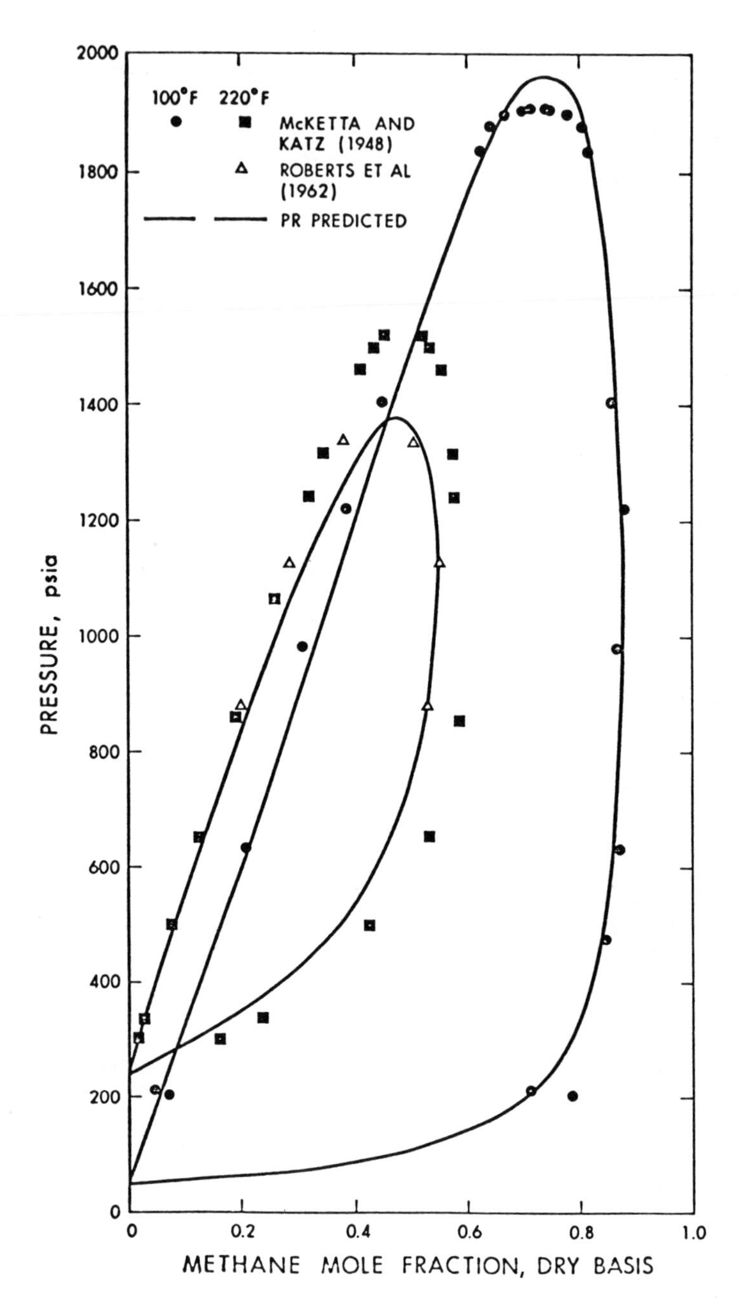

Figure 9. Hydrocarbon distribution in the hydrocarbon-rich liquid and vapor in three-phase system containing water, methane, and n-*butane*

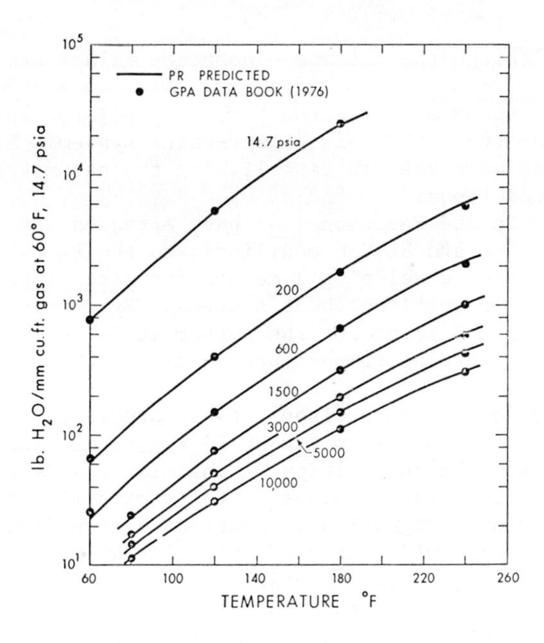

Figure 10. Predicted equilibrium water content of natural gas in contact with water liquid

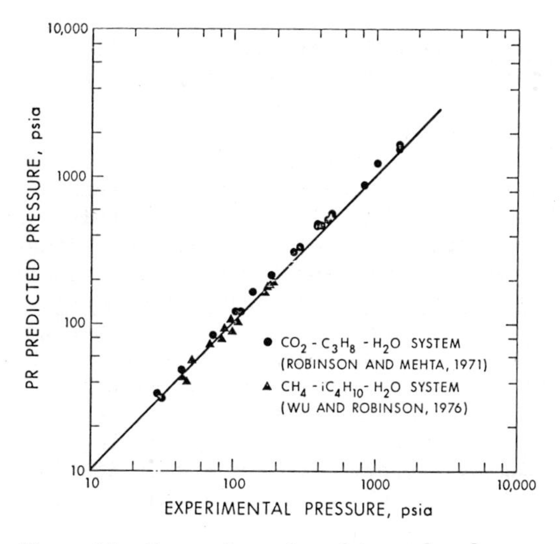

Figure 11. Comparison of experimental and predicted hydrate formation pressures in the HL_1G region (data from Refs. 17, 18)

throughout, although the methane--isobutane values are biased slightly high.

Figure 12 shows a comparison between predicted and experimental HL_1L_2G data for the methane-isobutane-water system. None of the four-phase data were used in establishing the parameters used in making the predictions.

Recently, Ng and Robinson (14) have extended the methods used above for the HL_1G and HL_1L_2G equilibria to the HL_1L_2 region. This was accomplished by developing a method for predicting the slope of the HL_1L_2 locus in multicomponent systems. By combining this with the predicted HL_1L_2G point for the system it is possible to calculate the hydrate forming temperature in the HL_1L_2 region at any pressure.

Figure 13 shows a comparison between predicted and experimental hydrate forming conditions in the HL_1L_2 region for three selected liquid mixtures. The composition of the mixtures is given in Table 4. These include a mixture of essentially paraffin hydrocarbons, a mixture containing a significant amount of carbon dioxide, and a mixture containing a wide range of molecular weights. The mixture containing the carbon dioxide shows a difference between the predicted and experimental hydrate formation temperature of about $2^{\circ}F$ although the slopes of the predicted HL_1L_2 locii agree very well with those of the experimental data in every case. The discrepancy between the experimental and predicted hydration temperatures probably results from a combination of small errors in both the bubble point and hydrate programs.

Future Work

Undefined Fractions. The example applications illustrated above show the ability of the PR equation to make acceptable fluid property predictions for a wide variety of situations when the composition of the system is fully defined. At the present time, a suitable method has not been developed for characterizing the critical properties and acentric factor for an undefined fraction such as C_{7+} or its equivalent. Work is currently in progress on this problem. The importance of being able to handle this kind of situation for industrial systems involving petroleum fractions can readily be appreciate

Water Content of Systems Containing CO_2 and H_2S. Frequently reservoir fluids and petroleum fractions contain significant concentrations of CO_2 and H_2S. When these fluids are in the presence of liquid water and an equilibrium gas and/or liquid phase it is of interest to be able to calculate the water content of the gas and hydrocarbon liquid. A suitable scheme for making these predictions is currently being developed. Preliminary indications are that the δ_{ij} values for water with CO_2 or H_2S will have to be temperature dependent.

Nomenclature

a - attraction parameter in PR equation
b - Van der Waals covolume

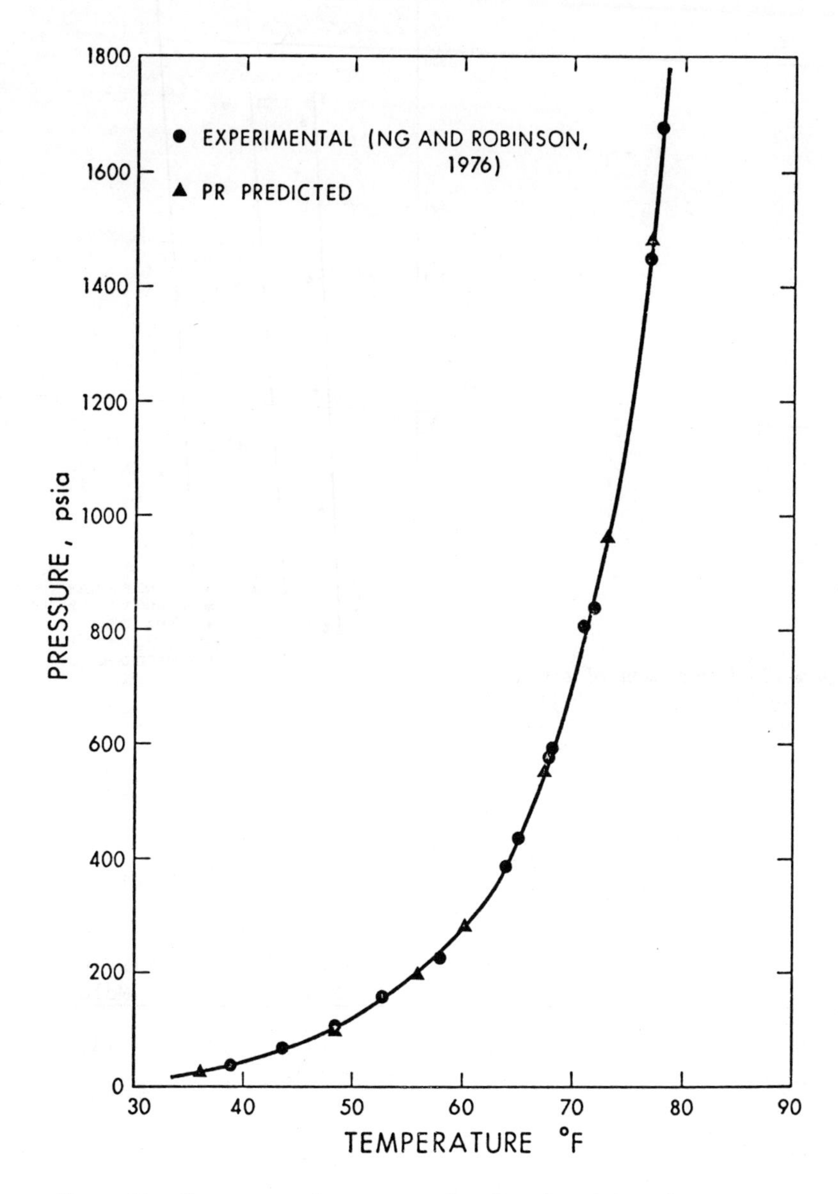

Figure 12. Comparison of experimental and predicted four-phase HL_1L_2G equilibrium in the methane–isobutane–water system

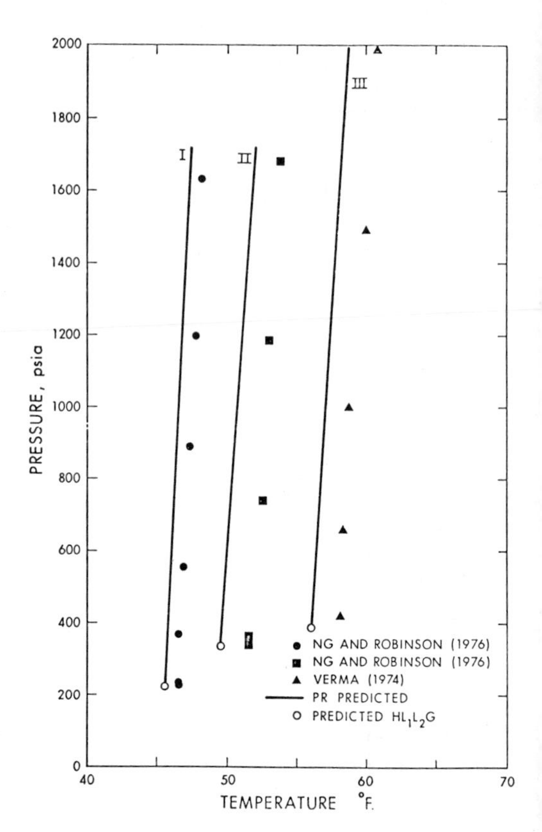

Figure 13. Comparison of experimental and predicted hydrate-forming conditions in selected liquid mixtures in the HL_1L_2 region

Table 4

Composition of Mixtures Used for Predicting Hydrate Formation in the H_1L_2 Region

Component	Mixture I[a] Mole %	Mixture II[a] Mole %	Mixture III[c] Mole %
M Methane	2.2	--	14.5
Ethane	30.6	17.0	--
Propane	50.8	38.6	27.1
Isobutane	16.2	18.9	--
Nitrogen	0.2	--	--
Carbon Dioxide	--	25.5	--
n-Decane	--	--	58.4

[a] Ng and Robinson (13)

[c] Verma (16)

G - gas phase
H - solid hydrate phase
L_1 - water-rich liquid phase
L_2 - liquid phase rich in hydrate former(s)
P - pressure
R - gas constant
T - temperature
v - molar volume
x - mole fraction
Z - compressibility factor

Greek Letters

α - scaling factor in PR equation
δ - interaction parameter
γ - characteristic constant used for PR parameters
ω - acentric factor

Subscripts

c - critical property
i,j - component identification
m - mixture property

Literature Cited

1. Peng, D.-Y., and Robinson, D. B., Ind. Eng. Chem. Fundamentals, (1976) 15, 59.
2. Soave, G., Chem. Eng. Sci. (1972) 27, 1197.
3. Redlich, O., and Kwong, J. N. S., Chem. Rev., (1949) 44, 233.
4. Chen, R. J., Chappelear, P. S., and Kobayashi, R., J. Chem. Eng. Data, (1974) 19, 53.
5. Chen, R. J., Chappelear, and Kobayashi, R., Chem. Eng. Data, (1974) 19, 58.
6. Peng, D.-Y, and Robinson, D. B., "A rigorous Method for Predicting the Critical Properties of Multicomponent Systems Using an Equation of State, " AIChE Journal, (1977) 23, 137.
7. Gibbs, J. Willard, "On the Equilibrium of Heterogeneous Substances," (October 1876 - May 1877). Collected Works, Vol. 1, p. 55, Yale University Press, New Haven, Connecticut, 1928.
8. Peng, D.-Y., and Robinson, D. B., "Two- and Three Phase Equilibrium Calculations for Systems Containing Water," Can. Journal of Chem. Eng.,(1976) 54, 595.
9. McKetta, J. J., and Katz, D. L., Ind. Eng. Chem., (1948) 40, 853.
10. Wehe, A. H., and McKetta, J. J., J. Chem. Eng. Data, (1961) 6, 167.
11. Roberts, L. R., Wang, R. H., Azarnoosh, A., and McKetta, J. J., J. Chem. Eng. Data, (1962) 7, 484.
12. Parrish, W. R., and Prausnitz, J. M., Ind. Eng. Chem. Process Des. Develop., (1972) 11, 26.

13. Ng, H.-J., and Robinson, D. B., Ind. Eng. Chem. Fundamentals (1976) 15, 293.
14. Ng, H.-J., and Robinson, D. B., "Hydrate Formation in Condensed Systems," AIChE Journal, (1977), In Press.
15. Yarborough, L., and J. Chem. Eng. Data, (1972) 17, 129.
16. Gonzales, M. H., and Lee, A. L., J. Chem. Eng. Data, (1968) 40, 853.
17. Robinson, D. B., and Mehta, B. R., and J. Can. Pet. Tech., (1971 10, 33.
18. Wu, B.-J., Robinson, D. B., and Ng, H.-J., J. Chem. Thermo., (1976) 8, 461.

9

Application of Equations of State in Exxon's Production Operations

S. W. HOPKE

Exxon Production Research Co., Houston, TX 77001

Equations of state are the heart of Exxon's computer simulation of phase behavior in field separation facilities, pipelines, gas plants and condensate or volatile oil reservoirs. These simulation programs are executed scores of times each working day. There are several state equation methods available to our engineers. One of them, the BWRS equation, (1 - 3), is more accurate than the others and is used for almost all of our simulations that employ state equations.

The BWRS equation is Starling's modification of the Benedict-Webb-Rubin equation of state. It contains eleven adjustable pure component parameters plus a binary interaction parameter for each component pair. Thus, a typical 20 component mixture would be characterized by 220 pure component parameters and 180 different binary interaction parameters--a total of 400 constants. Exxon's set of constants were determined by multi-property regression, a procedure in which parameters are adjusted until available data on density, enthalpy, vapor pressure, K-values, sonic velocity, and specific heats are all matched simultaneously. The large number of constants to be determined requires that these data be accurate and that they cover a wide range of conditions. Nearly 20,000 data points were used to determine our set of constants. The large amount of data required limits the components that can be handled to the relatively few for which such data exist and may also place a practical limit on the number of parameters desirable in an equation of state.

To be useful for typical petroleum production applications, a state equation must be able to predict the phase behavior of mixtures containing hydrocarbon fractions. The BWRS equation can do this for mixtures containing fractions with molecular weights as high as 500. The BWRS parameters for these fractions are obtained from correlations of the parameters, made dimensionless by dividing by appropriate powers of the critical temperature and critical volume, with the acentric factor. Accurate predicted K-values for light components in oils result from correlations of binary interaction parameters. Reliable simulations of absorber plant performance are possible even when only the molecular weight and density of the absorption oils are known.

In condensate and volatile oil reservoirs, in high pressure separation facilities, and in pipelines associated with production of reservoirs in remote locations, phase behavior is dominated by hydrocarbons heavier than C_{10}. Since the concentrations and properties of these heavy hydrocarbon fractions are not usually known, reliable phase behavior predictions for these produced fluids are impossible unless they are based upon experimental phase behavior data for the fluid in question. Consequently, to predict the phase behavior of these systems, we take the following approach. First, the composition of the fluid is measured to $C_{10}+$ and the molecular weight and density determined for each of the C_6 to $C_{10}+$ fractions. The $C_{10}+$ fraction is then split into several components--typically the C_{10}, C_{15}, C_{20}, and C_{25} fractions. This split and the binary interaction parameters of these heavy hydrocarbon fractions with other components are adjusted until the model matches experimental phase behavior data for the fluid. The minimum phase behavior data required is the volume percent liquid at several pressures and at a constant temperature.

Because of the large number of constants involved, the BWRS method requires slightly longer computation times than simpler, less accurate methods. Considerable attention to improving computation efficiency has enabled us to use the BWRS equation in tray-to-tray column calculations and in a one-dimensional compositional reservoir simulator, applications which involve large numbers of state equation calculations.

While the BWRS equation of state is adequate for most of our applications, there are problem areas. Some low temperature processing conditions approach the critical region, where calculations lose accuracy and are difficult to converge. To date, we have not systematically evaluated the accuracy near mixture critical points nor studied ways to improve convergence in this region.

We have not had great success modeling water with the BWRS equation. Water is present in most production systems and we have had to use separate correlations for water behavior.

Binary interaction parameters, especially for component pairs that greatly differ in molecular size, tend to be temperature dependent. This indicates that the mixing rules can be improved.

In the future, we plan to continue incorporating new data into our correlations as they become available. We also await with interest studies currently underway in the universities that might lead to fast, generalized equations of state that can handle water and polar compounds while retaining adequate accuracy for engineering applications.

Literature Cited

1. Lin, C. J. and Hopke, S. W., AIChE Symposium Series (1974), 70, No. 140, 37.
2. Hopke, S. W. and Lin C. J. presented at 76th National AIChE meeting, Tulsa, Oklahoma (1974).
3. Hopke, S. W. and C. J. Lin, Proc. 53rd Gas Processors Assn. Annual Conv. (1974) p. 63.

10

Phase Equilibria from Equations of State

T. E. DAUBERT

Pennsylvania State University, University Park, PA 16802

As I am last to make opening remarks I have decided to take a slightly different approach to the problem. The previous six speak have alluded to the meaning and application of various equations--t advantages, disadvantages and accuracy. As co-director of the API Technical Data Book project my major interest is in ascertaining wh equation is the most generally interpolatable and extrapolatable wi reasonable accuracy over a wide range rather than what is the most accurate equation within a limited range of applicability.

My major premise is: <u>No equation of state is necessarily bett than the data used to validate it</u>. In order to be generally valid any equation must be tested over a wide range of molecular type and weight of components, a wide range of compositions, and the entire temperature and pressure span. We've evaluated many of the equatio available and conclude that a decision as to the universally "best' possible equation cannot be made with current data. In addition, I personally would suggest that until more data are available further refinements of existing base equations may be empty exercises.

Thus, what I would like to do is show some examples of the data that are available and then point out where the largest gaps exist and what must be done about it. My discussion will be limite to hydrocarbon systems and systems of hydrocarbons and the industrially important gases--H_2, N_2, H_2S, CO_2, and CO. It will become apparent that the easy data already have been taken and the diffic data are yet to be determined.

Discussion

Table 1 shows the data which have been gathered for our work from a comprehensive survey of the literature. All binary data were rigorously tested for thermodynamic consistency using the meth of Van Ness and co-workers (1, 2) using the general coexistence equation reduced to constant temperature or constant pressure according to the data source. It is readily apparent that less tha

Table 1

Types and Amounts of VLE Data Available

Type System	Points of Data	
	Available	"Consistent"
Binary Hydrocarbon	2836	1850
Ternary Hydrocarbon	670	----
Quaternary Hydrocarbon	134	----
Five and Six Component Hydrocarbon	60	----
Binary Hydrocarbon - Nonhydrocarbon		
- Hydrogen	750	430
- Hydrogen Sulfide	350	240
- Carbon Dioxide	700	323
- Carbon Monoxide	240	156
- Nitrogen	450	184
Ternary Hydrocarbon - Nonhydrocarbon		
- Hydrogen	209	----
- Nitrogen	97	----
- Hydrogen + Hydrogen Sulfide	54	----
Quaternary Hydrocarbon - Nonhydrocarbon	52	----
Hydrocarbon - Water	----	----
Petroleum Fractions	----	----

two-thirds of the binary data were consistent. Thus, only the consistent points were used for data evaluation. The method was extended to ternary systems. However, since necessary parameters for testing often were not available for many data sets, all available data which were not obviously in error were used for testing.

From this table it is apparent that data on ternary hydrocarbon-non-hydrcarbon systems are in short supply as are data on all four and higher component systems. However, the former need is the most critical for testing purposes as estimating equations which are accurate for ternaries, in general, are accurate for higher multicomponent systems. The table also shows no entries for hydrocarbon-water or petroleum fraction systems. Such systems will be discussed later.

Table 2 breaks down the consistent binary hydrocarbon data by

Table 2

Consistent Binary Hydrocarbon VLE Data

System Types		Number of Points	Carbon Number Range	
A	B		A	B
Paraffin -	Paraffin	727	C_1-C_8	C_2-C_{10}
Paraffin -	Olefin	160	C_2-C_7	C_2-C_6
Olefin -	Olefin	29	C_3	C_4
Paraffin -	Naphthene	133	C_1-C_8	C_6-C_7
Paraffin -	Aromatic	450	C_1-C_8	C_6-C_9
Olefin -	Aromatic	128	C_2-C_8	C_6-C_8
Naphthene -	Aromatic	160	C_5-C_8	C_6-C_8
Naphthene -	Naphthene	3	C_6	C_7
Aromatic -	Aromatic	60	C_6-C_9	C_7-C_{10}
	Total	1850		

molecular type systems and carbon number range. Of systems of real interest, date is very sparse for naphthene-naphthene and aromatic-aromatic systems. Data are nonexistent above C_{10} for any of the system types.

Table 3 shows the distribution of pressure ranges of the data

Table 3

Pressure Range of Binary Hydrocarbon Data

Pressure Range (psia)	Percent of Data Points
$P \leq 14.7$	37
$14.7 \leq p \leq 100$	6
$100 \leq p \leq 1000$	44
$1000 \leq p$	13

for the total binary hydrocarbon data set. Few data exist in the important range of 1 - 7 atm. or above 65 atm.

Table 4 lists the consistent binary hydrocarbon-nonhydrocarbon VLE data. Data are again essentially nonexistent above hydrocarbons containing ten carbon atoms. Aromatic data is essentially limited to benzene and naphthenic data are almost absent.

Table 5 summarizes data for multicomponent systems. For hydrocarbons, the same deficiencies exist as were present for binaries--only worse. Aromatic and naphthenic data are nonexistent. For nonhydrocarbon-hydrocarbon systems no data are available for CO or CO_2 and other data are extremely sketchy.

Two areas listed in Table 1 were not discussed--hydrocarbon-water binaries and petroleum fractions. For the former systems only eight systems of experimental VLE data (ethylene, propylene, 1-butene, 1-hexene, n-hexane, cyclohexane, benzene, and n-nonane) were available. Use of solubility data and calculated VLE data added only five more systems (propane, propyne, cyclopropane, n-butane, and 1,3-butadiene). For petroleum fraction systems only three sources and seven systems have been characterized well enough to use in analytical correlations to be tested. These systems include three naphtha-fuel oil systems, two hydrogen-hydrocrackate fractions, and two hydrogen-hydrogen sulfide-hydrocrackate fractions. No reasonable work can be done without additional data.

Conclusions

The major conclusion of this survey is obvious--more data are necessary in order to validate an extended range, completely general equation of state. Particularly important areas for which new data must be taken or proprietary data must be released are listed below in no particular order as priorities depend on the users needs.

1. All systems with hydrocarbon components above C_{10}.

Table 4

Consistent Binary Hydrocarbon - Nonhydrocarbon VLE Data

Nonhydrocarbon	Hydrocarbon Type and Carbon Number Range	Approximate Number of Data Points	
Hydrogen	Paraffins ($C_1 \rightarrow C_{12}$)	280	
	Olefins ($C_2 \rightarrow C_3$)	60	
	Naphthenes ($C_6 \rightarrow C_7$)	50	
	Aromatic (C_6)	40	
			430
Hydrogen Sulfide	Paraffins ($C_1 \rightarrow C_{10}$)	220	
	Naphthene (C_9)	13	
	Aromatic (C_8)	7	
			240
Carbon Dioxide	Paraffins ($C_1 \rightarrow C_{10}$)	282	
	Olefin (C_3)	24	
	Naphthene (C_3)	8	
	Aromatic (C_6)	9	
			323
Carbon Monoxide	Paraffins ($C_1 \rightarrow C_8$)	105	
	Aromatic (C_6)	51	
			156
Nitrogen	Paraffins ($C_1 \rightarrow C_{10}$)	168	
	Aromatic (C_6)	16	
			184

Table 5

3, 4, 5, and 6 Component Systems

Type System	No. of Systems	No. of Data Points	
Hydrocarbon			
Ternary Hydrocarbon			
Paraffin (C_1 - C_{10})	12	435	
Paraffin - Olefin (C_1 - C_3	1	13	
Paraffin - Naphthene - Olefin (C_6 - C_7)	2	67	
Miscellaneous Systems with Paraffins, Olefins, Acetylenes	5	152	
			668
Quaternary Hydrocarbon			
Paraffin (C_1 - C_7)	4	15	
Paraffin - Olefin - Diolefin - Acetylene (C_3)	1	53	
Paraffin - Olefin - Diolefin - Acetylene (C_5)	1	66	
			134
Five and Six Carbon Atom Hydrocarbon			
5 Paraffin (C_1 - C_{10})	8	28	
6 Paraffin (C_1 - C_{10})	1	17	
6 Paraffin - Olefin (C_4)	1	15	
			60
Nonhydrocarbon			
Ternary Nonhydrocarbon			
Hydrogen - Methane - Ethane	1	82	
Hydrogen - Methane - Propane	1	30	
Hydrogen - Methane - Ethylene	1	97	
Nitrogen - Methane - Ethane	1	44	
Nitrogen - Methane - Hexane	1	53	
Hydrogen - Hydrogen Sulfide - C_9 aromatic or paraffin or naphthene	3	54	360
Quaternary Nonhydrocarbon			
Hydrocarbon Nitrogen - Methane - Ethane	1	7	
Hydrogen - Benzene - Cyclohexane - Hexane	1	36	
Hydrogen - Hydrogen Sulfide - Methane - Isopropylcyclohexane	1	9	
			52

2. Aromatic and naphthenic binary and ternary hydrocarbon systems.
3. Nonhydrocarbon-naphthenic and aromatic binary and ternary systems, especially hydrogen-aromatic-naphthenic systems.
4. Water-hydrocarbon binary and ternary systems of all types.
5. Water-hydrocarbon-nonhydrocarbon ternary systems.
6. Petroleum fraction and petroleum fraction-nonhydrocarbon systems.

Progress in obtaining new data is essential if the users of equations of state ever hope to adopt a reliable, dependable, and accurate predictor.

References Cited

1. Van Ness, H. C., "Classical Thermodynamics of Non-Electrolyte Solutions," Pergamon Press, Oxford (1964).
2. Van Ness, H. C., Byer, S. M., Gibbs, R. E. AIChE Journal, (1973) 2, 238.

Acknowledgment

The Refining Department of the American Petroleum Institute is gratefully acknowledged for financial support of this work.

11

Equations of State in the Vapor-Liquid Critical Region

P. T. EUBANK

Texas A&M University, College Station, TX 77843

Equations of state (ES) may be divided between those that are analytic and those that are not. Analytic equations of the form $P(\rho,T,[Z_i])$ cannot provide an accurate description of thermodynamic properties in the critical region whether for the pure components or their mixtures. Scaled ES are non-analytic in the usual $P(\rho,T)$ coordinates but assume analyticity in $\mu(\rho,T)$ for pure components. The choice of variables for a scaled ES for a mixture is not well-defined although Leung and Griffiths (1) have used $P(T,[\mu_i])$ with success on the ^3He-^4He system. Phase diagrams are simplier in such coordinates as the bubble-point surface and dew-point surface collapse into a single sheet.

Analytical equations operating outside the critical region generally correlate P-ρ-T-$[Z_i]$ data with greater difficulty as one of the components is exchanged for a compound of greater acentricity or, particularly, polarity. Disclaimers usually accompany classical equations to discourage use with highly polar compounds. Steam is a good example--both the ES and correlation procedures used to produce steam tables differ considerably from those used with hydrocarbons. Scaled ES operating in the critical region, where intermolecular forces are not so important, do not incur additional difficulty with highly polar compounds.

I. Correlation of Fluid Properties with Analytic ES

Angus (2) has recently surveyed modern ES including

1. virial-type
2. extended BWR including Strobridge, Bender, Gosman-McCarty-Hust, and Stewart-Jacobsen
3. Helmholtz free energy equations as used by Pollak and by Keenan-Keyes-Hill-Moore for water/steam
4. orthogonal polynomials as in the NEL steam tables of 1964
5. spline functions as used by Schot in 1969 for steam
6. the Goodwin ES (non-analytic)
7. the Kazavchinskii ES.

The form of the ES plus the number and type of constraints to be placed upon its constants depends on the compound and the region of reduced pressure and temperature over which the equation is to be used. First, we will examine compounds that are not highly polar--i.e., dipole moment < 0.5D.

A. Non-Polar Compounds. If we wish to fit the data for nitrogen (T_c = 126.26K) above 200K, a Bender (3) ES,

$$(P/\rho RT) = 1 + B\rho + C\rho^2 + D\rho^3 + E\rho^4 + F\rho^5 + (G + H\rho^2)\rho^2 e^{-n_{20}\rho^2}, \quad (1)$$

where $B = n_1 - n_2/T - n_3/T^2 - n_4/T^3 - n_5/T^4$

$$C = n_6 + n_7/T + n_8/T^2$$

$$D = n_9 + n_{10}/T$$

$$E = n_{11} + n_{12}/T$$

$$F = n_{13}/T$$

$$G = n_{14}/T^3 + n_{15}/T^4 + n_{16}/T^5$$

$$H = n_{17}/T^3 + n_{18}/T^4 + n_{19}/T^5,$$

could be used with only the following constraints:

1. the equation for B containing five constants should accurately reproduce the literature second virial coefficients over the desired temperature range.
2. the constants in the equation for C may also be constrained to fit any C (T) data and/or to provide a C vs T curve in agreement with the Lennard-Jones pair-potential (4).
3. at high densities the Bender equation could be further constrained to reproduce a hard-sphere ES.

However, if we wish to fit both the liquid and vapor PVT surface for nitrogen at temperatures above 80K, then additional constraints must be placed on the Bender constants. As most of the properties in the two-phase region are functions only of temperature, are related by the Clapeyron equation, and often represent measurements (i.e., heats of vaporization) independent of the single-phase density data, it is generally best to fix the coexistence dome first and later constrain the PVT surface to consistently pass through it.

Douslin and coworkers (5) provide excellent examples as to the determination of accurate and consistent two-phase properties. The critical constants, the vapor pressure curve, and the saturation densities are found graphically from the two-phase data. The heat of vaporization is calculated from the Clapeyron equation and checked for power-law behavior against the temperature variable

$\tau = 1 - T_R$ as discussed by Hall and Eubank (6). This reference also provides a method for accurate determination of the vapor pressure slope at the CP,

$$\psi_c \equiv [dP_R/dT_R]|_{T_R = 1}, \tag{2}$$

which ranges from about four for ^4He to over eight for water.

The following additional constraints may now be placed on our Bender ES constants:

4. slope of critical isochore at CP equal to ψ_c.
5. $(\partial^m P/\partial \rho^m)_{T_c} = 0$ at CP, with m = 1, 2, 3, 4.
6. obey Maxwell's equal area rule on each subcritical isotherm spaced 10K.

At this point there is the danger that our twenty-constant Bender equation has been constrained to the point where there are no free constants left to fit the single-phase density data. Bender (2) generally used constraints 1, 4, 5 (m = 1,2) and 6 (about 5 isotherms) which leaves seven free constants.

B. Polar Compounds. The same procedures can be applied to highly polar compounds except that it is difficult to accurately represent the single-phase density data without a large number of unconstrained constants. For the water example, the Pollak equation uses 40 constants, Keenan-Keyes-Hill-Moore used 50 and Jůsa used over 100 as did Schot in her bicubic spline fit.

The major difficulty concerning ES for polar compounds is the lack of accurate, comprehensive data or, in most cases, the lack of any data. Steam is probably the only highly polar compound for which there is a large body of data for a variety of properties such as vapor pressure, critical constants, saturation densities, heats of vaporization, single-phase densities in both liquid and vapor, calorimetric, Joule-Thomson, sonic velocities and heat capacity measurements (7,8,9). Recent data (10) and compilation work (11) have improved the picture for ammonia. There is also a reasonable body of data for methanol but it is not altogether thermodynamically consistent (12).

Inconsistencies are more likely to occur between sets of PVT data from different laboratories due primarily to surface effects --physical adsorption and chemical reaction. Likewise, calculated enthalpy deviations with pressure at constant temperature from density measurements are not so likely to agree with the calorimetric data as for non-polar compounds.

II. Non-Analytic ES

The reader may note that there is now a tendency to use some of the results consistent with scaling theory and the data in conjunction with an assumed analytic ES. Examples are (1) a non-analytic vapor pressure equation such as that of Goodwin or of Wagner and (2) m = 3,4 in constraint 5, above. While theoretically inconsistent, an analytic ES so constrained generally provides a better approximation in the critical region although incorrect at the CP. Two difficulties that arise in using non-analytic ES are that (1) they are not easily tractable and (2) they are not desirable outside the critical region. By tractable, we mean the ease with which one can derive an equation for a measureable quantity, such as the heat of vaporization, from a reasonably simple scaled ES, such as that due to Vicentini-Missoni, Levelt Sengers and Green (13), in the form of $\mu(\rho,T)$. One often derives an expression involving a power series as in Ref. 6 for the vapor pressure.

Attempts to blend analytic and non-analytic ES along an arbitrary critical region face the usual difficulty of blending different analytic equations--discontinuities are likely to occur in the second derivatives of the PVT surface (i.e., the heat capacity) plus the dissimilarity of the form of the two equations. A successful example is the blending of the Bender ES with Schofield's parametric scaled ES by Chapela and Rowlinson (14) for methane and carbon dioxide. One alternate procedure suggested by Levelt Sengers is to use the analytic ES as the background term and add on the necessary non-analytical terms in a manner such that they only contribute in the critical region. A second alternative is just the reverse--take the CP as reference and add analytic terms to the scaled ES as one moves away from the CP.

Literature Cited

1. Leung, S. S. and Griffiths, R. B., Phys. Rev. A (1973), 8, 2670.
2. Angus, S., "Guide to the Correlation of Experimental Thermodynamic Data on Fluids", IUPAC Thermodynamic Tables Project Centre, Department of Chemistry, Imperial College, London. (May 1975).
3. Bender, E., Kaltetechnik-Klimatisierung 23, 258 (Sept. 1971).
4. Mason, E. A. and Spurling, T. H., "The Virial Equation of State", p. 13, Pergamon Press, 1969.
5. Douslin, D. R. and Harrison, R. H., J. Chem. Thermodynamics (1973), 5, 491.
6. Hall, K. R. and Eubank, P. T., I&EC Fundam. (1976), 15, 323.
7. Keenan, J. H., Keyes, F. G., Hill, P. G., and Moore, J. G., "Steam Tables", Wiley and Sons, Inc., New York, 1969.
8. Levelt Sengers, J. M. H. and Greer, S. C., Int. J. Heat and Mass Transfer (1972), 15, 1865.
9. Baehr, H. D. and Schomaecker, H., Forsch. Ingenieurw. (1975) 41 (2), 43.
10. Baehr, H. D., Garnjost, H., and Pollak, R., J. Chem. Thermodynamics (1976), 8, 114.
11. Haar, L. and Gallagher, J. S., "Thermodynamic Properties of Ammonia", Nat. Bur. of Stds., Washington, D. C. (to be published).
12. Eubank, P. T., Chem. Engr. Sym. Ser. (1970), 66 (98), 16.

13. Vicentini-Missoni, M., Levelt Sengers, J. M. H., and Green, M. S., J. Res. Natl. Bur. Stand. (1969), 73A, 563.
14. Chapela, G. A. and Rowlinson, J. S., J. Chem. Soc. (Faraday Trans.I) (1974), 70, 584.

12

On the Development of an Equation of State for Vapor-Liquid Equilibrium Calculations

M. G. KESLER and B. I. LEE

Mobil Oil Corp., Princeton, NJ 08540

Mobil's VLE and related thermodynamic correlations find their use in the modeling of gas and crude reservoirs, natural gas plants, refinery units, as well as petrochemical plants. Because of the diversity and overlap of these needs, we favor the use of an equation of state (EOS) for most thermodynamic calculations in support of process modeling. Our recent experience in the development of the Lee-Kesler generalized correlation and in the use of a modified Soave-Redlich-Kwong equation (SRK) give us reason to believe that the above goal is attainable, at least to meet current needs. We would like to outline a few criteria that might be helpful in guiding future work in this area.

Desired Characteristics of an EOS for VLE Calculations

An important requirement of an EOS, to fill the needs of the petroleum industry, is that it apply to a wide range of hydrocarbons, from methane to components boiling above 1500F and to wide boiling mixtures of these components, including also H_2, H_2O N_2, O_2, CO_2, CO and H_2S. The EOS would need to cover temperature ranges of -250 to 1000F and pressures from vacuum to several thousand psi, correspondi to reduced temperatures (T_r) of 0.3 to 30 and to reduced pressures (P_r) of 0.01 to 10.

As a corollary of the above, the EOS should conform with the principle of corresponding states, that is, it should be generalizab The advantages of a generalizable form are in the ease of developing an EOS and of extending its application. A generalized form of the coefficients should also help the development of mixing rules. It is worth noting that the BWR EOS shows excessive and erroneous dependence on composition that can be traced to the specificity of the constants.

A good starting point for the development of an EOS for VLE application is reasonable representation of the compressibility. In this connection we have plotted in Figure 1, representation of the compressibility factor of n-octane for several isotherms, using the Lee-Kesler, Soave and Peng-Robinson (P-R) EOS's (1, 2, 3). As is

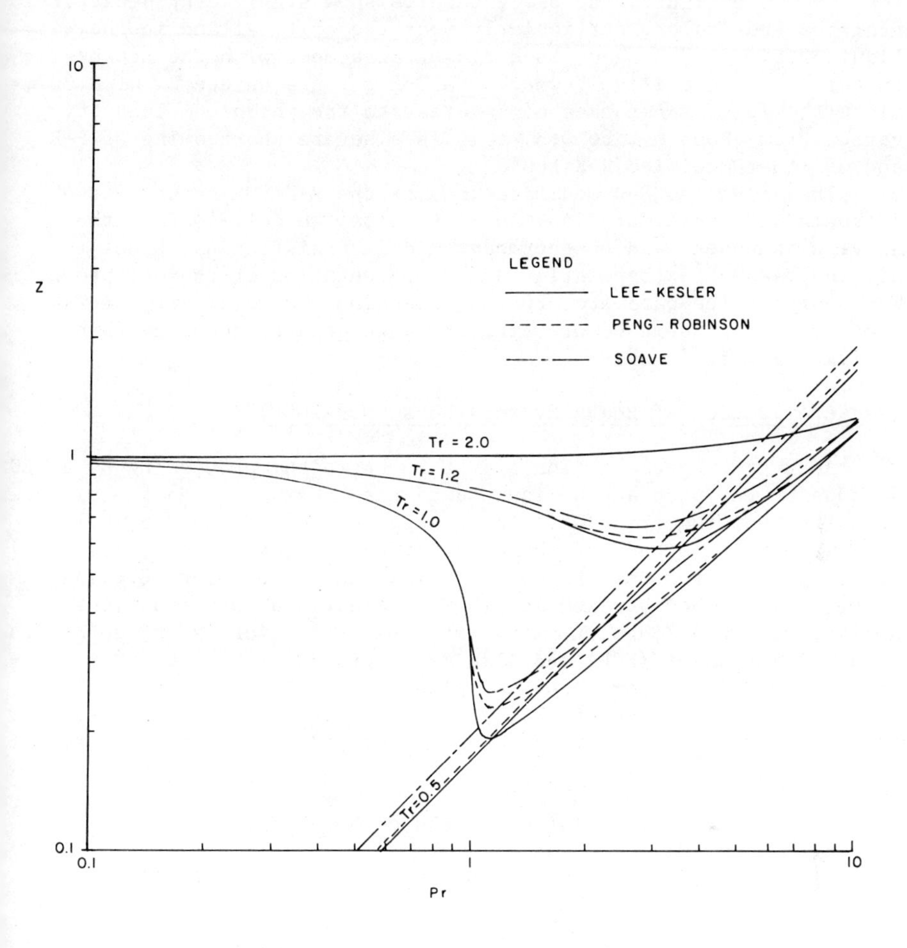

Figure 1. Compressibility factor comparison

seen from the figure, the Soave results show significant departure from the Lee-Kesler, particularly near the critical and in the liquid region. P-R shows much closer agreement with the Lee-Kesler except near the critical, where the P-R gives a unique Z_c of 0.307. Since the Z_c of substances of interest to the petroleum industry varies from about 0.2 to 0.3, this is a severe shortcoming of P-R and of other modified R-K EOS's.

The Soave and P-R modifications of the R-K EOS have remarkably improved VLE representation. Nevertheless, we believe that the above weaknesses lead to thermodynamic inconsistencies, hence to difficulties of extrapolation and representation of thermal properti We favor a three-parameter EOS, approaching the complexity of BWR, that would represent compressibility with greater accuracy than modified R-K EOS's do.

Characterization of Heavy Hydrocarbons (NBP>200F)

Accurate representation of VLE in petroleum processing by an EO depends on the accuracy of input data that characterizes the compone or narrow-boiling fraction. In this connection it would be helpful review current methods of characterizing a typical crude. The starting point for that is a crude assay shown in Figure 2. The crude, boiling between 100 and 1000F, is divided into fractions boiling within a range of, say, 25F. Each fraction is characterized by a boiling point (NBP) and a specific gravity (SG). Current practice is to derive from these two properties, the other properties needed for the EOS such as P_c, T_c, W, MW, etc. The current state-of the-art based on the corresponding states principle can be represent by:

(Correlations based on NBP and SG)

$$f(T_c, P_c, \omega) = f[T_c(NBP,SG),P_c(NBP,SG),\omega(NBP,SG)]=\emptyset(NBP,SG) \quad (1)$$

We believe that the above mapping can be improved, particularly for hydrocarbons derived from coal, shale, etc., by introducing a third parameter, such as viscosity. The proposed can be represented by:

(Correlations based on NBP, SG, and μ)

$$f(T_c,P_c,\omega) = f[T_c(NBP,SG,\mu),P_c(NBP,SG,\mu),\omega(NBP,SG,\mu)]=\emptyset(NBP,SG,\mu) \quad (2)$$

Two of the reasons for the suggested relation (2) are:

1. The ideal-vapor specific heats of naphthenes differ significantl from the specific heats of a mixture of n-paraffins and aromatic with the same NBP and SG as those of the naphthenes (4).
2. The viscosities of petroleum fractions with the same NBP and SG are different from cracked than for virgin stocks (5). In fact,

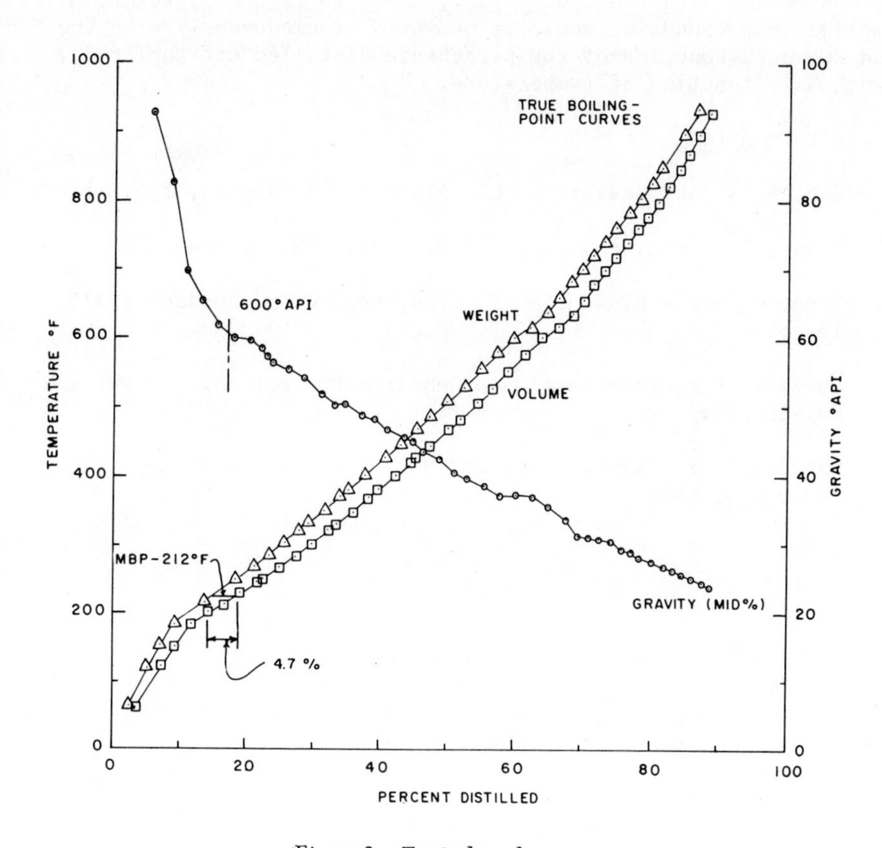

Figure 2. Typical crude assay

viscosity is suggested as a third parameter because it reflects well structural differences between substances.

In connection with the crude assay, shown in Figure 2, relation (2) implies that viscosity would be measured simultaneously with the NBP and SG, as a function of the percentage distilled off the crude, hence as a function of temperature.

Literature Cited

1. Lee, B. I. and Kesler, M. G., AIChE J., (1975) 21, 510.

2. Soave, G., Chem. Eng. Sci., (1972) 27, 1197.

3. Peng, D. and Robinson, O. B., Ind. Eng. Chem. Fundam. (1976) 15, 59.

4. American Petroleum Institute Research Project 44, Texas A & M University.

5. Watson, K. M., Wien, J. L. and Murphy, G. B., Ind. Eng. Chem. (1936) 28, 605.

13

Oil Recovery by CO_2 Injection

RALPH SIMON

Chevron Oil Field Research Co., LaHabra, CA 90631

In the U.S. (excluding Alaskan North Slope) ∿420 billion barrels of oil have been discovered (Figure 1). Of this amount 110 billion have already been produced and an additional 30 billion can be economically recovered at present prices as primary or secondary production. Of the ∿280 billion unrecoverable by primary and secondary methods only a fraction can be recovered. Published studies suggest that CO_2 might recover 5 to 10 billion barrels. This amount depends on economic parameters, the primary one being the price of crude oil.

The suitability of a specific reservoir for CO_2 injection can be estimated from phase behavior measurements, physical property calculations, and displacement tests in various porous media. Reservoirs that are candidates for CO_2 injection generally have pressures exceeding 1500 psia (100 atmospheres) and crude oils with specific gravity <0.87 (>30° API).

In 1976 there are 6 commercial CO_2 injection projects (1), and more are planned. It's doubtful that enough good-quality naturally occurring CO_2 can be found for all of them.

Displacement Mechanisms in CO_2--Reservoir Oil Systems

The physical phenomena that occur during CO_2 injection can be explained with the aid of a series of simplified network model drawings of a porous medium. Figure 2 represents a virgin reservoir containing oil as the continuous phase and connate water in isolated clusters of pores. Figure 3 shows the reservoir after a waterflood, indicating the residual oil not displaced by the water and the continuous water phase reaching from inlet to outlet. Figure 4 illustrates the invasion of CO_2 following the waterflood.

The advancing CO_2 dissolves in the residual reservoir oil, causing it to swell, decreasing its viscosity, and vaporizing the more volatile components in the oil. The advancing vapor phase is also able to enter and displace pores containing residual oil

	BILLION BARRELS
US DISCOVERIES*	420
PRODUCED TO DATE	110
PRODUCIBLE WITH EXISTING PROCESSES AND ECONOMICS	30
FROM CO_2	5–10

* EXCLUDING ALASKAN NORTH SLOPE

Figure 1

Figure 2. Oil (continuous) with connate water

Figure 3. Oil after waterflood

Figure 4. Oil with CO_2 injection

because of low interfacial tension between the advancing gas and residual liquid. The combination of these phenomena enables CO_2 to displace oil in either secondary or tertiary conditions.

Phase and Flow Behavior Characteristics of CO_2 --Reservoir Oil Systems

The mechanisms described in the previous section are expressed quantitatively in Figures 5 through 10.

Figure 5 shows CO_2 solubility in oil as a function of pressure and temperature. Note that at typical reservoir conditions e.g. 2,000 psia (135 atmospheres), and 170°F (77°C) the solution contains 65 mol percent CO_2. At these conditions the density of the

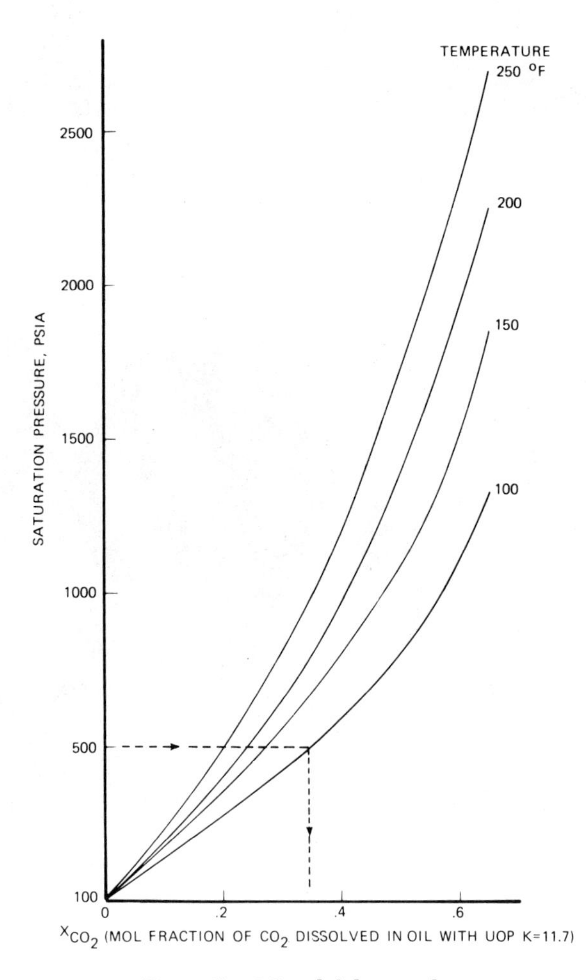

Figure 5. CO_2 solubility in oil

CO_2 and some reservoir oils are nearly equal, thus minimizing CO_2 gravity override.

Figure 6 shows oil phase volumetric expansion. With 65 mol percent CO_2 in solution the swelling can be 25 percent. As the oil swells it displaces adjacent oil toward the producing wells.

Figure 7 displays the viscosity of CO_2-crude oil mixtures. The data show that dissolved CO_2 can reduce viscosity as much as 100 fold. This improves the mobility ratio and decreases the tendency of CO_2 to finger in the high permeability paths. A complete discussion of Figures 5, 6, and 7 and related measurements is in Reference 2.

Figure 8 shows the two phase boundary and critical point for a CO_2-reservoir oil system and notes that a solid precipitated at

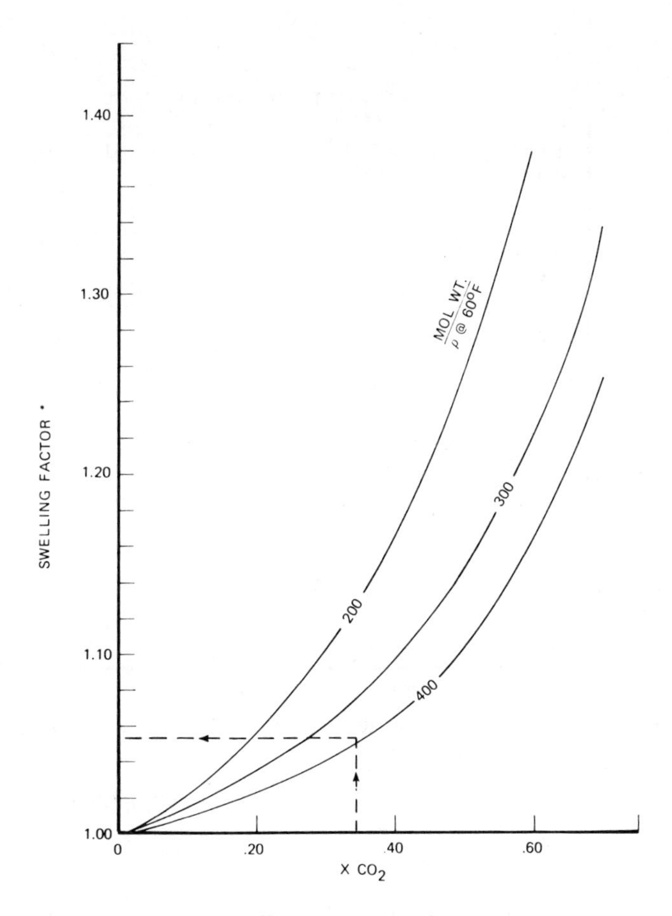

Figure 6. Swelling factor vs. mol fraction CO_2

$$*\frac{\text{vol @ sat'n press. \& temp}}{\text{vol @ 1 atm \& temp}}$$

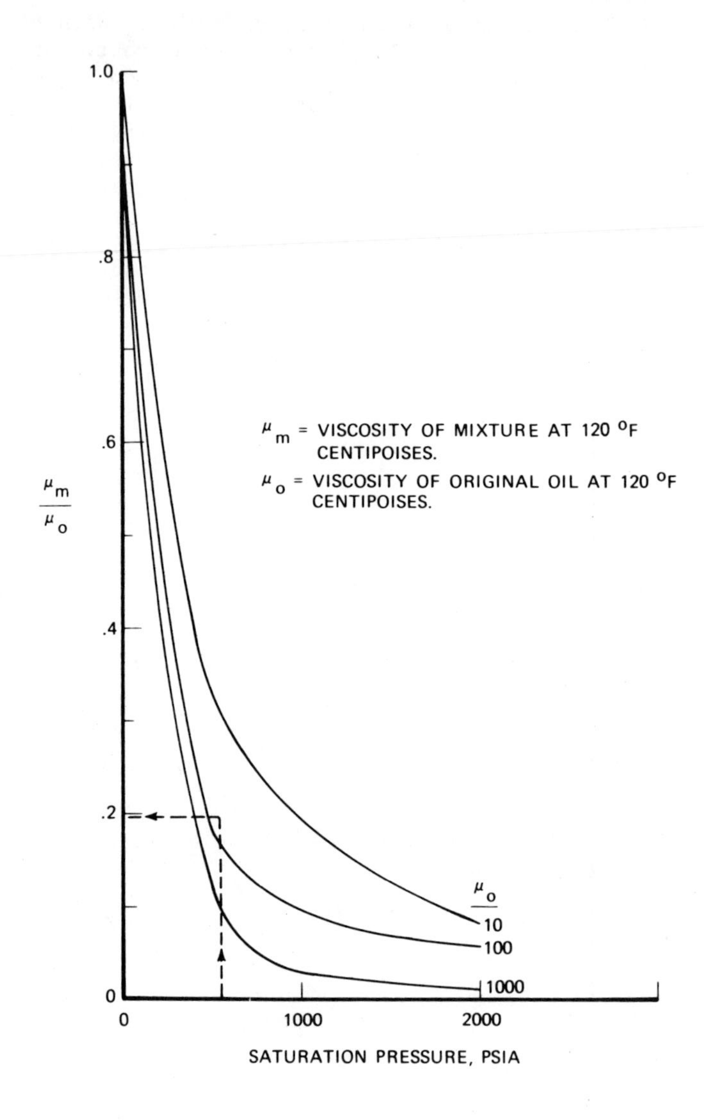

Figure 7. Viscosity of CO_2–crude oil mixtures at 120°F

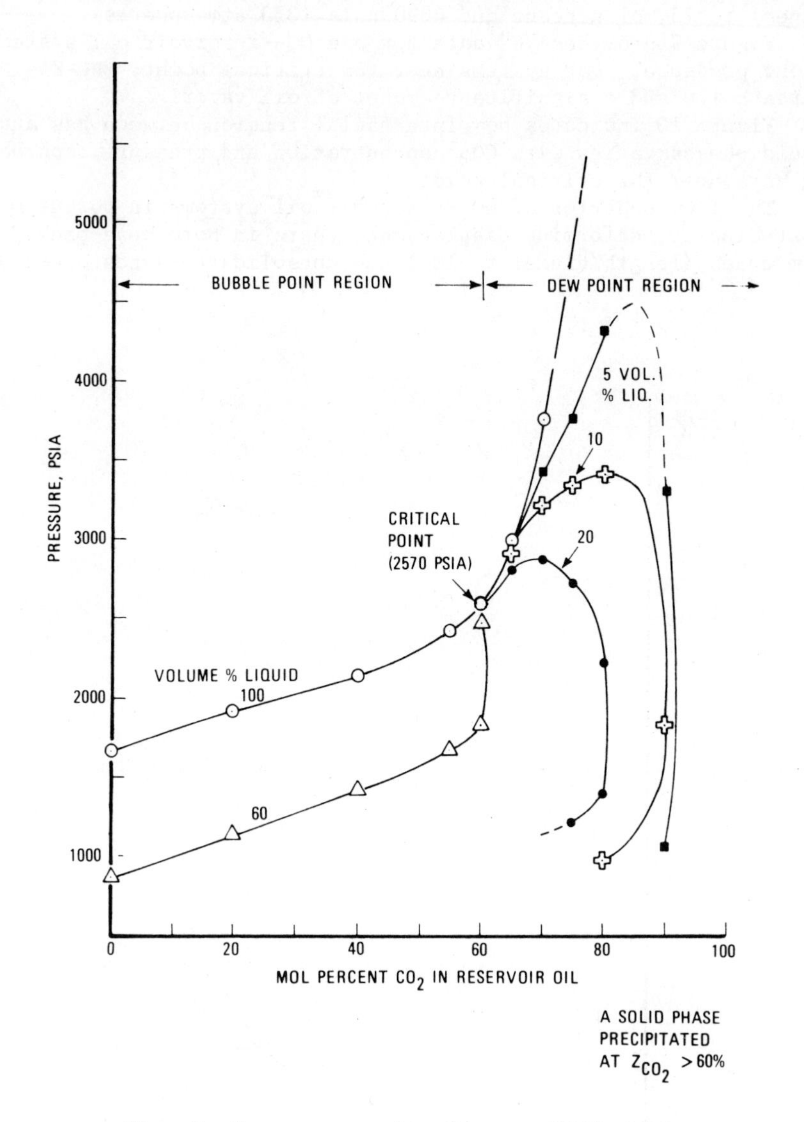

Figure 8. Pressure–composition diagram at 130°F: Oil A

CO_2 concentrations >60 mol percent. Critical points of systems studied ranged from 60 mol percent CO_2 and 2570 psia (175 atmospheres) to 75 mol percent and 4890 psia (330 atmospheres).

Figure 9 provides "K" data for one CO_2-reservoir oil system at one pressure. For systems near the critical point, the K's approach 1.0 and a significant amount of oil vaporizes.

Figure 10 indicates how interfacial tension between gas and liquid phases varies with CO_2 concentration and pressure, approaching zero near the critical point.

The flow behavior of CO_2-reservoir oil systems in porous media is studied by performing displacement tests in both bead-packed slim tubes (length/diameter >100) and consolidated sandstone and

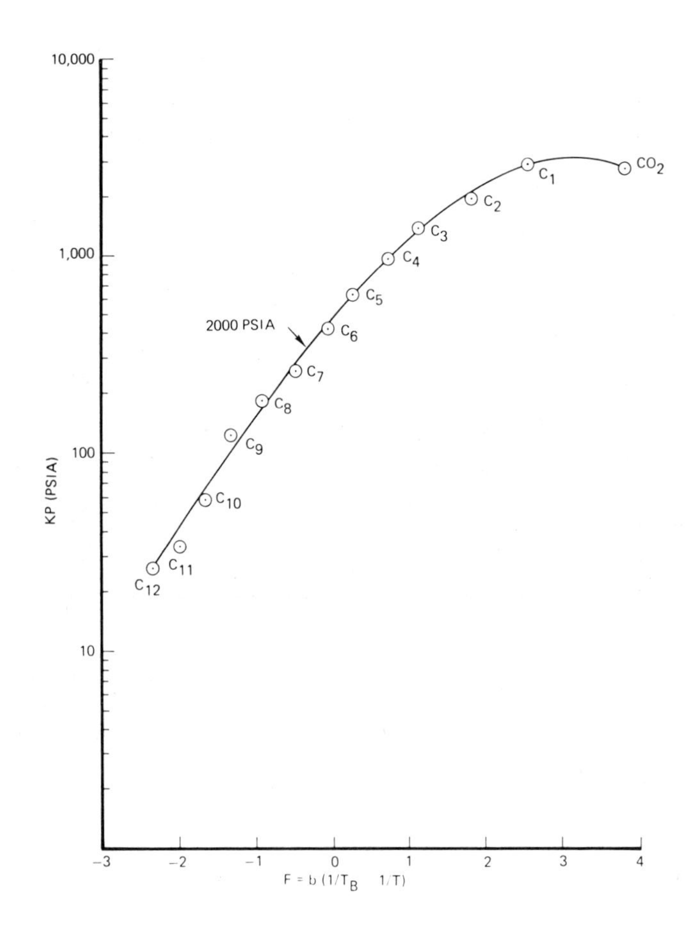

Figure 9. KP vs. F plot for 55 mol % CO_2, 45 mol % reservoir oil at 130°F

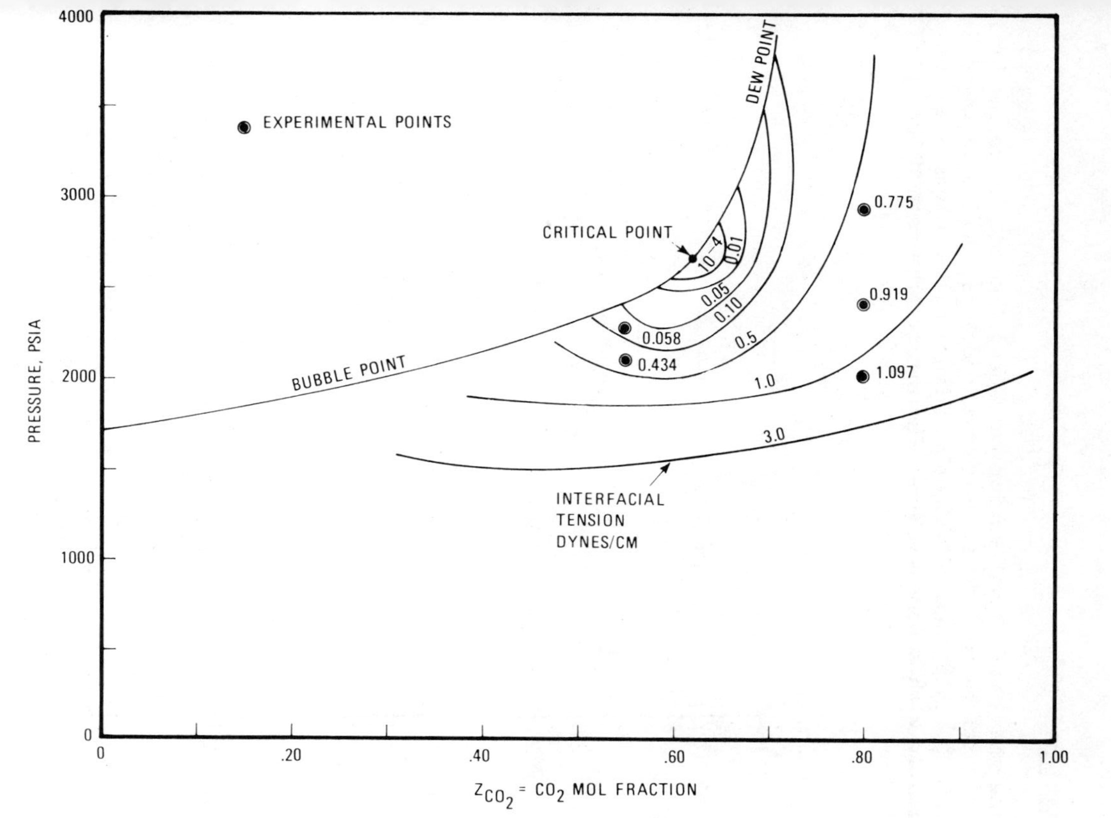

Figure 10. Calculated interfacial tension vs. pressure and CO_2 concentration: Oil A

limestone cores. The tubes are used to determine the optimum pressure for oil displacement (Figure 11 (3)); the cores to measure oil recovery versus pore volumes injected. Figure 12 indicates the potential of CO_2 to recover oil from a previously waterflooded reservoir.

Calculations

The phase and flow behavior data described above are the basis for calculations used to evaluate new projects, design approved ones, and operate them efficiently. These calculations are done principally with a Compositional Simulator. The main calculation

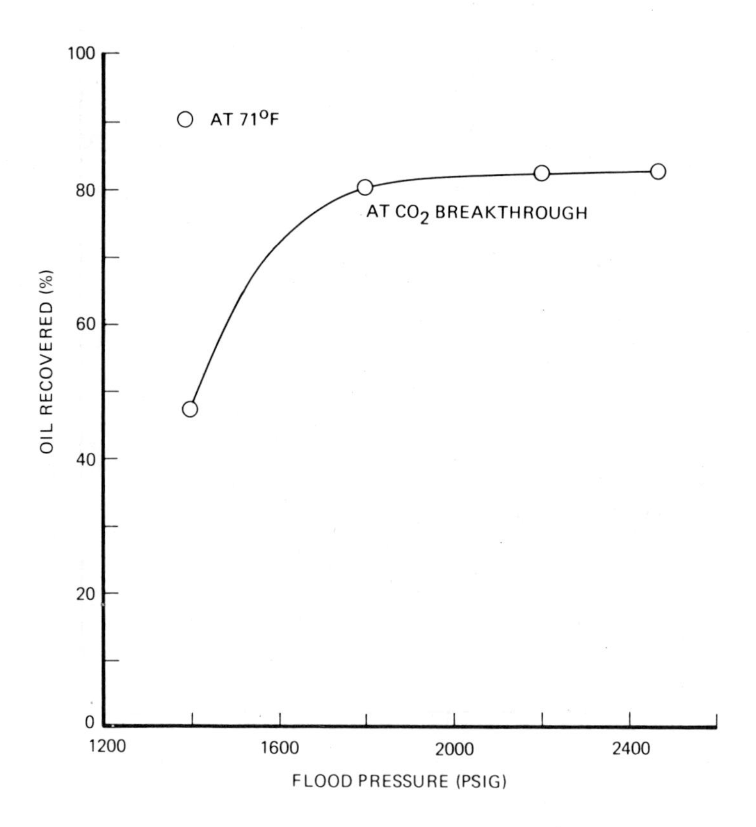

Figure 11. Oil recovered from CO_2 floods of 48-ft long sand pack

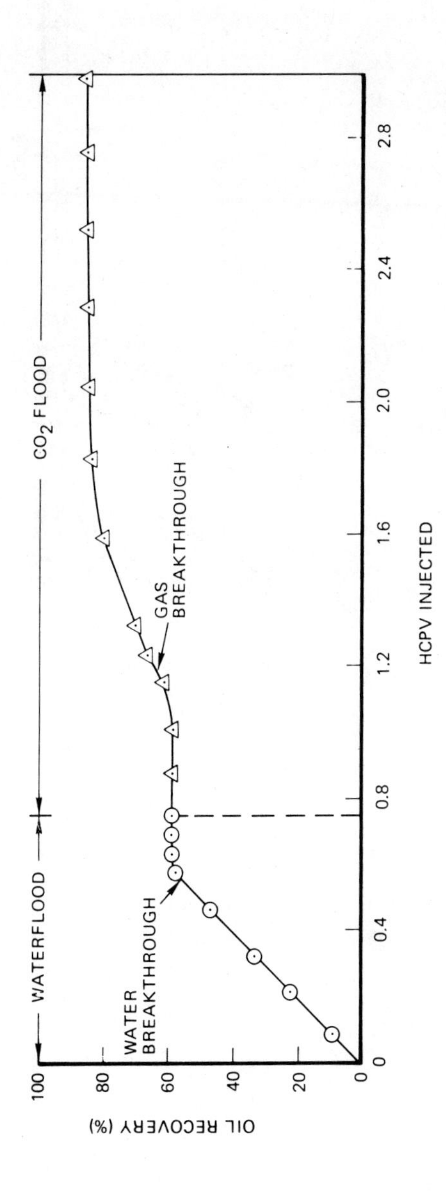

Figure 12. Oil recovery vs. HCPVI

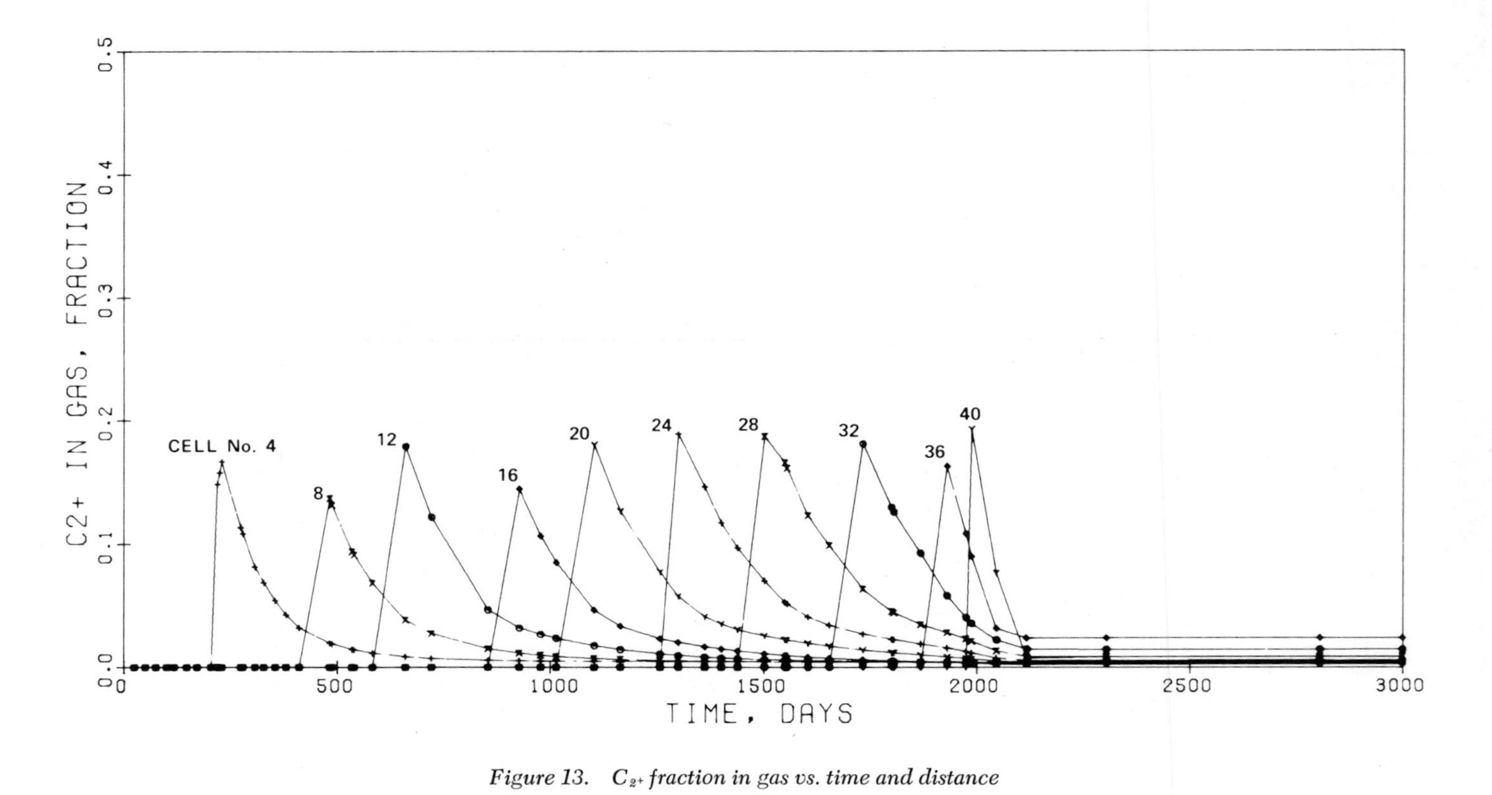

Figure 13. C_{2^+} *fraction in gas vs. time and distance*

is a performance prediction that shows recovery versus time for a specified number and location of wells, injection rate, reservoir pressure and temperature, etc.

Compositional Simulator calculations are also used to help understand the effects of mass transfer as CO_2 advances through the reservoir and displaces oil. Two examples:

Figure 13 shows the C_2+ in the gas phase versus time and distance. Note how the gas phase enriched with 20 mol percent C_2+ moves through the reservoir.

Figure 14 displays interfacial tension versus time and distance. Here a low (essentially zero) interfacial tension front is shown moving through the reservoir. Under these conditions CO_2 provides a miscible displacement.

Research

In recent years there have been significant advances in obtaining data on CO_2-reservoir oil systems. However, more data are needed in order to evaluate, design, and operate individual CO_2 injection projects without first making extensive, costly, and time-consuming experimental studies. Additional information is needed primarily in two categories:

1. Predicting the properties of two-phase mixtures--specifically, more reliable K correlations and interfacial tension equations.
2. Understanding mass transfer and predicting its effect versus time and distance as a CO_2 front moves through the reservoir.

Abstract

Various authors have estimated that five to ten billion barrels of U.S. crude oil (excluding Alaskan North Slope) may be potentially recoverable by CO_2 injection. This recovery will result when CO_2 dissolves in the oil, swells it, reduces its viscosity and lowers interfacial tension between the injected gas and the residual oil thus enhancing oil displacement. Also CO_2 vaporizes the volatile part of the oil, carrying it from the reservoir in the gas phase.

Physical property correlations and displacement tests with CO_2-reservoir oil systems are needed to calculate the effect of CO_2 injection. The calculations are used in evaluating new projects, designing approved ones, and operating them efficiently.

Research is needed to improve the reliability of present calculation methods, particularly for predicting the properties of two-phase mixtures.

Literature Cited

(1) Oil & Gas Journal, April 5 (1976).

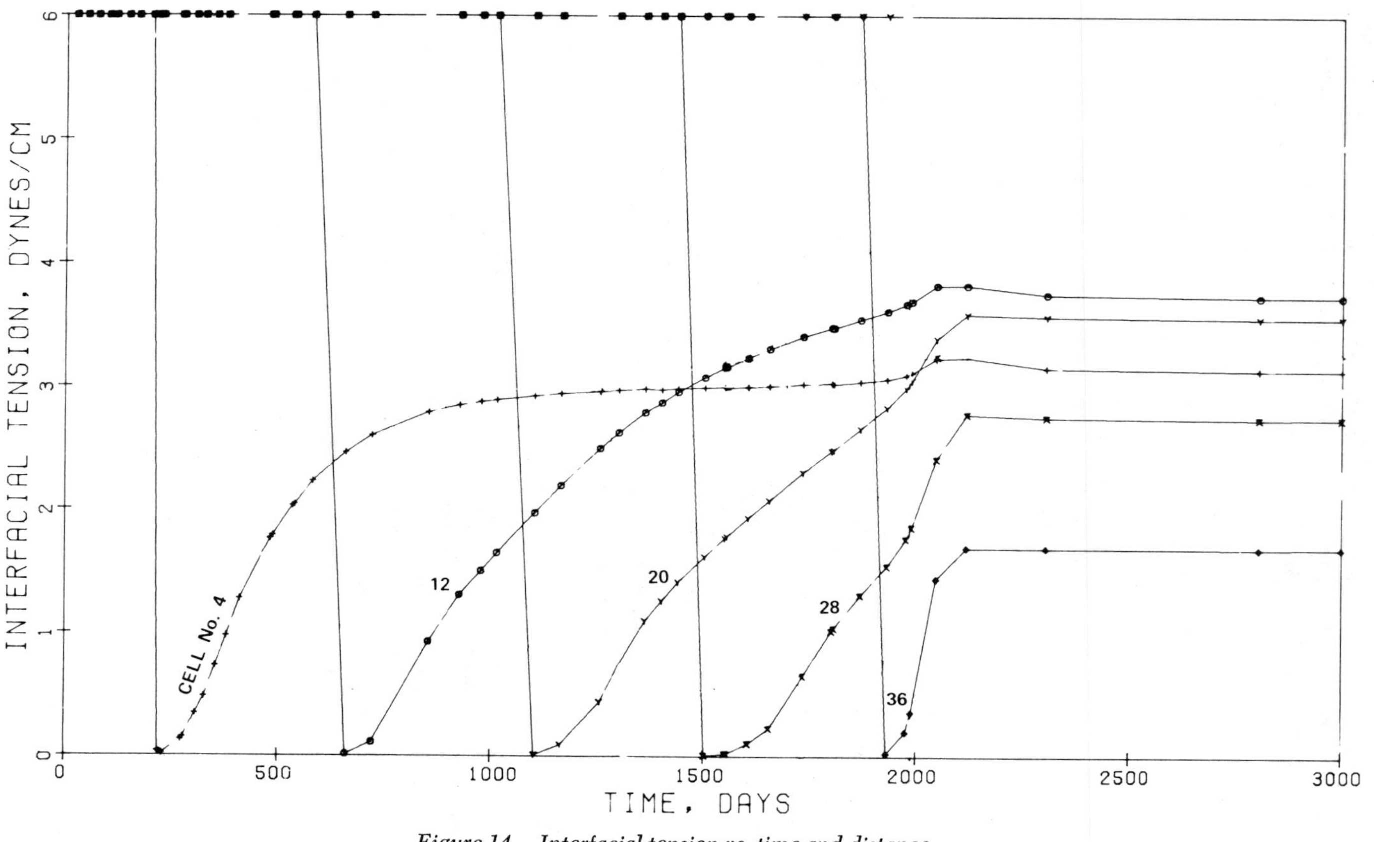

Figure 14. Interfacial tension vs. time and distance

(2) Simon, R., and Graue, D. J., J. Pet. Tech., (1965).
(3) Holm, L. W., and Josendahl, V. A., J. Pet. Tech., (1974).

14

A Computer Model for Calculating Physical and Thermodynamic Properties of Synthetic Gas Process Streams

GRANT M. WILSON and MARK D. WEINER

Brigham Young University, Provo, UT 84602

This paper describes our progress to date on development of a computer model based on a thermodynamically-consistent correlation that will accurately and reliably predict enthalpies, entropies, densities and ultimately K-values for synthetic gas systems over the range of typical synthetic gas processing conditions.

A review of various gasification processes is given in Table This table shows that process operating temperatures range from 550° to 3000°F. The gasification of liquid hydrocarbons is done at temperatures ranging from 550°F to 1000°F, while coal gasification processes operate at temperatures from 500°F to 3000°F. Gasification pressures range from ambient to 1500 psia, and may be extended to 4000 psia.

The principal chemical reactions of these processes can be characterized by the following equations.

$$C_xH_y + H_2O \rightarrow H_2 + CO \quad (1)$$

$$CO + H_2O \rightarrow H_2 + CO_2 \quad (2)$$

$$CO + 3H_2 \rightarrow CH_4 + H_2O \quad (3)$$

In some cases hydrogen is produced by reactions 1 and 2, and then the feedstock is hydrogenated to produce gas and oil. High operating temperatures are achieved either by pre-heating the reactants or by partial combustion using oxygen or air. Reactor products and by-products therefore consist of the following compounds.

Hydrogen	Argon
Carbon monoxide	Hydrogen sulfide
Carbon dioxide	Carbonyl sulfide

Table 1, Process Conditions of Various Gasification Methods

Process (5)	Feedstock	Temp. ^{o}F	Pressure, psia	Product
CRG Hydrogasification	liquid hyd.	600-1000	375	97% CH_4
Gasynthan	liquid hyd.	550-750	300-500	98% CH_4
JGC Methane-Rich Gas (MRG)	liquid hyd.	not given	not given	98% CH_4
Shell Gasification	liquid hyd.	not given	not given	50% H_2, 45% CO
Agglomerating Burner Gasification	coal	1600-2100	100-400	H_2/CO
Bi-Gas	coal	1700-3000	1000-1500	SNG, 900 Btu/ft^3
CO_2 Acceptor	coal	500-1870	150	SNG, 950 + Btu/ft^3
Coal Solution Gasification	coal	not given	not given	SNG, 1000 + Btu/ft^3
Hydrane	coal	1450-1800	1000	95% CH_4
Lurgi-SNG/Coal	coal	600-1850	1500	SNG, 950 + Btu/ft^3
Molten Salt Coal Gasification	coal or coke	1700	1200	SNG, 950 Btu/ft^3
Koppers-Totzek	coal or coke	2900	not given	H_2/CO
U-Gas	coal	700-1900	300-350	H_2/CO
Winkler Gasification	coal	2000	not given	H_2/CO

Water	Ammonia
Methane	Light hydrocarbons
Nitrogen	Heavy hydrocarbons
Oxygen	

The prediction and correlation of energy requirements, heat exchanger duty, chemical equilibria, and separation equilibria requires a knowledge of the enthalpy, compressibility, fugacity, and vapor-liquid equilibrium properties of these compounds and their mixtures.

Existing prediction methods apply primarily to hydrocarbon mixtures at relatively low temperatures and their application to mixtures containing significant concentrations of water, CO_2, H_2S, and COS at high temperatures is questionable. Thus, accurate prediction methods are needed either by modification of existing methods or by development of new methods.

A discussion of various alternatives for development of a new prediction method is given in the next section of this report.

Possibilities for New Prediction Methods

Equation-of-state methods appear to be the most likely candidates for reliable accurate data. They are capable of predicting enthalpy, entropy, density, fugacity, vapor-liquid, and liquid-liquid data from one equation in regions where both low and high densities are encountered. Rapid computation requires that the equation be simple yet accurate without computational difficulties in areas surrounding the critical point.

Possible candidates for this purpose are the following.

1. Modify existing correlations based on the Redlich-Kwong equation of state.
   a) Mark V (P-V-T, Inc., Houston, Texas)
   b) Soave (1) method
2. Adapt the BWR equation
3. Develop new equations of state

Correlations based on the Redlich-Kwong equation of state have a "built in" volume-dependence limitation. The equation has the right form for non-polar compounds but not for polar compounds. For example, if one correlates the solubility of CO_2 in water at $77^{\circ}F$; then one would expect that the correlation should also predict the solubility of water in CO_2. This is not the case, since the predicted solubility of water in CO_2 at liquid-liquid saturation is 2.1 mole percent instead of the measured value of 0.25 mole percent (2). This represents a factor-of-eight prediction error. No amount of change of the temperature parameters wil correct this error because it is related to the volume dependence of the equation. Since water is a principal component of syntheti gas processes, an accurate prediction of this mixture seems essential.

The BWR equation has not been adequately tested for predictin the properties of polar and non-polar compounds. The author has some doubts whether it would be suitable. A second disadvantage is that the BWR equation has so many constants that it takes about five times the computer time required by the Redlich-Kwong equation.

What is needed then is a new equation comparable in computation speed to existing Redlich-Kwong correlations, but with the capability of correlating the properties of both polar and non-polar mixtures. A new modified Van der Waals equation which meets these requirements is described in the next section of this report.

A New Modified Van der Waals Equation for Both Polar and Non-Polar Compounds and Their Mixtures

Van Laar (3,4) used the Van der Waals equation to derive the now widely used Van Laar equation for calculating activity coefficients in non-ideal mixtures. His derived equation is as follows.

$$\ell n\ \gamma_1 = \left(\frac{x_2 b_2}{x_j b_1 + x_2 b_2}\right)^2 b_1 a' \qquad (1)$$

$$\ell n\ \gamma_2 = \left(\frac{x_1 b_1}{x_1 b_1 + x_2 b_2}\right)^2 b_2 a' \qquad (2)$$

where b = Van der Waals b

$$a' = \frac{1}{RT}\left(\frac{a_1}{b_1} - \frac{a_2}{b_2}\right)^2$$

a_1, a_2 = Van der Waals a

As derived, the equation gave only approximate results; but it was found that by making the b's and a's empirical parameters a large number of mixtures could be accurately correlated. Of course, by this method the parameters lose their physical significance as parameters in the Van der Waals equation. Nevertheless, the equation has proved to be a very useful equation, and is now more appropriately written in the following form.

$$\ell n\ \gamma_1 = \left(\frac{x_2 S_2}{x_1 S_1 + x_2 S_2}\right)^2 S_1 B_{12} \qquad (3)$$

$$\ell n\ \gamma_2 = \left(\frac{x_1 S_1}{x_1 S_1 + x_2 S_2}\right)^2 S_2 B_{12} \qquad (4)$$

where S, B = empirical parameters.

If the original Van Laar equation came from the Van der Waals equation, then what equation of state corresponds to the empirical Van Laar equation? If this equation of state were known, one would have an equation of state with empirical parameters to adjus for assymetric non-ideal behavior in mixtures.

The authors believe that such an equation is derivable by assuming that void spaces in a fluid can be considered as additional component of a mixture. When this method is used with the empirical Van Laar equation, then the following modified Van der Waals (M-VDW) equation is produced. (See the appendix for details.)

$$\frac{P\overline{V}}{RT} = \frac{\overline{V}}{\overline{V} - b} - \frac{\overline{V} \sum_j \sum_k x_j S_j x_k S_k (A_{jk}/RT)}{(\overline{V} + S - b)^2} \tag{5}$$

where b = molecular-volume parameter analogous to Van der Waals b

A_{jk} = energy parameter analogous to Van der Waals A_{jk}

S = symmetry parameter from the empirical Van Laar equation

$B = \sum_i x_i b_i$

$S = \sum_i x_i S_i$

This equation reduces to the form of the Van der Waals equation when S equals b. Thus the model is in agreement with the original Van Laar equation when S = b; this confirms the method by which it was derived. The parameters S_j and A_{jj}/RT have been assumed to be temperature dependent as follows.

$$\ell n(S_j/b_j) = C_j \left(\frac{Tc_j}{T}\right) \tag{6}$$

$$S_j A_{jj}/RT = \alpha + \beta \left(\frac{Tc_j}{T}\right) + \gamma \left(\frac{Tc_j}{T}\right)^2 + \delta \left(\frac{Tc_j}{T}\right)^3 + \sigma \left(\frac{Tc_j}{T}\right)^4 \tag{7}$$

Equation (6) is significant because it shows that as $T \to \infty$ then $S_j \to b_j$. This means that the equation approaches the Van der Waals form at high temperatures. The parameter S can be considered to be the effective volume of a molecule resulting from interaction forces between the molecules. The trajectory of a molecule passing a central molecule is changed as a result of molecular interactions, and passing molecules thus collide more frequently with the central molecule than would otherwise be expected. When this happens, the effective size of the molecule is larger than its actual size. At high temperatures, the molecular interactions become small compared to the kinetic energy of the molecules and the trajectory of a molecule is negligibly altered. When this happens, the effective size of a molecule is the same as the actual size of the molecule.

The parameter A_{jj} has the units of energy per unit volume and is analogous to the square of the solubility parameter in the Scatchard-Hildebrand equation. Numerous tests of the Scatchard-Hilderbrand equation have shown that there is significance to the solubility parameter. Previously, there has been no way to interpret this type of parameter from an equation of state. This new M-VDW equation could provide this opportunity.

Preliminary studies show that the M-VDW equation can be adjusted to accurately predict the physical properties of water, hydrogen, carbon dioxide and methane. Comparisons with experimental data are given in Tables 2 to 21. Tables 2 to 5 give comparisons between experimental and predicted liquid molar volume, vapor pressure, and saturated vapor compressibility factor data for these compounds. These are preliminary results, nevertheless the agreement between experimental and predicted data from the M-VDW equation are quite good. With some exceptions, liquid molar volumes are predicted within about ±2%. This could be improved by assuming a small temperature dependence of the b parameter. Vapor pressure data appear to be predicted to better than ±1%, and saturated vapor compressibility factors are accurately predicted except at conditions close to the critical temperature where this property changes quite drastically with only small changes in temperature. The prediction of these properties presumably can be improved by adjusting either the A or S parameters. One important conclusion from these comparisons is that water properties can be predicted with accuracy comparable to other non-polar compounds. This result is important in determining the suitability of the M-VDW for correlating and predicting the properties of mixtures containing polar and non-polar compounds.

Tables 2 to 5 compare properties at temperatures below the critical temperature of the compounds. Tables 6 to 9 extend the comparisons to temperatures above the critical temperatures. The tables compare experimental and predicted compressibility factor data. Deviations between experimental and predicted data are small except in regions near the critical point. All equations of state tend to deviate in this region, and the authors believe the accuracy is comparable to predictions from other equations of state.

From past experience, the authors have found that if other physical properties such as vapor pressure and compressibility factor are accurately predicted; then enthalpy will be accurately predicted. This was found to be so in this case as is shown in Tables 10 to 15. Tables 20, 11, and 12 compare published and calculated enthalpy data in the saturation regions for water, CO_2

Table 2, WATER, Comparison of Predicted and Experimental Liquid Molar Volume, Vapor Pressure, and Saturated Vapor Compressibility Factor

	Liquid Molar Volume ft^3/lb-mole			Vapor Pressure psia			Saturated Vapor Compressibility Factor		
Temperature	Exp[6]	M-VDW	%Diff	Exp[6]	M-VDW	%Diff	Exp[6]	M-VDW	%Diff
100°F	.2906	.3026	4.13	.949	.948	-0.11	.9987	.9978	-0.09
200	.3000	.2993	-0.23	11.53	11.50	-0.30	.9881	.9872	-0.09
300	.3144	.3082	-1.97	67.01	67.46	0.67	.9574	.9573	-0.01
400	.3358	.3280	-2.32	247.3	250.07	1.12	.8986	.9012	0.29
500	.3675	.3623	-1.41	680.8	686.26	0.80	.8010	.8143	1.66
600	.4252	.4231	-0.49	1543	1538.07	-0.32	.6468	.6914	6.90

Table 3, NORMAL HYDROGEN, Comparison of Predicted and Experimental Liquid Molar Volume, Vapor Pressure, and Saturated Vapor Compressibility Factor

	Liquid Molar Volume ft^3/lb-mole			Vapor Pressure psia			Saturated Vapor Compressibility Factor		
Temperature	Exp (7)	M-VDW	%Diff	Exp (7)	M-VDW	%Diff	Exp (7)	M-VDW	%Diff
-430.87°F	.4278	.4059	-5.12	2.966	2.966	0.00	.9678	.9662	-0.17
-423.67	.4521	.4326	-4.31	13.07	13.04	-0.23	.9128	.9124	-0.04
-416.47	.4876	.4727	-3.06	37.41	37.50	0.24	.8278	.8304	0.31
-409.27	.5454	.5400	-0.99	83.69	84.36	0.80	.7077	.7164	1.23
-402.07	.6920	.6902	-0.26	160.67	161.45	0.49	.5161	.5446	5.52

Table 4, CARBON DIOXIDE, Comparison of Predicted and Experimental Liquid Molar Volume, Vapor Pressure, and Saturated Compressibility Factor

	Liquid Molar Volume ft^3/lb-mole			Vapor Pressure psia			Saturated Vapor Compressibility Factor		
Temperature	Exp[(8)]	M-VDW	%Diff	Exp[(8)]	M-VDW	%Diff	Exp[(8)]	M-VDW	%Diff
$-69.6^{o}F$	.5985	.5887	-.164	75.15	74.7	-0.60	.9142	.9175	0.36
-40	.6324	.6253	-1.12	145.87	147.2	0.91	.8713	.8698	-0.17
0	.6914	.6916	0.03	305.8	308.2	0.78	.7923	.7865	-0.73
40	.7860	.7949	1.13	567.3	561.9	-0.95	.6714	.6796	1.22

Table 5, METHANE, Comparison of Predicted and Experimental Liquid Molar Volume, Vapor Pressure, and Saturated Vapor Compressibility Factor

	Liquid Molar Volume ft^3/lb-mole			Vapor Pressure psia			Saturated Vapor Compressibility Factor		
Temperature	Exp(9)	M-VDW	%Diff	Exp(9)	M-VDW	%Diff	Exp(9)	M-VDW	%Diff
-280°F	.5839	.5795	-0.75	4.90	4.90	0.00	.980	.9817	0.20
-250	.6160	.6098	-1.01	21.71	21.56	-0.69	.946	.9483	0.21
-220	.6561	.6524	-0.56	64.5	64.65	0.23	.898	.8933	-0.56
-190	.7107	.7144	0.52	150.0	151.83	1.22	.819	.8142	-0.61
-160	.7925	.8135	2.65	297.0	301.91	1.65	.713	.7082	-0.70
-130	.9625	1.0142	5.37	527.0	534.54	1.43	.554	.5564	0.36

Table 6, WATER, Comparison of Predicted and Experimental Compressibility Data in Super Heat Region Data Obtained From (6)

700°F a)				1200°F			
PSIA	EXP	M-VDW	%DIFF	PSIA	EXP	M-VDW	%DIFF
500	.9443	.9422	-0.22	500	.9866	.9865	-0.01
1000	.8809	.8801	-0.09	1000	.9727	.9733	0.06
1600	.7915	.7969	0.68	1600	.9560	.9561	0.01
2000	.7207	.7341	1.86	2000	.9445	.9452	0.07
2500	.6103	.6400	4.87	2500	.9302	.9315	0.14
3000	.4274	.5000	16.99	3000	.9159	.9179	0.22
4000	.1662	.1582	-4.81	4000	.8870	.8910	0.45
5000	.1940	.1834	-5.46	5000	.8579	.8644	0.76
900°F				1600°F			
PSIA	EXP	M-VDW	%DIFF	PSIA	EXP	M-VDW	%DIFF
500	.9703	.9704	0.01	500	.9953	.9946	-0.07
1000	.9390	.9382	-0.09	1000	.9901	.9894	-0.07
1600	.8996	.8996	0.00	1600	.9841	.9832	-0.09
2000	.8723	.8732	0.10	2000	.9800	.9792	-0.08
2500	.8366	.8395	0.35	2500	.9750	.9743	-0.07
3000	.7998	.8035	0.46	3000	.9699	.9695	-0.04
4000	.7221	.7289	0.94	4000	.9596	.9587	-0.09
5000	.6397	.6488	1.42	5000	.9493	.9492	-0.01

a) This is slightly below the critical temperature of 706°F.

Table 7, NORMAL HYDROGEN, Comparison of Predicted and Experimental Compressibility Data in Super Heat Region Data Obtained From (7)

-398.47°F [a]				-351.67°F			
PSIA	EXP	M-VDW	%DIFF	PSIA	EXP	M-VDW	%DIFF
17.75	.9708	.9729	0.22	19.24	.9941	.9948	0.07
150.93	.6882	.7056	2.53	183.11	.9462	.9514	0.55
221.91	.2890	.2784	-3.67	454.69	.8808	.8913	1.19
443.52	.3913	.3872	-1.05	876.18	.8488	.8582	1.11
833.26	.6512	.6302	-3.22	1540.14	.9552	.9582	0.31
-391.27°F				-189.67°F			
PSIA	EXP	M-VDW	%DIFF	PSIA	EXP	M-VDW	%DIFF
19.94	.9757	.9777	0.20	80.90	1.003	1.0028	-0.02
159.45	.7803	.7935	1.69	1385.83	1.074	1.0668	-0.67
235.87	.6415	.6635	3.43	2192.64	1.132	1.1221	-0.87
368.28	.4292	.4360	1.58	3640.20	1.254	1.2398	-1.13
612.24	.5167	.5122	-0.87	5894.57	1.462	1.4552	-0.47
-380.47°F				80.33°F			
PSIA	EXP	M-VDW	%DIFF	PSIA	EXP	M-VDW	%DIFF
14.04	.9887	.9897	0.10	162.39	1.007	1.0059	-0.11
163.27	.8623	.8726	1.19	3311.01	1.140	1.1325	-0.66
310.38	.7287	.7465	2.44	6126.76	1.266	1.2596	-0.51
467.33	.6174	.63.26	2.46	9005.71	1.396	1.3989	0.21
747.00	.6315	.6347	0.51	12526.87	1.553	1.5748	1.40

a) This is slightly above the critical point at -399.93°F.

Table 8, CARBON DIOXIDE, Comparison of Predicted and Experimental Compressibility Data in Super Heat Region Data Obtained From (8)

100°F a)				400°F			
PSIA	EXP	M-VDW	%DIFF	PSIA	EXP	M-VDW	%DIFF
200	.9260	.9386	1.36	200	.9885	.9886	0.01
400	.8471	.8717	2.90	400	.9763	.9775	0.12
600	.7585	.7961	4.96	600	.9647	.9658	0.11
800	.6695	.7080	5.75	800	.9534	.9547	0.14
1000	.5789	.5921	2.28	1000	.9427	.9439	0.13
140°F				600°F			
PSIA	EXP	M-VDW	%DIFF	PSIA	EXP	M-VDW	%DIFF
200	.9406	.9522	1.23	200	.9955	.9966	0.11
400	.8727	.9016	3.31	400	.9912	.9934	0.22
600	.8043	.8477	5.40	600	.9865	.9904	0.40
800	.7337	.7882	7.43	800	.9822	.9875	0.54
1000	.6853	.7230	5.50	1000	.9777	.9848	0.73
240°F				1000°F			
PSIA	EXP	M-VDW	%DIFF	PSIA	EXP	M-VDW	%DIFF
200	.9718	.9737	0.20	200	.9992	1.0020	0.28
400	.9433	.9455	0.23	400	.9986	1.0042	0.56
600	.9145	.9175	0.33	600	.9979	1.0063	0.84
800	.8859	.8891	0.36	800	.9972	1.0085	1.13
1000	.8564	.8605	0.48	1000	.9964	1.0108	1.45

a) This is slightly above the critical temperature at 88°F.

Table 9, METHANE, Comparison of Predicted and Experimental Compressibility Data in Super Heat Region Data Obtained From (9)

-100°F				500°F			
PSIA	EXP	M-VDW	%DIFF	PSIA	EXP	M-VDW	%DIFF
100	.9567	.9567	---	100	1.0016	1.0000	-0.16
300	.8588	.8620	0.37	300	1.0023	1.0030	0.07
600	.6723	.6820	1.44	600	1.0047	1.0064	0.17
1000	.3217	.3045	-5.35	1000	1.0095	1.0115	0.20
2000	.4646	.4350	-6.37	2000	1.0252	1.0275	0.22
-60°F				1000°F			
PSIA	EXP	M-VDW	%DIFF	PSIA	EXP	M-VDW	%DIFF
100	.9694	.9706	0.12	100	1.0016	1.0021	0.05
300	.9026	.9061	0.39	300	1.0062	1.0065	0.03
600	.7924	.8009	1.07	600	1.0126	1.0131	0.05
1000	.6250	.6378	2.05	1000	1.0215	1.0222	0.07
2000	.5206	.4931	-5.28	2000	1.0440	1.0438	-0.02
0°F				1500°F			
PSIA	EXP	M-VDW	%DIFF	PSIA	EXP	M-VDW	%DIFF
100	.9808	.9818	0.10	100	1.0023	1.0018	-0.05
300	.9405	.9439	0.36	300	1.0062	1.0056	-0.06
600	.8788	.8867	0.90	600	1.0124	1.0112	-0.12
1000	.7974	.8097	1.54	1000	1.0206	1.0188	-0.18
2000	.6777	.6696	-1.20	2000	1.0420	1.0371	-0.47
100°F				2000°F			
PSIA	EXP	M-VDW	%DIFF	PSIA	EXP	M-VDW	%DIFF
100	.9906	.9915	0.09	100	1.0022	1.0012	-0.10
300	.9711	.9752	0.42	300	1.0057	1.0038	-0.19
600	.9440	.9502	0.66	600	1.0115	1.0076	-0.39
1000	.9108	.9200	1.01	1000	1.0192	1.0192	-0.64
2000	.8584	.8639	0.64	2000	1.0378	1.0260	-1.14

Table 10

Comparison of Calculated Water Enthalpy

Data with Literature Data in the Saturation Region (10)

Temp °F	Pressure PSIA	Enthalpy, BTU/LB					
			LIQUID			VAPOR	
		Calc	Lit	Diff	Calc	Lit	Diff
32.02	.089	40.8	0	40.8	1080.5	1075.5	5.0
213.03[a]	15.0	181.2	181.2	--	1156.6	1150.9	5.7
320.28	90.0	292.2	290.7	1.5	1191.8	1185.3	6.5
377.53	190.0	354.7	350.9	3.8	1204.9	1197.6	7.3
414.25	290.0	395.9	390.6	5.3	1210.8	1202.6	8.2
444.60	400.0	430.6	424.2	6.4	1213.8	1204.6	9.2
503.08	700.0	500.0	491.6	8.4	1214.4	1201.8	12.6
567.19	1200.0	581.4	571.9	9.5	1205.3	1184.8	20.5
613.13	1700.0	645.1	636.5	8.6	1190.5	1158.6	31.9
662.11	2400.00	722.0	719.0	3.0	1163.2	1103.7	59.5

a) Enthalpy base adjusted to fit the liquid enthalpy at 213.03°F

Table 11

Comparison of Calculated Carbon Dioxide Enthalpy

Data with Literature[1] Data in the Saturation Region (11)

Temp °F	Pressure PSIA	Enthalpy, BTU/LB					
		LIQUID			VAPOR		
		Calc	Lit	Diff	Calc	Lit	Diff
-69.9	75.1	-19.9	-13.7	-6.2	138.2	136.0	2.2
-50.0	118.3	-6.4	-4.6	-1.8	139.6	137.3	2.3
-40.0[a]	145.9	0.0	0.0	--	140.1	137.9	2.2
-20.0	215.0	12.2	9.2	3.0	140.8	138.7	2.1
0.0	305.8	23.8	18.8	5.0	140.9	138.9	2.0
20.0	421.8	35.3	29.6	5.7	140.2	138.5	1.7
40.0	567.3	47.0	41.8	5.2	138.5	136.8	1.7
60.0	747.4	59.5	55.7	3.8	135.5	132.2	3.3
80.0	969.3	73.9	74.0	-0.1	129.8	119.0	10.8

a) Enthalpy base adjusted to fit the liquid enthalpy at -40°F.

Table 12

Comparison of Calculated Methane Enthalpy Data with Literature Data in the Saturation Region (9)

Temp °F	Pressure PSIA	Enthalpy, BTU/LB					
		LIQUID			VAPOR		
		Calc	Lit	Diff	Calc	Lit	Diff
-280	4.90	-1934.0	-1934.0	--	-1706.1	-1705.8	-0.3
-260	13.8	-1918.1	-1917.4	-0.7	-1697.9	-1697.6	-0.3
-258.68[a]	14.7	-1917.0	-1917.0	--	-1697.4	-1697.5	0.1
-235	39.0	-1897.4	-1896.4	-1.0	-1689.2	-1689.2	--
-210	87.6	-1875.6	-1875.2	-0.4	-1682.8	-1683.0	0.2
-185	169.7	-1852.2	-1851.2	-1.0	-1679.5	-1679.2	-0.3
-160	297.0	-1825.9	-1826.2	0.3	-1680.3	-1679.2	-1.1
-135	482.0	-1797.0	-1795.3	1.3	-1688.3	-1686.8	-1.5
-116.5[b]	673.1	-1760.1	-1730.0	-30.1	-1708.6	-1730.0	21.4

a) Enthalpy base adjusted to fit the liquid enthalpy at -258.68°F.

b) Critical point of methane.

Table 13

Comparison of Calculated Water Enthalpy Data with Literature Data in the Superheat Region (10)

Temp °F	Pressure PSIA	Enthalpy, BTU/LB Calc	Lit	Diff
750[a]	1000	1361.8	1358.7	3.1
	1500	1332.4	1328.0	4.4
	2000	1299.6	1292.6	7.0
	2500	1262.6	1250.6	12.0
	3000	1218.1	1197.9	20.2
	3500	1158.9	1127.1	31.8
	4000	933.8	1007.4	-73.6
	6000	819.4	822.9	-3.5
	8000	796.1	796.6	-0.5
	10000	783.8	783.8	--
	15000	769.6	769.7	0.1
1500	250	1806.4	1800.2	6.2[b]
	500	1803.2	1796.9	6.3
	750	1799.9	1793.6	6.3
	1000	1796.7	1790.3	6.4
	1500	1790.2	1783.7	6.5
	2000	1783.8	1777.1	6.7
	2500	1777.1	1770.4	6.7
	3000	1770.6	1763.8	6.8
	3500	1764.1	1757.2	6.9
	4000	1757.6	1750.6	7.0
	6000	1731.8	1724.2	7.6
	8000	1706.4	1698.1	8.3
	10000	1681.7	1672.8	8.9
	15000	1624.6	1615.9	8.7

a) Slightly above critical temperature of water at 706°F.

b) The error of 6 BTU/lb is an error in the ideal gas enthalpy at 1500°F referred to the liquid at 213°F. This error is caused by a small error in the ideal gas heat capacity of water. This problem will be corrected.

Table 14

Comprison of Calculated Carbon Dioxide Enthalpy Data with Literature Data in the Superheat Region (11)

Temp °F	Pressure PSIA	Enthalpy, BTU/LB Calc	Lit	Diff
-20	1.0	152.5	151.5	1.0
	10.0	152.1	151.1	1.0
	25.0	151.4	150.4	1.0
	50.0	150.2	149.0	1.2
	100.0	147.6	146.0	1.6
	200.0	141.9	139.9	2.0
300	1.0	219.6	218.8	0.8
	10.0	219.5	218.8	0.7
	25.0	219.3	218.7	0.6
	50.0	218.9	218.4	0.5
	100.0	218.3	217.8	0.5
	200.0	216.9	216.4	0.5
	400.0	214.2	214.0	0.2
	600.0	211.4	211.4	0
	1000.0	205.7	205.9	-0.2
1000	1.0	399.9	399.0	0.9
	10.0	399.9	399.0	0.9
	25.0	399.8	399.0	0.8
	50.0	399.8	398.9	0.9
	100.0	399.6	398.9	0.7
	200.0	399.3	398.8	0.5
	400.0	398.7	398.4	0.3
	600.0	398.1	398.0	0.1
	1000.0	397.0	397.2	-0.2

Table 15

Comparison of Calculated Methane Enthalpy Data with Literature Data in the Superheat Region (9)

Temp °F	Pressure PSIA	Enthalpy, BTU/LB Calc	Lit	Diff
-60	1.0	-1595.7	-1595.6	-0.1
	25.0	-1596.9	-1596.8	-0.1
	50.0	-1598.2	-1597.9	-0.3
	100.0	-1600.7	-1600.4	-0.3
	200.0	-1606.0	-1605.7	-0.3
	400.0	-1617.3	-1617.0	-0.3
	600.0	-1630.0	-1629.8	-0.2
	1000.0	-1662.0	-1661.9	-0.1
	2000.0	-1727.3	-1720.8	-6.5
700	1.0	-1110.0	-1107.8	-2.2
	25.0	-1110.1	-1107.9	-2.2
	50.0	-1110.3	-1108.0	-2.3
	100.0	-1110.4	-1108.3	-2.1
	200.0	-1110.8	-1108.9	-1.9
	400.0	-1111.5	-1109.8	-1.7
	600.0	-1112.3	-1110.3	-2.0
	1000.0	-1113.7	-1112.6	-1.1
	2000.0	-1117.0	-1116.3	-0.7
2200	1.0	518.5	518.6	-0.1
	50.0	518.4	518.6	-0.2
	100.0	518.7	518.8	-0.1
	200.0	518.8	519.2	-0.4
	400.0	519.0	519.8	-0.8
	600.0	519.2	520.4	-1.2
	1000.0	519.7	521.8	-2.1
	2000.0	521.2	525.3	-4.1

Table 16, WATER, Comparison of Predicted and Experimental Second Virial Coefficients

T^oK	M-VDW	Literature (12)	
		G. S. Kell	M. P. Vukalovich
353.16	-561.4		-844.4
373.16	-465.75		-453.6
423.16	-308.59	-326	-283.3
473.16	-217.35	-209	-196.1
523.16	-160.18	-152.5	-145.4
573.16	-122.2	-117.1	-112.9
623.16	-95.67	-92.38	-90.2
673.16	-76.53	-73.26	-72.4
723.16	-62.24	-59.36	-60.6
773.16	-51.29		-50.4
823.16	-42.72		-42.0
923.16	-30.33		-29.4
1073.16	-18.75		-17.0
1173.16	-13.68		-11.6

Table 17, CARBON DIOXIDE, Comparison of Predicted and Experimental Second Virial Coefficients

°K	M-VDW	Literature (12, 13)						
		1*	2	3	4	5	6	7
203.83	-316.61					-330		
209.03	-296.24					-302		
233.34	-223.46					-210		
258.15	-174.05							-173.5
273.15	-151.70	-147.4	-147.4	-156.36	-142			-149.3
323.15	-100.80	-100.7		-102.63	-104.3		-103.1	-104.5
373.15	-69.99	-69.5	-70.2	-71.85	-73.9		-73.1	
398.15	-58.74		-58.4		-59.4		-61.7	
423.15	-49.35	-46.3		-50.59	-52.6		-52.5	
473.15	-34.54	-29.1		-34.08			-36.8	
573.15	-14.64			-13.58			-15.9	
673.15	-1.82			-1.58			-3.4	
773.15	7.15			6.05			4.0	
873.15	13.81			12.11			10.4	
1023.15	21.10						15.8	

Primary Data Source:

1. E. G. Butcher (12)
2. R. S. Dadson (12, 13)
3. K. E. MacCormack (12, 13)
4. A. Perez Masia (12, 13)
5. K. Schafer (12, 13)
6. M. P. Vukalovich (12, 13)
7. B. L. Turlington (12)

Table 18, HYDROGEN, Comparison of Predicted and Curve Fit Experimental Second Virial Coefficients

T^oK	M-VDW	Literature[(1]	
15	-239.2	-230	±5
25	-106.0	-111	±3
50	-30.3	-35	±2
100	-1.6	-1.9	±1
200	9.7	11.3	±0
300	13.0	14.8	±0
400	14.5	15.2	±0.

Table 19, METHANE, Comparison of Predicted and Curve Fit Experimental Second Virial Coefficients

T^oK	M-VDW	Literature[(1]	
110	-415.5	-344	±10
150	-204.8	-191	±6
200	-107.3	-107	±2
250	-62.7	-67	±1
300	-37.7	-42	±1
400	-11.1	-15.5	±1
500	2.6	-0.5	±1
600	11.0	8.5	±1

Table 20, Predicted Solubility of Water in Carbon Dioxide at Liquid-Liquid Saturation from the M-VDW and Mark V Equations of State at 25°C

	Solubility Mole %	Error %
Measured (2)	0.25	--
Predicted[b] M-VDW	0.20	20
Predicted[b] Mark V	2.1	840

b) The CO_2-H_2O interaction coefficient was adjusted to fit the measured solubility of CO_2 in H_2O which is 2.55 mole % as reported by Francis.

Table 21, WATER-CO_2, Comparison of Predicted and Experimental Compressibility Data at Temperatures from 840°F to 1470°F

	Exp (14)	M-VDW	%Diff	Exp (14)	M-VDW	%Diff	Exp (14)	M-VDW	%Diff
0% CO_2									
842°F	.8919	.8912	-0.08	.6037	.6182	2.40	.3746	.3409	-9.00
1022	.935	.9365	0.11	.801	.8031	0.25	.672	.6681	-0.60
1202	.9621	.9605	-0.17	.8872	.8822	-0.56	.8168	.8086	-1.00
1472	.9801	.9797	-0.04	.9756	.9397	-0.62	.9191	.9037	-1.68
30% CO_2									
842°F	.9454	.9387	-0.71	.8288	.8235	-0.64	.7755	.7377	-4.87
1022	.9687	.9693	0.06	.9158	.9128	-0.33	.8762	.8720	-0.48
1202	.9750	.9849	1.02	.9527	.9597	0.73	.9381	.9433	0.55
1472	.9755	1.0000	2.51	.9863	1.0000	1.39	.9894	1.0060	1.68
70% CO_2									
842°F	.9859	.9843	-0.16	.9761	.9676	-0.87	1.0101	.9722	-3.75
1022	.9918	1.0000	0.83	1.0076	1.0090	0.14	1.0436	1.0283	-1.47
1202	.9953	1.0092	1.40	1.0193	1.0320	1.25	1.0601	1.0604	0.03
1472	.9940	1.0180	2.41	1.0332	1.0541	2.02	1.0719	1.0918	1.86
100% CO_2									
842°F	1.0047	1.0072	0.25	1.0419	1.0341	-0.75	1.1017	1.0761	-2.32
1022	1.0171	1.0171	0.00	1.0543	1.0568	0.24	1.1167	1.1043	-1.11
1202	1.0165	1.0226	0.60	1.0608	1.0691	0.78	1.1285	1.1196	-0.79
1472	1.0152	1.0265	1.11	1.0615	1.0776	1.52	1.1328	1.1297	-0.27

and methane respectively. Water deviates on the average about 7 BTU/lb in both the liquid and vapor regions except at high pressures in the vapor region. This bias could be corrected by adjusting the enthalpy base, and the error at high pressures in the vapor could be minimized by simultaneous fitting of multiple properties. We consider the agreement here to be quite good considering that no enthalpy data have yet been used in developing the model. Enthalpy data on CO_2 in Table 11 show similar behavior except that a smaller adjustment needs to be made. Data on methane in Table 12 show almost perfect agreement except at the critical point where the computation method does not prove reliable. Superheat data in Tables 13, 14, and 15 also show good agreement. Data on water in Table 13 deviate near the critical point of water. Thus at 750°F and 4000 psia the largest deviation of -73.6 BTU/lb occurs. At higher and lower pressures the agrement is much better. At 1500°F there appears to be a constant bias in the calculated enthalpy of water, and this error can be corrected by use of a better ideal-gas heat capacity equation. Data in Table 14 on CO_2 agree very well with published data. Data on methane in Table 15 also show good agreement except at 700°F where deviations of about -2 BTU/lb occur. This error is probably due to ideal-gas enthalpy error, and can be corrected.

We suspect some variance between calculated and measured data is due to inconsistencies between various sets of data. This is shown in Tables 16 to 19 where calculated second virial coefficients are compared with literature data from various sources on water, CO_2, hydrogen, and methane respectively. Calculated data in all four tables appear to agree with the experimental data almost within the scatter of the data. This demonstrates two things. One is that the comparison depends on whose data you compare with and the other is that predictability almost within experimental accuracy has been made without fitting virial coefficient data directly for both polar and non-polar compounds. One problem in predicting values at 3000°F is the lack of experimental data. This means that theoretical models for the second virial coefficient will have to be used in order to extend to these higher temperatures.

Predicted data for mixtures of carbon dioxide and water have been made at low temperature and at high temperatures. These comparisons are given in Tables 20 and 21. Table 20 compares the predicted solubility of water in liquid carbon dioxide at liquid-liquid saturation at 25°C from the M-VDW and Mark V equations of state. This comparison shows a very large difference between the two equations of state. The M-VDW equation predicts the solubility within 20% while the Mark V is in error by a factor of eight. Because of the adjustable symmetry parameter, the M-VDW equation proves to be significantly superior to the Mark V. Presumably the Soave equation would give results comparable to the Mark V because they both assume the Redlich-Kwong volume dependence.

Table 21 compares experimental and predicted compressibility data on CO_2-water mixtures at high temperature from the M-VDW equation of state. This comparison shows good agreement with measured data where deviations on the order of ±1% depending on the temperature and pressure.

Parameters used in predicting data given in Tables 2 to 21 are summarized in Table 22.

Summary and Future Work

The results of this study show the following.

1. The M-VDW equation can be used to correlate both low-temperature and high-temperature physical property data.
2. Polar compounds such as water can be correlated with accuracy comparable to non-polar compounds.
3. Accurate liquid-liquid equilibria are predictable from the M-VDW equation of state.
4. The assymetric activity coefficient behavior of non-ideal mixtures is predictable.
5. This new equation should be comparable in computation speed to existing Redlich-Kwong modifications such as Mark V or Soave.

Although not presented here, it can be shown that the M-VDW parameters can be calculated from group contributions. This makes possible the calculation of heavy hydrocarbon parameters without a knowledge of the critical constants. By this method, paraffin, napthene, and aromatic solvent effects can be taken into account without knowing the actual compounds present in a given hydrocarbon fraction.

Work is in progress to generalize the parameters in the M-VDW model so that they will be predictable from the acentric factor and perhaps a polar parameter. Also work is planned to extend the model for predicting enthalpy, heat capacity, entropy, density, and phase behavior (including liquid-liquid-vapor phase behavior) of water, hydrogen, carbon dioxide, carbon monoxide, nitrogen, methane, and other light hydrocarbons for temperatures from $0^{\circ}F$ to $3000^{\circ}F$ and pressures from atmospheric to 4000 psia. Subsequent to this, the correlation will be extended, according to present plans, to include systems containing intermediate and heavy oils and tars of the types encountered in synthetic gas processing. The model is expected to take into account the aromatic, naphthenic, and parafinic nature of these oils and tars. Ultimately it is planned to also extend the M-VDW model to include low-temperature processing for hydrogen purification.

Nomenclature

Symbol	Definition
a_i	Van der Waals a
a_i'	derived from a_i and b_i, Eqns. 1-2

Table 22, Critical and Derived Constants Used in Computer Program

	Water	Carbon Dioxide	Methane	Hydrogen
	Critical Constants			
Temperature ^{o}R	1165.1	547.57	343.2	59.74
Pressure Psia	3206.2	1072.81	673.08	190.75
Acentric Factor	0.348	0.231	0.013	-0.214
	Derived Constants			
α	-0.256358	-1.56706	-0.704135	0.00595060
β	1.71075	6.05555	3.56806	2.46398
γ	2.49569	-2.09764	0.676355	1.87177
δ	-0.374112	1.12534	-0.0876456	-1.08136
σ	0	0	0	0.195429
B	0.149	0.372	0.414	0.301
C	1.05	0.54	0.44	0.268

Symbol	Definition
a_{ij}	derived from B_{ij}, A_{ii}, and A_{jj}, Eqn. a-6
a_{hj}	a_{ij} interaction parameter when i = void spaces
a_{HH}	a_{ij} when both i and j = void spaces
A	Helmholtz free energy
A_I	Ideal Helmholtz free energy of mixing
A_i^o	Helmholtz free energy of pure component
A_{ij}	M-VDW parameter in Eqn. 5
A_{ii}	interaction parameter defined by Eqn. a-4
b	$b = \Sigma_i x_i b_i$
b_i	b_i = Van der Waals volume
b_H	volume of "one mole" of void spaces
B_{ij}	Van Laar parameter in Eqn. a-1
B'_{ij}	Analogous Van Laar parameter in Helmholtz free energy Eqn., a-3
C	Temperature parameter for S_j in Eqn. 6
G	Gibbs free energy
G^E	Excess Gibbs free energy of mixing
G^I	Ideal Gibbs free energy of mixing
G_i^o	Free energy of pur component
n_H	moles of void spaces given by Eqn. a-7
n_i	moles of component i
P	pressure

Symbol	Definition
S_i	Van Laar parameter
S_H	Van Laar S_i for void spaces
S	$S = \sum_i x_i S_i$
T	Absolute temperature
T_c	Critical temperature (absolute)
V	Volume
$\overline{V}$	Molar volume
x_i	mole fraction of component i
Z	compressibility factor
α	
β	
γ	Parameters for calculating
δ	A_{jj} in Eqn. 7
σ	
γ_i	Activity coefficient

Appendix

Derivation of M-VDW Model

The excess free energy of mixing given by the Van Laar equation is given as follows.

$$\frac{G^E}{RT} = \sum_i n_i \ln\gamma_i = \frac{\Sigma_i \Sigma_j n_i S_i n_j S_j B_{ij}}{\Sigma_k n_k S_k} \qquad \text{(a-1)}$$

where $B_{ii} \equiv 0$.

The total free energy is given as the free energy of the components plus the ideal free energy of mixing plus the excess free energy of mixing as follows.

$$\frac{G}{RT} = \sum_i \frac{n_i G_i^o}{RT} + \frac{G^I}{RT} + \frac{G^E}{RT} \qquad \text{(a-2)}$$

If the same analytical form is assumed for the Helmholtz free energy, one obtains the following.

$$\frac{A}{RT} = \Sigma_i n_i A_i^o + \frac{A^I}{RT} + \frac{\Sigma_i \Sigma_j n_i S_i n_j S_j B'_{ij}}{\Sigma_k n_k S_k} \qquad \text{(a-3)}$$

where the prime is used to show that the B'_{ij} parameters have different values than B_{ij}.

The pure component free energy A_i^o can be defined in terms of parameters S_i and A'_{ii} as follows.

$$A_i^o = S_i A'_{ii} \qquad \text{(a-4)}$$

This definition can be substituted into Equation (a-3) to give the following.

$$\frac{A}{RT} = \frac{A^I}{RT} + \frac{\Sigma_i \Sigma_j n_i S_k n_j S_j a_{ij}}{\Sigma_k n_k S_k} \qquad \text{(a-5)}$$

where

$$a_{ij} = B'_{ij} + \frac{A'_{ii} + A'_{jj}}{2} \qquad \text{(a-6)}$$

In the next step, we assume that void spaces in a fluid can be considered as an additional component in calculating the Helmholtz free energy. This is done by assuming a molar volume for void spaces so that the number of moles of void spaces can be calculated as follows.

$$n_H = \frac{V - \Sigma_k n_k b_k}{b_H} \qquad \text{(a-7)}$$

where b_k = molar volume of molecules excluding void spaces

b_H = molar volume of void spaces

Substitution into Equation (a-5) give the following.

$$\frac{A}{RT} = \frac{A^I}{RT} + \frac{V}{[(1 - \sum_k \frac{n_k b_k}{V}) \frac{S_H}{b_H} + \sum_k \frac{n_k S_k}{V}]} \qquad \sum_{ij}\sum \frac{n_k S_k n_j S_j a_{ij}}{V^2}$$

$$+ \frac{2(V - \Sigma_k n_k b_k) \frac{S_H}{b_H} \sum_j n_j S_j a_{Hj}}{V^2}$$

$$+ \frac{(V - \Sigma_k n_k b_k)^2 (\frac{S_H}{b_H})^2 a_{HH}}{V^2} \tag{a-8}$$

The free energy of a void space is zero, so

$$a_{HH} \equiv 0 \tag{a-9}$$

Also at infinte volume, the free energy equals the free energy of an ideal gas at unit pressure:

$$\left.\frac{A}{RT}\right)_{V=\infty} = 2\frac{S_H}{b_H} \sum_i n_j S_j a_{Hj} \tag{a-10}$$

$$\text{or} \quad A_i^o = 2 \frac{S_H}{b_H} S_i a_{Hi} \tag{a-11}$$

When Equation (a-9) is substituted into (a-8) and the equation is rearranged algebraically we obtain the following.

$$\frac{A}{RT} = \frac{A^I}{RT} + \frac{\sum_{ij}\sum n_i S_i n_j S_j a_{ij} + 2(V - \Sigma k_k n_k b_k) \frac{S_H}{b_{Hj}} \Sigma n_j S_j a_{Hj}}{(V - \Sigma_k n_k b_k) \frac{S_H}{b_H} + \Sigma_k n_k S} \tag{a-12}$$

In the Van Laar equation, one of the S's is arbitrary; so we define as follows.

$$S_H \equiv b_H \tag{a-13}$$

With this change our final Helmholtz free energy equation is as follows.

$$\frac{A}{RT} = \frac{A^I}{RT} + \frac{\sum\limits_{ij}\sum n_i S_i n_j S_j a_{ij} + 2(V - \Sigma_k n_k b_k)\frac{S_H}{b_H}\sum\limits_j n_j S_j a_{Hj}}{V - \sum\limits_k n_k b_k + \sum\limits_k n_k S_k} \tag{a-14}$$

The equation of state is derived from the derivative of A with respect to volume as follows.

$$P = -\left(\frac{\partial A}{\partial V}\right)_{T,n} \tag{a-15}$$

The resulting equation is as follows.

$$\frac{P}{RT} = -\left(\frac{\partial A^I/RT}{\partial V}\right)_{T,n} + \frac{\Sigma_i \Sigma_j n_i S_i n_j S_j A_{ij}/RT}{(V - \Sigma_k n_k b_k + \Sigma_k n_k S_k)^2} \tag{a-16}$$

where A_{jk}/RT is defined as follows.

$$\frac{A_{jk}}{RT} = -a_{ij} + a_{Hi} + a_{Hj} \tag{a-17}$$

The derivative of the ideal Helmholtz free energy of mixing with respect to volume is assumed to given by the $V - \Sigma_i n_i b_i$ term in the Van der Waals equations; or

$$-\left(\frac{\partial A^I/RT}{\partial V}\right)_{T,n} = \frac{\Sigma_i n_i}{V - \Sigma_i n_i b_i} \tag{a-18}$$

Thus our final equation, called the Modified Van der Waals equation, is as follows.

$$\frac{P}{RT} = \frac{\Sigma_i n_i}{V - \sum\limits_i n_i b_i} - \frac{\Sigma_i \Sigma_j n_i S_i n_j S_j (A_{ij}/RT)}{(V + \Sigma_k n_k S_k - \Sigma_k n_k b_k)^2} \tag{a-19}$$

or $$Z = \frac{\overline{V}}{\overline{V} - b} - \frac{\overline{V}\Sigma_i \Sigma_j x_k S_i x_j S_j (A_{ij}/RT)}{(\overline{V} + \underline{S} - b)^2} \tag{a-20}$$

This equation is the same as Equation (5) in this paper.

This equation is unique because the mixture rules are already defined. Also Van Laar parameters derived from fitting binary and multicomponent vapor-liquid equilibrium data relate directly to parameters in the equation of state. By simple substitution it can be shown that A_{jk}/RT in Equation (a-17) is related to B_{ij} in Equation (a-3) by the equation:

$$\frac{A_{jk}}{RT} = -B'_{jk}, \; j \neq k \qquad \text{(a-21)}$$

This substitution is not quite as simple as it appears because the B'_{jk} parameters are defined as Van Laar interaction parameters when no void spaces are present; this is different from the usual definition.

Although not shown here, it can be shown that the S_i, and A_{jk} parameters can be calculated from group contributions as in the calculation of activity coefficients from group contributions.

Literature Cited

1. Soave, G., Chem. Eng. Sci. (1977) 27, 1197.
2. Francis, A., J. Phys. Chem. (1954) 58, 1099.
3. van Laar, Z. Physik. Chem. (1910) 72, 723.
4. Prausnitz, J. M., "Molecular Thermodynamics of Fluid Phase Equilibria," Prentice Hall, Englewood Cliffs, New Jersey, 1969, pp. 264-269.
5. "Gas Processing Handbook" Section in Hydrocarbon Processing (1975) 54, No. 4, 112-125.
6. Perry, J. H., "Chemical Engineer's Handbook," 4th Edition, McGraw-Hill, New York, 1968.
7. NBS Technical Note 120 (November 1968).
8. Sweigert, R. L., Weber, P. and Allen, R. L., Ind. & Eng. Chem. (1946) 38, 185-200.
9. Conjar, L. N. and Mannuig, F. S., "Thermodynamic Properties and Reduced Correlations for Gases," Gulf Publishing Co., Houston, 1967.
10. Meyer, C. A., McClintock, R. B., Silvestri, G. J. and Spencer, R. C., Jr., Editors, "Thermodynamic and Transport Properties of Steam," ASME, United Engr. Center, New York, New York, 1967.
11. "ASHRHE Thermodynamic Properties of Refriguants," Am. Soc. of Heating and Refrigerating and Air Conditioning Engineers, Inc., New York, 1969.
12. Dymond, J. H. and Smith, E. B., "The Virial Coefficients of Gases--A Critical Computation," Clarendon Press, Oxford, 1969.
13. Vukalovich, M. P. and Altunin, V. V., "Thermophysical Properties of Carbon Dioxide," Collets Publishing Ltd., London and Wellingborough, 1969.
14. Greenwood, H. J., Am. J. Sci., Schairer (1969) 267-A, 191-208.

15

Thermophysical Properties: Their Effect on Cryogenic Gas Processing

D. G. ELLIOT, P. S. CHAPPELEAR, R. J. J. CHEN, and R. L. MCKEE

McDermott Hudson Engineering, Houston, TX 77036

The changing economic climate in the natural gas processing industry has precipitated the design and construction of gas processing plants to recover 80 percent or more of the contained ethane. Typically, high ethane recoveries are attained by employing a cryogenic process utilizing a turboexpander. Historically, the availability of fundamental thermophysical property data required to design gas processing plants has lagged behind the design and construction of such facilities (1). The cryogenic process is no exception. Due largely to the efforts of the Gas Processors Association, pertinent experimental data has been taken and correlated by several methods. These thermophysical property correlations are generally available to the gas processing industry. It is the authors' intent to illustrate the relationship between process design and accurate prediction of K-values, enthalpy, entropy, and CO_2 solubility.

Scope

In 1970, White et al (2) demonstrated the importance of accurate K-value correlations in cryogenic plant design by comparing predicted product recoveries for a typical cryogenic plant using several correlations available at that time. The various K-value correlations gave predicted recoveries for ethane from 25.7% to 45.0% and for propane from 66.7% to 90.7%. Since that time much progress has occurred in correlations. Existing correlations have been refined and new correlations have been introduced.

It is the authors' intent to demonstrate the importance of accurate thermophysical properties on plant design and to compare the properties predicted by generally available correlations. This has been done by comparing predictions both from K-value correlations and enthalpy/entropy correlations for the major components in a typical cryogenic process. In addition, the predicted CO_2 concentrations near the conditions of CO_2 solids formation, a condition limiting ethane recovery in many plants, are presented.

The correlations examined, listed in Table I, are of three

TABLE I

CORRELATIONS

Type	Symbol	Description	Reference
Empirical:	GPA CONV	K-values from Hadden convergence pressure concept (1972 edition) with CH_4, C_2H_6, C_3H_8 from binary data at low temperatures. K-values (convergence) computer calculated from GPA Coefficient curve fit. K-values (binary) hand calculated.	(4)
	HUDSON	McDermott Hudson proprietary correlation for K-values of CH_4, C_2H_6, and C_3H_8. Uses GPA CONV for heavier components.	
Equation of State:			
van der Waals	MARK V	Wilson modification of the Redlich-Kwong equation of state; computer program marketed by PVT, Inc.	(5) (6)
	LEE	Lee et al. correlation as programmed in the GPA computer package K & H.	(7)
	SOAVE	Soave modification of the Redlich-Kwong equation of state, as programmed in the GPA computer package K & H.	(8)
	P-R	Peng-Robinson modification of the Redlich-Kwong equation of state. Values computer calculated from correlation by P-R distributed by GPA.	(9)
BWR	SH BWR	Starling-Han modification of the Benedict-Webb-Rubin equation of state as programmed in the GPA computer package K & H.	(10) (11)

TABLE I (Continued)

Type	Symbol	Description	Reference
Conformal Solution:	K VAL	1972 program for K-value by CHEMSHARE for high methane content streams. Based on binary K data for ten reference systems.	
	PROP-75	Property-75 Rice University shape factor corresponding states program for physical properties using methane and pentane as reference. 1975 revision of computer program distributed by GPA.	(12)
	K DELTA	T. W. Leland corresponding states correlation using data for binary systems as reference. Computer program marketed by Simulation Science Inc. (SSI).	(13)

types: empirical, equation of state, and conformal solution. Empirical correlations are graphs of experimental data plotted aga a correlating parameter. Equation of state correlations are divid into two types: one is similar to the van der Waals equation with small number of parameters; the other is based on the multiple con stant BWR equation. All equation of state correlations involve th computation of the parameters for the equation of state which will represent the multicomponent mixture. These computations are bas on parameters determined for the pure components, combining rules, and possibly binary intraction parameters.

The conformal solution or corresponding states correlation co putes pseudo-reduced conditions for the mixture to conform to the reference substance. The reference substance may be a pure compon or a mixture; its properties can be given in tabular or equation format.

The computer programs used in this study are current versions Other programming of the same equations may give slightly differen results. The abbreviations listed in Table I are used throughout the text.

Design Basis and Process Description

Properties predicted by these correlations determine not only the economic feasibility of gas processing installations; they als dictate the design (size) of its major components. These componen are usually the compressor, turboexpander, heat exchangers, extern refrigeration (if any), and demethanizer. A typical arrangement f a cryogenic gas plant of these components is shown in Figure 1.

The following process conditions are selected:

1. Inlet flow rate of 70 MMscfd
2. Inlet gas pressure of 250 psia
3. Inlet gas temperature of 120°F
4. High pressure separator at -80°F and 800 psia
5. Expander outlet pressure at 200 psia
6. Expander isentropic efficiency at 80% (commonly called adiabatic)
7. Compressor isentropic efficiency at 75% (commonly called adiabatic)
8. Inlet gas composition in Table II.

In this typical plant, the inlet gas is compressed from 250 p to 815 psia. It is then cooled by an air cooler to 120°F followed exchanging heat with the demethanizer reboiler and the plant resid gas. Any additional refrigeration required to cool the inlet gas -80°F before entering the high pressure separator is furnished fro an external source. The liquid separated in the high pressure separator is fed to the demethanizer where the methane is stripped out. Energy is recovered from the high pressure separator vapor b reducing the pressure through a turboexpander. The expander outle is much colder and partially condensed. This energy is converted

TABLE II - COMPOSITIONS USED IN CALCULATIONS

Component	Inlet Gas Mol %	Inlet Gas Mol/Day	Residue Gas Mol %	Residue Gas Mol/Day	Liquid Product Mol %	Liquid Product Mol/Day
Nitrogen	0.40	738	0.44	738		0
Carbon Dioxide	0.52	959	0.30	498	2.91	461
Methane	90.16	166,308	98.52	166,124	1.16	184
Ethane	4.69	8,651	0.70	1,181	47.15	7,470
Propane	1.85	3,412	0.04	73	21.07	3,339
Isobutane	0.79	1,457			9.20	1,457
Normal Butane	0.51	941			5.94	941
Isopentane	0.27	498			3.14	498
Normal Pentane	0.18	332			2.10	332
Hexane Plus*	0.63	1,162			7.33	1,162
	100.00	184,458	100.00	168,614	100.00	15,844

*All calculations used n-Heptane for the Hexane Plus fraction.

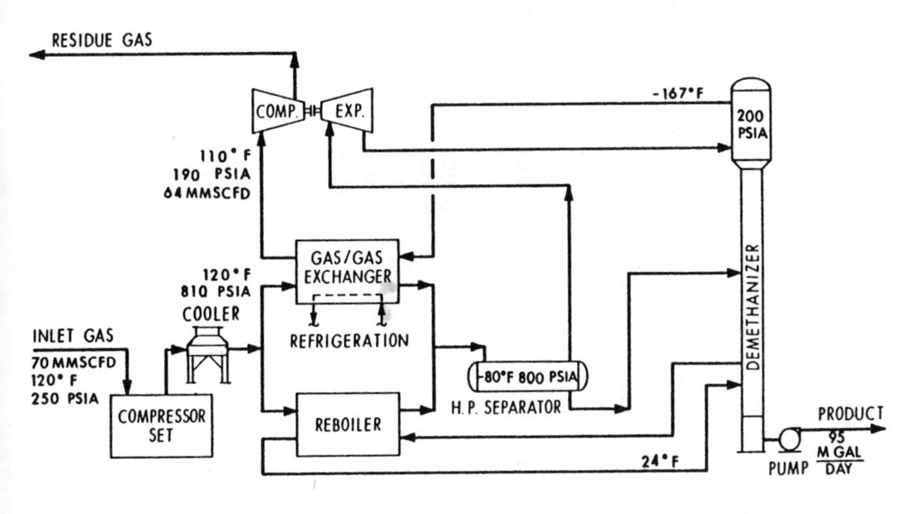

Figure 1. Schematic for typical turboexpander gas processing plant

TABLE III EFFECT OF K-VALUE CORRELATION IN HIGH PRESSURE SEPARATOR (PROPERTY 75 FOR THERMODYNAMIC PROPERTIES)

K-VALUE CORRELATION	GPA CONV 1050 PSIA	HUDSON	P-R	SOAVE	SH BWR	LEE	K VAL	K DELTA	MARK V
K-Value at -80F, 800 psia									
Nitrogen	3.954	3.954	3.616	3.644	5.262	2.723	3.374	4.019	3.512
Carbon Dioxide	0.673	0.572	0.543	0.574	0.598	0.414	0.706	0.645	0.599
Methane	1.367	1.310	1.321	1.337	1.672	1.161	1.336	1.378	1.327
Ethane	0.331	0.250	0.277	0.269	0.241	0.298	0.336	0.266	0.267
Propane	0.131	0.110	0.0889	0.0834	0.0665	0.127	0.0932	0.0797	0.0876
Isobutane	0.0614	0.0614	0.0400	0.0366	0.0262	0.0889	0.0386	0.0392	0.0407
Normal Butane	0.0437	0.0437	0.0287	0.0261	0.0176	0.0475	0.0271	0.0295	0.0293
Isopentane	0.0161	0.0161	0.0131	0.00114	0.00686	0.0366	0.0097	0.0098	0.0142
Normal Pentane	0.0173	0.0173	0.0096	0.00838	0.00497	0.0193	0.0064	0.00628	0.0107
Hexane Plus	0.0086	0.0086	0.0012	0.00092	0.00052	0.00363	0.0003	0.000302	0.0018
Liquid Condensed (Mol/Day)									
Methane	20,290	30,804	29,919	28,646	12,635	68,692	25,511	24,588	29,458
Ethane	3,154	4,704	4,427	4,390	3,136	6,345	3,615	4,095	4,482
Propane	2,011	2,490	2,619	2,619	2,306	2,951	2,472	2,564	2,619
Total Liquids	29,421	42,333	41,429	40,157	22,283	83,061	35,969	35,600	41,023
Cooling Duty (Mbtu/hr)	24,099	24,550	24,459	24,409	23,564	26,447	24,275	24,330	24,455

useful work by the turboexpander driven compressor which compresses the residue gas to the delivery pressure. The turboexpander outlet stream is then fed to a low temperature separator (normally in the top of the demethanizer) where condensed liquid is used to reflux the tower. The cold vapor goes overhead with the demethanizer residue gas.

Effects of K-Value Predictions

High pressure separator and turboexpander designs are highly dependent on vapor-liquid equilibrium K-values. Variation in K-values is examined by selecting a common basis (Property-75) for enthalpy/entropy calculations.

High Pressure Separator. The compressed inlet gas, after being cooled to 120°F by an air cooler, flows through the cryogenic heat exchanger system. For the sake of comparison, the gas is assumed to enter this system at 120°F and 810 psia and is cooled to -80°F at 800 psia, the conditions of the high pressure separator.

At the fixed conditions of the high pressure separator, the liquid knockdown and its composition are only dependent on the K-value correlation.

The results of the high pressure separator flash calculations are tabulated in Table III. For ease in comparing results, K-values predicted by the Hudson correlation as defined by Table I are used as a common basis. The range of K-values and amount of liquid knock-down is summarized as follows:

	Deviation from Hudson			
Methane K-value	-11%	to	28%	(1.161 to 1.672)
Ethane K-value	-4%	to	34%	(0.241 to 0.336)
Propane K-value	-40%	to	19%	(0.0665 to 0.131)
Liquid Condensed	-47%	to	96%	(22,283 to 83,061 mols/day)

The large variation in the amount of liquid condensed is mainly caused by the variation in the predicted methane K-value. This variation becomes increasingly pronounced as the conditions of the gas approach the critical point where most of the correlations are inadequate or fail altogether. Unfortunately, this condition occurs frequently in gas plant design. In such cases, one should be extremely cautious in using these correlations.

Turboexpander. Vapor from the high pressure separator feeds the expander. Table IV was generated using the different K-value correlations with the following common basis:

Expander Inlet Temperature	-80°F
Expander Inlet Pressure	800 psia
Expander Discharge Pressure	200 psia
Expander Isentropic Efficiency	80%

The gas flow rate was determined from the high pressure separator flash calculations. Otherwise, the expander inlet gas would not be at its dewpoint for the correlation being used. This fact precludes

TABLE IV EFFECT OF K-VALUE CORRELATIONS ON EXPANDER
(PROPERTY 75 FOR THERMODYNAMIC PROPERTIES)

K-Value Correlations	GPA* CONV	HUDSON	P - R	SOAVE	SH BWR	LEE	K VAL	K DELTA	MARK V
Expander Discharge Temperature, ^{o}F	-165.1	-166.1	-166.2	-167.1	-165.8	-164.4	-163.2	-166.4	-166.5
K-Value at 200 psia									
Nitrogen	9.189	9.068	8.674	8.539	8.491	12.28	8.337	9.417	7.735
Carbon Dioxide	0.229	0.332	0.1654	0.186	0.229	0.0131	0.299	0.271	0.219
Methane	1.335	1.268	1.283	1.290	1.316	1.241	1.263	1.251	1.283
Ethane	0.0524	0.0446	0.0421	0.0406	0.0431	0.0424	0.0556	0.0455	0.0383
Propane	0.0039	0.0026	0.0032	0.0029	0.0045	0.0042	0.0042	0.0041	0.0033
Isobutane	0.0006	0.0006	0.0005	0.0004	0.0009	0.0009	0.0006	0.00051	0.0006
Normal Butane	0.0003	0.0003	0.0002	0.0005	0.0005	0.0004	0.0003	0.0002	0.0003
Inlet Gas Flow Rate (Mol/Day)	155,037	142,125	143,029	144,301	162,175	101,397	148,489	148,858	143,435
Expander Horsepower	1,130	1,052	1,051	1,061	1,195	752	1,088	1,153	1,054

*K-values for methane, ethane and propane are obtained from interpolation of binary chart and infinite dilution chart in GPA 1972 Handbook. K-values for the rest of components are obtained by convergence method using 800 psi convergence pressure.

the isolation of the expander calculation from the high pressure separator calculation when comparing overall ethane recovery.

Again, with the Hudson correlation as the common basis, the variation in predicted K-values and recoveries are summarized as follows:

	Deviation from Hudson	
Methane K-value	-2.1% to 5.3%	(1.241 to 1.335)
Ethane K-value	-14% to 25%	(0.0382 to 0.0556)
Propane K-value	11% to 73%	(0.0029 to 0.0045)
Liquid Condensed	-36% to 21%	(13,576 to 25,584 mols/day)

The total effect of the high pressure separator and the turbo-expander flash calculations is shown by the estimated ethane recovery in Table IV which varies from 82.1% to 88.4%. The estimated propane recovery for all cases is 99+%.

The methane K-value shows very good agreement for all the correlations, but the ethane and propane K-values exhibit a large variation. Variation in the amount of liquid condensed at the expander outlet and the expander horsepower are directly related to the amount of vapor separated at high pressure separator. The magnitude of this variation can affect the expander design quite significantly.

The liquids condensed by the expansion are listed in Table V along with the total liquids. The total liquid condensed is the sum of expander outlet liquid and high pressure separator liquid. Since the ethane recovery is directly proportional to the total liquids condensed, the wide variation is significant. Also, the total amount of liquids has a large effect on the design of the demethanizer.

In the final analysis, the most important number in the plant is product rate. Here, the calculated overall variation in ethane recovery is 6.3%. This corresponds to 5,520 gallons per day of ethane. If one assumes ethane is worth $0.15 per gallon, the incremental product revenue is $828 per day, less gas shrinkage of $176 per day based on gas at $0.85 per MMscf. The resulting net income is $652 per day or more than $200,000 per year. Obviously, this variation in anticipated revenue is significant enough to affect investment decisions. Moreover, history has shown that it is not unusual for a competitive bid to be awarded on the basis of 1% higher guaranteed ethane recovery.

Effects of Thermophysical Property Correlations for Enthalpy and Entropy

To examine enthalpy effects of the various correlations, K-values from the Soave correlation were selected as a common basis. Comparisons are presented here in terms of the segments of the design calculations for an expander plant in the order of the flow shown on Figure 1.

Comparison of Predictions for Compressors. There are two compression schemes available for the design of a cryogenic

TABLE V

EFFECT OF K-VALUE CORRELATION ON TURBOEXPANDER LIQUIDS AND TOTAL RECOVERY
(PROPERTY 75 FOR THERMODYNAMIC PROPERTIES)

K-Value Correlations	GPA* CONV	HUDSON	P-R	SOAVE	SH BWR	LEE	K VAL	K DELTA	MARK V
Turboexpander Liquids Condensed (Mol/Day)									
Methane	18,815	16,380	15,900	15,421	18,575	10,809	19,091	19,073	15,790
Ethane	4,353	3,136	3,376	3,412	4,445	1,808	3,929	3,693	3,394
Propane	1,383	904	775	775	1,088	443	922	830	775
Total Liquids	25,584	21,102	20,696	20,198	24,865	13,576	24,551	24,238	20,567
Total Liquids (Mol/Day)									
Methane	39,105	47,184	45,819	44,067	31,210	79,501	44,602	43,661	45,248
Ethane	7,507	7,840	7,803	7,802	7,581	8,153	7,544	7,788	7,876
Propane	3,394	3,394	3,394	3,394	3,394	3,394	3,394	3,394	3,394
Total Liquids	55,005	63,435	62,125	60,355	47,148	96,637	60,520	59,838	61,590
*Estimated Ethane Stripped	387	354	355	337	272	503	446	338	319
Estimated Ethane Recovery, Percent	82.3	86.5	86.1	86.3	84.5	88.4	82.1	86.1	87.4

*Estimated Ethane Stripped = Total mol of methane condensed x mol ratio of ethane to methane in the

turboexpander plant. Since free pressure drop is usually not available to supply the energy of separation, either compression of the feed gas or recompression of the residue gas must be used (or some combination of the two). Selection is dependent upon the feed gas pressure level. Usually, precompression is used for low pressure feed and recompression is used for high pressure feed.

Precompressor. The example presented here in Figure 1 uses precompression. Since compressors are normally the most expensive pieces of equipment in the entire plant, the process is usually designed to fit the closest compressor frame size available.

The design basis is:

Flow Rate:	70 MMscfd
Gas Composition:	Inlet gas as shown in Table II
Suction Pressure:	250 psia
Suction Temperature:	120°F
Horsepower Available:	6,000 hp site rated
Isentropic Efficiency:	75%

Compressor discharge conditions are tabulated in Table VI.

The variation in discharge pressure is about 12 psi pressure from 809.5 to 821.5 psia. Although a higher discharge pressure results in greater recovery, the 12 psi is not a significant difference. Even a hand calculation by the k-method (3) yielded similar results.

The variation in discharge temperature is 6.8°F. For an air cooler cooling the compressed gas to 120°F, this results in a design duty variation up to 3.5% depending on which correlation is chosen.

Recompressor. Although not shown here, a similar comparison can be made for a case using recompression by choosing an estimated residue gas composition. The design basis is:

Flow Rate:	70 MMscfd
Gas Composition:	Residue Gas as shown in Table II
Suction Pressure	250 psia
Suction Temperature:	120°F
Horsepower Available:	6,000 hp site rated
Isentropic Efficiency:	75%

Calculated discharge conditions are tabulated in Table VI.

All the correlations agree reasonably well in predicting discharge pressure and temperature.

Inlet Gas Cooling. For this comparison, the inlet gas conditions are set as 120°F and 810 psia. The high pressure separator conditions are set as -80°F and 800 psia. Calculated cooling duties are given in Table VII. The variation in inlet gas cooling duty is -0.9% to 1.8% of the Property-75 calculation. (All comparisons in this section are made with the Property-75 results).

TABLE VI

CALCULATED COMPRESSOR DISCHARGE CONDITIONS

INLET: 70 MMSCFD AT 120°F 250 PSIA
6000 HP AT 75% EFFICIENCY

	Inlet Gas Compression		Residue Gas Compression	
	PSIA	$^{\circ}$F	PSIA	$^{\circ}$F
PROP-75	811.8	319.6	792.2	335.8
P-R	821.5	320.8	799.6	337.2
SOAVE	809.7	316.7	793.9	333.9
LEE	809.5	315.5	794.7	332.8
SH BWR	813.8	316.7	798.6	334.9
GPA-k	817.3	314.0	798.7	331.8

TABLE VII - COMPARISON OF ENTHALPY AND ENTROPY CORRELATIONS
(K-VALUES FROM GPA K&H SOAVE)

	SOAVE	P-R	LEE	PROP-75
Feed Gas Cooling Duty, MBtu/Hr	24,727	24,203	24,858	24,409
Expander Horsepower	1,050	1,018	942	1,061
Expander Discharge Temperature, $^{\circ}$F	-167.8	-168.2	-167.0	-167.1
Reboiler Duty, MBtu/Hr	5,727	5,858	5,662	5,223
Residue Heating Duty, MBtu/Hr	17,937	17,512	18,394	18,057
Refrigeration Duty, MBtu/Hr	1,063	833	802	1,129

Expander. All gas quantities and compositions are held constant by using a fixed K-value correlation since the expansion process is isentropic. The effect of the enthalpy-entropy correlations appears in the calculated horsepower and final temperature. The results in Table VII show from -1.0% to -11.2% deviation in horsepower and from +0.1°F to -1.1°F in expander discharge temperature.

If the same basis for ethane stripped as Table V is used, the estimated ethane recovery is:

LEE	86.0% @ -167.0°F
PROP-75	86.1% @ -167.1°F
SOAVE	87.0% @ -167.8°F
P-R	87.5% @ -168.2°F

This variation is primarily the effect of the K-value variation in this narrow temperature range. The methane K-values are 1.260 to 1.290; the ethane K-values are 0.0398 to 0.0406.

Demethanizer Reboiler. A fixed recovery was assumed by setting compositions of the residue gas and product as given in Table II. The feed to the demethanizer is set by the calculations from steps B and C above with a minor variation in the feed composition. With the column operating at 200 psia, the residue gas at a dew point of -167°F, and the product at a bubble point of 24.1°F, enthalpy calculations were made around the column for the various correlations. The reboiler duty was calculated as follows:

$$\text{Reboiler Duty} = \text{Enthalpy of the Residue} + \text{Enthalpy of the Product} - \text{Enthalpy of the Feed}$$

The calculated reboiler duty varies from +8.4% to 12.2%. A higher reboiler duty signifies a higher cooling duty available for the feed gas, which would result in a higher recovery.

Residue Gas Heating. Results of the calculations for heating the residue gas from -167°F to 110°F are also given in Table VII. This is the available duty for chilling the inlet gas in the gas-gas exchanger. The variation here is small, from -3.0% to +1.9%. This is not surprising since the residue gas is mostly methane and only sensible heat is involved.

External Refrigeration. The external refrigeration duty required for the plant is calculated from an overall heat balance of the entire plant.

$$\text{Refrigeration Duty} = \text{Feed Gas Cooling Duty} - \text{Reboiler Duty} - \text{Residue Gas Heating Duty}$$

The results listed in Table VII vary from -29.0% to -5.8%. The major contributor to these differences is the reboiler duty. The calculated low temperature (expander outlet) conditions are markedly different.

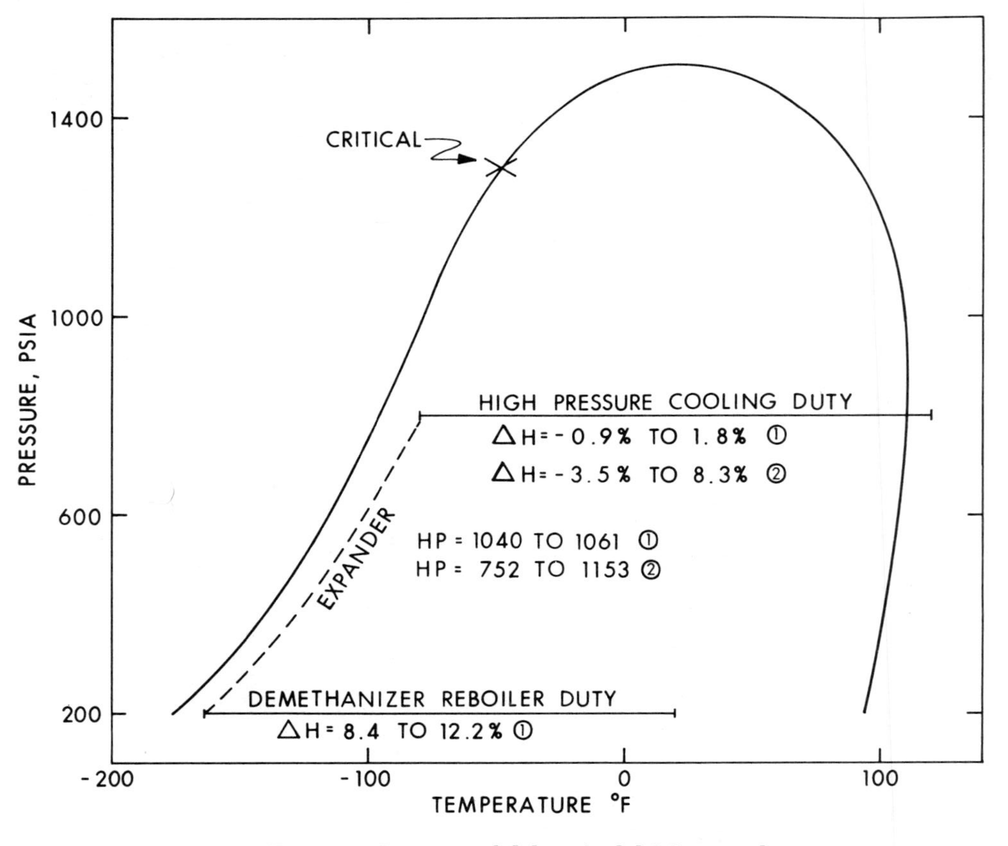

Figure 2. Limiting solubility of solid CO_2 in methane

Solids Formation by Carbon Dioxide

The possibility for freeze-up exists when carbon dioxide is present in the feed gas to an expander plant. For the calculations a feed composition of 0.52 mol% carbon dioxide is used. At fixed conditions, variations in the calculations for the carbon dioxide concentration arise solely from K-value correlations. The results from Section III of this paper, where thermodynamic properties are fixed by the Property-75 correlation and compositions are calculated for various K-value correlations, are selected as the basis for the comparison.

Any freeze-up will occur in the demethanizer. The feeds to this column are the expander outlet to the top of the column and the liquid from the high pressure separator at the fourth theoretical tray. A detailed tray-to-tray calculation was made for 8 theoretical trays using the Soave K-value correlation. At tray conditions, the limiting CO_2 content in the liquid phase was found from Figure 2. These results, presented in Table VIII, show the top three trays of the column are close to freeze-up conditions.

Flash calculations were made at the same pressure and temperature as the Soave calculations for the top three trays for all of the K-value correlations. Feed composition on each tray flash was set by the total composition from the tray-to-calculation. The amount of CO_2 was adjusted by the ratio of CO_2 knockdown in the expander outlet for the K-value correlation to the Soave case. The results in Table IX show that the Peng-Robinson correlation predicts that trays 2 and 3 will be only 0.21 to 0.25 mol% below freeze-up conditions. This is too close for operating control. A plant design based on the Peng-Robinson correlation without treating facilities would necessitate higher operating pressure for the demethanizer and a higher temperature and probably a lower recovery, in order to avoid solids formation. Other correlations such as K VAL and K DELTA would predict safe operation at 200 psia.

The carbon dioxide K-value has a definite effect on plant operation and recovery, since the carbon dioxide freezing conditions must be avoided. No data exist in the region of interest for the ternary methane-ethane-carbon dioxide system; such data would be useful for evaluation of these K-value correlations.

Conclusions

The effects of correlations on turboexpander plant design have been examined from two views: K-values and enthalpy/entropy. A summary of the results is given in Figure 3 for three portions of the plant. Detailed results are given in Tables III through IX.

K-Value Correlations. The K-values used have been shown to have a significant effect on plant design, more, in fact; than enthalpy/entropy. The total value of the plant is expressed in the total ethane recovery; this varied from 82.1% to 88.4%. The difference is equivalent to more than $200,000 per year in anticipated revenue.

TABLE VIII

TRAY-TO-TRAY DEMETHANIZER

SOAVE K-VALUES
PROP-75 THERMOPHYSICAL PROPERTIES
200 PSIA

Tray Number	Temperature °F	Carbon Dioxide mol %		Limiting Solubility mol %
		Vapor	Liquid	CO_2 Liquid
1	-166	0.32	1.72	2.45
2	-164	0.43	2.23	2.62
3	-158	0.62	2.85	3.27
Feed 4	-126	1.20	2.52	8.5
5	-121	1.80	3.38	-
6	-104	4.19	5.46	-
7	-66	10.5	7.29	-
8	-21	15.5	6.21	-
Reboiler	24.1	11.3	2.92	-

TABLE IX

SOLIDS FORMATION DEMETHANIZER

PROP-75 THERMOPHYSICAL PROPERTIES
200 PSIA

K-Value Correlation	Tray No. 1 -166°F K_{CO_2}	Tray No. 1 -166°F mol % CO_2 Liq.	Tray No. 2 -164°F K_{CO_2}	Tray No. 2 -164°F mol % CO_2 Liq.	Tray No. 3 -158°F K_{CO_2}	Tray No. 3 -158°F mol % CO_2 Liq.
GPA CONV	0.261	1.45	0.274	1.87	0.306	2.30
P-R	0.166	1.82	0.171	2.37	0.195	3.06
SOAVE	0.187	1.75	0.193	2.26	0.220	2.85
SH BWR	0.223	1.39	0.229	1.83	0.262	2.30
MARK V	0.219	1.55	0.225	2.00	0.257	2.54
K VAL	0.28	1.01	0.294	1.44	0.335	1.96
K DELTA	0.271	1.26	0.291	1.57	0.338	2.17
Figure 2		2.45		2.62		3.27

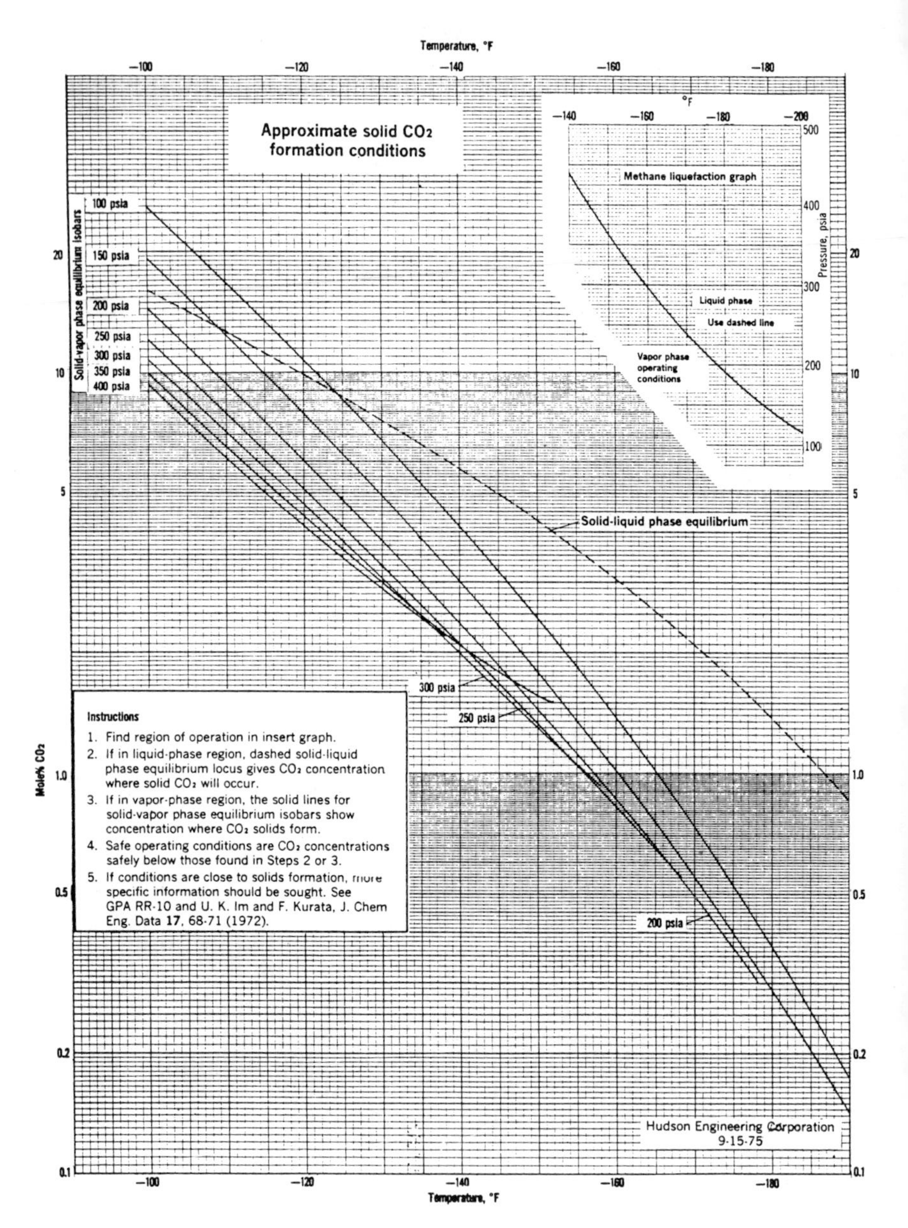

Figure 3. Phase envelope for feed gas with three process paths: Variations in predictions for different enthalpy correlations at a fixed K-value correlation denoted by ①; variations in predictions for different K-value correlations at a fixed enthalpy correlation indicated by ②. Approximate critical conditions are indicated by an X.

The high pressure cooling duty varies from -4.0% to 7.7% of the selected base. This 12% difference affects sizing of heat exchangers, amount of external refrigeration, and size of expander.

The expander horsepower varies from 752 to 1153. This calculation is coupled with the high pressure separator calculation. If in actual plant operations the design expander discharge temperature calculated is not reached, the total recovery will not meet design.

The design of the demethanizer tower, the reboiler, and the refrigeration requirement is a function of the total liquids condensed. The largest calculated value is twice the smallest.

Freeze-up due to carbon dioxide solid formation can occur in the cold portions of the process. The CO_2 K-value is important at conditions approaching the formation of solids, but no ternary experimental data now exist for comparison with the various correlations.

Enthalpy/Entropy Correlations. The most expensive item in a turboexpander plant is the compressor which is designed from enthalpy/entropy calculations. For a fixed horsepower the results are almost the same for Property-75, Peng-Robinson, GPA K&H Soave, GPA K&H Lee, GPA K&H Starling-Han BWR, and GPA-k method. The largest variation predicted is in the discharge temperature (maximum differences of 6.8^o and 5.4^oF) which would affect the discharge cooler sizes (assuming 120^oF) by 3%.

The effect on the high pressure cooling duty and the expander horsepower is rather small for the different H & S correlations as shown in Figure 3.

The major effect is seen in the calculated demethanizer reboiler duty, which varies from 8.4% to 12.2%. This is largely a result of the large differences in calculated enthalpy at the expander outlet conditions. If the enthalpy balance is in error, sufficient refrigeration will not be available and the expected recovery will not be obtained.

Acknowledgment

The authors thank Patricia J. Castillo, Charlotte L. Eberwein, and Deborah L. Woods for their efforts in the preparation of the manuscript. Also, the authors are indebted to Dr. J. Randall Johnson for his helpful comments.

Literature Cited

1. Elliott, D. G., McKee, R. L., and White, W. E., Gas Processors Assn. Proc. (1976) 55, 81.
2. White, W. E., Wilson, G., and Kobayashi, R., Nat. Gas Processors Assn. Proc. (1970) 49, 126.
3. GPSA Engineering Data Book, Ninth Edition, Gas Processors Assn., Tulsa, Okla., 1972, Section 5, "Centrifugal Compressors".

4. GPSA Engineering Data Book, Ninth Edition, Gas Processors Assn., Tulsa, Okla., 1972, Section 18, "Equilibrium Ratios".
5. Wilson, G., Adv. Cryog. Eng. (1964) 9, 168.
6. Redlich, O., and Kwong, I.N.S., Chem. Rev. (1949) 44, 233.
7. Lee, B. L., and Edmister, W. C., Nat. Gas Processors Assn. Proc. (1971) 50, 56.
8. Soave, G., Chem. Eng. Sci. (1976) 15, 59.
9. Peng, D.-Y., and Robinson, D. B., Ind. Eng. Chem. Fundamentals (1976) 15, 59.
10. Starling, K. E., and Han, M. S., Hyd. Proc. (May 1972) 51 (5), 129.
11. Benedict, M., Webb, G. B., and Rubin, L. C., J. Chem. Phys. (1940) 8, 334.
12. Leach, J. W., Chappelear, P. S., and Leland, T. W., AIChE J. (1968) 14, 568.
13. Chapela-Castañares, B. A., and Leland, T. W., AIChE Sym. Series (1975) 70, No. 140, 48.

Discussion

LYMAN YARBOROUGH

Comments:

Two constant equations of state can be fitted to specific fluid properties in the two-phase region up to and including the critical. However, if this is done, considerable accuracy is lost when calculating other fluid properties not used in the fitting process; i.e., if pure component vapor pressures and liquid phase densities up to the critical point and equality of vapor and liquid phase fugacities are used to determine the two constants, then properties such as enthalpy correction and the derivaties of $(\partial P/\partial v_T)$ at or near the critical point will be incorrect. These equations of state do not have a sufficient number of independent variables to fit all properties correctly.

Too many equations of state and modifications of equations of state are being used. The number of equations should be reduced by selecting the best ones, and those equations of state should be further developed.

Q: Because of environmental regulations, the concentration of dissolved components in water is of importance. Can the solubility of dissolved gaseous and liquid components in water be estimated by equations of state?

A: Presently available equations of state can be used to accurately calculate the water content of gas and hydrocarbon liquid phases but not the gaseous or liquid hydrocarbon content of the water phase. Equations of state can predict only qualitative results for the water phase.

Q: At cryogenic temperatures some natural gas mixtures can form two liquid phases, and liquid-liquid-vapor phase equilibria occurs. Can this behavior be calculated?

A: Yes, equilibrium for three-phase systems can be calculated for non-aqueous systems; but little data are available to compare against and good initial values are required to start the calculation.

Comments:

Calculations employing free energy minimization often show less sensitivity to starting values than other methods.

Additional equilibrium data on liquid-liquid-vapor phase systems are needed.

Q: How is the acentric factor calculated for petroleum fractions? What kind of values are used in equations of state?

A: The acentric factor can be predicted by adjusting it until the calculated normal boiling point for the fraction matches the one measured experimentally. Values of one and greater have been obtained in this manner.

Comments:

The acentric factor was developed as a perturbation from the spherical molecules and was never intended for use with large molecules. An alternative could be a parameter employed by Prigogine.

A large acentric factor doesn't make sense physically but is used to correlate vapor pressures for the petroleum fractions. Perhaps a vapor pressure correlation for petroleum fractions could supplant the use of acentric factor.

The Lee-Kesler method for correlating gas phase properties (reference equation and corresponding states approach) has been extended to vapor-liquid equilibria calculations. The calculation of accurate phase equilibria results is more difficult than for the properties of single phase fluids.

For highly asymmetric mixtures (small and large molecules), recent studies have shown that mixing rules which provide parameters intermediate between those of Van der Waal's method and Kay's method should be used. This may be helpful in the corresponding states approach.

It is not really possible to represent mixture behavior by a pseudo pure component approach. Possibly should separate correlation equation into repulsive and attractive potentials. The repulsive potential could be represented by the hard core potential and only the attractive potential correlated by the pseudocritical approach.

Comment by H. T. Davis:

In the emerging theory of interfacial thermodynamics, the properties of the thermodynamic functions (PVT or free energy quantities) at densities corresponding to thermodynamically unstable states play a crucial role. For example, the interfacial tension is computed from a formula calling for Helmholtz free energy data at every density (composition) between the compositions of the bulk phases in equilibrium. A particular implication of the theory is that Method (a) of Professor Prausnitz's classification is useful for interfacial calculations of vapor-liquid systems, but Method (b) is not. For those interested in the interfacial theory to which I refer, the following papers and references therein may be useful:

1. Bongiorno, V., and Davis, H. T., _Phys. Rev_.
2. Bongiorno, V., Scriven, L. E., and Davis, H. T., _J. Colloid and Interface Science_.

Comment by John P. O'Connell:

There are a number of aspects of improving equations of the "van der Waal" type (e.g., R.K.) which did not appear in the presentations and discussions, three of which I want to address. (I also feel that present knowledge of the critical state scaling laws should be included in future developments.)

The form of the equation is

$$Z = \frac{pv}{RT} = f(v/b) - \frac{a}{vRT}\, g(v/b)$$

where a, b can be functions of T, $\underline{x}$.

1. Excluded Volume

The f(v/b) represents excluded volume effects of the liquids by a rigid-body formula, the van der Waals form

$$f(v/b) = \frac{v/\Sigma\, x_i b_i}{v/\Sigma\, x_i b_i - 1}$$

is not a good one. A better one is that of Mansoori, et al. (_1_) known as the "Carnahan-Starling" form

$$f(v/b) = [1 + v/b + (v/b)^2 - (v/b)^3]/(1 - v/b)^3$$

2. Second Virial Coefficients

These equations yield expressions

$$\beta(T,x) = \sum_{ij}\sum x_i x_j \beta_{ij}(T) = b(T,\underline{x}) - a(T,\underline{x})/RT$$

Since highly accurate forms for β_{ij} are known, The T,$\underline{x}$ behavior of a and b should be chosen appropriately. This can be quite important even for the liquid since second virial contributions to the integrals for fugacity and enthalpy are not negligible.

3. Thermal Pressure Coefficient and Compressibility of Liquids

For liquids where $T < T_c$, $v < v_c/2$

$\left.\frac{\partial p}{\partial T}\right)_{v\underline{x}}$ and $\left.\frac{1}{T}\frac{\mu p}{\alpha v}\right)_{T\underline{x}}$ are very weak functions of T over a wide range of T,v.

Thus, setting $\left(\frac{\partial^2 pv/R}{\partial T^2}\right)_{v,\underline{x}} = 0$ we find a relation for the temperature variation of a and b

$$\left(T\frac{df}{df} - \frac{a}{Rv}\frac{dg}{db}\right)\left(\frac{\partial^2 b}{\partial T^2}\right)_{\underline{x}} + 2\left[\frac{df}{db} - \frac{1}{Rv}\left(\frac{\partial a}{\partial T}\right)_{\underline{x}}\frac{dg}{db}\right]\left(\frac{\partial b}{\partial T}\right)_{\underline{x}} = \frac{g}{Rv}\left(\frac{\partial^2 a}{\partial T^2}\right)_{\underline{x}}$$

If a has a T dependence which yields a second derivative, b must also depend on T. Since liquid results depend strongly on the value of b, this could be significant.

Brelvi (2,3) showed that this result could be generalized to

$$\ell n\left[1 + \frac{v}{K_R RT}\right] = \ell n\left[1 + \frac{\partial(p/RT)}{\partial(1/v)}\right]_{T,\underline{x}} = A + B(v^*/v) + C(v^*/v)^2 + D(v^*/v)^3$$

where $v^* = v^c$ for nonpolar substances and $v^* < v^c$ for polar substances. For this 5% correlation of K_T, the substances included Ar, alcohols, hydrocarbons from CH_4 to those with MW > 400, Amines, etc.

Gubbins (3) showed how the generalization to all fluids should be expected (though not why the T dependence is absent). It was shown that for $T > T_c$ a 2-parameter corresponding states plot for $K_T RT/v$ of H_2O and Ar is successful over wide ranges of T/T^* and v/v* with the values of T* and v* corresponding to intermolecular force parameters of the spherically symmetric forces. These should be related to the a and b parameters of the equation of state since compressibility is easily measured.

References

1. Mansoori, G. A., Carnahan, N. F., Starling, K. E., and Leland, T. W., J. Chem. Phys. (1971) 54, 1523.
2. Brelvi, S. W. and O'Connell, J. P., AIChE J., (1972) 18, 1239; (1975) 21, 171, 1024.
3. Gubbins, K. E. and O'Connell, J. P., J. Chem. Phys. (1974) 60, 3449.

Q: Approximately what proportion of the total fluid in your pvT cell was the solid precipitate, and what were the API gravities of the two oils tested?

A: The solid precipitate varied from 2% to 5% of the total fluid, and the gravity for the oil studied at 130F was 40°API while the gravity for the oil studied at 255F was 26°API.

Q: What is the minimum value of the interfacial tension that can be measured using the pendant drop apparatus? Has this been checked against other experimental techniques?

A: Experience has shown the pendant drop and spinning drop techniques agree down to interfacial tensions of 0.01 dynes/cm. The spinning drop apparatus is not applicable at pressures up to 2000 psia.

Comment:

The Weinaug and Katz correlation was based on surface tension and is not designed to correlate interfacial tension.

Q: Does water affect the process of CO_2 displacing oil?

A: No, the water is nearly immiscible in both phases. The interfacial tension between the CO_2-water and water-oil phases is high. Water is often injected alternately with the CO_2 for mobility control.

Q: Is it possible to use any activity coefficient model and derive a consistent equation of state in a manner similar to that presented?

A: No. One must be careful of the reference state. In the derivation shown in the paper, the reference state is the pure component.

Q: What parameters are required to characterize the heavy fractions which are expected to be rich in aromatic hydrocarbons?

A: I don't know.

Q: This problem of characterization is still being studied by both API and GPA for petroleum base fluids. At least one easily measured parameter is required in addition to normal boiling point and specific gravity.

Comments:

Some Russian articles have suggested refractive index as a characterizing parameter.

Refractive index correlates too closely with density or specific gravity to be considered as another independent parameter.

A parameter suggested by Prigogine for large molecules combines the density, the coefficient of thermal expansion, and the isothermal compressibility of the fluid. This might be useful for large petroleum molecules also.

Q: What is the significance of the assumed hole in the liquid state in your development?

A: It is really just an adjustable parameter, a sort of free volume.

Q: In the derivation, what assumptions were made concerning the size of the holes?

A: No assumptions were necessary because the hole volume cancelled out.

Q: When comparing the phase equilibria prediction methods you used against experimental data, which one appears to be best?

A: The Soave Redlich-Kwong marketed by GPA. However, one correlation may be good in a certain region but not as good in another region.

Q: Why show comparisons with all these correlations? Why not throw out some that are known not to be applicable?

A: All correlations used in process simulators which are available on a time-sharing basis, or which are marketed by GPA, were tested to show which are poor and which are good for this type of plant. This allows the engineer to see why some correlations should not be used.

Comments:

Some of the two constant equations of state predict good phase equilibria results at cryogenic temperatures. Experimental data measured on an LNG plant were used to compare against predictions from both the Soave Redlich-Kwong and the Peng-Robinson equations of state marketed through GPA, and both calculation methods compared well with the data.

Phase equilibria calculations from both an Orye-type BWR and the Soave Redlich-Kwong equation of state were compared with data from an LNG plant. Both correlations showed good comparisons with the plant data and experimental data measured by P-V-T, Inc. The enthalpies predicted using the Soave Redlich-Kwong equation of state were poorer than those predicted using the Lee-Kesler correlation.

The calculated ethane recovery is very sensitive to the method used to characterize the heptanes plus fraction. If this fraction contained primarily naphthenic and aromatic components, the calculated ethane recovery could be several percent different than if the fraction contained only n-paraffin hydrocarbons. The characterization of the heavy fraction is important even when that fraction contsists of less than 0.5 mol % of the inlet feed.

16

Prediction of Thermodynamic Properties

JOHN S. ROWLINSON

Physical Chemistry Laboratory, Oxford, England

No chemical engineer has ever relied on directly measured values for all the thermodynamic properties he uses. It is one of the great virtues of classical thermodynamics (indeed, it has been said that it is its only virtue (1)) that it allows us to calculate one physical or chemical property from another, for example, a latent heat of evaporation from the change of vapour pressure with temperature. Less obvious, but almost equally secure calculations can be made by fitting measured values to an empirical equation and calculating other properties by a subsequent manipulation of that equation. Thus we can fit the pressure of a gas to a Redlich-Kwong (RK) or Benedict-Webb-Rubin (BWR) equation, from which the other properties such as changes of enthalpy and entropy are then derived. Similarly we fit activity coefficients to Wilson's equation and calculate K-values.

None of this, necessary and useful though it is, is what is meant by *prediction*, for that term is usually reserved for calculations which go beyond the comfortable security of classical thermodynamics (2,3). They may still be empirical, for example, the extrapolation of a BWR equation to a range remote from that of the experimental evidence used to determine its parameters, or the estimation of BWR parameters for a mixture from those for the pure components. It is, however, becoming more common to restrict the word prediction to methods of calculation which are based, at least in part, on the two theoretical disciplines of quantum and statistical mechanics, and it is that field which is the subject of this review.

The distinction between empiricism and theory is not sharp, nor is it unchanging. Thus before 1938-39 we should have said that the principle of corresponding states was empirical, but after the work of de Boer and of Pitzer (4,5,6) we could see its basis in statistical mechanics, and it now ranks as a theoretically-based principle. Moreover it was only after its theoretical basis, and hence its theoretical limitations, had become evident that we were able to discuss with success systematic departures from the principle. These departures are now handled by means of the acentric factor, as Professor Pitzer has described in the opening paper of this meeting. The sequence of empiricism, followed by theory, followed by an extension which is empirical in form but which is guided by theory, is one which is, I believe, likely to be followed increasingly in the development of methods of prediction of value to the chemical engineer.

There is no need to stress the confidence we all feel in calculations that are securely founded on classical thermodynamics, but it is important to emphasize that those based on statistical mechanics are not inherently any less secure. This science is now about 100 years old, if we reckon Maxwell, Boltzmann and Gibbs to be its founders, and that is ample time for any faults in the foundations to have revealed themselves. In practice, the calculations may be more speculative because of approximations that we have introduced, but the existence of these approximations is always evident, even if their consequences are not fully known. When the approximations are negligible our confidence in the calculations should be high (2,3).

The Dilute Gas

Statistical calculations are most accurate in the calculations of the properties of the dilute or perfect gas, where they provide the link between quantal and spectroscopic determinations of molecular energy levels, on the one hand, and molar thermodynamic properties on the other. Even twenty years ago (7) it was accepted that statistical calculations were more accurate than experimental measurements for the heat capacities, energies and entropies of such simple gases as Ar, N_2, H_2, CO, CO_2 and H_2O; indeed it would now be hard to find a recent measurement of these properties. The case for statistical calculation, while still strong, is not so overwhelming once we go to more complicated molecules. The limiting factor is the ability of the spectroscopist to assign energy levels to molecules, an ability which depends on the symmetry and ridigity of the molecule. If these desirable properties are absent, as in the alcohols, for example, then his job is difficult, but even here his accuracy is now not necessarily worse than that of all but the very best measurements. The position has improved considerably over that described by Reid and Sherwood (8) ten years ago, as is shown by the methods and results for organic molecules described in a review by Frankiss and Green (9). The dilute gas, its mixtures, and chemical reactions occurring therein, can now be regarded as a solved problem.

The Virial Expansion

When we go to the next level of difficulty, the imperfect gas, then the claims made for statistical mechanics, although still strong, must be moderated. We note, first, that it tells us the appropriate form of the equation of state, namely that the compression factor $Z = pV/nRT$, has a power series expansion in the molar density n/V, where n is the amount of substance. This expansion is convergent at the low densities and divergent at high. We do not know the limit of convergence but it must be below the critical density at temperatures at and below T^c. Perhaps fortunately, the mathematical range of convergence is unimportant in practice since the expansion is useless when it needs to be taken to more than two or three terms.

More important is that statistical mechanics does tell us the composition dependence of each of the virial coefficients of this expansion. The nth virial coefficient is determined by the forces between a group of n molecules, and so is a polynomial of degree n in the mole fractions y_i. Thus for the second virial coefficient

$$B = \sum_i \sum_j y_i \, y_j \, B_{ij} \qquad (B_{ij} = B_{ji}), \tag{1}$$

which is a result we owe to statistical not classical thermodynamics.

Intermolecular Forces

When we come to calculate the second virial coefficients of pure and mixed gases we come to one of the principal difficulties of all methods of prediction based on statistical mechanics - what do we know of intermolecular forces? Here caution is needed, for twenty years ago we thought that we knew more about them than we did. The Lennard-Jones potential is

$$u(r) = 4\varepsilon[(\sigma/r)^{12} - (\sigma/r)^6] \tag{2}$$

where ε is the depth at the minimum and σ, the collision diameter: i.e. $u(\sigma) = 0$. Because of some cancellation of errors, which need not be discussed here, and under the influence of an important book by Hirschfelder, Curtiss and Bird (10), it was thought that the Lennard-Jones potential was an accurate representation of the forces between simple molecules such as Ar, N_2, O_2, CH_4, etc. The accepted value of ε/k for the Ar-Ar interaction was 120K; we know now that the potential has a different functional form and that ε/k is 141 $\pm$ 2 K. Moreover it is now only for the inert gases (11) and for some of their mixtures that we are confident that we know ε, σ, and the functional form of $u(r)$. Nevertheless this salutary shock to our confidence should not lead us into the opposite error, since for many pairs of molecules, like or unlike, we still know much about the strength of the intermolecular potential relative to that of another pair, and such relative rather than absolute values often suffice for prediction. Fortunately the second virial coefficient and many other thermodynamic properties of gases and liquids are not very sensitive to the exact form of the potential (indeed our error with the

Lennard-Jones potential could not have arisen if they were), and so predictions can be better than the potentials they depend on. Thus the editions of the American Petroleum Institute Research Project 44 published since the 1950's have relied on the Lennard-Jones potential for the calculation of the properties of methane to 1500 K and 100 atm. The temperature is 1000 K higher than any measurements of B but I think that the results are more reliable than any that could be obtained by the extrapolation of an empirical equation of state, however complicated, that had been fitted to the p-V-T properties at low temperatures.

When we come to mixtures even relative knowledge of the strength of intermolecular forces is harder to obtain. In practice we have to back-calculate from one or more observed properties in order to obtain the strengths of the unlike forces which we then use to calculate another property which we wish to know. Thus in one calculation a knowledge of the diffusion coefficient, D_{12} for the system CO_2 + N_2 was used to aid the calculation of the virial coefficients B_{12} and C_{112} and so to predict the solubility of solid CO_2 in compressed air (12). More usual is the use of one thermodynamic property (B_{12}, T^c as a function of y, etc.) to calculate another (K-values, enthalpy of liquid mixtures, etc.). In each case, however, we need an experimental measurement on which to base our estimate of the strength of the 1-2 forces if we are to have a secure base for our calculations. Intermolecular forces are, to a good approximation, forces between pairs of molecules, and so experimental evidence on the properties of binary mixtures will always be the foundation of satisfactory methods of prediction. Fortunately there are now rapid, indeed almost automatic, methods of measuring heats of mixing and vapour pressures at least for binary mixtures at and near room temperatures. It is only the need for data at inconvenient pressures and temperatures and, above all, the need for the properties of multi-component mixtures, that justifies the great effort put into the development of methods of prediction.

Liquid and Dense Gas Mixtures

In using the virial equation of state our problems are those of knowing enough of the intermolecular forces; the statistical mechanics we are using is essentially exact. When we come to dense gases and liquids the intermolecular force problem is still with us but we have now the additional problem of having to introduce approximations in the statistical mechanics. This is not the place to attempt to review progress in the theory of liquids and liquid mixtures (13, 14, 15, 16), although this has been gratifyingly rapid in the last decade, but simply to state that the methods of prediction of value to engineers are still based on the state of the statistical art of about ten years ago. No doubt the more recent work on theory will, in time, be used, but this has not yet happened to any appreciable extent. This section of this review if therefore restricted to methods based on the principle of corresponding states, on its extension by means of the acentric factor to the less simple fluids, and to mixtures by the best pseudo-critical approximation available to us.

The principle was first used to relate the p-V-T properties of one pure substance to those of another by writing:

$$P_i(V,T) = (h_{ii}/f_{ii})[P_o(V/h_{ii}, T/f_{ii})] \quad (3)$$

where f_{ii} and h_{ii} are the ratios of the critical constants:

$$f_{ii} = T_i^c/T_o^c \text{ and } h_{ii} = V_i^c/V_o^c \quad (4)$$

The subscripts are doubled because the size of the parameters f and h depends on the strengths of the potential energies $u_{ii}(r)$ and $u_{oo}(r)$ between two molecules of species i and two of species o.

$$f_{ii} = \varepsilon_{ii}/\varepsilon_{oo} \qquad h_{ii} = \sigma_{ii}^3/\sigma_{oo}^3 \quad (5)$$

These equations enable the (unknown) pressure of substance i to be calculated from the (presumed known) pressure of substance o at a different volume and temperature. A more generally useful equation is that relating the configurational parts of the Helmholtz free energy (17)

$$A_i(V,T) = f_{ii}[A_o(V/h_{ii}, T/f_{ii})] - nRT\ell nh_{ii} \quad (6)$$

from which equation (3) follows at once, since:

$$P = -(\partial A/\partial V)_T.$$

Equations (3) to (6), although simple, are not accurate except for closely related pairs of substances, for example Kr from Ar, or i-C_4H_{10} from n-C_4H_{10} that is, pairs for which u(r) can be expected to have geometrically similar shapes. They have the consequence that for both substances, i and o, the reduced vapour pressure (P^σ/P^c) is the same function of the reduced temperature (T/T^c), and that both substances have the same value for the critical compression ratio $Z^c = P^cV^c/nRT^c$. If we choose argon as a reference substance, subscript o, then $(P^\sigma/P^c)_o$ is 0.100 at $(T/T^c) = 0.7$, and $Z_o{}^c = 0.293$. The same ratios, namely 0.100 and 0.293, are found for krypton and xenon, but lower values for all other substances other than the "quantal" fluids, hydrogen, helium and neon. Moreover, the departures of other substances from the reduced behavior of argon are not random but can be well correlated by a third parameter in addition to f and h, which is a measure of the increasing departure of u(r) from the spherical nonpolar form it has for argon.

If, following Pitzer (18, 19, 20), we define an acentric factor, ω, by:

$$\omega_{ii} = -1.000 - \log_{10}(P_i^{\sigma}/P_i^{c}) \tag{7}$$

then ω is zero for Ar, Kr, and Xe and takes the following values for some other substances:

Substance	ω
O_2	0.021
N_2	0.040
CH_4	0.013
C_2H_6	0.105
C_3H_8	0.152
C_6H_6	0.215
H_2O	0.348

The change of Z^c from the value of 0.293 is found to be related to the value of ω:

$$Z^c = 0.293/(1 + 0.375\ \omega). \tag{8}$$

Either ω or $(Z^c - 0.293)$ can be used as a third parameter with which to extend the principle of corresponding states to substances which depart from strict agreement with equations (3) to (6). The first was used by Pitzer and his colleagues, (18-20) and the second by Hougen, Watson, and Ragatz (21). The first, has the practical advantage that the vapour pressure at $(T/T^c) = 0.7$ is known for most substances since this temperature is not far above the normal boiling point.

Pitzer wrote, for any thermodynamic function Y:

$$Y_i(V/V_i^c,\ T/T_i^c) = Y_o(V/V_o^c,\ T/T_o^c) + \omega\ Y_o'(V/V_o^c,\ T/T_o^c) \tag{9}$$

where Yo is the same function as for argon and Y_o' is an empirically determined correction function. A more convenient way of expressing the same result was devised by Leland and his colleagues (22, 23) who introduced what they called shape factors, θ and ϕ. These are functions of the reduced density and temperature which are defined as follows. Let us suppose that, for a pair of substances, i and o, which do not necessarily conform mutually to a principle of corresponding states, we use (1) and (4) to define functions f_{ii} and h_{ii}. It can be shown (17, 24) that these equations can be satisfied simultaneously and consistently only by one pair of values of f_{ii} and h_{ii} at each V and T of the substance i. Clearly if i and o do conform to the principle (i.e. if $\omega_{ii} = \omega_{oo}$) then f and h, so defined, would be constants given by equation (4). If i and o do not conform, then they are slowly varying functions of V and T. Leland therefore writes:

$$f_{ii,oo}(V/V_i^c,\ T/T_i^c) = (T_i^c/T_o^c) \times \theta_{ii,oo}(V/V_i^c,\ T/T_i^c)$$

$$h_{ii,oo}(V/V_i^c,\ T/T_i^c) = (V_i^c/V_o^c) \times \phi_{ii,oo}(V/V_i^c,\ T/T_i^c) \qquad (10)$$

where θ and ϕ are the shape factors of substance i, with reference to substance o. They can be expressed:

$$\theta_{ii,oo} = 1 + (\omega_{ii} - \omega_{oo}) \times F_\theta(V/V^c,\ T/T^c)$$

$$\phi_{ii,oo} = 1 + (\omega_{ii} - \omega_{oo}) \times F_\phi(V/V^c,\ T/T^c) \qquad (11)$$

where within the ambit of Pitzer's formulation, F_θ and F_ϕ are universal functions, determinable empirically by comparing a few fluids with, say, argon (25, 26).

With this apparatus we are equipped to calculate the configurational free energy, and hence the other thermodynamic properties, for a single component, i, from a knowledge of $A_o(V,T)$ for argon (or other reference substance of our choice), from the universal functions F_θ and F_ϕ, and the constants V_i^c, T_i^c, and ω_{ii}. The limitation of the method is that the departure of i from the reduced properties of o should be sufficiently small and regular to be described by the one parameter ω. In practice this means that if our reference substance is one for which ω is approximately zero, we are restricted to substances for which ω is less than about 0.25. This includes most simple molecules except NH_3 and H_2O, and the lower hydrocarbons, but excludes such technically important substances as the alcohols and the lower amines, ammonia and water.

Many attempts have been made to extend the recipe above to mixtures by introducing parameters f_x and h_x (or, what is equivalent, T_x^c and V_x^c) which are averaged parameters ascribed to a single hypothetical substance, subscript x, which is chosen to represent the mixture. That is, we write for the configurational free energy of a mixture of C components:

$$A_{mixt}(V,T) = A_x(V,T,) + \sum_{i=1}^{c} n_i\ RT\ln x_i \qquad (12)$$

$$A_x(V,T) = f_x \cdot A_o(V/h_x,\ T/f_x) - n\ RT\ln h_x \qquad (13)$$

where $\Sigma\, n_i = n$ and $\sum_{i=1}^{c} x_i = 1.$ (14)

The parameters f_x and h_x clearly depend on all molecular interactions in the mixture, that is, on both those between like molecules, f_{ii} and h_{ii}, and those between unlike molecules, f_{ij} and h_{ij}. Since intermolecular forces are, at least approximately, pair-wise additive, we do not need to introduce three-body parameters, f_{ijk}, etc. The most successful recipe for combining the parameters, and the one for which there are the best theoretical arguments (27) is that usually called the van der Waals approximation:

$$f_x h_x = \sum_{i=1}^{c} \sum_{j=1}^{c} x_i x_j f_{ij} h_{ij}$$

$$h_x = \sum_{i=1}^{c} \sum_{j=1}^{c} x_i x_j h_{ij}. \tag{15}$$

These equations, together with equations (12) to (14), are a solution to the problem of calculating the thermodynamic properties of a multi-component mixture, provided:

(1) That the change of f and h with reduced volume and temperature can again be incorporated into shape factors, as for pure substances in equation (10) and (11).

(2) That we can assign values to the cross-parameters f_{ij} and h_{ij} with i not equal to j.

The first problem has been solved, but needs care if thermodynamic consistency is to be preserved (17, 22-26). The second is usually solved by writing:

$$f_{ij} = \xi_{ij}(f_{ii} f_{jj})^{\frac{1}{2}} \tag{16}$$

$$h_{ij} = \eta_{ij}(\tfrac{1}{2} h_{ii}^{1/3} + \tfrac{1}{2} h_{jj}^{1/3})^3 \tag{17}$$

$$\omega_{ij} = \tfrac{1}{2}(\omega_{ii} + \omega_{jj}). \tag{18}$$

If nothing is known of the binary system formed from species i and j then it is usual to adopt the Lorentz-Berthelot assumption:

$$\xi_{ij} = \eta_{ij} = 1. \tag{19}$$

Measurements of almost any thermodynamic property of the binary mixture allows at least one of these parameters to be determined more precisely. It is usual to retain $\eta_{ij} = 1$ and to use the binary measurements to estimate ξ_{ij}. Typical values are: (28-31)

Substances	ξ
$Ar + O_2$	0.99
$N_2 + CO$	0.99
$N_2 + CH_4$	0.97
$CO + CH_4$	0.99
$CH_4 + C_2H_6$	0.99
$CH_4 + C_3H_8$	0.97
$C_2H_2 + C_3H_8$	1.00
$C_3H_8 + n - C_4H_{10}$	0.99
$CO_2 + CH_4$	0.94
$CO_2 + C_2H_6$	0.91
$CH_4 + CF_4$	0.92
$C_6H_6 + c - C_6H_{12}$	0.97

Such figures are subject to an uncertainty of at least 0.01, since the theory does not exactly correspond with reality and so different properties of the binary mixture lead to slightly different values of ξ. The properties most commonly used to estimate ξ are the second virial coefficient and T^c as functions of x, and the excess Gibbs free energy and enthalpy. There is now abundant evidence that ξ is usually significantly less than unity, particularly in mixtures of chemically different type. The use of values determined from binary mixtures leads to more accurate predictions for multi-component mixtures than the universal use of equation (19).

The principle of corresponding states, extended as above to mixtures of acentric molecules, has been applied to the calculation of many of the properties needed for the design of separation equipment. The examples reviewed briefly here are taken from our own work on cryogenic fluids, liquified natural gas (LNG), mixtures of hydrocarbons, and mixtures of carbon dioxide with hydrocarbons. In all this work methane was used as the reference substance.

Liquid-Vapour Equilibrium. For mixtures of $Ar + N_2 + O_2$ the method described above yields excellent K-values (28). Similar results were obtained by Mollerup and Fredenslund for $Ar + N_2$, using a method almost identical with that described here (32). The principal error in the work on the ternary system arose in the pure components. Pitzer's acentric factor, although it provides an exact fit to the vapour pressure at $(T/T^c) = 0.7$, leads to small errors in the vapour pressures at other temperatures. The boiling point of each component is almost certainly one of the experimental facts known before the calculations are started for the multi-component mixture, and so it is natural to modify the procedure above in such a way as to use this information. This can be done by treating ω as a floating parameter which, at each pressure, is chosen afresh so as to reproduce exactly the boiling point of each component. If this is done, (29) then the liquid vapour equilibrium of $Ar + N_2 + O_2$ is reproduced with a mean error of 0.14% which is almost within experimental error, and excellent representations are obtained of the systems: $C_3H_8 + C_3H_6$: $C_3H_8 + n\text{-}C_4H_{10}$: $C_2H_6 + C_6H_6$: and $CO_2 + n\text{-}C_4H_{10}$ (29).

Compression factors Z, and enthalpies of one-phase system. For one-phase systems comparison of theory with experiment is hindered by the lack of good experimental results for multi-component systems. However, Z is given (29) to 0.1 - 1.0% for pressures up to 200 to 300 bar for air, for $CH_4 + H_2S$, and for $CH_4 + CO_2$. The enthalpy of air is equally well predicted (29).

Azeotropic Lines. The calculation of liquid-vapour equilibrium includes, in principle, the determination of the conditions for azeotropy. Nevertheless, in practice these are best formulated and tested for in a separate calculation (30). Values of ξ_{12} less than unity lead correctly to the appearance of azeotropes at high reduced temperatures for the systems: $CO_2 + C_2H_6$: $CO_2 + C_2H_4$: and $CO_2 + C_2H_2$ (30).

Critical Lines. The calculation of the thermodynamic conditions for critical points in even a binary system is a severe test of any method of prediction, because of the surprising variety of critical lines that can occur (2). The Redlich-Kwong equation has been used by chemical engineers (33, 34, 35) to represent the simplest class of binary system, when there is but one critical line in p-T-x space. This joins the gas-liquid critical point of the two pure components. The treatment described above can represent, quantitatively for simple systems, and qualitatively for more complex systems, the critical lines of many binary mixtures (30). In the more complex of these, the lines representing liquid-liquid critical states intrude into the gas-liquid critical region, giving rise to a topologically wide variety of behavior.

Density of LNG. An accurate knowledge of the density of LNG is needed for its metering and pumping. The densities of some natural and some "synthetic" mixtures have been measured but it is impossible to cover all compositions that might be encountered in practice. The problem of predicting densities in such a system is one that is ideally suited to the method described above.

The results obtained were incorporated in a program which specified first whether the conditions are those of: (1) the compressed liquid, (2) the liquid at a bubble point of specified temperature, (3) or the liquid at a bubble point of specified pressure. The program then yielded the liquid density with an accuracy usually better than 0.2%, and, for conditions (2) and (3), the composition, pressure, (or temperature) and density of the vapour phase (31).

This program was prepared by Mollerup (36) and was issued by him to many interested in testing this method of prediction. Saville and his colleagues (37) later modified and shortened programs of this kind, first by algebraic improvements which removed one layer of iteration, secondly by using more efficient programming and thirdly by using Stewart and Jacobsen's equation for nitrogen as the reference equation (38). (Vera and Prausnitz (39) had earlier pointed out the advantages of using an explicit equation of state: they chose Strobridge's equation. Mollerup used an explicit equation for methane.) It is difficult to compare rates of computation, but it may be useful to estimate costs. At the standard rates quoted by commercial bureaux in Britain, Saville estimates that the calculation of the equilibrium states of liquid and vapour in a multi-component mixture and the tabulation of their enthalpies, densities and fugacities, costs now 3 cents per point. (Within the university the rates are only 2% of this, but these costs are not economically realistic.) A sum of 3 cents is, of course, negligible for a small number of calculations. It is appreciable, but not prohibitive, if the program is to be run many times as, for example, in an iterative design calculation for a distillation or stripping column.

One point which is worth emphasizing here is the need in such calculations for high accuracy in "difficult" regions of the phase diagram. Separation processes may account for half the capital cost of a modern plant, and the capital cost of separation varies roughly as $(\ell n\ \alpha)^{-1}$, where α is the volatility ratio. Hence high accuracy is needed in those regions, often near infinite dilution or azeotropes, where α is close to unity. The testing of methods of prediction does not usually pay enough attention to this point. Moreover the cost of error in these fields make it almost certain that the design is tested experimentally before a final decision is taken. There are definite limits to what predictive methods can be expected to do.

Advantages and Disadvantages of the Principle of Corresponding States as a Method of Prediction

The method described above is truly predictive in that it uses thermodynamic information from one- and, if possible, two-component systems to calculate the same or different thermodynamic properties of multi-component mixtures. There are other methods which have been used for the same purpose of which one of the most powerful is the generalization to mixtures of the Benedict-Webb-Rubin equation, and of equations derived from it. For each substance we must determine a set of eight or more parameters from the p-V-T properties. If we use the BWR equation for mixtures by writing for each of its eight parameters X an equation of the form:

$$X_x^{1/n} = \sum_{i=1}^{c} x_i X_i^{1/n} \quad \text{where } (n = 1,2, \text{ or } 3) \tag{20}$$

then the equation has predictive power--we obtain the properties of the mixture from those of the pure components. But such use, although often of value, suffers from the same loss of accuracy as the use of the principle of corresponding states with all $\xi_{ij} = 1$.

If we are to get the best use out of equations such as the BWR then it is necessary both to allow some of the parameters to become themselves functions of density or temperature and to introduce parameters which represent the actual behaviour of the relevant binary system (40, 41). We have then a flexible and accurate method of predicting the properties of multi-component mixtures.

When comparative tests have been made, the accuracy of the BWR equation and of the principle of corresponding states has been about the same (29), but the extended BWR equations are more accurate for the systems to which they have been fitted. The use of the principle of corresponding states is now little more expensive in computing time.

The advantages of the principle are, first, that it is more closely tied to theory and so can more safely to extrapolated to new regions of pressure and temperature. Its second advantage is that it is much more easily extended to take in new components. All we need are T_i^c, V_i^c, ω_{ii}, and estimates of ξ_{ij} for the binary systems formed from the new and each of the old components. If the latter are not known then they can be put equal to unit or, better, guessed by analogy with chemically similar systems. To extend the BWR, or Stewart and Jacobsen's equation, or Bender's equation (41) to additional components requires a large body of p-V-T and, preferably, calorimetric information on the new components, and a program to fit between eight and 20 parameters. A similar new fit must be made for each of the binary systems involving the new component if the best accuracy is to be achieved.

The greatest limitation of both methods is their restriction to comparatively simple molecules. The problem of predicting with such high accuracy the properties of multi-component systems containing ammonia, water, alcohols, phenols, amines, etc. is still substantially unsolved. It is hard to say how rapid progress will be here. Undoubtedly there are regularities in the behaviour of such series of compounds as the alcohols, but for two reasons I do not want to speculate further on how mixtures containing highly polar substances can be handled. The first is that, at least in the first instance, such methods are likely to be more empirical than those I have described, perhaps of the kind called "molecular thermodynamics" by Prausnitz (42), and the second is that I do not know how far companies and other agencies have gone in developing such methods in schemes of prediction that are wholly or partly unpublished.

Abstract

If the prediction of thermodynamic properties is to be soundly based then it requires an understanding of the properties of molecules and the forces between them, and of the methods of statistical mechanics. This paper reviews the extent of our knowledge in these fields and of our ability to make useful predictions. Particular emphasis is placed on methods that derive from the principle of corresponding states.

Literature Cited

1. McGlashan, M. L., "Chemical Thermodynamics", Vol. 1, Chap. 1. Specialist Report, Chemical Society, London, 1973.
2. Bett, K. E., Rowlinson, J. S., and Saville, G., "Thermodynamics for Chemical Engineers", Chapt. 9, M.I.T. Press, Cambridge, Mass., 1975.
3. Reed, T. M. and Gubbins, K. E., "Applied Statistical Mechanics", McGraw-Hill, New York, 1973.
4. de Boer, J. and Michels, A., Physica, (1938) 5, 945.
5. Pitzer, K. S., J. Chem. Phys., (1939) 7, 583.
6. Guggenheim, E. A., J. Chem. Phys., (1945) 13, 253.
7. "Tables of Thermal Properties of Gases", National Bureau of Standards Circular 564, Washington, D.C., 1955.
8. Reid, R. C. and Sherwood, T. K., "The Properties of Gases and Liquids: Their Estimation and Correlation", Chapt. 5, 2nd edn., McGraw-Hill, New York, (1966).
9. Frankiss, S. G. and Green, J. H. S., "Chemical Thermodynamics", Vol. 1, Chap. 8, Specialist Report, Chemical Society, London, (1973).
10. Hirschfelder, J. O., Curtiss, C. F. and Bird, R. B., "Molecular Theory of Gases and Liquids", John Wiley, New York (1954).
11. Smith, E. B., Physica, (1974), 73, 211.
12. Rowlinson, J. S., Chem. Ind. (1961), 929.
13. Smith, W. R., "Statistical Mechanics", Vol. 1, Chap. 2, Specialist Reports, Chemical Society, London, 1973.
14. McDonald, I. R., Vol. 1, Chap. 3, Ibid.
15. Gray, C. G., "Statistical Mechanics", Vol. 2, Chap. 5, Specialist Reports, Chemical Society, London, 1975.
16. Hansen, J. P. and McDonald, I. R., "Theory of Simple Liquids", Academic Press, London, 1976.
17. Rowlinson, J. S. and Watson, I. D., Chem. Eng. Sci., (1969) 24, 1565.
18. Pitzer, K. S. and Curl, R. F., "Thermodynamic and Transport Properties of Fluids", p. 1, London, Institution of Mechanical Engineers, 1958.

19. Edmister, W. C., Petrol. Refiner, (1958) 37, 173.
20. See reference 8, Appendix A.
21. Hougen, O. A., Watson, K. M., and Ragatz, R. A., "Chemical Process Principles: Part II, Thermodynamics", p. 569, 2nd ed., John Wiley, New York, 1959.
22. Leland, T. W., Chappelear, P. S., and Gamson, B. W., AICHE J. (1962) 8, 482.
23. Reid, R. C. and Leland, T. W., AICHE J. (1965) 11, 228.
24. Canfield, F. B. and Gunning, A. J., Chem. Eng. Sci., (1971) 26, 1139.
25. Leach, J. W., Chappelear, P. S. and Leland, T. W., Proc. Am. Petrol. Inst., (1966) 46, 223.
26. Leach, J. W., Chappelear, P. S. and Leland, T. W., AICHE J. (1968) 14, 568.
27. Leland, T. W., Rowlinson, J. S. and Sather, G. A., Trans. Faraday Soc., (1968) 64, 1447.
28. Watson, I. D. and Rowlinson, J. S., Chem. Eng. Sci. (1969) 24, 1575.
29. Gunning, A. J. and Rowlinson, J. S., Chem. Eng. Sci. (1973) 28, 529.
30. Teja, A. S. and Rowlinson, J. S., Chem. Eng. Sci. (1973) 28, 521.
31. Mollerup, J. and Rowlinson, J. S., Chem. Eng. Sci. (1974) 29, 1373.
32. Mollerup, J. and Fredenslund, A., Chem. Eng. Sci. (1973) 28, 1295.
33. Joffe, J. and Zudkevitch, D., Chem. Eng. Prog. Symp. Ser., No. 81, (1967) 63, 43.
34. Spear, R. R., Robinson, R. L., and Chao, K. C., Ind. Eng. Chem. Fundam. (1969) 8, 2.
35. Hissong, D. W. and Kay, W. B., AICHE J. (1970) 16, 580.
36. Mollerup, J., Program ICLNG. Instituttet for Kemiteknik, Danmarks Tekniske Højskole, Lyngby, Denmark.
37. Saville, G., Department of Chemical Engineering, Imperial College of Science and Technology, London, England, private communication.
38. Stewart, R. B. and Jacobsen, R. T., Center for Applied Thermodynamic Studies, College of Engineering, Idaho University, Moscow, Idaho, private communication.
39. Vera, J. H. and Prausnitz, J. M., Chem. Eng. Sci. (1971) 26, 1772.
40. See examples and papers cited in, S. K. Sood and G. G. Haselden, AICHE J. (1970) 16, 891.
41. Bender, E., The Calculation of Phase Equilibrium from a Thermal Equation of State, applied to the Pure Fluids Argon, Oxygen, and Nitrogen, and their Mixtures, Müller, Karlsruhe, 1973.
42. Prausnitz, J. M., "Molecular Thermodynamics of Fluid-Phase Equilibria", Prentice-Hall, Englewood Cliffs, N.J., 1969.

17

Can Theory Contribute: Transport Properties

HOWARD J. M. HANLEY

National Bureau of Standards, Boulder, CO 80302

In this lecture we comment briefly on how molecular theory--that is statistical mechanics and kinetic theory--can describe transport coefficients, e.g., the viscosity (η), thermal conductivity (λ), and the diffusion (D), of fluids such as argon, nitrogen, oxygen, carbon dioxide, and the simple hydrocarbons.

Unfortunately it turns out that statistical mechanics is not always satisfactory in this context because of the serious conceptual and mathematical difficulties in describing a fluid in nonequilibrium. And the problems are not only theoretical: an assessment of any theory from the viewpoint of the engineer has been, and is, handicapped by lack of quantitative data. Reliable data for pure fluids over a wide temperature and pressure range were not available until about ten years ago, with only limited exceptions: few data are yet available for mixtures in the liquid. So, although in principle theory can contribute, in practice its contribution is not always obvious.

One should distinguish between formal statistical mechanics and approximate theories, based on statistical mechanics, which bypass in some way the detailed problems of the molecular theory for a fluid in nonequilibrium. Such theories will not be discussed here (1), with the exception of corresponding states.

Formal Theory

One can for convenience say that formal studies of transport phenomena in fluids are based on two principle approaches. The more general examines how a system responds to a force or gradient: the transport coefficients are a quantitative measure of the response, and can be expressed in terms of the decay of fluctuations induced by the force.

It can be shown that any transport coefficient, L, for a fluid which is not too far from equilibrium can be written as an integral of the corresponding correlation function:

$$L \sim \int_{\infty}^{o} \left\langle J_L(o) \cdot J_L(\tau) \right\rangle d\tau \tag{1}$$

where $J_L(o)$ and $J_L(\tau)$ is the flux associated with L at arbitrary time o and at time τ, respectively. For example the simplest expression is that for the self-diffusion coefficient, D_{11}:

$$D_{11} \sim \int_{\infty}^{o} \left\langle \underset{\sim}{v}(o) \cdot \underset{\sim}{v}(\tau) \right\rangle d\tau \tag{2}$$

where $\underset{\sim}{v}$ is the velocity of a molecule. Similarly for the viscosity, and the other transport properties. The expressions are standard (2).

The alternative approach is to examine the behavior of a dilute gas, i.e., a gas for which one only has to consider binary molecular collisions. This procedure is clearly far more restricted than the correlation function route but the corresponding transport expressions are standard and practical. For example, the Chapman-Enskog solution (3) of the Boltzmann equation gives the dilute gas viscosity, η_o, and the other coefficients, e.g.,

$$\eta_o = \frac{5}{16} \frac{(\pi mkT)^{1/2}}{\pi\sigma^2 \Omega^{(2,2)*}} \tag{3}$$

where m is the molecular mass, k is Boltzmann's constant, σ is a size parameter characteristic of the intermolecular pair potential, ϕ. $\Omega^{(2,2)*}$--the collision integral--depicts the dynamics of a binary collision and is a function of ϕ.

The relationship between the general correlation function approach and the dilute gas should, in principle, be clear-cut. In fact, Equation (3) can be derived from the corresponding equation (2). Many authors have attempted to generalize the Boltzmann equation for all densities; alternatively, many authors have attempted to write the fluxes of Equation (1) in terms of the dynamics of molecular collisions. But these are difficult problems.

A simplistic explanation of the difficulties, which perhaps should be compared to those encountered for a fluid in equilibrium, is as follows. In equilibrium, it turns out one is concerned with molecular configurations over distances of the order of the range of the intermolecular potential (see Figure 1a). In nonequilibrium, however, the characteristic distances are of the order of the mean free path, or greater (see Figure 1b). A description and understanding of the dynamics of the molecular motion in the fluid requires certain integrals to be evaluated over such distances. They have not been solved for a realistic potential, other than for the dilute gas.

The thrust of kinetic theory in the last few years has been to

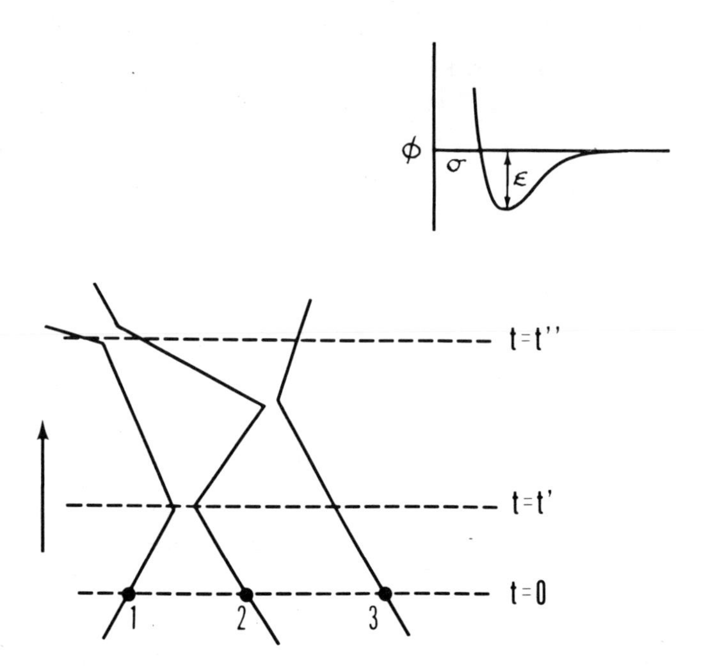

Figure 1. (a) General form of the intermolecular pair potential, ϕ, as a function of separation, r. The parameter σ, defined by $\phi(r) = 0$, corresponds to the molecular diameter: ϵ, is the maximum value of ϕ. (b) Schematic of a possible sequence of collisions between three molecules, 1, 2, and 3, for three times t = 0, t′, t″.

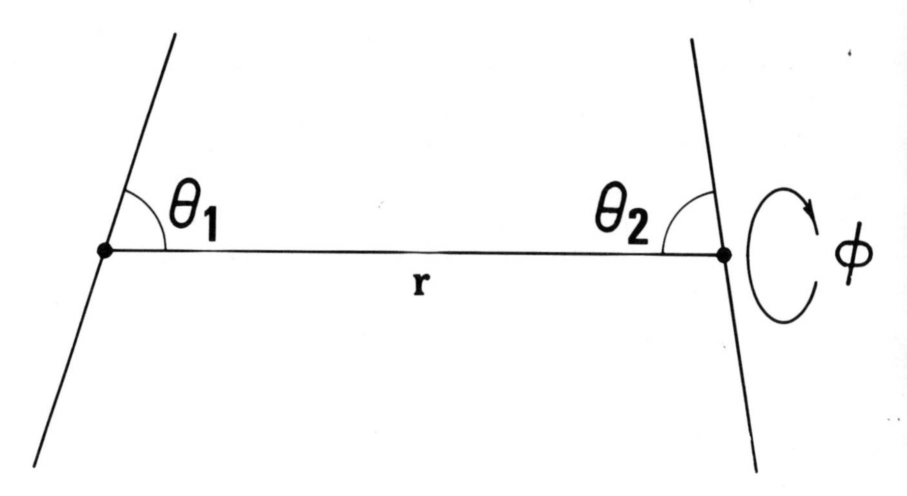

Figure 2. Coordinate system for two linear molecules interacting along the line of the centers

try to understand the nature of the dynamical events in a nonequilibrium system and, in particular, to explain the decay of the correlation function with time.

Comparison with Experiment for the Dilute Gas

Substantial progress has been made recently in applying the kinetic theory dilute gas expressions, both for the transport and equilibrium properties. For example, we now have a good grasp on how a model intermolecular potential can be used to relate theory to data. Also simple nonspherical molecules can be considered systematically.

Some results for carbon dioxide (4) are presented here graphically to illustrate the type of agreement one can expect between kinetic theory and experiment.

In our work with carbon dioxide we have assumed the intermolecular potential is of the form

$$\phi(r\theta_1\theta_2\phi) = \phi_s(r) + \phi_{ns}(r\theta_1\theta_2\phi) \tag{4}$$

and that the molecules interact according to the coordinate system shown in Figure 2. ϕ_s is the spherical, angle independent contribution given by the m-6-8 potential

$$\phi_s(r) = \varepsilon\left[\frac{6+2\gamma}{m-6}\left(\frac{\sigma}{r}\right)^m d^m - \frac{m-\gamma(m-8)}{m-6}\left(\frac{\sigma}{r}\right)^6 d^6 - \sigma\left(\frac{\sigma}{r}\right)^8 d^8\right] \tag{5}$$

where $\phi(\sigma) = 0$, $\phi(r_{min}) = -\varepsilon$ and $d = r_{min}/\sigma$. The parameters, m and γ, represent the strength of the repulsive term and an inverse eight attractive term, respectively. ϕ_{ns} is a function of a quadrupole moment, the polarizability and the relative angles of orientation.

Comprisons between theory and experiment for the viscosity, thermal conductivity, equilibrium second virial coefficient, and the dielectric second virial coefficient are shown in Figures 3-6. One observes that a wide range of independent thermophysical properties have been fitted quite well.

It should be stressed that the potential parameters are the same for all properties, and that the values of quadrupole moment and the polarizability were taken from independent experiments and were not treated as adjustable parameters. It should also be stressed that the fits depend on a proper choice of the spherical part of the potential, ϕ_s. Models for ϕ_s simpler than Equation (5) are inadequate.

Results similar to those of Figures 3-6 are available for other simple polyatomic molecules (F_2, O_2, N_2, CH_4, etc.) the rare

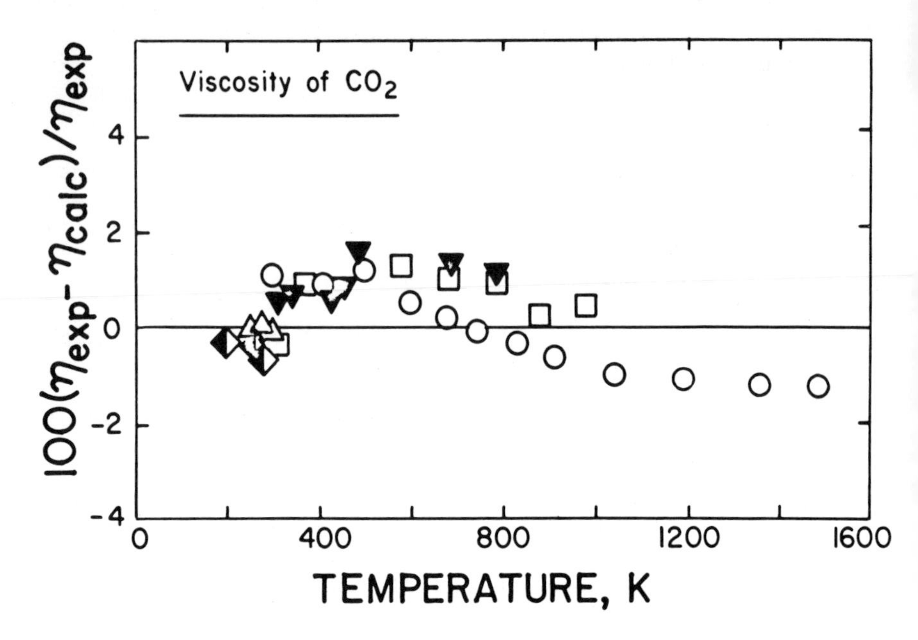

Figure 3. Viscosity of CO_2: Comparison between values determined from Equation 3, using the potential (4), and experiment (details in Ref. 4)

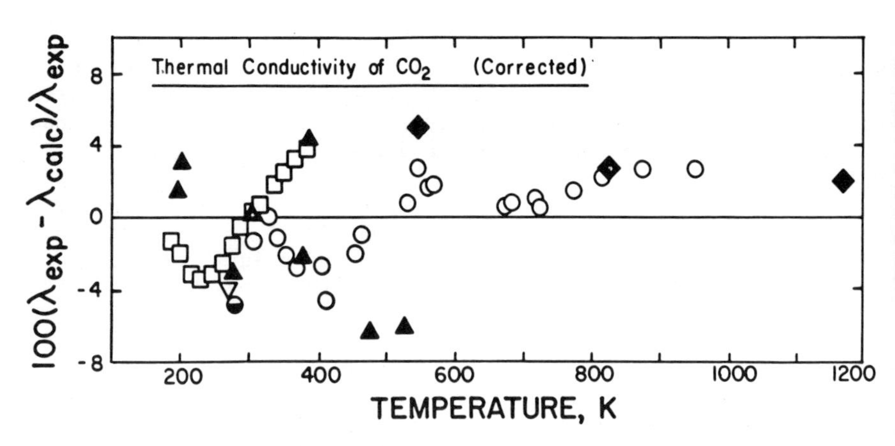

Figure 4. Thermal conductivity of CO_2: comparison between kinetic theory and experiment (details in Ref. 4)

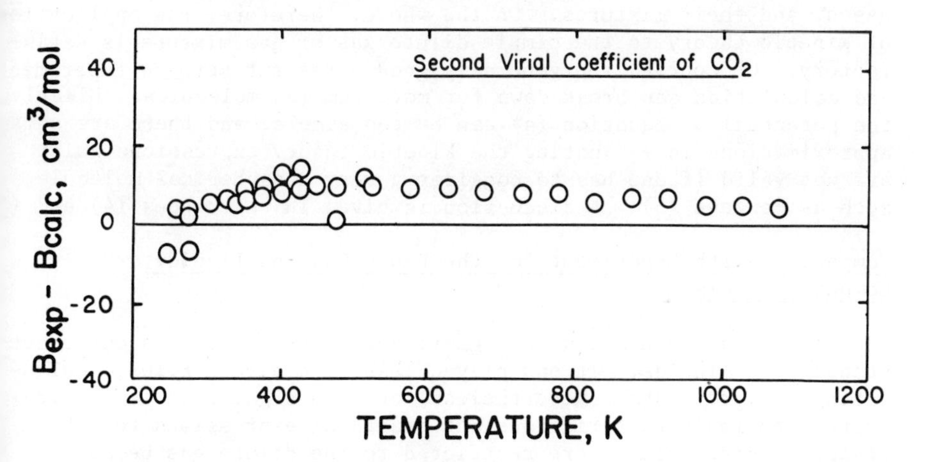

Figure 5. Second virial coefficient of CO_2 predicted from statistical mechanics using the potential of Equation 4 (Ref. 4)

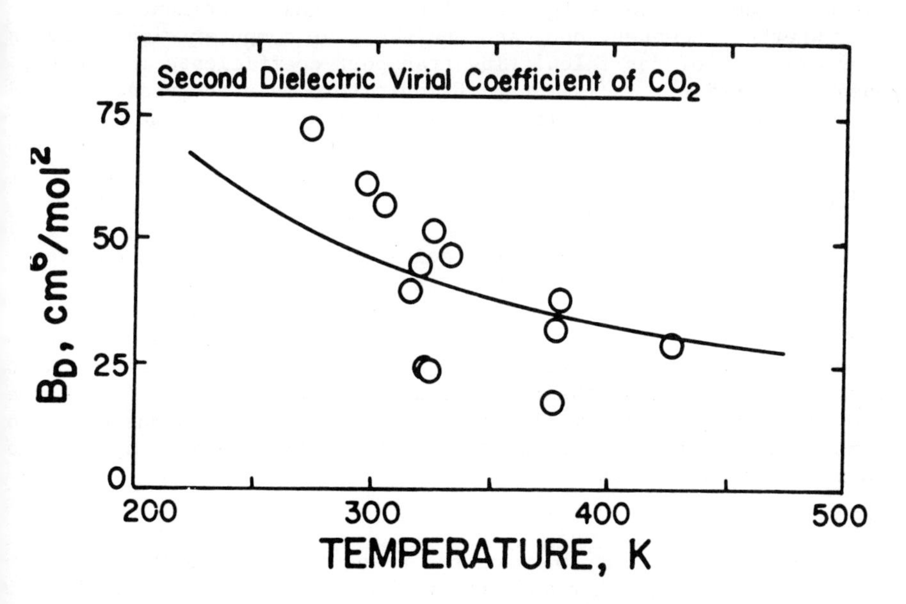

Figure 6. Dielectric virial, B_D, of CO_2. This is defined by the expansion

$$\frac{\epsilon - 1}{\epsilon + 2}\frac{1}{\rho} = A_D + B_D\rho + \ldots$$

where ϵ is the dielectric constant, ρ is the density (comparison between theory and experiment, Ref. 4)

gases, and their mixtures. On the whole, therefore, the applicatio of kinetic theory to the simple dilute gas or gas mixture is satisfactory. Of course, the generally good agreement between experiment and calculation can break down for more complex molecules. Clearly, the potential of Equation (4) can be too simple, and there are some approximations in evaluating the kinetic theory expressions which are not valid if one has to consider a very nonspherical molecule, such as benzene (5). A discussion is given in references (4) and (5

Comparison with Experiment for the Dense Gas and Liquid: Computer Simulation

The computer has led to significant contributions in our understanding of liquids. It has played less of a direct role in calculating property data. Nevertheless, one can argue that it is not correct to imply that rigorous but practical expressions for the transport coefficients are restricted to the dilute gas because Equation (2) can be evaluated numerically if the fluid is simulated on the computer. For example, it is relatively straightforward to calculate the self-diffusion coefficient via Equation (3). In fact Figure 7 shows the results obtained by us (6) for methane, assuming that methane interacts with the m-6-8 potential of Equation (5). Agreement between the computer calculation and experiment is good.

Whether, however, one can conclude that computer simulation is a practical tool for calculating transport coefficients in general open to question. It is more difficult to obtain the viscosity and thermal conductivity to within a reasonable error than it is to detmin and self-diffusion, and the latter is not a trivial calculation

Corresponding States

The contribution of statistical mechanics is not, of course, limited to the kind of results shown in Figures 3-7. Several approximate theories have been proposed which are based on statistical mechanics but which require some assumption, or assumptions, to avoid a detailed description of a fluid in nonequilibrium. A go example is the Enskog theory and its modifications (1). As remarke however, such theories will not be discussed here with one exceptio namely the theory of corresponding states. One specific applicatio is outlined as follows.

---

* However, a different approach has been introduced recently by Hoover (7), and by other authors, which could eventually make compu simulation a viable tool. Rather than evaluate the correlation fun tion, it is suggested that a laboratory experiment, which would lea to the transport coefficient, be simulated. For example, a flow system can be set up on the computer to represent a shear flux produced by a velocity gradient. Numerical evaluation of the flux for a given gradient gives the viscosity coefficient.

DIFFUSION OF METHANE

Density (ρ) kg/m^3	Temperature K	$10^6 \rho D$(calc) kg/ms	$10^6 \rho D$(exp) kg/ms
440.9	121.5	1.6	2.2
440.9	151.4	2.5	2.7
347.4	172.8	5.7	6.0
258.7	202.8	10.1	10.2
192.4	213.0	9.4	10.6
133.6	233.0	12.7	12.0

Molecular Physics

Figure 7. Table comparing experimental diffusion coefficients for methane with values calculated from Equation 3 using computer simulation (6)

We have developed (8) a corresponding states procedure the object of which is to predict the viscosity and thermal conductivity coefficient of a pure fluid or a mixture from thermodynamic (PVT) data.

The procedure incorporates the extended corresponding states approach to the thermodynamic properties of nonconformal fluids introduced by Leland, and by Rowlinson and their co-workers (9).

The thermodynamic properties of a fluid, α, can be related to the properties of a reference fluid, o, via (for example) the functions $f_{\alpha\alpha,o}$ and $h_{\alpha\alpha,o}$ where

$$f_{\alpha\alpha,\,o} = (T^c_{\alpha\alpha}/T^c_o)\,\theta_{\alpha\alpha,\,o}\;;\qquad h_{\alpha\alpha,\,o} = (\rho^c_o/\rho^c_{\alpha\alpha})\,\phi_{\alpha\alpha,\,o} \tag{6}$$

The shape factors, θ and ϕ, are weak functions of temperature (T) and density (ρ) and can be regarded as characteristic of the Pitzer acentric factors of fluids α and o, respectively. The superscript c denotes the critical point value.

Applying extended corresponding states ideas to the transport properties, one has for the viscosity and thermal conductivity coefficients of fluid, α, at a given density and temperature with respect to the equivalent coefficient of fluid, o (10)

$$\eta_\alpha(\rho,T) = \eta_o(\rho h_{\alpha\alpha,\,o}\,,\,T/f_{\alpha\alpha,\,o})\;\;FH^{\eta}_{\alpha,\,o} \tag{7}$$

and

$$\lambda_\alpha(\rho,T) = \lambda_o(\rho h_{\alpha\alpha,o}, T/f_{\alpha\alpha,o})\, FH^\lambda_{\alpha,o} \tag{8}$$

where

$$FH^\eta_{\alpha,o} = (M_\alpha/M_o)^{1/2}\, h^{-2/3}_{\alpha\alpha,o}\, f^{1/2}_{\alpha\alpha,o} \tag{9}$$

$$FH^\lambda_{\alpha,o} = (M_o/M_\alpha)^{1/2}\, h^{-2/3}_{\alpha\alpha,o}\, f^{1/2}_{\alpha\alpha,o} \tag{10}$$

For a mixture x, one has

$$\eta_x(\rho,T) = \eta_o(\rho h_{x,o}, T/f'_{x,o})\, FH^\eta_{x,o} \tag{11}$$

with a corresponding equation for $\lambda_x(\rho,T)$. Mixing rules, consistent with the Van der Waals one-fluid approximation (11) are

$$h_{x,o} = \sum_\alpha \sum_\beta x_\alpha x_\beta\, h_{\alpha\beta,o}$$
$$f_{x,o}\, h_{x,o} = \sum_\alpha \sum_\beta x_\alpha x_\beta\, f_{\alpha\beta,o}\, h_{\alpha\beta,o} \tag{12}$$

with

$$f_{\alpha\beta,o} = \xi_{\alpha\beta}\,(f_{\alpha\alpha,o}\, f_{\beta\beta,o})^{1/2}$$

and (13)

$$h_{\alpha\beta,o} = \psi_{\alpha\beta}\left(\frac{1}{2}\, h^{1/3}_{\alpha\alpha,o} + \frac{1}{2}\, h^{1/3}_{\beta\beta,o}\right)^3$$

where $\xi_{\alpha\beta}$ and $\psi_{\alpha\beta}$ are the binary interaction parameters. We use the expression $M_x = \sum_\alpha x_\alpha M_\alpha$ for the mixing rule for the molecular weight [but see Mo and Gubbins (11)].

It is important to note that Equations (7) and (8), and the equivalent equations for a mixture can fail to predict correctly the transport properties of a nonconformal fluid _if the corresponding states parameters are estimated from thermodynamic data_. The equations can break down, in particular, if $\rho/\rho^c \gtrsim 1$.

Modification of the Transport Expressions

It is proposed that the predictive capability of the extended corresponding states approach to transport phenomena would be improved significantly if one considers equations of the form (for the viscosity coefficient, for example)

$$\eta_\alpha(\rho,T) = \eta_o(\rho h_{\alpha\alpha,o}, T/f_{\alpha\alpha,o})\, FH^\eta_{\alpha,o}\, X^\eta_{\alpha,o}(\rho,T). \tag{14}$$

A correction factor, X^{η}, is introduced which should be unity if fluids α and o follow simple two-parameter corresponding states, but should be a strong function of density and a weak function of temperature otherwise. However, X^{η} should nevertheless approach unity as the density approaches zero.

In like manner, we introduce $X^{\eta}_{x,o}(\rho,T)$ for the mixture, and $X^{\lambda}_{\alpha,o}(\rho,T)$ and $X^{\lambda}_{x,o}(\rho,T)$ for the thermal conductivity coefficient of the pure fluid and mixture, respectively.

We have suggested expressions for X^{η} and X^{λ}. Based on the behavior of the transport coefficients according to the Modified Enskog Theory (1), one can derive for the pure fluid, (and similarly for the mixture)

$$X^{\eta}_{\alpha,o}(\rho,T) = Q_{\alpha,o}\, G^{\eta}_{\alpha,o} ; \quad X^{\lambda}_{\alpha,o}(\rho,T) = Q_{\alpha,o}\, G^{\lambda}_{\alpha,o} \tag{15}$$

$Q_{\alpha,o}$ is function of ρ and T defined as

$$Q_{\alpha,o} = [(b\rho)_{\alpha}/(b\rho)_{o}]^{[1 - 1/\exp\,(\rho^{c}/\rho)^{3}]} \tag{16}$$

where b is a term given by $b = B + TdB/dT$, with B the equilibrium second virial coefficient. $G^{\eta}_{\alpha,o}$ and $G^{\lambda}_{\alpha,o}$ are ratios given, respectively by $G^{\eta}_{\alpha,o} = [\,]^{\eta}_{\alpha}/[\,]^{\eta}_{o}$ and $[\,]^{\lambda}_{\alpha}/[\,]^{\lambda}_{o}$. The bracket expressions are

$$[\,]^{\eta} = [1/b\rho\chi + 0.8 + 0.761\, b\rho\chi]$$

$$[\,]^{\lambda} = [1/b\rho\chi + 1.2 + 0.755\, b\rho\chi] \tag{17}$$

with

$$b\rho\chi = \frac{1}{\rho R}\left(\frac{\partial P}{\partial T}\right)_{\rho} - 1 \tag{18}$$

for fluids α or o. R is the gas constant.

It should be noted that the calculation of the correction factors X^{η} and X^{λ} require only thermodynamic (PVT) information for fluid α or x with respect to the reference fluid; for example, b_{α} Equation (16) can be obtained from the second virial coefficient of the reference fluid, B_o, according to the expression

$$b_{\alpha}(T) = h_{\alpha\alpha,o}\, B_o + T\, \frac{d(h_{\alpha\alpha,o}\, B_o)}{dT} \tag{19}$$

and similarly for the term $b\rho\chi$ Equation (18).

Comparisons with Experimental Data Mixtures

Apart from the dilute and moderately dense gas, viscosity and thermal conductivity data for nonconformal mixtures are scarce and their reliability are often difficult to assesss. The modified equations, however, were checked as far as possible over a wide range of experimental conditions. Methane was chosen as the reference fluid [equation of state data from Goodwin (12), transport dat from the correlation of Hanley, McCarty and Haynes (13)].

Table 1 gives typical and representative results for the metha propane system. The data (estimated accuracy ∿ 3%) are those of Huang, Swift and Kurata (14). [Values from Equation (11) are given in parentheses and the agreement between calculation and experiment is seen to be poor.]

Table 2 shows a comparison between predicted and experimental values for an natural gas mixture (15) at 273 K. The mixture has the following composition: CH_4, 91.5%; C_2H_6, 1.8%; C_3H_8, 0.8%; C_4H_{10}, 0.6%; N_2, 5%; others 0.3%. We cannot evaluate the accuracy of the data but the prediction appears reasonable.

Results for a nitrogen-carbon dioxide mixture are shown in Table 3: data from reference (15). Note that neither nitrogen, or carbon dioxide is used as the reference fluid.

Table 4 compares our prediction with thermal conductivity data of a nitrogen-ethane mixture from Gilmore and Comings (16). Values from the unmodified equation are given in parentheses. Finally, Table 5 shows the results for a methane-n-butane mixture (17).

TABLE 1. VISCOSITY OF A METHANE (50%) - PROPANE (50%) MIXTURE AT 153.15 K. ($T^c \approx 319$ K, $\rho^c \approx 7.7$ mol/ℓ)

ρ mol/ℓ	η(exp)	η(calc)	
	μg cm^{-1} s^{-1}		
18.69	2780	2614	(2286)
18.78	2860	2710	(2360)
18.96	3030	2878	(2509)
19.12	3200	3002	(2659)
19.28	3380	3232	(2807)
19.42	3550	3407	(2952)

*Results from the extended corresponding states equation (11).

TABLE 2. NATURAL GAS VISCOSITY AT 273 K

P	η(exp)	η(calc)
atm	$\mu g\ cm^{-1}\ s^{-1}$	
20.	109.3	110.0
60.	132.2	123.6
100.	160.2	145.8
200.	244.5	227.7
300.	313.7	298.8
400	372.3	345.0

TABLE 3. NITROGEN-CARBON DIOXIDE VISCOSITY

Comparison between predicted viscosity coefficients for a nitrogen (62%) - carbon dioxide (38%) mixture at 289 K.

P	ρ	η(exp)	η(calc)
atm	mol/ℓ	$\mu g\ cm^{-1}\ s^{-2}$	
20.	0.9	167.0	167.0
60.	2.8	179.5	186.5
100.	4.9	200.5	213.6
120.	7.0	212.5	229.1

TABLE 4. THERMAL CONDUCTIVITY OF A NITROGEN (59.8%) - ETHANE (40.2%) MIXTURE AT 348.16 K ($T^c \approx$ 260 K, $\rho^c \approx$ 12.0 mol/ℓ)

ρ	λ(exp)	λ(calc)	
mol/ρ	mW m^{-1} K^{-1}		
7.34	46.0	46.11	(61.5)*
10.16	55.2	54.69	(72.65)
13.75	70.3	69.47	(92.4)
16.43	85.4	84.44	(113)

*Results from the extended corresponding states equation corresponding to equation (11).

TABLE 5. THERMAL CONDUCTIVITY OF A METHANE (39.4%) - n-BUTANE MIXTURE (60.6%) at 277.6 K ($T^c \approx$ 390 K, $\rho^c \approx$ 6.25 mol/ℓ)

ρ	λ(exp)	η(calc)
mol/ℓ	mW m^{-1} K^{-1}	
12.26	99.00	95.33
12.49	101.46	99.63
13.01	113.39	109.99
13.25	114.90	115.05

The representative results reported here indicate our procedure is satisfactory. In other words it appears, given the limited data available at this time, that transport properties can be predicted satisfactorily via thermodynamic data. Hence only one set of parameters ($\theta,\phi,\psi_{\alpha\beta}$,etc.) are required to fit both thermodynamic and transport properties.

Abstract

A very brief outline of theoretical calculations of the transport coefficients is given. A method to predict the viscosity and thermal conductivity coefficients of pure fluids and mixtures is presented.

Literature Cited

1. Hanley, H. J. M., McCarty, R. D. and Cohen, E. G. D., Physics (1972) 60, 322.
2. McQuarrie, D. G., "Statistical Mechanics" (Harper and Row, N.Y., 1976).
3. Chapman, S. and Cowling, T. G., "The Mathematical Theory of Non-Uniform Gases," (Cambridge Univ. Press, London, 1970).
4. Ely, J. F. and Hanley, H. J. M., Molecular Physics (1975) 30, 565.
5. Evans, D. J., Ph.D. Thesis, Australian National University, Canberra, Australia (1975).
6. Hanley, H. J. M. and Watts, R. O., Molecular Physics (1975) 29, 1907.
7. Ashurst, W. T. and Hoover, W. G., AIChE Journal (1975) 21, 410.
8. Hanley, H. J. M., Cryogenics (1976) 16, 643.
9. Rowlinson, J. S. and Watson, I. D., Chem. Eng. Sci. (1969) 24, 1565.
10. Haile, J. M., Mo, K. C. and Gubbins, K. E., "Advances in Cryogenic Eng." Vol. 21, Ed. K. D. Timmerhaus and D. H. Weitzel (Plenum Press, N.Y., 1975) 501.
11. Mo, K. C. and Gubbins, K. E., Chem. Eng. Commun. (1974) 1, 281.
12. Goodwin, R. D., Nat. Bur. Stand. (U.S.), Tech. Note No. 653 (1974).
13. Hanley, H. J. M., McCarty, R. D. and Haynes, W. M., Cryogenics (1975) 15, 413; J. Phys. Chem. Ref. Data (In press).
14. Huang, E. T. S., Swift, G. W. and Kurata, F., AIChE J. (1967) 13, 846.
15. Golubev, I. F., "Viscosity of Gases and Gas Mixtures," Israel Program for Scientific Translations (Jerusalem, 1970).
16. Gilmore, T. F. and Comings, W. E., AIChE Journal (1966) 12, 1172.
17. Carmichael, L. T., Jacobs, J. and Sage, B. H., J. Chem. Eng. Data (1968) 13, 489.

18

Polar and Quadrupolar Fluid Mixtures

K. E. GUBBINS and C. H. TWU

Cornell University, Ithaca, NY 14853

Phase equilibrium calculations may be made by either semiempirical or theoretical methods. In the semiempirical approach one uses an empirical equation of state for one or more of the phases involved; for the liquid phase one of the empirical equations for the excess Gibbs energy (Wilson, van Laar, etc.) is usually used. The semiempirical approach gives good results provided that one has a significant amount of experimental data available for the mixture. However, such methods are better suited to interpolation of existing data than to extrapolation or prediction. Theoretical methods are based in statistical thermodynamics, require less mixture data, and should be more reliable for prediction. Theoretically-based methods that have found extensive use by chemical engineers include regular solution theory (1), corresponding states methods (conformal solution theory) (2-4), and perturbation expansions based on a hard sphere fluid as reference system (3,4). For mixtures of simple nonpolar molecules these methods give good results, especially the corresponding states and perturbation expansion theories (3). However, all three theories are based on the assumption that the molecules are spherical, with intermolecular forces that are a function only of the intermolecular separation. This assumption is strictly valid only for mixtures of the inert gases (Ar, Kr, Xe) and for certain fused salts and liquid metals. In spite of this restriction the corresponding states and perturbation methods have been applied with success to mixtures in which the intermolecular forces depend on the molecular orientations, e.g., mixtures containing O_2, N_2, light hydrocarbons, etc. (see ref. 4 for review of applications up to 1973). The extension to weakly nonspherical molecules can be accomplished, for example, by introducing shape factors as suggested by Leland and his colleagues (5). These methods have been extensively exploited for both thermodynamic (2) and transport (6) properties. They are predictive only if equations are available for the composition dependence of the shape factors. The existing methods for doing this work well for relatively simple mixtures, but break down when constituents with strongly orientation-dependent forces are present (e.g., strongly polar or quadrupolar constituents).

In this paper we review a recently developed theoretical method for phase equilibrium calculation which explicitly accounts for strongly orientation-dependent forces. These anisotropic forces are taken into account through a perturbation scheme in which the reference fluid is composed of simple spherical molecules; in practice, the known properties of argon, or those of a Lennard-Jones fluid simulated on the computer, may be used. Such a perturbation scheme was first suggested by Pople (7) more than twenty years ago, but was not immediately used for liquid phase calculations because the reference fluid properties were not sufficiently well known. Since 1972 the theory has been extended and improved, and its successful application to liquids of strongly polar or quadrupolar molecules dates from 1974 (7).

The most successful form of the theory is briefly outlined in the Theory Section. In the Section on Results the theory is used to classify mixture phase diagrams in terms of the intermolecular forces involved, and also to predict vapor-liquid equilibria for several binary and ternary mixtures.

Intermolecular Forces

The intermolecular pair potential u between two axially symmetric molecules of components α and β can be written as a sum of parts

$$u^{\alpha\beta}(r\theta_1\theta_2\phi) = u_o^{\alpha\beta}(r) + u_{mult}^{\alpha\beta}(r\theta_1\theta_2\phi) + u_{dis}^{\alpha\beta}(r\theta_1\theta_2\phi) + u_{ov}^{\alpha\beta}(r\theta_1\theta_2\phi) + u_{ind}^{\alpha\beta}(r\theta_1\theta_2\phi) \quad (1)$$

where u_{mult}, u_{dis}, u_{ov}, and u_{ind} are multipolar (electrostatic), anisotropic dispersion, anisotropic charge overlap, and induction terms, respectively. Here r is the intermolecular separation, $(\theta_i\phi_i)$ are the polar angles giving the orientation of molecule i, the $\underline{r}$ direction being taken as the polar axis, and $\phi = \phi_1 - \phi_2$. We take the isotropic central potential $u_o(r)$ to be an (n,6) model (8),

$$u_o^{\alpha\beta}(r) = \frac{n_{\alpha\beta}}{n_{\alpha\beta} - 6}\left(\frac{n_{\alpha\beta}}{6}\right)^{6/(n_{\alpha\beta} - 6)} \varepsilon_{\alpha\beta}\left[\left(\frac{\sigma_{\alpha\beta}}{r}\right)^{n_{\alpha\beta}} - \left(\frac{\sigma_{\alpha\beta}}{r}\right)^{6}\right] \quad (2)$$

where $\varepsilon_{\alpha\beta}$, $\sigma_{\alpha\beta}$ and $n_{\alpha\beta}$ are potential parameters. The multipolar, dispersion, overlap and induction potentials are usually approximated by the first few terms in a spherical harmonic expansion.+

+In practice not all of these contributions are equally important, and some may be neglected for particular fluids. In calculating thermodynamic properties for the fluids considered in this review, the multipolar forces make the largest anisotropic contribution.

Equations for these terms are given in several reviews (7-10). The leading multipole term in u_{mult} is the dipole-dipole potential in the case of polar fluids, and the quadrupole-quadrupole potential for fluids of linear symmetrical molecules (CO_2, N_2, etc.). These are given by

$$u_{\mu\mu} = \frac{\mu_\alpha \mu_\beta}{r^3} (s_1 s_2 c - 2c_1 c_2) \tag{3}$$

$$u_{QQ} = \frac{3Q_\alpha Q_\beta}{4r^5} (1 - 5c_1^2 - 5c_2^2 + 17c_1^2 c_2^2 + 2s_1^2 s_2^2 c^2 - 16s_1 s_2 c_1 c_2 c) \tag{4}$$

where $s_i = \sin\theta_i$, $c_i = \cos\theta_i$, $c = \cos\phi$, and μ and Q are the dipole and quadrupole moments. Experimental values of μ and Q have been tabulated by Stogryn and Stogryn (11).

The anisotropic overlap potential may be approximated for symmetrical linear molecules (CO_2, N_2, Br_2, etc.) by (7)

$$u_{ov}^{\alpha\beta} = 4\,\varepsilon_{\alpha\beta} \left(\frac{\sigma_{\alpha\beta}}{r}\right)^{12} \delta_2^{\alpha\beta} (3\,c_1^2 + 3\,c_2^2 - 2) \tag{5}$$

while for unsymmetrical linear molecules (HCl, NO, N_2), etc.) one has

$$u_{ov}^{\alpha\beta} = 4\,\varepsilon_{\alpha\beta} \left(\frac{\sigma_{\alpha\beta}}{r}\right)^{12} \delta_1^{\alpha\beta} (c_1 - c_2) \tag{6}$$

where n in Eq. (2) is taken to be 12, and where δ_1 and δ_2 are dimensionless overlap parameters that must lie within the ranges $-0.5 \leq \delta_1 \leq 0.5$ and $-0.25 \leq \delta_2 \leq 0.5$. The anisotropic dispersion potential is given to a good approximation by the London expression (7,9,10).

Theory

In thermodynamic perturbation theory the properties of the real system, in which the pair potential is $u^{\alpha\beta}$, are expanded about the values for a reference system, in which the pair potential is $u_0^{\alpha\beta}$. Here we take the reference potential to be the (n,6) potential, so that the anisotropic parts of the potential in Eq. (1) are the perturbation. Expanding the Helmholtz free energy A in powers of the perturbing potential about the value A_o for the reference system gives

$$A = A_o + A_2 + A_3 + \ldots \tag{7}$$

where A_2, A_3, . . . are the second-, third-order, etc.

perturbation terms (the first-order term vanishes). General expressions for A_2 and A_3 for an arbitrary intermolecular potential have been worked out (12,13). Equations for the other thermodynamic properties (pressure, internal energy, etc.) may be obtained by applying the usual thermodynamic identities to (7). When (7) is terminated at the third-order term it is found to give good results for moderately polar fluids (e.g., HCl) but to fail for strong dipoles (e.g., H_2O, NH_3, acetone). This is shown for a liquid state condition in Figure 1.+ Similar results are found for quadrupole forces. The slow convergence of (7) for strong multipole strengths led Stell et al. (14) to suggest the following simple Padé approximation for the free energy.

$$A = A_o + A_2\left[\frac{1}{1 - \frac{A_3}{A_2}}\right] \tag{8}$$

Eq. (8) is found to give excellent results, even for the strongest dipole or quadrupole moments observed in nature (Figure 1). For the anisotropic overlap potential of Eq. (5), the Padé approximation gives good results for $-0.2 \leq \delta_2$ _ +0.3. This range of δ_2 includes molecules such as HCl, CO_2, ethane, ethylene, methanol, etc., for which δ_2 is 0.2 or less.

The expressions for A_2 and A_3 involve the state variables, intermolecular potential parameters, and certain integrals J and K for the reference fluid. Thus, if the only anisotropic part of the potential in (1) is the dipole-dipole term of Eq. (3), then

$$A_2 = -\frac{2\pi N\rho}{3kT} \sum_{\alpha\beta} x_\alpha x_\beta \left(\left(\frac{1}{\sigma_{\alpha\beta}^3}\right) \right) \mu_\alpha^2 \mu_\beta^2 \; J_{\mu\mu}^{\alpha\beta} \tag{9}$$

$$A_3 = \frac{32\pi^3}{135} \left(\frac{14\pi}{5}\right)^{1/2} \frac{\rho^2 N}{(kT)^2} \sum_{\alpha\beta\gamma} x_\alpha x_\beta x_\gamma \left(\left(\frac{1}{\sigma_{\alpha\beta}\sigma_{\alpha\gamma}\sigma_{\beta\gamma}}\right) \right) \mu_\alpha^2 \mu_\beta^2 \mu_\gamma^2 \; K_{\mu\mu\mu}^{\alpha\beta\gamma} \tag{10}$$

where N is Avogadro's number, $\rho = N/V$ is number density, k is Boltzmann constant, T is temperature, and $x_\alpha = N_\alpha/N$ is mole fraction of component α. The summations are over the components of the mixture. $J_{\mu\mu}$ and $K_{\mu\mu\mu}$ are integrals over the two- and three-

+For HCl μ*0.8, while for acetone and H_2O μ^* is about 1.2 and 2-3, respectively; for N_2 and CO_2 the reduced quadrupole moment Q* is about 0.5 and 0.9, respectively. Here $\mu^* = \mu/(\epsilon\sigma^3)^{1/2}$ and $Q^* = A/(\epsilon_\sigma^5)^{\frac{1}{2}}$. The influence of polar or quadrupolar forces on the thermodynamic properties is determined by μ^* and Q*, rather than by μ and Q themselves. Thus the effects of these forces are greatest for small molecules with relatively large μ or Q values.

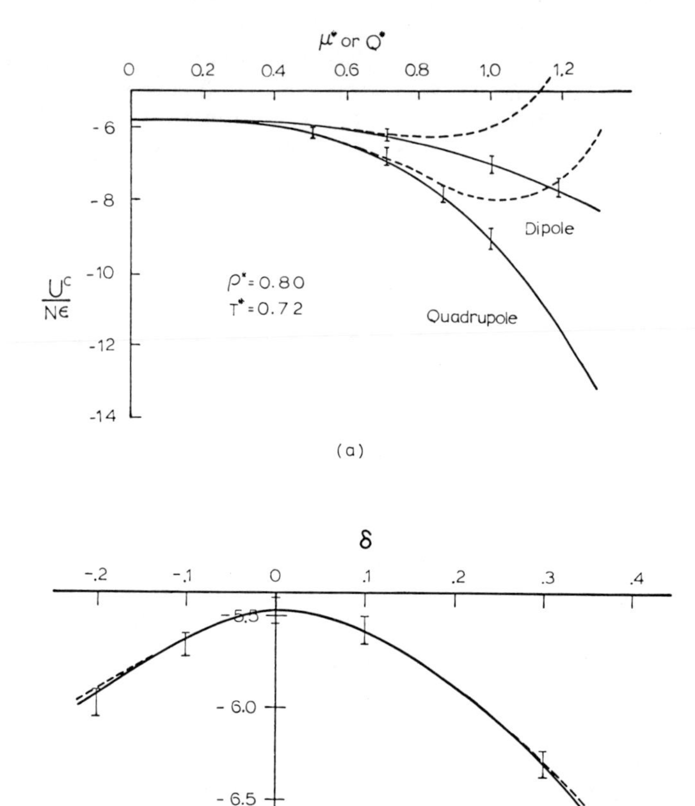

Figure 1. Comparison of the perturbation expansion to third order (– – –), Equation 7, with the Padé approximant to the series (——), Equation 8, with molecular dynamics computer simulation data (15, 16). U^c is the configurational contribution to the internal energy. The potential is of the form $u_o + u_a$, where u_o is the Lennard-Jones (12,6) potential and u_a, is given by either Equation 3 or 4 in Figure (a), and is the anisotropic overlap model of Equation 5 in Figure (b). Here $\mu^* = \mu/(\epsilon\sigma^3)^{\frac{1}{2}}$ and $Q^* = Q/(\epsilon\sigma^5)^{\frac{1}{2}}$.

molecule correlation functions for the reference fluid. These integrals have been calculated for the case n = 12 from molecular dynamics results for a pure Lennard-Jones (12,6) fluid (12,13, 17), and are given as a function of reduced density $\rho^* = \rho\sigma^3$ and temperature $T^* = kT/\varepsilon$ by

$$\ell n\ J_{\mu\mu} = -0.488498\ \rho^{*2}\ \ell n\ T^* + 0.863195\ \rho^{*2} + 0.761344\ \rho^* \times \ell n\ T^* - 0.750086\ \rho^* - 0.218562\ \ell n\ T^* - 0.538463 \quad (11)$$

$$\ell n\ K_{\mu\mu\mu} = -1.050534\ \rho^{*2}\ \ell n\ T^* + 1.747476\ \rho^{*2} + 1.769366\ \rho^* \times \ell n\ T^* - 1.999227\ \rho^* - 0.661046\ \ell n\ T^* - 3.028720 \quad (12)$$

The equations for A_2 and A_3 for other types of anisotropic forces are listed elsewhere (12,13,17).

In order to make mixture calculations using the above equations, methods for calculating A_o, $J^{\alpha\beta}_{\mu\mu}$, and $K^{\alpha\beta\gamma}_{\mu\mu\mu}$ (together with J and K integrals for any other parts of the anisotropic potential) are necessary. Accurate methods for these calculations have been described by Twu et al. (13,17). The thermodynamic properties of the reference mixture were related to those of a pure, conformal reference fluid by using the van der Waals 1 form of the corresponding states theory (2,3,13). The properties of this pure reference fluid were obtained from experimental data for argon, by using the multiparameter equations of Gosman et al. (18) in reduced form. Mixture values of $J^{\alpha\beta}$ and $K^{\alpha\beta\gamma}$ were also related to pure fluid values using the van der Waals 1 theory (17).

Results

In this section we give as examples two uses of the theory outlined above: (a) the classification of binary phase diagrams, critical behavior, etc. in terms of the intermolecular forces involved, and (b) comparison with experiment for some binary and ternary systems.

Classification of Binary Phase Diagrams. Figure 2 shows a classification of binary phase diagrams suggested by Scott and Van Konynenburg (19), with examples of each class. This classification is based on the presence or absence of three-phase lines and the way critical lines connect with these; this is best seen on a P,T projection. In classes I, II and VI the two components have similar critical temperatures, and the gas-liquid critical line passes continuously between the pure component critical points; class II and VI mixtures differ from class I in that they are more nonideal and show liquid-liquid immiscibility. Class II behavior is common, whereas class VI, in which closed solubility

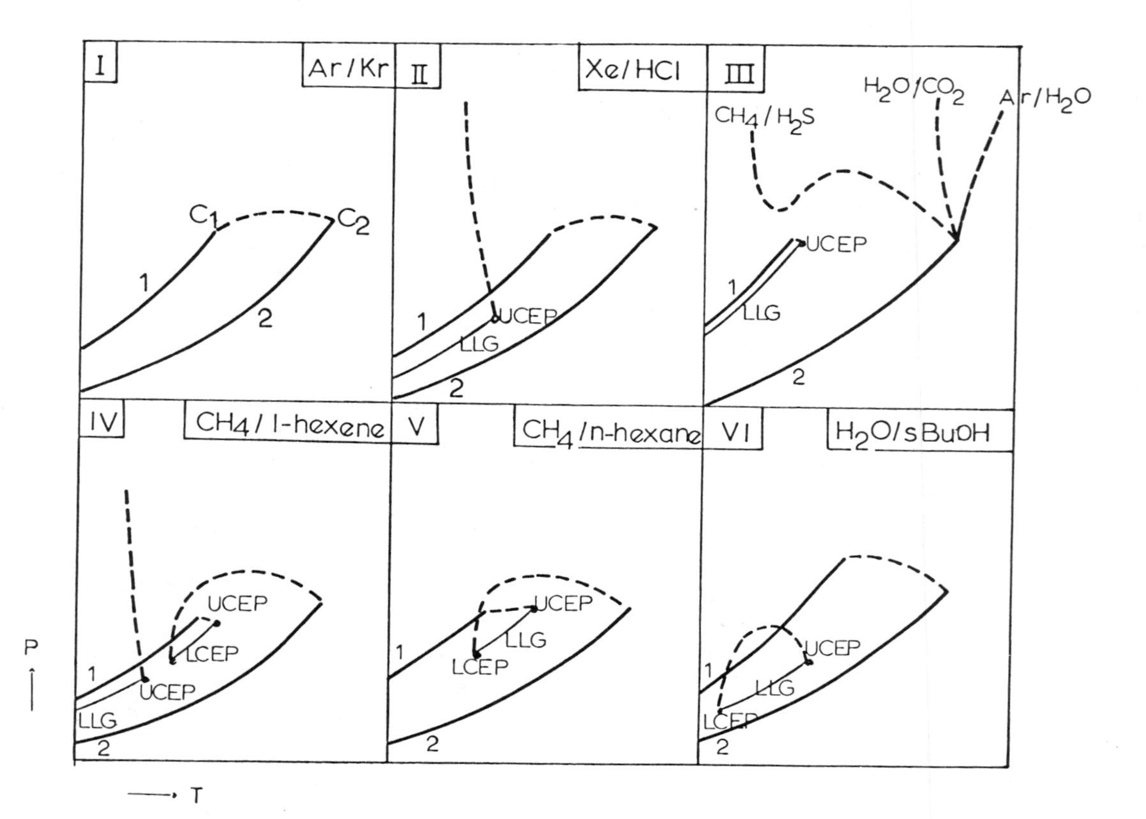

Figure 2. Classification of binary fluid phase diagrams (19) with examples of each class. Vertical and horizontal axes are pressure and temperature, respectively. Solid lines labelled 1 and 2 are vapor pressure curves for the pure components; solid lines labelled LLG are three-phase (liquid–liquid–gas) lines; dashed lines are critical loci.

loops occur, arises in only a relatively few cases (usually water mixed with alcohols or amines). Mixtures of classes III, IV and V are usually composed of components with widely different critical temperatures ($T_1^c/T_2^c \geq 2$), and the gas-liquid critical curve does not pass continuously from one pure component to the other (e.g., because the liquid-liquid immiscibility region extends to that of the gas-liquid critical curve). Included in class III are systems that exhibit "gas-gas" immiscibility. These six classes may be further subdivided, according to whether or not azeotropes are formed, etc. Detailed discussion of the experimental behavior of binary systems is given in several reviews (20-22).

In this section we report calculations for fluids in which one of the constituents interacts with a Lennard-Jones (12,6) potential,

$$u_o(r) = 4\varepsilon[(\frac{\sigma}{r})^{12} - (\frac{\sigma}{r})^6] \quad (13)$$

while the other constituent interacts with a potential $u_o + u_{mult}$, where u_o is again the (12,6) model and u_{mult} is either the dipole-dipole or quadrupole-quadrupole potential (Eq. (3) or (4)). In particular, we investigate the relationship between the intermolecular forces and the resulting class of binary phase diagram. Vapor-liquid, liquid-liquid, and "gas-gas" equilibrium surfaces, three phase lines, critical lines, and azeotropic lines are calculated. For a more detailed discussion of the methods and results, the papers of Twu *et al*. (13,17,23) should be consulted. The unlike pair parameters $\sigma_{\alpha\beta}$ and $\varepsilon_{\alpha\beta}$ are related to the like pair parameters by the usual combining rules,

$$\sigma_{\alpha\beta} = \frac{1}{2} \eta_{\alpha\beta} (\sigma_{\alpha\alpha} + \sigma_{\beta\beta}) \quad (14)$$

$$\varepsilon_{\alpha\beta} = \zeta_{\alpha\beta} (\varepsilon_{\alpha\alpha} \varepsilon_{\beta\beta})^{1/2} \quad (15)$$

where $\eta_{\alpha\beta}$ and $\zeta_{\alpha\beta}$ are close to unity.

Figures 3 to 5 illustrate the effects of adding a dipole-dipole or quadrupole-quadrupole interaction (Eq. (3) or (4)) to one of the components in a binary Lennard-Jones mixture. In this case the Lennard-Jones parameters are chosen so that the reference system approximates an argon-krypton mixture (24):

$\varepsilon_{ArAr}/k = 119.8$ K $\quad \varepsilon_{KrKr}/k = 167.0 \quad \zeta_{ArKr} = 0.989$

$\sigma_{ArAr} = 3.405$ Å $\quad \sigma_{KrKr} = 3.633 \quad \eta_{ArKr} = 1.000$

The argon-krypton reference system is a class I system with no azeotrope,and argon is the more volatile component. If a dipole moment is now added to the argon component, so that the mixture

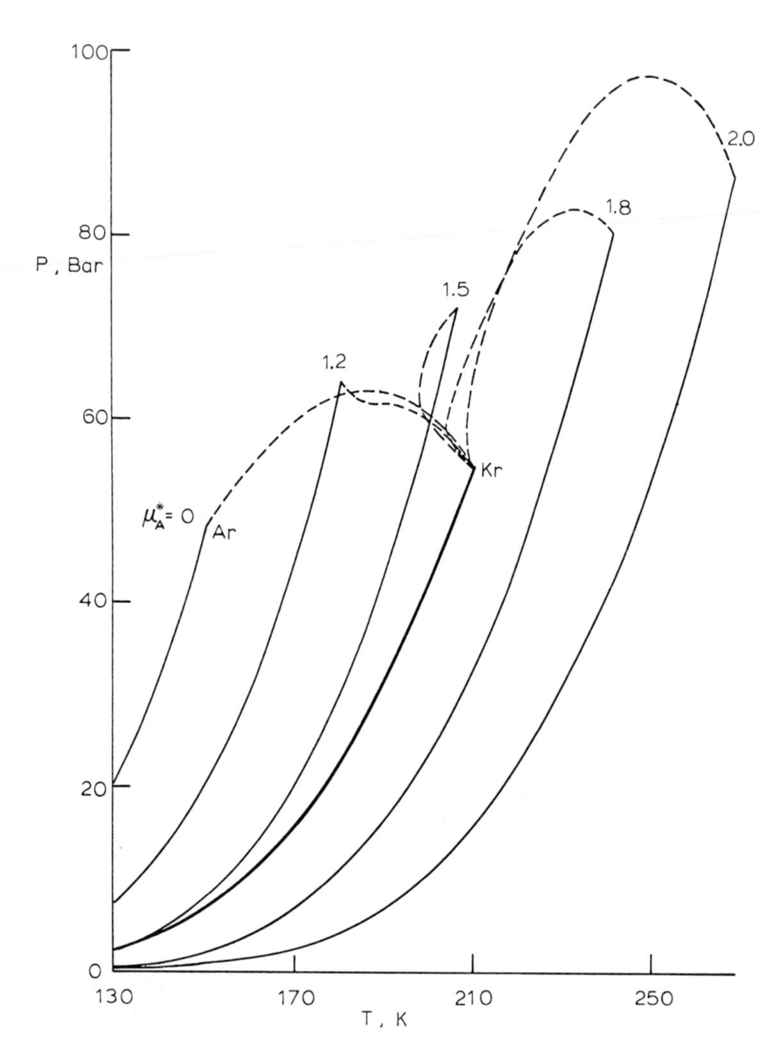

Figure 3. Effect of dipole moment μ_A^* on gas–liquid critical lines (– – –) and on pure component vapor pressures (——) for class I and II systems. Reference system is an argon–krypton mixture.

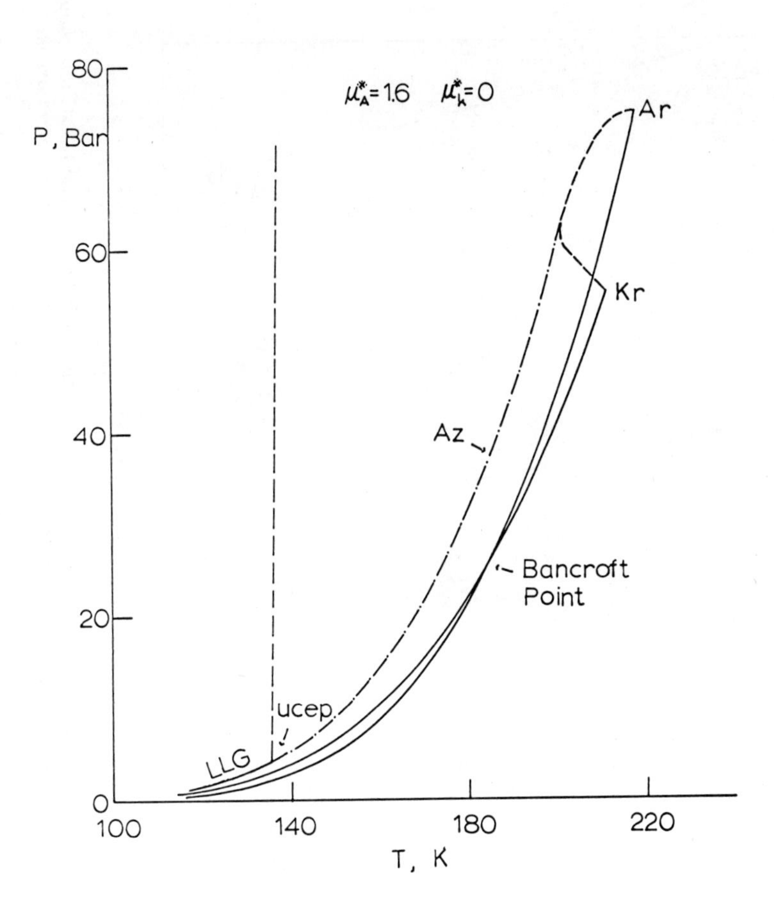

Figure 4. Class II behavior for a polar–nonpolar mixture (argon–krypton reference system) shown as a P,T *projection*

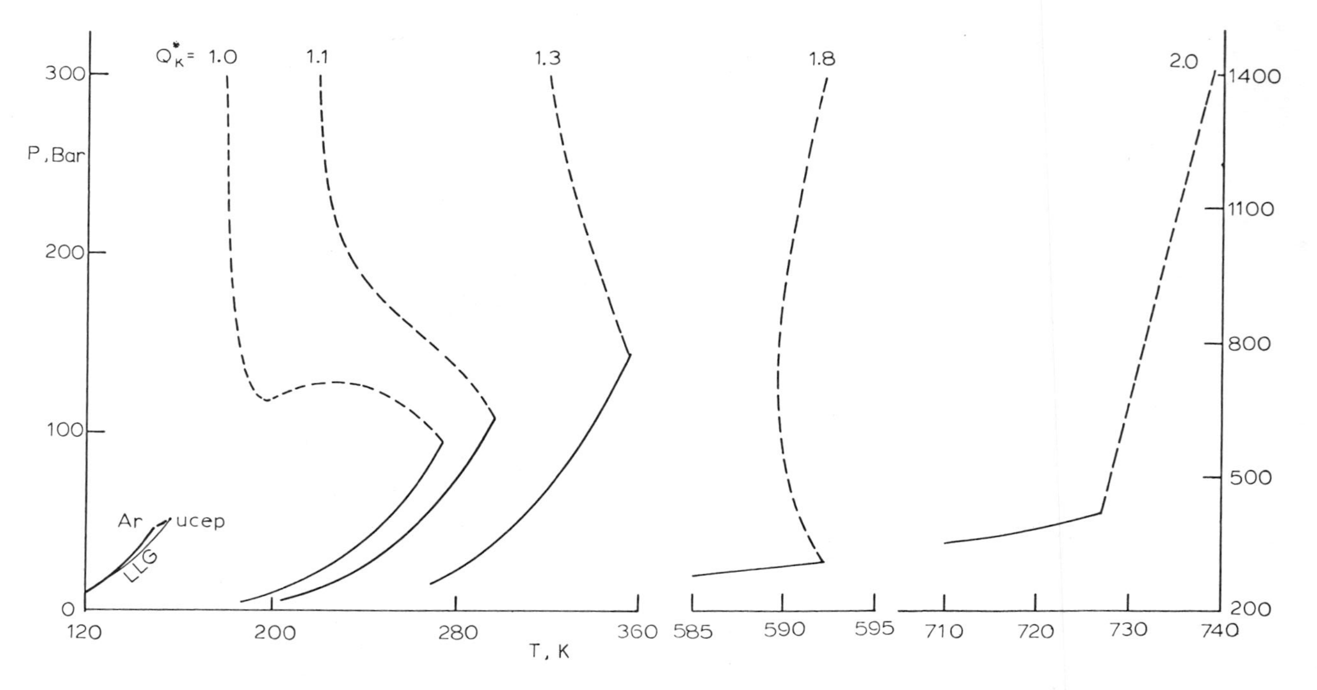

Figure 5. P,T *projection for quadrupolar–nonpolar mixtures ($Q_A^* = 0$) showing class III behavior. The L_1L_2G line shown is for the case $Q_K^* = 1.0$. For higher Q_K^* values this line lies closer to the vapor pressure curve for pure argon.*

becomes a polar/nonpolar one, the effect is to make the Ar component less volatile relative to the Kr component. Figure 3 shows the effect of the dipole moment on the gas-liquid critical curve and on the pure component vapor pressures. At a value of μ_A^* of 1.6 the volatility of the two components is about the same and the system has a positive azeotrope. Despite the similarity in the volatilities and boiling points of the two components, their intermolecular forces are very different, and the system is of class II, showing liquid-liquid immiscibility at lower temperatures. Figure 4 shows the three-phase line, the azeotropic line, and the liquid-liquid critical line for this case. If the value of μ_A^* is further increased the Kr component becomes the more volatile one, and for sufficiently large μ_A^* the system passes to class III. Similar results are found when a quadrupole is added to Ar instead of a dipole (17,23).

If the dipole is added to the less volatile Kr component, leaving the Ar interaction unchanged, the effect is to increase the relative volatility, and the system soon changes from class I to IV, and then quickly to III (Table 1). This behavior is shown for the case of quadrupole interactions in Figure 5. For $Q_k^* > \sim 1.7$ the slope of the critical curve becomes positive at high pressures, so that the critical temperature exceeds that of either pure component at sufficiently high pressures. Such systems exhibit "gas-gas" immiscibility (20,21,25).

Table 1 summarizes the classes of phase behavior found for these polar/nonpolar systems, using an argon-krypton reference system, and compares it with the behavior for simple nonpolar Lennard-Jones systems. An important difference between the two types of systems is that the Lennard-Jones mixtures do not form azeotropes, and appear to exhibit class II behavior only when the components have very different vapor pressures and critical temperatures ($T_b^c/T_a^c > 2$). In practice, the liquid ranges of the two components would not overlap in such cases, so that liquid-liquid immiscibility (and hence class II behavior) would not be observed in Lennard-Jones mixtures (the only exception to this statement seems to be when the unlike pair interaction is improbably weak). Thus, the use of theories based on the Lennard-Jones or other isotropic potential models cannot be expected to give good results for systems of class II, and will probably give poor results for most systems of classes III, IV and V also.

The polar/nonpolar mixtures studied here exhibit four of the six classes of behavior shown in Figure 2. Class V is presumably present also, but is indistinguishable from class IV because the location of the solid-fluid boundary is not calculated. For polar/nonpolar mixtures in which the reference system is a weakly nonideal Lennard-Jones mixture, increasing the dipole moment of the polar molecule causes a continuous transition among the classes,

$$I \rightarrow II \rightarrow (V) \rightarrow (IV) \rightarrow III$$

Table 1

CLASSIFICATION OF BINARY PHASE DIAGRAMS

A. Lennard-Jones Mixtures ($\zeta_{ab} = 1$)

$(\sigma_{bb}/\sigma_{aa})^3$	$\varepsilon_{bb}/\varepsilon_{aa}$	Class	T_b^c/T_a^c	Azeotrope
1	1-2.0	I	1-2	None
	2.0-2.2	II	2.0-2.2	None
	2.2-4.0	III	2.2-4.0	None
2	1-2.2	I	1-2.2	None
	2.2-2.7	II	2.2-2.7	None
	2.7-4.0	III	2.7-4.0	None

B. Nonpolar/Polar Mixtures (Ar/Kr Reference)

μ^*_{Ar}	μ^*_{Kr}	Class	T_{Kr}^c/T_{Ar}^c	Azeotrope
0-1.2	0	I	1.39-1.16	None
1.2-1.3	0	I	1.16-1.11	Positive (limited above)
1.3-1.5	0	I	1.11-1.01	Positive (absolute)
1.5-2.14	0	II	1.01-0.74	Positive (absolute)
2.14-2.5	0	III	0.74-0.58	Positive (absolute)
0	0-1.2	I	1.39-1.67	None
0	1.2-1.37	II	1.67-1.81	None
0	1.37-1.45	IV	1.81-1.87	None
0	1.45-2.5	III	1.87-3.35	None

Classes IV and V are observed only for certain values of the Lennard-Jones parameters. The same sequence is found for quadrupolar/nonpolar mixtures. The potential models considered here fail to account for the class VI systems, i.e., those with low temperature lower critical solution points. Such behavior is believed to arise from strong unlike pair forces, and presumably requires different potential models than those used here.

It is interesting to note that it is possible to observe a tricritical point+ in binary polar/nonpolar (or quadrupolar/nonpolar) systems of the type considered above. Thus if the polar component is a and μ_a is increased, the tricritical point is observed as an intermediary stage in the transition from class II to class III behavior. This is shown in Figure 6, the tricritical point occurring where the vapor-liquid critical curve, liquid-liquid critical curve, and the liquid-liquid-gas curve meet. (It should be noted that the formation of a tricritical point in this binary mixture does not violate the phase rule, since μ_a acts as an additional degree of freedom).

Comparison with Experiment. We consider systems involving the polar constituents HCl and HBr, and the quadrupolar constituents CO_2, ethane, ethylene and acetylene, together with the nonpolar monatomic fluid xenon. For the like pair interaction the intermolecular potential models used were:

$$\text{Xe:} \quad u(r) = u_o(n,6) \tag{16}$$

$$\text{HCl:} \quad u(r\theta_1\theta_2\phi) = u_o(n,6) + u_{mult}\,(\mu\mu + \mu Q + QQ) + u_{ov} + u_{dis} \tag{17}$$

$$\text{HBr:} \quad u(r\theta_1\theta_2\phi) = u_o(n,6) + u_{mult}\,(\mu\mu + \mu Q + QQ) \tag{18}$$

$$CO_2,\ C_2H_2,\ C_2H_4,\ C_2H_6: \quad u(r\theta_1\theta_2\phi) = u_o(n,6) + u_{mult}\,(QQ) + u_{ov} + u_{dis} \tag{19}$$

where u_o is the (n,6) potential, $\mu\mu$ is dipole-dipole, μQ is dipole-quadrupole, and QQ is quadrupole-quadrupole interaction, u_{dis} is the London model for anisotropic dispersion, and u_{ov} is given by Eq. (6) for HCl and by (5) for the other cases. The unlike pair potential model for CO_2/ethane, CO_2/ethylene, CO_2/acetylene, and ethane/ethylene was Eq. (19). For Xe/HCl and Xe/HBr the models

+A normal (bi) critical point occurs when two phases (gas-liquid, liquid-liquid, or "gas-gas")become indistinguishable, i.e., their intensive properties become identical. A tricritical point occurs when three phases become identical; for the binary systems considered here, one of these phases is gaseous and the other two are liquid.

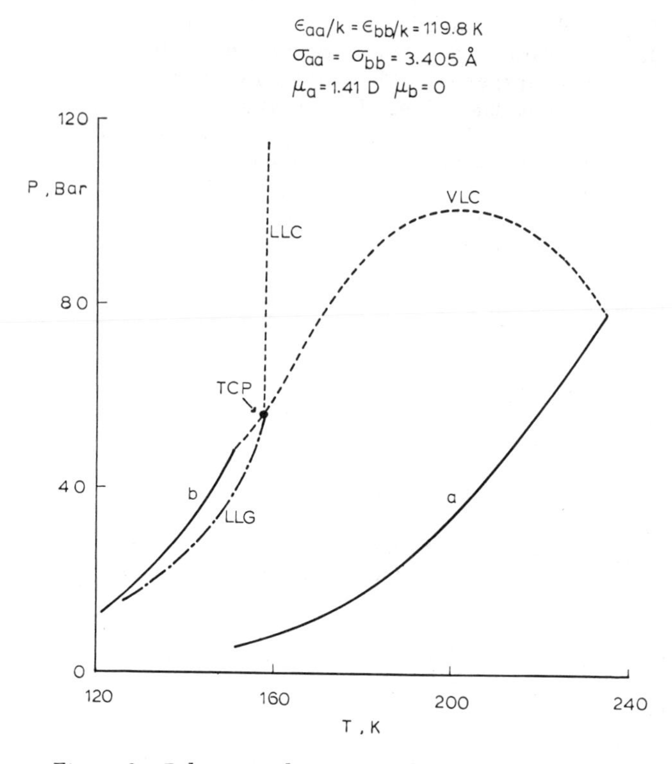

Figure 6. Polar–nonpolar mixture showing tricritical point

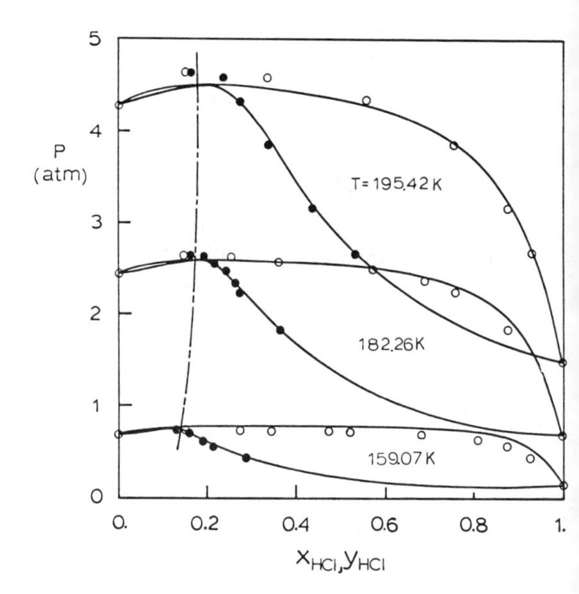

Figure 7. Vapor–liquid equilibria for the system Xe–HCl from theory (lines) and experiment (27) (points). The dash–dot line is the azeotropic locus.

were

$$\text{Xe/HCl} \quad u(r\theta_1\theta_2\phi) = u_o(n,6) + u_{ind} + u_{ov} + u_{dis} \tag{20}$$

$$\text{Xe/HBr} \quad u(r\theta_1\theta_2\phi) = u_o(n,6) + u_{ind} \tag{21}$$

For the Xe/HCl case the overlap model corresponding to Eq. (6) is

$$u_{ov}^{\alpha\beta} = 4\varepsilon_{\alpha\beta}\left(\frac{\sigma_{\alpha\beta}}{r}\right)^{12} \delta_1^{\alpha\beta} \cos\theta_1 \tag{22}$$

and θ_1 is the polar angle for the HCℓ molecule. The term u_{ind} in (20) and (21) includes dipole-induced dipole, dipole-induced quadrupole, quadrupole-induced dipole, and quadrupole-induced quadrupole terms.

The above potential models were arrived at by considering all possible terms in Eq. (1) in each case, and omitting terms that were found to be negligible. Multipole moments and polarisabilities were available from independent experimental measurements. For the like-pair interaction the remaining parameters (ε, σ, n, δ_1, or δ_2) were determined as those giving the best fit to the saturated liquid density and vapor pressure data (13). The like-pair parameters are shown in Table 2. For the unlike-pair interactions one must also know the parameters $\varepsilon_{\alpha\beta}$, $\sigma_{\alpha\beta}$, $n_{\alpha\beta}$ and $\delta_i^{\alpha\beta}$, where i is 1 or 2. The last two parameters were estimated from the combining rules

$$n_{\alpha\beta} = (n_{\alpha\alpha}n_{\beta\beta})^{1/2} \tag{23}$$

$$\delta_1^{\alpha\beta} = 1/2\ (\delta_1^{\alpha\alpha} + \delta_1^{\beta\beta}) \tag{24}$$

while $\varepsilon_{\alpha\beta}$ and $\sigma_{\alpha\beta}$ are given by (14) and (15). The quantities $\eta_{\alpha\beta}$ and $\zeta_{\alpha\beta}$ were taken to be the values giving the best fit to vapor-liquid equilibrium data for the binary mixture at a single temperature. The unlike-pair parameters are given in Table 3.

Figures 7 and 8 show a comparison of theory and experimental data of Calado et al. (27,28) for vapor-liquid equilibrium in the systems Xe/HCl and Xe/HBr. Figure 9 shows a comparison for the excess volume for Xe/HBr (similar agreement is obtained for Xe/HCl). These systems are highly nonideal, and both the polar and quadrupolar forces are important. The Xe/HCl system has a positive azeotrope which is well reproduced by the theory, and apparently exhibits liquid-liquid immiscibility at low temperatures (27). The theory predicts the vapor-liquid compositions and pressures well at each temperature, the maximum error being 0.1 bar, and also reproduces the excess volume. It is interesting to note that the V^E composition curve for Xe/HBr is cubic. Such behavior

Table 2

POTENTIAL PARAMETERS

Substance	$\varepsilon/k^{(a)}$ (K)	$\sigma^{(a)}$ (Å)	$n^{(a)}$	$\delta_1^{(a)}$	$\delta_2^{(a)}$	α (Å^3)	$\kappa^{(d)}$	$\mu \times 10^{18}$ esu cm	$Q \times 10^{26}$ esu cm^2
Xe	231.52	3.961	12			$4.10^{(b)}$			
HCl	153.23	3.670	9	0.10				1.07 (c)	3.80 (c)
HBr	248.47	3.790	12					0.788(c)	4.0 (c)
CO_2	244.31	3.687	16		-0.1		0.257		-4.30 (c)
C_2H_2	253.66	3.901	16		+0.3		0.270		5.01 (d)
C_2H_4	229.32	4.097	13		0.1		0.158		3.85 (d)
C_2H_6	238.99	4.324	13		-0.2		0.132		-0.65 (c)

(a) From saturated liquid density and pressure.
(b) Isotropic polarizability, from Reed and Gubbins.[8]
(c) From Stogryn and Stogryn.[11]
(d) From Spurling and Mason.[26] $\kappa \equiv (\alpha_{\parallel} - \alpha_{\perp})/(\alpha_{\parallel} + 2\alpha_{\perp})$, where $\alpha_{\parallel}$ and $\alpha_{\perp}$ are the polarizabilities parallel and perpendicular to the symmetry axis, respectively.

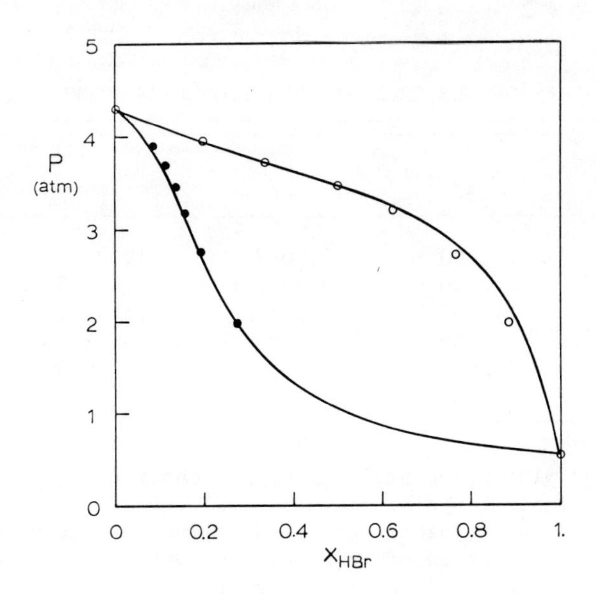

Figure 8. Vapor–liquid equilibria for Xe–HBr from theory (lines) and experiment (28) (points)

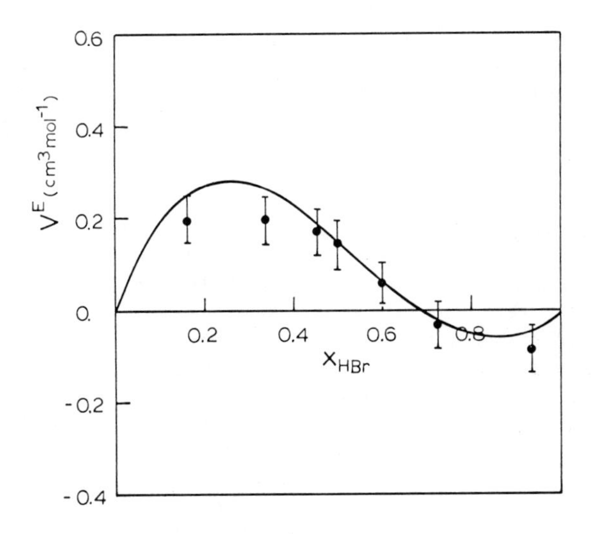

Figure 9. Excess molar volumes for Xe–HBr from theory (lines) and experiment (28) (points)

Table 3

VALUES OF THE CROSS-INTERACTION PARAMETERS

System	$\eta_{\alpha\beta}$	$\zeta_{\alpha\beta}$	$n_{\alpha\beta}$	$\delta_i^{\alpha\beta}$
HCl/Xe	1.0050	1.0042	10.3923	+0.10
HBr/Xe	0.9996	0.9880	12.0000	-
CO_2/C_2H_6	1.0000	0.9635	14.4222	-0.15
CO_2/C_2H_4	1.0169	0.9038	14.4222	0.00
CO_2/C_2H_2	1.0000	0.8380	16.0000	+0.10
C_2H_2/C_2H_6	1.0000	1.0197	13.0000	-0.05

is common for mixtures containing polar constituents, but is not observed in simple nonpolar mixtures (20).

Figure 10 shows theoretical predictions for the system CO_2/ethane over a wide range of temperatures, pressures and compositions. The system has a positive azeotrope over the whole of the liquid range. Excellent agreement is obtained between theory and experiment for the vapor-liquid equilibria at all temperatures, although small discrepancies appear in the critical line. Calculated azeotropic compositions agree with experimental values within 1% over the entire liquid range; calculated azeotropic pressures agree within 0.1 bar. Figure 11 shows the predicted vapor-liquid equilibrium curves for CO_2/acetylene. This system is of particular interest because it forms a negative azeotrope; according to the theory this azeotrope is bounded above. The negative azeotrope in this system is believed to arise because the quadrupole moments of CO_2 and acetylene are of opposite sign (20). The theoretical predictions are in good agreement with the very meager experimental data for this system. These are limited to a measurement of the boiling point of the azeotrope at 1 bar, (32) and to measurements of the critical point for an equimolar mixture (29,33). For the systems CO_2/ethylene and ethane/ethylene, the agreement between theory and experiment is similar to that for CO_2/ethane.

Figure 12 shows a test of the theory for the ternary system CO_2/ethane/ethylene. Such a test is of particular interest, since no new potential parameters need be introduced. For temperatures ranging from 263 to 293K, and pressures from 20 to 60 bar, the mean deviation between theory and experiment was only 0.003 in mole fraction and 0.16 bar in pressure. The results for the ternary system are as accurate as those for the binary systems. Thus the binary systems are adequate building blocks for calculating the properties of the multicomponent systems. There is no need to invoke any characteristically three-component parameter.

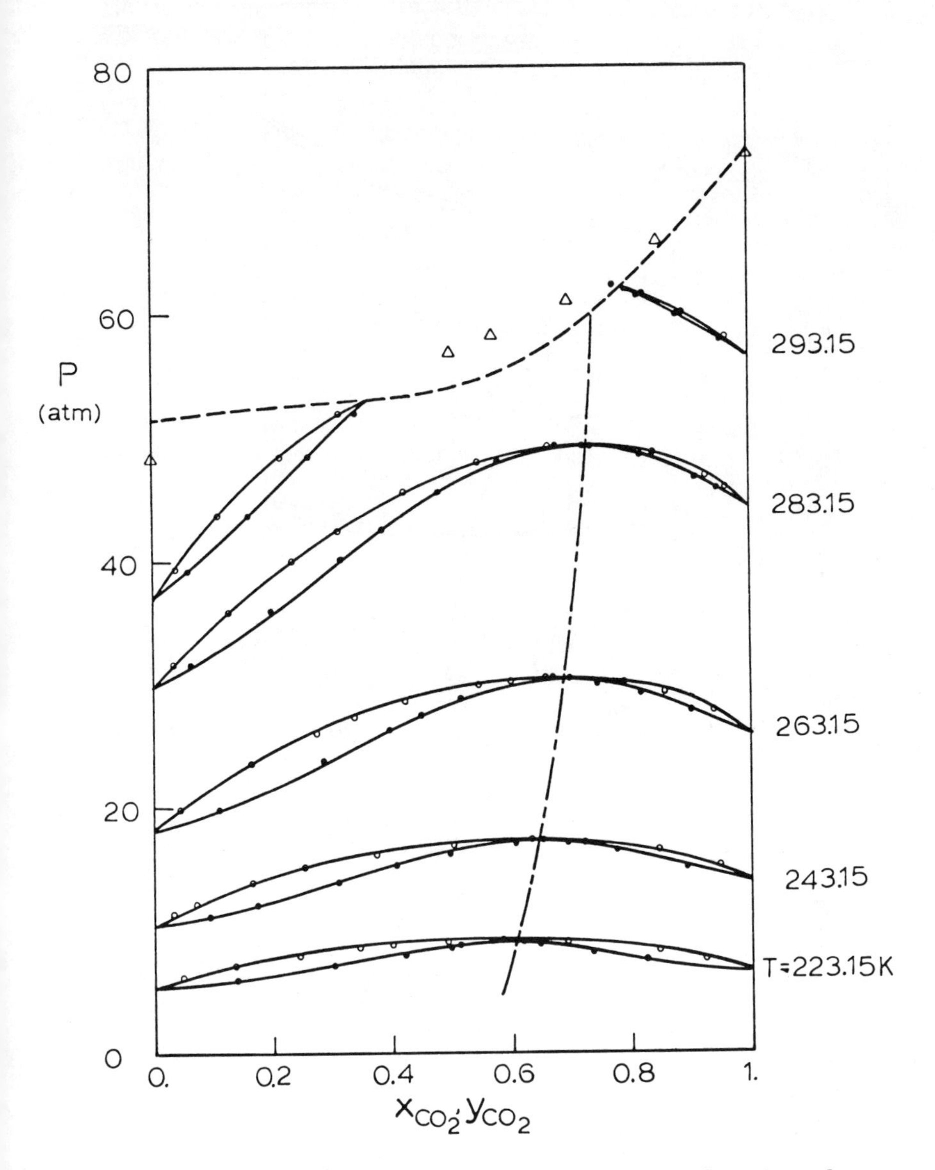

Figure 10. Vapor–liquid equilibria and critical line for CO_2–ethane from theory (lines) and experiment (29–31) (points)

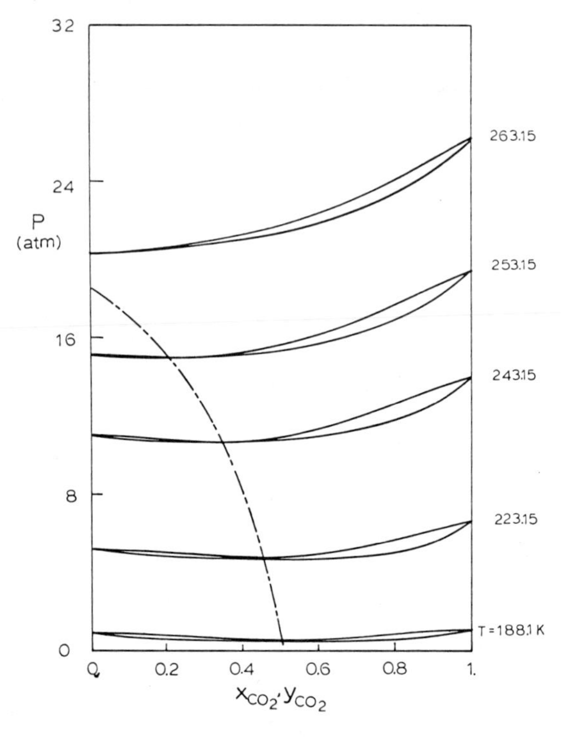

Figure 11. Calculated vapor–liquid equilibria for CO_2–acetylene

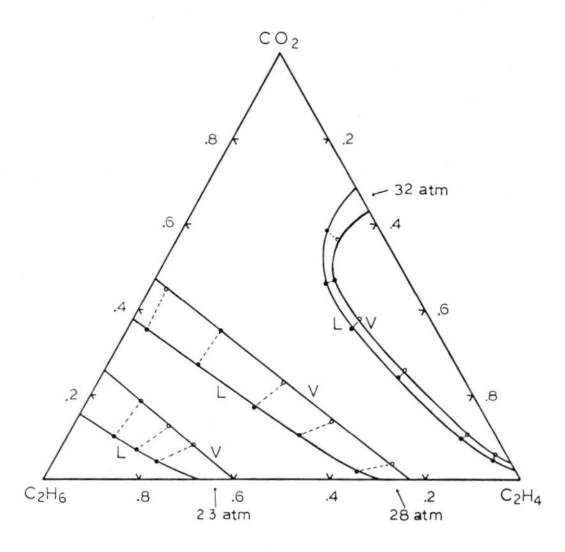

Figure 12. Vapor–liquid equilibria for the system CO_2–ethane–ethylene at 263°K from theory (lines) and experiment (34) (points)

Acknowledgments

We are grateful to the National Science Foundation, the American Gas Association, and the Petroleum Research Fund (administered by the American Chemical Society) for financial support of this work. It is a pleasure to thank L. A. K. Staveley, J. C. G. Calado, A. Fredenslund and J. Mollerup for sending papers prior to publication, and B. Widom for a helpful discussion.

Abstract

A method based on thermodynamic perturbation theory is described which allows strong directional intermolecular forces to be taken into account when calculating thermodynamic properties. This is applied to the prediction of phase equilibrium and critical loci for mixtures containing polar or quadrupolar constituents. Two applications of the theory are then considered. In the first, the relation between intermolecular forces and the type of phase behavior is explored for binary mixtures in which one component is either polar or quadrupolar. Such systems are shown to give rise to five of the six classes of binary phase diagrams found in nature. The second application involves comparison of theory and experiment for binary and ternary mixtures.

References

1. Hildebrand, J. H., Prausnitz, J. M., and Scott, R. L., "Regular and Related Solutions," Van Nostrand Reinhold Co., New York 1970.
2. J. S. Rowlinson, this volume.
3. Smith, W. R., "Statistical Mechanics, Vol. 1, Specialist Periodical Report," ed. K. Singer, Chemical Society, London (1973).
4. Gubbins, K. E., AIChE Journal, (1973) 19, 684.
5. T. W. Leland, P. S. Chappelear and B. W. Gamson, A.I.Ch.E. Journal, (1962), 8,482; J. W. Leach, P. S. Chappelear and T. W. Leland, Proc. Am. Petrol. Inst., (1966), 46, 223.
6. K. C. Mo and K. E. Gubbins, Molec. Phys., (1976),31,825; J. M. Haile, K. C. Mo and K. E. Gubbins, Adv. Cryogen. Engng., (1976), 21,501; S. Murad and K. E. Gubbins, Chem. Eng. Science, (1977), 32, 499.
7. For a review see Egelstaff, P. A., Gray, C. G., and Gubbins, K. E., "Molecular Structure and Properties," International Review of Science. Physical Chemistry, Series 2, Volume 2, ed. A. D. Buckingham, Butterworths, London 1975.
8. Reed, T. M., and Gubbins, K. E., "Applied Statistical Mechanics," McGraw Hill, New York 1973.
9. Gray, C. G., and Van Kranendonk, J., Canad. J. Phys., (1966) 44 2411.

10. Armstrong, R. L., Blumenfeld, S. M. and Gray, C. G., Canad. J. Phys. (1968) 46, 1331.
11. Stogryn, D. E. and Stogryn, A. P., Molec. Phys. (1966) 11, 371.
12. Flytzani-Stephanopoulos, M., Gubbins, K. E., and Gray, C. G., Molec. Phys., (1975) 30, 1649.
13. Twu, C. H., Ph.D. Dissertation, University of Florida (1976).
14. Stell, G., Rasaiah, J. C., and Narang, H., Molec. Phys., (1974) 27, 1393.
15. Wang, S. S., Egelstaff, P. A., Gray, C. G., and Gubbins, K. E., Chem. Phys. Lett., (1974) 24, 453.
16. Haile, J. M., Ph.D. Dissertation, University of Florida (1976).
17. Twu, C. H., Gubbins, K. E., and Gray, C. G., J. Chem. Phys., (1976) 64, 5168.
18. Gosman, A. L., McCarty, R. D., and Hust, J. D., Nat. Stand. Ref. Data Serv. Natl. Bur. Stand. 27 (1969).
19. Scott, R. L., and Van Konynenburg, P. H., Discuss. Faraday Soc., (1970) 49, 87.
20. Rowlinson, J. S., "Liquids and Liquid Mixtures," Butterworths, London, 2nd edition 1969.
21. Schneider, G. M., Adv. Chem. Phys., (1970) 17, 1.
22. Scott, R. L., Ber. Bunsenges. Phys. Chem. (1972) 76, 296.
23. Twu, C. H., and Gubbins, K. E., to be published.
24. McDonald, I. R., "Statistical Mechanics, Vol. 1 Specialist Periodical Report," ed. K. Singer, Chemical Society, London (1973).
25. Streett, W. B., Canad. J. Chem. Eng., (1974) 52, 92.
26. Spurling, T. H., and Mason, E. A., J. Chem. Phys., (1967) 46, 322.
27. Calado, J. C. G., Kozdon, A. F., Morris, P. J., da Ponte, N., Staveley, L. A. K., and Woolf, L. A., J. Chem. Soc., Faraday Trans. I, (1975) 71, 1372.
28. Calado, J. C. G., and Staveley, L. A. K., private communication.
29. Kuenen, J. P., Phil. Mag. (1897) 44, 174.
30. Fredenslund, A., and Mollerup, J., J. Chem. Soc. Faraday Trans. I., (1974) 70, 1653.
31. Mollerup, J., J. Chem. Soc. Faraday Trans. I, (1975) 71, 2351.
32. Clark, A. M., and Din, F., Trans. Faraday Soc., (1950) 46, 901.
33. Dewar, J., Proc. Roy. Soc. London, (1880) 30, 542.
34. Fredenslund, A., Mollerup, J., and Hall, K. R., J. Chem. Eng. Data, (1976), 21, 301.

Discussion

TRUMAN S. STORVICK

Progress on the theoretical description at the properties of moderately dense and dense fluids systems has traditionally been limited by the lack of an equation of state for dense fluid phases. The past decade has seen rapid progress on this problem using electronic computers to obtain ensemble average properties for fluid systems modelled with simple particles. The equilibrium and non-equilibrium statistical mechanical and kinetic theory relationships admit the direct calculation of the fluid properties without identifying the mathematical form of the equation of state. This application of computers has revolutionized the fluid properties studies in ways as profound as their application to process design problems.

The determination of dense fluid properties from *ab initio* quantum mechanical calculations still appears to be some time from practical completion. Molecular dynamics and Monte Carlo calculations on rigid body motions with simple interacting forces have qualitatively produced all of the essential features of fluid systems and quantitative agreement for the thermodynamic properties of simple pure fluids and their mixtures. These calculations form the basis upon which perturbation methods can be used to obtain properties for polyatomic and polar fluid systems. All this work has provided insight for the development of the principle of corresponding state methods that describe the properties of larger molecules.

The perturbation methods were discussed in detail and applications to polar-nonpolar mixtures described the potential power of these techniques. The energy and distance scaling parameters for each component are obtained from pure component vapor pressure data. Dipole or quadrupole moments are obtained from independent measurements. An energy interaction parameter is evaluated from vapor-liquid equilibrium data at one temperature. An effective equation of state for these mixtures is obtained from the formalizm using this small set of data. More work must be done to improve the accuracy of the calculations to provide design data but it clearly shows promise and continued effort should be productive.

The calculation of the transport properties of moderately dense fluids using kinetic theory and an equilibrium equation of state shows remarkable agreement between predictions and experimental measurements. This success suggests that the theoretical structure required to obtain the functional forms that produced these results should be studied. The verification of this procedure for a wider range of conditions and mixtures could provide a way to obtain transport properties of fluids.

The question and answer session focused on problems that must be treated in order to make the theoretical advances available to the design people. It takes great attention to the detailed structure of the theory to understand what the essential features of the models are and what experimental information must be supplied to make quantitative calculations.
There are some special areas where work should now be done:

1. Experimental data to test theoretical procedures is often not available. Thermodynamic and transport data on mixtures containing hydrogen and acid gases (HCl, NH_3, SO_2, etc.) would be very useful. Dense gas data are generally not available. Accurate measurements over wide ranges of temperature and density are especially useful.
2. Requests for data from field engineers to company data centers are currently distributed about 70% phase equilibrium, 25% enthalpy and 5% all other data. Design decisions are made on the basis of the distribution coefficients between phases and on the process energy requirements. The requests reflect this demand. Procedures to accurately predict these data would be widely adopted and are in great demand.
3. Polyatomic molecules are difficult to treat because the size, shape, flexibility, and charge distribution are all important factors in the fluid equation of state. Complex molecules do not satisfy random mixing criteria and the energy and entropy functions must contain proper accounting of these non-random effects.
4. The extended principle of corresponding state methods are being further developed. Wider experience with this approach might produce earlier adoption of new methods of prediction because most practicing engineers have a better intuitive understanding of these methods.

The rapid development of the theory of dense fluid systems will require a number of workers who can transform these results into practical properties prediction schemes. This is often a difficult step to accomplish but it is essential if we are to reduce the time between the theoretical description of fluid behavior and the use of this understanding to help make design decisions.

19

Criteria of Criticality

MICHAEL MODELL

Massachusetts Institute of Technology, Cambridge, MA 02139

Over a century ago, Gibbs (1) developed the mathematical criteria of criticality. He defined the critical phase as the terminal state on the binodal surface and reasoned that it has one less degree of freedom than the binodal surface (i.e., f = n-1). He then developed two equations as the criteria of criticality; these two equations, when imposed upon the Fundamental Equation, reduce the degrees of freedom from n+1 to n-1.

Gibbs presented three alternate sets of the two criticality criteria. In terms of the variable set $T,\underline{V},\mu_1,\ldots,\mu_{n-1},N_n$, they are:

$$\left(\frac{\partial \mu_n}{\partial N_n}\right)_{T,\underline{V},\mu_1,\ldots,\mu_{n-1}} = 0 \tag{1}$$

$$\left(\frac{\partial^2 \mu_n}{\partial N_n^2}\right)_{T,\underline{V},\mu_1,\ldots,\mu_{n-1}} = 0 \tag{2}$$

To ensure stability of the critical phase, Gibbs noted that the following relation must also be satisfied:

$$\left(\frac{\partial^3 \mu_n}{\partial N_n^3}\right)_{T,\underline{V},\mu_1,\ldots,\mu_{n-1}} > 0 \tag{3}$$

Gibbs noted that Eq. (1) may lead to an indeterminant form. When this occurs, he suggested permutting the component subscripts or using "another differential coefficient of the same general form. He noted that Eq. (1) is the criterion of the limit of stability (i.e., the spinodal surface). In an earlier discussion on that subject, Gibbs showed that all of the following partial derivatives vanish on the spinodal surface.

$$\left(\frac{\partial T}{\partial \underline{S}}\right)_{\underline{V},\mu_1,\ldots,\mu_n}, \left(\frac{\partial \mu_1}{\partial N_1}\right)_{T,\underline{V},\mu_2,\ldots,\mu_n}, \ldots, \left(\frac{\partial \mu_n}{\partial N_n}\right)_{T,\underline{V},\mu_1,\ldots,\mu_{n-1}} \tag{4}$$

Thus, any one of these forms could be used in place of Eq. (1), with corresponding changes made in Eqs. (2) and (3).

As an alternative to Eqs. (1) and (2), Gibbs suggests "for a perfectly rigorous method there is an advantage in the use of $\underline{S}$,$\underline{V}$, $N_1,\ldots,N_n$ as independent variables." In this variable set the criticality conditions become:

$$R_{n+1} = 0 \tag{5}$$

$$S = 0 \tag{6}$$

where
$$R_{n+1} = \begin{vmatrix} U_{SS} & U_{N_1S} & \cdots & U_{N_nS} \\ U_{SN_1} & U_{N_1N_1} & \cdots & U_{N_nN_1} \\ \cdots & \cdots & \cdots & \cdots \\ U_{SN_n} & U_{N_1N_n} & \cdots & U_{N_nN_n} \end{vmatrix} \tag{7}$$

and S is the determinant formed by replacing one of the rows of R_{n+1} by $(R_{n+1})_S$ $(R_{n+1})_{N_1}$ $\cdots$ $(R_{n+1})_{N_n}$

$$\text{e.g., } S = \begin{vmatrix} U_{SS} & U_{N_1S} & \cdots & U_{N_nS} \\ U_{SN_1} & U_{N_1N_1} & \cdots & U_{N_nN_1} \\ \cdots & \cdots & \cdots & \cdots \\ U_{SN_{n-1}} & U_{N_1N_{n-1}} & \cdots & U_{N_nN_{n-1}} \\ (R_{n+1})_S & (R_{n+1})_{N_1} & \cdots & (R_{n+1})_{N_n} \end{vmatrix} \tag{8}$$

where the notation is $U_{ij} = (\partial^2\underline{U}/\partial i\partial j)$ and the variables held constant in the differentiation are n+1 of the set $\underline{S},\underline{V},N_1,\ldots,N_n$. Similarly, $(R_{n+1})_{N_1} = (\partial R_{n+1}/\partial N_1)_{\underline{S},\underline{V},N_2,\ldots,N_n}$ Gibbs states that Eqs. (7) and (8) "will hold true of every critical phase without exception," implying that this set of criticality criteria is more stringent than Eqs. (1) and (2).

Gibbs presents a third alternative set with T,P, $N_1,\ldots,N_n$ as independent variables:

$$B = \begin{vmatrix} G_{N_1N_1} & G_{N_2N_1} & \cdots & G_{N_{n-1}N_1} \\ G_{N_1N_2} & G_{N_2N_2} & \cdots & G_{N_{n-1}N_2} \\ \cdots & \cdots & \cdots & \cdots \\ G_{N_1N_{n-1}} & G_{N_2N_{n-1}} & \cdots & G_{N_{n-1}N_{n-1}} \end{vmatrix} = 0 \tag{9}$$

$$\text{and} \quad C = 0 \tag{10)*$$

where C is the determinant formed by replacing one of the rows of the B-determinant by $B_{N_1}\ B_{N_2}\ \cdots\ B_{N_{n-1}}$. Gibbs does not specify if Eqs. (9) and (10) are true without exception or if there are restrictions to their use.

Development of the General Criticality Criteria

Each of the criticality criteria presented by Gibbs can be shown

*Gibbs uses the symbols U and V for the B- and C-determinants. We use B and C here to avoid confusion with internal energy and volume.

to be special cases of more general forms. The development of the general form follows from the unique feature of the critical phase which distinguishes it from other phases in the vicinity; namely, the critical phase is the stable condition on the spinodal surface. This statement is equivalent to Gibbs' argument that the critical phase lies on the binodal surface and, therefore, is stable (with respect to continuous changes) and also satisfies the condition of the limit of stability (which defines the spinodal curve).

Starting with the Fundamental Equation in the internal energy representation,

$$\underline{U} = f_U(\underline{S},\underline{V},N_1,\ldots,N_n) \tag{11}$$

we can test the stability of a substance by expanding $\underline{U}$ in a Taylor series and examining the change in $\underline{U}$ for small perturbations while holding the total entropy, volume, and mass constant(2). That is,

$$\Delta\underline{U} = \delta\underline{U} + \frac{1}{2!}\,\delta^2\underline{U} + \frac{1}{3!}\,\delta^3\underline{U} + \ldots \tag{12}$$

For equilibrium to exist,

$$\delta\underline{U} = 0 \tag{13}$$

and for the system to be stable,

$$\delta^P\underline{U} > 0 \tag{14}$$

where $\delta^P\underline{U}$ is the lowest order non-vanishing variation. The spinodal surface represents the conditions where the second-order variation first loses the positive, definite character of the stable, single phase (3). The second-order variation can be expressed as

$$\begin{aligned}\delta^2\underline{U} = {} & U_{SS}(\delta\underline{S})^2 + 2U_{SV}(\delta\underline{S})(\delta\underline{V}) + U_{VV}(\delta\underline{V})^2 \\ & + 2\sum_{j=i}^{n}(U_{SN_j}\,\delta\underline{S} + U_{VN_j}\,\delta\underline{V})\delta N_j \\ & + \sum_{j=1}^{n}\sum_{k=1}^{n} U_{N_jN_k}\,\delta N_j\,\delta N_k\end{aligned} \tag{15}$$

or, in closed form, as a sum of squares,

$$\delta^2\underline{U} = U_{SS}\,\delta Z_1^{\,2} + \frac{\begin{vmatrix} U_{SS} & U_{SV} \\ U_{VS} & U_{VV} \end{vmatrix}}{U_{SS}}\,\delta Z_2^{\,2}$$

$$
+ \frac{\begin{vmatrix} U_{SS} & U_{SV} & U_{SN_1} \\ U_{VS} & U_{VV} & U_{VN_1} \\ U_{N_1S} & U_{N_1V} & U_{N_1N_1} \end{vmatrix}}{\begin{vmatrix} U_{SS} & U_{SV} \\ U_{VS} & U_{VV} \end{vmatrix}} \delta Z_3^{\,2} + \ldots
$$

$$
+ \frac{\begin{vmatrix} U_{SS} & U_{SV} & U_{SN_1} & \cdots & U_{SN_n} \\ U_{VS} & U_{VV} & U_{VN_1} & \cdots & U_{VN_n} \\ U_{N_1S} & U_{N_1V} & U_{N_1N_1} & \cdots & U_{N_1N_n} \\ \cdots & \cdots & \cdots & \cdots & \cdots \\ U_{N_nS} & U_{N_nV} & U_{N_nN_1} & \cdots & U_{N_nN_n} \end{vmatrix}}{\begin{vmatrix} U_{SS} & U_{SV} & U_{SN_1} & \cdots & U_{SN_{n-1}} \\ U_{VS} & U_{VV} & U_{VN_1} & \cdots & U_{VN_{n-1}} \\ U_{N_1S} & U_{N_1V} & U_{N_1N_1} & \cdots & U_{N_1N_{n-1}} \\ \cdots & \cdots & \cdots & \cdots & \cdots \\ U_{N_{n-1}S} & U_{N_{n-1}V} & U_{N_{n-1}N_1} & \cdots & U_{N_{n-1}N_{n-1}} \end{vmatrix}} \delta Z_{n+2}^{2} \qquad (16)
$$

Using the shorthand notation,

$$A_1 = U_{SS}$$

$$A_2 = \begin{vmatrix} U_{SS} & U_{SV} \\ U_{VS} & U_{VV} \end{vmatrix}$$

$$A_{i+2} = \begin{vmatrix} U_{SS} & U_{SV} & U_{SN_1} & \cdots & U_{SN_i} \\ U_{VS} & U_{VV} & U_{VN_1} & \cdots & U_{VN_i} \\ U_{N_1S} & U_{N_1V} & U_{N_1N_1} & \cdots & U_{N_1N_i} \\ U_{N_iS} & U_{N_iV} & U_{N_iN_1} & \cdots & U_{N_iN_i} \end{vmatrix}$$

Eq. (16) takes the form,

$$\delta^2\underline{U} = A_1\delta Z_1{}^2 + \sum_{j=2}^{n+2} \frac{A_j}{A_{j-1}} \delta Z_j{}^2 \qquad (17)^*$$

The last term in Eqs. (16) or (17) can be shown to be zero as a virtue of the fact that only n+1 intensive variables are independent for a single phase. The proof follows.

Let us expand each of the n+2 intensive variables T, P, $\mu_1,\ldots,\mu_n$ as functions of $\underline{S},\underline{V},N_1,\ldots,N_n$.

$$dT = \left(\frac{\partial T}{\partial \underline{S}}\right)_{\underline{V},N} d\underline{S} + \left(\frac{\partial T}{\partial \underline{V}}\right)_{\underline{S},N} d\underline{V} + \sum_{i=1}^{n} \left(\frac{\partial T}{\partial N_i}\right)_{\underline{S},\underline{V},N_j[i]} dN_i$$

*For the form of the δZ_j variations, see (3); since they always appear as squared terms, we need not consider them further here.

or $$dT = U_{SS}\, d\underline{S} + U_{SV} d\underline{V} + \sum_{i=1}^{n} U_{SN_i}\, dN_i \quad (18)$$

$$dP = U_{VS}\, d\underline{S} + U_{VV}\, d\underline{V} + \sum_{i=1}^{n} U_{VN_i}\, dN_i \quad (19)$$

$$d\mu_1 = U_{N_1S}\, d\underline{S} + U_{N_1V}\, d\underline{V} + \sum_{i=1}^{n} U_{N_1N_i}\, dN_i \quad (20)$$

$$\cdots \quad \cdots \quad \cdots \quad \cdots \quad \cdots$$

$$d\mu_{n-1} = U_{N_{n-1}S}\, d\underline{S} + U_{N_{n-1}V}\, d\underline{V} + \sum_{i=1}^{n} U_{N_{n-1}N_i}\, dN_i \quad (21)$$

$$d\mu_n = U_{N_nS}\, d\underline{S} + U_{N_nV}\, d\underline{V} + \sum_{i=1}^{n} U_{N_nN_i}\, dN_i \quad (22)$$

Since the n+2 differentials of $T, P, \mu_1, \ldots, \mu_n$ are related by the Gibbs-Duhem equation, any one from this set can be expressed as a function of the other n+1. Thus, if $P, \mu_1, \ldots, \mu_n$ are held constant (i.e., $dP = d\mu_1 = \ldots = d\mu_n = 0$), then T must also be constant (i.e., $dT = 0$). It follows that the determinant of the matrix of coefficients in Eqs. (18) to (22) must be zero. Since this determinant is A_{n+2}, it follows that $A_{n+2} = 0$. Therefore, the general form of Eq. (17) is

$$\delta^2\underline{U} = A_1 \delta Z_1^{\,2} + \sum_{j=2}^{n+1} \frac{A_j}{A_{j-1}} \delta Z_j^{\,2} \quad (23)$$

The limit of stability is defined as the condition under which $\delta^2\underline{U}$ loses its positive, definite character. That is, one of the coefficients in Eq. (23) vanishes. As has been shown previously (3), when the spinodal surface is approached from a stable single phase region, the coefficient of δZ_{n+1} is among the first to reach zero (i.e., if another coefficient also vanishes, it does so simultaneously with the coefficient of δZ_{n+1}). Therefore, on the spinodal surface

$$A_{n+1} = 0 \quad (24)$$

This equation is one of the criteria of criticality. It is equivalent to reducing the degrees of freedom from n+1 for a stable single phase to n on the spinodal surface. Since A_{n+1} is the determinant of the matrix of coefficients of Eqs. (18) to (21) at constant N_n, the requirement of $A_{n+1} = 0$ is equivalent to $dT = dP = d\mu_1 = \ldots = d\mu_{n-1} = 0$. That is, if n variables from the set $T,P,\mu_1,\ldots,\mu_{n-1}$ are held constant, the remaining variable will also be constant.

The second criterion of criticality is that the differential of A_{n+1} must vanish for all possible variations. Thus,

$$dA_{n+1} = \left(\frac{\partial A_{n+1}}{\partial \underline{S}}\right)_{\underline{V},N} d\underline{S} \left(\frac{\partial A_{n+1}}{\partial \underline{V}}\right)_{\underline{S},N} d\underline{V} + \sum_{i=1}^{n} \left(\frac{\partial A_{n+1}}{\partial N_i}\right)_{\underline{S},\underline{V},N_j[i]} dN_i \tag{25}$$

Alternatively, Eq. (25) can be satisfied simultaneously with n equations from the set Eqs. (18) to (21) at constant N_n. Choosing $dT = dP = d\mu_1 = \ldots = d\mu_{n-2} = 0$ from this set, the second criteria of criticality becomes:

$$\mathcal{B} = \begin{vmatrix} U_{SS} & U_{SV} & U_{SN_1} & \cdots & U_{SN_{n-1}} \\ U_{VS} & U_{VV} & U_{VN_1} & \cdots & U_{VN_{n-1}} \\ U_{N_1S} & U_{N_1V} & U_{N_1N_1} & \cdots & U_{N_1N_{n-1}} \\ \cdots & \cdots & \cdots & \cdots & \cdots \\ U_{N_{n-2}S} & U_{N_{n-2}V} & U_{N_{n-2}N_1} & \cdots & U_{N_{n-2}N_{n-1}} \\ (A_{n+1})_S & (A_{n+1})_V & (A_{n+1})_{N_1} & \cdots & (A_{n+1})_{N_{n-1}} \end{vmatrix} = 0 \tag{26}$$

Alternate Criteria in the U-Representation

The two criteria of criticality developed in the last section similar to, but no identical with Eqs. (5) and (6) which were state by Gibbs. We shall now show that the two are equivalent and that t are particular sets from a more general form of the criteria.

We began the derivation of Eq. (24) by expanding $\underline{U}$ in terms $\underline{S},\underline{V},N_1,\ldots N_n$, Eq. (15), and then closed the sum of squares, Eq. (16 In the process, we maintained the ordering of the independent varia as $\underline{S},\underline{V},N_1,\ldots N_n$. If we had chosen the order as $\underline{S},N_1,\ldots N_n,\underline{V}$, then the numerator of the coefficient of δZ_{n+1}^2, which was A_{n+1}, would ha been R_{n+1}, as expressed by Eq. (7). Thus, Eqs. (5) and (24) differ only in the ordering of variables in closing the sum of squares and hence, are of equal validity. The same statement also applies to

Eqs. (6) and (26).

It follows immediately that there are n+2 equivalent forms of Eqs. (5) or (24) and (6) or (26), each formed by omitting one of the n+2 variables $\underline{S}, \underline{V}, N_1, \ldots, N_n$.

The general form can be stated in terms of $y^{(o)} = f(z_1, \ldots, z_m)$ where $y^{(o)}$ is $\underline{U}$ and $z_1, \ldots, z_m$ is any ordering of the n+2 variables $\underline{S}, \underline{V}, N_1, \ldots, N_n$ (i.e., m=n+2). The general form of Eq. (23) becomes

$$\delta^2 y^{(o)} = \mathcal{D}_1 \ \delta Z_1{}^2 + \sum_{j=2}^{n+1} \frac{\mathcal{D}_j}{\mathcal{D}_{j-1}} \ \delta Z_j{}^2 \tag{27}$$

$$\text{where } \mathcal{D}_k = \begin{vmatrix} y_{11}^{(o)} & y_{12}^{(o)} & \cdots & y_{1k}^{(o)} \\ y_{21}^{(o)} & y_{22}^{(o)} & \cdots & y_{2k}^{(o)} \\ \cdots & \cdots & \cdots & \cdots \\ y_{k1}^{(o)} & y_{k2}^{(o)} & \cdots & y_{kk}^{(o)} \end{vmatrix} \tag{28}$$

and $y_{ij}^{(o)} = \partial^2 y^{(o)} / \partial z_i \partial z_j$.

In this terminology, the two general criteria of criticality, which apply without exception, are:

$$\mathcal{D}_{n+1} = \begin{vmatrix} y_{11}^{(o)} & y_{12}^{(o)} & \cdots & y_{1(n+1)}^{(o)} \\ y_{21}^{(o)} & y_{22}^{(o)} & \cdots & y_{2(n+1)}^{(o)} \\ \cdots & \cdots & \cdots & \cdots \\ y_{(n+1)1}^{(o)} & y_{(n+1)2}^{(o)} & \cdots & y_{(n+1)(n+1)}^{(o)} \end{vmatrix} = 0 \tag{29}$$

and

$$\mathcal{E}_{n+1} = \begin{vmatrix} y_{11}^{(o)} & y_{12}^{(o)} & \cdots & y_{1(n+1)}^{(o)} \\ y_{21}^{(o)} & y_{22}^{(o)} & \cdots & y_{2(n+1)}^{(o)} \\ \cdots & \cdots & \cdots & \cdots \\ y_{n1}^{(o)} & y_{n2}^{(o)} & \cdots & y_{n(n+1)}^{(o)} \\ (\mathcal{D}_{n+1})_1 & (\mathcal{D}_{n+1})_2 & \cdots & (\mathcal{D}_{n+1})_{n+1} \end{vmatrix} = 0 \tag{30}$$

Alternate Criteria in Other Potential Functions

In addition to the criteria of criticality in the U-representation, Gibbs presented two alternative sets; namely Eqs. (1) and (2) and Eqs. (9) and (10). These are two of many sets which we shall show derive from representations of the Fundamental Equation in alternate potential functions. We shall use the methodology of Legendre transformations to shift from one set of independent variables to another.*

It has been shown that the common potential functions of enthalpy, $\underline{H}$, Helmholtz free energy, $\underline{A}$, and Gibbs free energy, $\underline{G}$, can be viewed as Legendre transforms of $\underline{U}$ in the following independent variable sets:

$$\underline{H} = f_H(\underline{S},P,N_1,\ldots,N_n) \tag{31}$$

$$\underline{A} = f_A(T,\underline{V},N_1,\ldots,N_n) \tag{32}$$

$$\underline{G} = f_G(T,P,N_1,\ldots,N_n) \tag{33}$$

The functions f_H, f_A and f_G are entirely equivalent to f_U in Eq. (11 That is, the information content of the Fundamental Equation, Eq. (11), is maintained in the Legendre transformation.

For multicomponent substances, additional potential functions may be defined wherein one or more mole numbers, N_i, is transformed to its conjugate coordinate, μ_i. Following the notation of Beegle, et al., (3), we shall call these the prime potential functions, wherein the number of primes indicate the number of N_i that have been so transformed. That is,

$$\underline{U}' = f_{U'}(\underline{S},\underline{V},\mu_1,N_2,\ldots,N_n) \tag{34}$$

$$\underline{U}'' = f_{U''}(\underline{S},\underline{V},\mu_1,\mu_2,N_3,\ldots,N_n) \tag{35}$$

$$\underline{H}' = f_{H'}(\underline{S},P,\mu_1,N_2,\ldots,N_n) \tag{36}$$

$$\underline{A}' = f_{A'}(T,\underline{V},\mu_1,N_2,\ldots,N_n) \tag{37}$$

$$\underline{G}' = f_{G'}(T,P,\mu_1,N_2,\ldots,N_n) \tag{38}$$

*Refer to Modell and Reid (2) or Beegle, et al. (3) for a description of the Legendre technique and for a more detailed explanation of terminology.

Note that A, H and U' are all single variable Legendre transforms; G, A', H' and U" are double transforms; and G', A", H" and U"'are triple transforms. Defining ξ_i as the conjugate coordinate of z_i,

$$\xi_i \equiv \left(\frac{\partial y^{(o)}}{\partial z_i}\right)_{z_j[i]} \tag{39}$$

the k-variable Legendre transform is

$$y^{(k)} = f(\xi_1,\ldots,\xi_k,z_{k+1},\ldots,z_m) \tag{40}$$

In terms of the $y^{(k)}$ transform, the criticality criteria can be simplified by reducing the $\mathcal{D}$- determinant forms in Eq. (27) to single partial derivatives. It has been shown (3) that

$$y_{kk}^{(k-1)} = \frac{\mathcal{D}_k}{\mathcal{D}_{k-1}} \tag{41}$$

Substituting Eq. (41) into Eq. (27),

$$\delta^2 y^{(o)} = y_{11}^{(o)}\,\delta Z_1^{\,2} + y_{22}^{(1)}\,\delta Z_2^{\,2} + y_{33}^{(2)}\,\delta Z_3^{\,2} + \ldots + y_{(n+1)(n+1)}^{(n)}\,\delta Z_{n+1}^{\,2} \tag{42}$$

Instead of Eq. (29), we obtain for the criterion of the spinodal surface:

$$y_{(n+1)(n+1)}^{(n)} = \left(\frac{\partial^2 y^{(n)}}{\partial z_{n+1}^{\,2}}\right)_{\xi_1,\ldots,\xi_n,z_{n+2}} = 0 \tag{43}$$

The second criterion of criticality becomes

$$y_{(n+1)(n+1)(n+1)}^{(n)} = \left(\frac{\partial^3 y^{(n)}}{\partial z_{n+1}^{\,3}}\right)_{\xi_1,\ldots,\xi_n,z_{n+2}} = 0 \tag{44}$$

To ensure stability of the critical phase, we further require

$$y^{(n)}_{(n+1)(n+1)(n+1)(n+1)} = \left(\frac{\partial^4 y^{(n)}}{\partial z_{n+1}^4}\right)_{\xi_1,\ldots,\xi_n,z_{n+2}} \geq 0 \tag{45}$$

Using the G-prime notation,

$$y^{(n)} = \underline{G}^{(n-2)'} = f(T,P,\mu_1,\ldots,\mu_{n-2},N_{n-1},N_n) \tag{46}$$

and

$$y^{(n)}_{n+1} = \left(\frac{\partial \underline{G}^{(n-2)'}}{\partial N_{n-1}}\right)_{T,P,\mu_1,\ldots,\mu_{n-2},N_n} = \mu_{n-1} \tag{47}$$

Thus, Eqs. (43) to (45) become:

$$y^{(n)}_{n+1)(n+1)} = \left(\frac{\partial \mu_{n-1}}{\partial N_{n-1}}\right)_{T,P,\mu_1,\ldots,\mu_{n-2},N_n} = 0 \tag{48}$$

$$y^{(n)}_{(n+1)(n+1)(n+1)} = \left(\frac{\partial^2 \mu_{n-1}}{\partial N_{n-1}^2}\right)_{T,P,\mu_1,\ldots,\mu_{n-2},N_n} = 0 \tag{49}$$

and

$$y^{(n)}_{(n+1)(n+1)(n+1)(n+1)} = \left(\frac{\partial^3 \mu_{n-1}}{\partial N_{n-1}^3}\right)_{T,P,\mu_1,\ldots,\mu_{n-2},N_n} \geq 0 \tag{50}$$

If we choose to work in the A-prime system, we have

$$y^{(n)} = \underline{A}^{(n-1)'} = f(T,\underline{V},\mu_1,\ldots,\mu_{n-1},N_n) \tag{51}$$

Two equivalent criteria are obtained because z_{n+1} could be taken as $\underline{V}$ or N_n, the two extensive variables of $\underline{U}$ which are not transformed to obtain $\underline{A}^{(n-1)'}$. Therefore,

$$y^{(n)}_{(n+1)} = A_V^{(n-1)'} = \left(\frac{\partial \underline{A}^{(n-1)'}}{\partial \underline{V}}\right)_{T,\mu_1,\ldots,\mu_{n-1},N_n} = -P \tag{52}$$

$$\text{or} \quad A^{(n-1)'}_{N_n} = \left(\frac{\partial \underline{A}^{(n-1)'}}{\partial N_n}\right)_{T,\underline{V},\mu_1,\ldots,\mu_{n-1}} = \mu_n \tag{53}$$

Equations (43) to (45) then take either of two equivalent forms:

$$\left(\frac{\partial P}{\partial \underline{V}}\right)_{T,\mu_1,\ldots,\mu_{n-1},N_n} = 0 \tag{54}$$

$$\left(\frac{\partial^2 P}{\partial \underline{V}^2}\right)_{T,\mu_1,\ldots,\mu_{n-1},N_n} = 0 \tag{55}$$

$$-\left(\frac{\partial^3 P}{\partial \underline{V}^3}\right)_{T,\mu_1,\ldots,\mu_{n-1},N_n} \geq 0 \tag{56}$$

$$\text{or} \quad \left(\frac{\partial \mu_n}{\partial N_n}\right)_{T,\underline{V},\mu_1,\ldots,\mu_{n-1}} = 0 \tag{57}$$

$$\left(\frac{\partial^2 \mu_n}{\partial N_n^2}\right)_{T,\underline{V},\mu_1,\ldots,\mu_{n-1}} = 0 \tag{58}$$

$$\left(\frac{\partial^3 \mu_n}{\partial N_n^3}\right)_{T,\underline{V},\mu_1,\ldots,\mu_{n-1}} \geq 0 \tag{59}$$

We have now reached another one of our objectives: to show that the Gibbs criteria of Eqs. (1) to (3) are a special case of a broader generality. Equations (57) to (59) are identical to Eqs. (1) to (3). By our development herein, we also see that Eqs. (1) to (3) are but one of many equivalent sets which obtain from the n-variable Legendre transform. The particular set for Eqs. (1) to (3) derives from the $\underline{A}^{(n-1)'}$ system. Other equivalent first criticality criteria for binary, ternary and quaternary systems are given in Table I.

Table I. The First Criticality Criteria in $y^{(n)}$

Binary	Ternary	Quaternary
$y_{33}^{(2)} = 0$	$y_{44}^{(3)} = 0$	$y_{55}^{(4)} = 0$
$G_{N_1N_1} = \left(\frac{\partial \mu_1}{\partial N_1}\right)_{T,P,N_2} = 0$	$G'_{N_2N_2} = \left(\frac{\partial \mu_2}{\partial N_2}\right)_{T,P,\mu_1,N_3} = 0$	$G''_{N_3N_3} = \left(\frac{\partial \mu_3}{\partial N_3}\right)_{T,P,\mu_1,\mu_2,N_4} = 0$
$A'_{VV} = -\left(\frac{\partial P}{\partial \underline{V}}\right)_{T,\mu_1,N_2} = 0$	$A''_{VV} = -\left(\frac{\partial P}{\partial \underline{V}}\right)_{T,\mu_1,\mu_2,N_3} = 0$	$A'''_{VV} = -\left(\frac{\partial P}{\partial \underline{V}}\right)_{T,\mu_1,\mu_2,\mu_3,N_4} = 0$
$A'_{N_2N_2} = \left(\frac{\partial \mu_2}{\partial N_2}\right)_{T,\underline{V},\mu_1} = 0$	$A''_{N_3N_3} = \left(\frac{\partial \mu_3}{\partial N_3}\right)_{T,\underline{V},\mu_1,\mu_2} = 0$	$A'''_{N_4N_4} = \left(\frac{\partial \mu_4}{\partial N_4}\right)_{T,\underline{V},\mu_1,\mu_2,\mu_3} = 0$
$H'_{SS} = \left(\frac{\partial T}{\partial \underline{S}}\right)_{P,\mu_1,N_2} = 0$	$H''_{SS} = \left(\frac{\partial T}{\partial \underline{S}}\right)_{P,\mu_1,\mu_2,N_3} = 0$	$H'''_{SS} = \left(\frac{\partial T}{\partial \underline{S}}\right)_{P,\mu_1,\mu_2,\mu_3,N_4} = 0$
$H'_{N_2N_2} = \left(\frac{\partial \mu_2}{\partial N_2}\right)_{\underline{S},P,\mu_1} = 0$	$H''_{N_3N_3} = \left(\frac{\partial \mu_3}{\partial N_3}\right)_{\underline{S},P,\mu_1,\mu_2} = 0$	$H'''_{N_4N_4} = \left(\frac{\partial \mu_4}{\partial N_4}\right)_{\underline{S},P,\mu_1,\mu_2,\mu_3} = 0$
$U''_{SS} = \left(\frac{\partial T}{\partial \underline{S}}\right)_{\underline{V},\mu_1,\mu_2} = 0$	$U'''_{SS} = \left(\frac{\partial T}{\partial S}\right)_{\underline{V},\mu_1,\mu_2,\mu_3} = 0$	$U''''_{SS} = \left(\frac{\partial T}{\partial \underline{S}}\right)_{\underline{V},\mu_1,\ldots,\mu_4} = 0$
$U''_{VV} = -\left(\frac{\partial P}{\partial \underline{V}}\right)_{\underline{S},\mu_1,\mu_2} = 0$	$U'''_{VV} = -\left(\frac{\partial P}{\partial \underline{V}}\right)_{\underline{S},\mu_1,\mu_2,\mu_3} = 0$	$U''''_{VV} = -\left(\frac{\partial P}{\partial \underline{V}}\right)_{\underline{S},\mu_1,\ldots,\mu_4} = 0$

The second criticality criteria follow directly by a second differentiation.

Gibbs noted that Eq. (1) may, in some instances, be indeterminate, while Eq. (5) holds true without exception. Eq. (1) is a special case of Eq. (43) while Eq. (5) is a special case of Eq. (29). The difference between the two criteria can be seen by comparing Eqs. (27) and (42). The coefficient of the last differential term in each of the equations defines the first criticality criterion. They are equivalent; i.e.,

$$y^{(n)}_{(n+1)(n+1)} = \frac{\mathcal{D}_{n+1}}{\mathcal{D}_n} \tag{60}$$

If, under some circumstances, $\mathcal{D}_n$ and $\mathcal{D}_{n+1}$ vanish simultaneously, Eq. (29) will be satisfied while Eq. (43) will be indeterminate. Thus, Eq. (29) is the more general criterion.

On the other hand, when Eq. (43) is indeterminate, Gibbs states that we could always alter the ordering of the independent variables to remove the indeterminate condition. This procedure would be equivalent to choosing another form of $y^{(n)}_{(n+1)(n+1)}$, such as illustrated in Table I.

As an alternative to reordering variables when $y^{(n)}_{(n+1)(n+1)}$ is indeterminate, the following procedure can be used. Applying Eq. (60) successively n times,

$$\mathcal{D}_{n+1} = y^{(n)}_{(n+1)(n+1)} \mathcal{D}_n = y^{(n)}_{(n+1)(n+1)} y^{(n-1)}_{nn} \mathcal{D}_{n-1} = \ldots$$

or $$\mathcal{D}_{n+1} = \prod_{k=1}^{n+1} y^{(k-1)}_{kk} \tag{61}$$

Thus, $\mathcal{D}_{n+1}$ is equal to the product of the coefficients in Eq. (42). When $y^{(n)}_{(n+1)(n+1)}$ is indeterminate (i.e., $\mathcal{D}_n = 0$), then $y^{(n-1)}_{nn}$ should be zero or indeterminate; if $y^{(n-1)}_{nn}$ is also indeterminate, then $y^{(n-2)}_{(n-1)(n-1)}$ should be zero or indeterminate; etc. It then follows that when $y^{(n)}_{(n+1)(n+1)}$ is indeterminate, we could choose progressively smaller values of $y^{(k-1)}_{kk}$ until we obtain one which is zero. In this manner, we can satisfy the first criticality criterion without reordering variables.

Let us examine the implications of these procedures in ternary and binary systems.

For a ternary system, Eq. (42) is

$$\delta^2 y^{(o)} = y^{(o)}_{11}\, \delta Z_1{}^2 + y^{(1)}_{22}\, \delta Z_2{}^2 + y^{(2)}_{33}\, \delta Z_3{}^2 + y^{(3)}_{44}\, \delta Z_4{}^2 \tag{62}$$

In terms of the normal ordering of variables for $\underline{U}$ [see Eq. (11)], Eq. (62) can be written in terms of potential functions as

$$\delta^2\underline{U} = U_{SS}\ \delta Z_1^{\ 2} + A_{VV}\ \delta Z_2^{\ 2} + G_{N_1N_1}\ \delta Z_3^{\ 2} + G'_{N_2N_2}\ \delta Z_4^{\ 2} \tag{63}$$

where $\underline{G} = f(T,P,N_1,N_2,N_3)$ and $\underline{G}' = f(T,P,\mu_1,N_2,N_3)$. If we are dealing with a special case in which $\mathcal{D}_3$ and $\mathcal{D}_4$ vanish simultaneously then $G_{N_1N_1} = \mathcal{D}_3/\mathcal{D}_2 = 0$ and $G'_{N_2N_2} = \mathcal{D}_4/\mathcal{D}_3$ is indeterminate. A simple reordering of variables to, e.g., $\underline{G} = f(T,P,N_2,N_1,N_3)$ and G' $f(T,P,\mu_2,N_1,N_3)$ should result in a finite value of $G_{N_2N_2}$ and a zero value of $G'_{N_1N_1}$. Alternatively, we could test the first criticality criterion by using Eq. (61):

$$\mathcal{D}_4 = y_{44}^{(3)}\ y_{33}^{(2)}\ y_{22}^{(1)}\ y_{11}^{(o)} \tag{64}$$

or
$$\mathcal{D}_4 = G'_{N_2N_2}\ G_{N_1N_1}\ A_{VV}\ U_{SS} \tag{65}$$

That is, we can test for $y_{33}^{(2)}$ when $y_{44}^{(3)}$ is indeterminate. As pointed out by Heidemann (4), in testing the stability of liquid-liquid coexisting phases in a ternary system, there are conditions within the unstable region where $y_{33}^{(2)}$ vanishes (i.e., where $y_{44}^{(3)}$ is negative). Thus, when testing transforms of lower order than $y_{(n+1)(n+1)}^{(n)}$, we must be sure that we have approached the condition under which $y_{(n+1)(n+1)}^{(n)}$ is indeterminate from a region of proven stability.

For a binary system, Eqs. (42), (62) and (63) are truncated aft the third term. If we are dealing with the special case in which $\mathcal{D}_2$ and $\mathcal{D}_3$ vanish simultaneously, then $A_{VV} = \mathcal{D}_2/\mathcal{D}_1 = 0$ and $G_{N_1N_1} = \mathcal{D}_3/\mathcal{D}_2$ is indeterminate. This special case is known to occur when a binary mixture exhibits azeotropic behavior in the critical region; that is, when the locus of azeotropic conditions intersects the locus of critical conditions (5).

For such a system, the reordering of variables might take the following form. Starting with $\underline{U} = f(\underline{S},N_1,N_2,\underline{V})$, we would have $y^{(1)} = f(T,N_1,N_2,\underline{V})$ and $y^{(2)} = f(T,\mu_1,N_2,\underline{V})$. The truncated binary form of Eq. (62) would then become

$$\delta^2\underline{U} = U_{SS}\ \delta Z_1^{\ 2} + A_{N_1N_1}\ \delta Z_2^{\ 2} + A'_{N_2N_2}\ \delta Z_3^{\ 2} \tag{66}$$

If this form is successful in removing the indeterminate condition, then we should find $A_{N_1N_1}$ non-vanishing and $A'_{N_2N_2} = 0$.* If it is unsuccessful, then another choice of reordered variables should be pursued.

As an alternative to reordering the independent variables, we can use Eq. (61) for the binary:

$$\mathcal{D}_3 = y_{33}^{(2)}\, y_{22}^{(1)}\, y_{11}^{(o)} \tag{67}$$

or $$\mathcal{D}_3 = G_{N_1N_1}\, A_{VV}\, U_{SS} \tag{68}$$

Thus, when azeotropic behavior is suspected in the critical region, we could test A_{VV} and $G_{N_1N_1}$ simultaneously or use the product of the two to determine the vanishing of $\mathcal{D}_3$. This latter procedure was applied successfully by Teja and Kropholler (6).

Criteria in Gibbs Free Energy

The third criteria put forth by Gibbs, Eqs. (9) and (10), are the forms frequently quoted in the current literature. These criteria, which were presented by Gibbs for $\underline{G} = f_G(T,P,N_1,\ldots,N_n)$, are the special case of $y^{(2)} = f(\xi_1,\xi_2,z_3,\ldots,z_{n+2})$.

The form offered by Gibbs can be developed directly from the procedure of the last section. Using Eq. (61) to evaluate $\mathcal{D}_2$,

$$\mathcal{D}_2 = y_{22}^{(1)}\, y_{11}^{(o)} \tag{69}$$

Eq. (61) can be rewritten as

$$\frac{\mathcal{D}_{n+1}}{\mathcal{D}_2} = \prod_{k=3}^{n+1} y_{kk}^{(k-1)} \tag{70}$$

---

*Note that for a pure material, the first criticality criterion is $y_{22}^{(1)} = 0$ where $y_{22}^{(1)} = A_{VV}$ or A_{NN} (3). A reordering of variables does not remove the criterion of mechanical instability. On the other hand, for the binary, A_{VV} may vanish while some other form of $y_{22}^{(1)}$ (e.g.,$A_{N_1N_1}$) should not vanish. Thus, to test mechanical stability of a binary, one would have to evaluate A_{VV}. However, in the general case the region of mechanical instability will lie within the region of material instability and, thus, the condition of mechanical instability will be academic. It is always the vanishing of the last term in Eq. (27), or its equivalent, Eq. (42), that defines the limit of intrinsic stability.

Each term in the product of Eq. (70) can be expressed in terms of derivatives of $y^{(2)}$ by using the step-down operator [Eq. (20) of Beegle, et al., (3)]:

$$y_{kk}^{(k-1)} = y_{kk}^{(k-2)} - {y_{k(k-1)}^{(k-2)}}^2 / y_{(k-1)(k-1)}^{(k-2)} \tag{71}$$

Repeated use of the step-down operator leads to the following equation:

$$y_{kk}^{(k-1)} = \frac{\begin{vmatrix} y_{33}^{(2)} & y_{34}^{(2)} & \cdots & y_{3k}^{(2)} \\ y_{34}^{(2)} & y_{44}^{(2)} & \cdots & y_{4k}^{(2)} \\ \cdots & \cdots & \cdots & \cdots \\ y_{3k}^{(2)} & y_{4k}^{(2)} & \cdots & y_{kk}^{(2)} \end{vmatrix}}{\begin{vmatrix} y_{33}^{(2)} & y_{34}^{(2)} & \cdots & y_{3(k-1)}^{(2)} \\ y_{34}^{(2)} & y_{44}^{(2)} & \cdots & y_{4(k-1)}^{(2)} \\ \cdots & \cdots & \cdots & \cdots \\ y_{3(k-1)}^{(2)} & y_{4(k-1)}^{(2)} & \cdots & y_{(k-1)(k-1)}^{(2)} \end{vmatrix}} \tag{72}$$

Substituting Eq. (72) for each of the terms in the product of Eq. (70), we obtain:

$$\frac{\mathcal{D}_{n+1}}{\mathcal{D}_2} = \begin{vmatrix} y_{33}^{(2)} & y_{34}^{(2)} & \cdots & y_{3k}^{(2)} \\ y_{34}^{(2)} & y_{44}^{(2)} & \cdots & y_{4k}^{(2)} \\ \cdots & \cdots & \cdots & \cdots \\ y_{3(n+1)}^{(2)} & y_{4(n+1)}^{(2)} & \cdots & y_{(n+1)(n+1)}^{(2)} \end{vmatrix} \tag{73}$$

Eq. (73) is the determinant of Eq. (9) when $y^{(2)}$ is taken as $\underline{G}$. Thu for a quarternary system (n=4), Eq. (73) is

$$\frac{\mathcal{D}_5}{\mathcal{D}_2} = \begin{vmatrix} y^{(2)} & y_{34}^{(2)} & y_{35}^{(2)} \\ y_{34}^{(2)} & y_{44}^{(2)} & y_{45}^{(2)} \\ y_{35}^{(2)} & y_{45}^{(2)} & y_{55}^{(2)} \end{vmatrix} \tag{74}$$

$$= \begin{vmatrix} G_{N_1N_1} & G_{N_1N_2} & G_{N_1N_3} \\ G_{N_1N_2} & G_{N_2N_2} & G_{N_2N_3} \\ G_{N_1N_3} & G_{N_2N_3} & G_{N_3N_3} \end{vmatrix} \tag{75}$$

From this analysis, it should be clear that the criterion offered by Gibbs in Eq. (9) is equivalent to the requirement that the product of Eq. (70) should vanish. That is, for a quarternary system, for example,

$$\frac{\mathcal{D}_5}{\mathcal{D}_2} = y_{44}^{(4)}\ y_{44}^{(3)}\ y_{33}^{(2)} = 0 \tag{76}$$

Following the discussion in the last section, this form may be indeterminant if $y_{22}^{(1)}$ vanishes. Thus, for a system in which an azeotrope intersects the critical locus, we would expect Eq. (76) to be indeterminant. In such cases, the alternate procedures outlined in the preceding section should be followed.

Concluding Remarks

In this paper, we have developed the criteria of criticality for multicomponent, classical systems. The development utilized Legendre transform theory described in the first two papers of this set (Beegle, et al., 1974a,b). It was shown that the criteria can be expressed in terms of (n+1)-order determinants involving second order partial derivatives of U [Eqs. (29) and (30)], single second order partial derivatives of n-variable Legendre transforms of U [Eqs. (43) and (44)], or determinants of two-variable Legendre transforms [Eq.(73)]. The alternate sets presented by Gibbs are special cases of the general criteria presented herein.

Abstract

The general criteria of criticality for a classical thermodyna system are developed in terms of the internal energy representation the Fundamental Equation, $\underline{U}, (\underline{S},\underline{V},N_1,\ldots,N_n)$. Alternative criteria in other variable sets are derived using Legendre transformations. The criteria of criticality as stated by Gibbs are shown to be spec cases of the general criteria. The development utilizes extensive potential functions with mole number variables rather than mole fractions.

Notation

$\underline{A}$ = total Helmholtz free energy
$\mathcal{A}_{i+2}$ = determinant defined above Eq. (17)
B = determinant defined in Eq. (9)
$\mathcal{B}$ = determinant defined in Eq. (26)
$\mathcal{D}_k$ = determinant defined in Eq. (28)
$\mathcal{E}_{n+1}$ = determinant defined in Eq. (30)
$\underline{G}$ = total Gibbs free energy
$\underline{H}$ = total enthalpy
N_j = moles of component j
n = number of components in mixture
P = pressure
R_{n+1} = determinant defined in Eq. (7)
S = determinant defined in Eq. (8)
$\underline{S}$ = total entropy
T = temperature
$\underline{U}$ = total internal energy
$\underline{V}$ = total volume
$y^{(k)}$ = k-variable Legendre transform
Z_k = extensive parameter defined in Eq. (16)
z_k = extensive independent variable

Greek Letters

μ_j = chemical potential of component j
ξ_j = intensive independent variable that is the conjugate coordinate of z_j

Literature Cited

1. Gibbs, J. W., "On the Equilibrium of Heterogeneous Substances," Trans. Conn. Acad. III, 108 (1876).
2. Modell, M., and R. C. Reid, Thermodynamics and Its Applications,

Ch. 5, 7, Prentice-Hill, Englewood Cliffs, N. J. (1974).

3. Beegle, B. L., M. Modell and R. C. Reid: (a) "Legendre Transformations and Their Application in Thermodynamics," AIChE J., 20, 1194 (1974); (b) "Thermodynamic Stability Criterion for Pure Substances and Mixtures," Ibid., 1200 (1974).
4. Heidemann, R. A., "The Criteria for Thermodynamic Stability," AIChE J., 21, 824 (1975).
5. Rowlinson, J. S., Liquids and Liquid Mixtures, 2nd Ed., Ch. 6. Butterworth, London (1969).
6. Teja, A. S., and H. W. Kropholler, "Critical States of Mixtures in which Azeotropic Behaviour Persists in the Critical Region," Chem. Engng Sci., 30, 435 (1975).

20

Thermodynamic Data Needs in the Synthetic Fuels Industry

HOWARD G. HIPKIN

Research and Engineering, Bechtel Corp., San Francisco, CA 94119

The petroleum industry, which is responsible for producing most of the energy used in modern society, has produced, as a by-product over the past 40 years, a broad base of thermodynamic data for the hydrocarbons it processes. Even though the petroleum industry is mature, the data development and correlation effort has not slacked off, and indeed has accelerated in the last two decades. A similar, but more proprietary, effort has been carried on by the chemical industry for nonhydrocarbons. In view of these long-lived programs, the question arises,--does the new synthetic fuels industry, which promises to become important in the last quarter of the century, need specific data programs or are the present data systems adequate for its needs? This paper looks at that question and attempts to outline areas where work is needed. As in the petroleum industry, the need is for thermodynamic properties of a small number of pure compounds, and for mathematical procedures to predict the properties of mixtures of those compounds.

What are the Synthetic Fuels?

The synthetic fuels comprise a spectrum of gaseous, liquid, and solid fuels. Gaseous fuels include high-, medium, and low-heating values gas from coal, similar gases from fermentation or pyrolysis of biomass, low-heating value gas from retorting shale oil, and hydrogen. Liquified natural gas, while not a synthetic, is included here because it promises to be an important answer to the near-term energy shortage, and because the data needs for LNG are specific.

High-Btu gas from coal is methane, made by reacting carbon monoxide and hydrogen. It can replace natural gas in all its applications without any equipment changes. Its heat of combustion is about 950 Btu per standard cubic foot. Medium-Btu gas is the carbon monoxide and hydrogen from which high-Btu gas is made. Its heat of combustion is about 300 Btu per standard cubic foot. Since methanation is not required, it is cheaper to produce than high-Btu gas; but is not interchangeable with natural gas, and it is more expensive to transport because of its larger volume. Low-Btu gas is carbon monoxide and hydrogen diluted with atmospheric nitrogen. It is made by using air instead of oxygen in the coal gasifier. It is the cheapest gas to make from coal, but its heat of combustion is only 150 Btu per standard cubic foot, and the very large volume required prevents its transportation much away from the gasifier. The major use for low-Btu gas is expected to be for combined cycle power generation (gas turbine followed by steam turbine).

Gas made by pyrolysis of biomass, or of municipal solid waste, resembles coal syntehsis gas in composition, and can be produced in the same heating values. Gas made by anaerobic fermentation of biomass is predominantly methane and carbon dioxide. When the carbon dioxide is removed, it is a replacement for natural gas.

Depending on the method of retorting, a gas of variable composition and quantity is made from oil shale. The gas is usually burned as in-plant fuel, but it could be processed for syngas. An interesting new development is the production of high-Btu gas by hydrotreating oil shale. This process, under development by the Institute of Gas Technology, provides more complete utilization of the organic carbon content than simple retorting.

Hydrogen, from decomposition of water, may become the dominant gaseous fuel well after the turn of the century. This concept is the basis for the so-called "hydrogen economy", which visualizes a virtually inexhaustible supply of non-polluting gaseous energy from this source. The hydrogen economy suffers from three problems; the high-energy, high-temperature heat source required, the relatively low conversion efficiency of that energy, and the difficulty of storing hydrogen. Since this form of synthetic fuel is unlikely to be important in the next 20 or 30 years, it will not be considered further in this paper.

Liquified natural gas has been produced in commercial quantities for a number of years in the Middle East and Alaska. A number of projects for liquifying waste natural gas in the Middle East, Indonesia, and Alaska, shipping the LNG to the United States, and regasifying it to augment our diminishing supplies are under active development at present. LNG appears to be the cheapest way to supply gaseous fuel to the American market. The energy required to liquefy and transport the gas is less than 15% of the heating value of the initial gas; as a result, the energy efficiency of this process is higher than that of any of the other processes under consideration.

Liquid synthetic fuels include shale oil and its various fractions; coal liquids by cooking, by hydrogenation, and by Fischer-tropsch synthesis; methyl fuel from coal or from overseas natural gas; and liquid products from biomass, probably by gasification and Fischer-Tropsch synthesis.

Oil shale is one of the most plentiful resources in the United States. It is well known that the reserves of oil shale in the Piceance Creek Basin of Colorado and Utah exceed all the known petroleum reserves in the world. It is less well known that the reserves of oil shale in the deeper and leaner deposits of the eastern states are even larger (1). Shale oil is produced by retorting oil shale, either in-situ, or on the surface after mining. The viscous shale oil requires a light hydrogenation to remove sulfur, nitrogen, and oxygen, and then can be refined like a crude oil.

Simple pryolysis of coal produces light oils and tars, both predominantly aromatic in character, together with some gas and a residue of solid coke (usually high in sulfur). The liquid yield can be increased by hydrogenating the coal under pressure. The hydrogenation also reduces the sulfur content of the solid residue, and eliminates the nitrogen content as ammonia. An alternate processing scheme is to gasify the coal with steam and oxygen to ma hydrogen and carbon monoxide, which can then be catalytically reacted to a spectrum of hydrocarbons and oxygenated compounds. This mixture is then refined to make various grades of fuels and chemicals. At present, only the Sasol plant in South Africa uses this Fischer-Tropsch process commercially.

A special case of Fischer-Tropsch is the production of methanol, either from gasified coal or from reformed natural gas (overseas). Crude methanol can be used as a fuel in gas turbines, industrial boilers, fuel cells, and internal combustion engines. It represents a commercially available way to convert coal to a liquid fuel, and has the advantage (as compared to conventional coal liquids) of eliminating the refining step.

A number of proprietary processes produce liquid products by pyrolysis of biomass. Alternatively, the biomass can be gasified and reacted to liquid products by Fischer-Tropsch.

Solid fuels are, in most cases, the undesirable residues from coal processes. Obviously, if a solid fuel is needed, coal itself would be the cheapest material. The residue from most gasification and liquification processes is high in ash and usually contains a higher percentage of sulfur than the original coal. If the combustible content of the residue is low, it is discarded. If, as in coke, it is too large to be economically discarded, the solid can be gasified to produce the hydrogen required for the process.

An exception is Solvent Refined Coal, a process which produces a low-sulfur solid fuel with a melting point of about 350°F. The process also produced a high-sulfur, carbonaceous residue.

The solid residue left from retorting oil shale consists of large amounts of rock with enough carbon to support combustion. The carbon is usually burned to supply the heat required for retorting. Under some circumstances, this material could be a solid fuel for other purposes.

Most of these processes have a common characteristic--they are increasing the hydrogen content of a raw fuel that is hydrogen-deficient compared to petroleum fuels. There are three ways in which this is done:

The hydrogenation processes add hydrogen directly

$$C + 2H_2 = CH_4$$

The gasification processes add hydrogen from water. The oxygen in the water is rejected to the atmosphere as carbon dioxide. A simplified representation of the coal-to-synthetic natural gas processes is

$$2C + 2H_2O = CH_4 + CO_2.$$

It can be seen that about half the carbon in the coal is rejected in order to upgrade the other half.

The simple pyrolysis processes create a high-carbon residue and a high-hydrogen product from the original low-hydrogen fuel. The coking of coal can be represented in an oversimplified way as

$$10.71\ (CH_{0.56}) = C_6H_6 + 4.71C.$$

The good yield of benzene indicated by this equation is deceptive, because the coal also contains oxygen, nitrogen, and sulfur, and these elements are driven off as water, ammonia, and hydrogen sulfide, thus reducing the hydrogen available for forming hydrocarbons.

It is not possible to follow the technology of all the synthetic fuel processes in a single paper. Since coal gasification for the production of high-BTU gas is the process on which most effort is currently concentrated, and since coal gasification is the first step in several liquification processes, and since coal or char gasification is the most likely way of producing hydrogen for a variety of other processes, the various coal gasification processes will be followed in some detail. In addition, this paper takes a shorter look at in-situ shale retorting.

How are Synthetic Fuels Made?

This look at synthetic fuels is restricted to coal gasification, as mentioned above. The present generation of commercial gasification processes includes three, all German, and all designed for chemical feedstock, rather than for fuel.

Lurgi. The Lurgi gasifier has had more installations than any other, possibly because it has been the only pressure gasifier (up to 500 psi). It is the gasifier used in several of the coal gasification facilities proposed for American SNG production. It is a fixed bed gasifier, in which coal is fed in at the top through lock hoppers, and ash is removed at the bottom, also through lock hoppers. Steam and oxygen enter at the bottom and pass up counter-

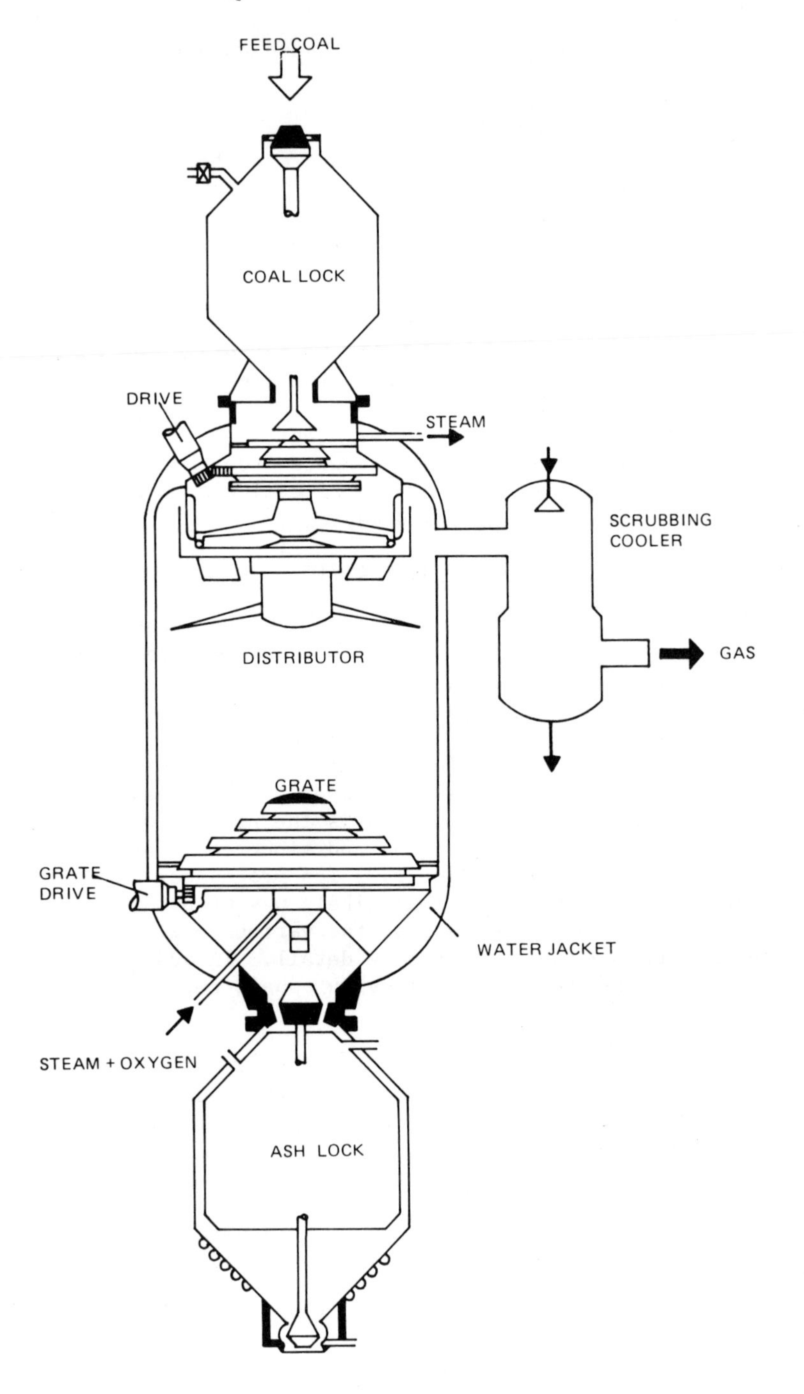

Figure 1. Lurgi reactor

current to the coal. The synthesis gas formed at the bottom of the reactor hydrogenates the coal in the middle sections, then cokes and preheats the incoming coal. Because of the countercurrent operation, the gas is cooled to about 1000°F before it leaves the reactor. The Lurgi reactor is shown schematically in Figure 1.

Since the coal is heated gradually as it passes slowly down through the bed, the Lurgi reactor produces light oils, tars, and phenols, as well as the usual impurities of hydrogen sulfide and ammonia. A typical Lurgi gas from bituminous coal has the following dry composition:

CO_2	29 volume%
CO	20
H_2	38
CH_4	11
H_2S, N_2, Ar, etc.	2

Purification of the gas from any coal gasifier is a major task for several reasons:

The gas contains up to 50% of unreacted water which, when condensed, is contaminated with solid flyash and carbon, dissolved hydrogen sulfide, ammonia, and carbon dioxide, possibly with phenols. The water must be treated to remove these contaminants.

The finely divided solids, some of them sub-micron in size, are particularly difficult to remove and, when removed, are a disposal problem.

It is always necessary to remove hydrogen sulfide, and usually necessary to remove carbon dioxide. Any clean-up process which removes both acid gases delivers a stream which is too dilute in H_2S to be handled in a Claus plant for elemental sulfur production. As a result, the gas is usually processed twice, once for hydrogen sulfide and later for carbon dioxide, at increased cost.

The various purification steps often require the gas to be heated and cooled several times in succession. Good heat economy requires fancy heat exchange between these streams, at increased capital cost.

With the Lurgi gasifier, the problem is further complicated by the necessity to remove oils (lighter than water), tars (heavier than water), phenols and cyanides (dissolved in water), and other sulfur compounds (mercaptans, carbonyl sulfide, and carbon disulfide). As a result, the Lurgi gas purification section is complicated and correspondingly expensive. Figures 2 through 8 show the purifica-

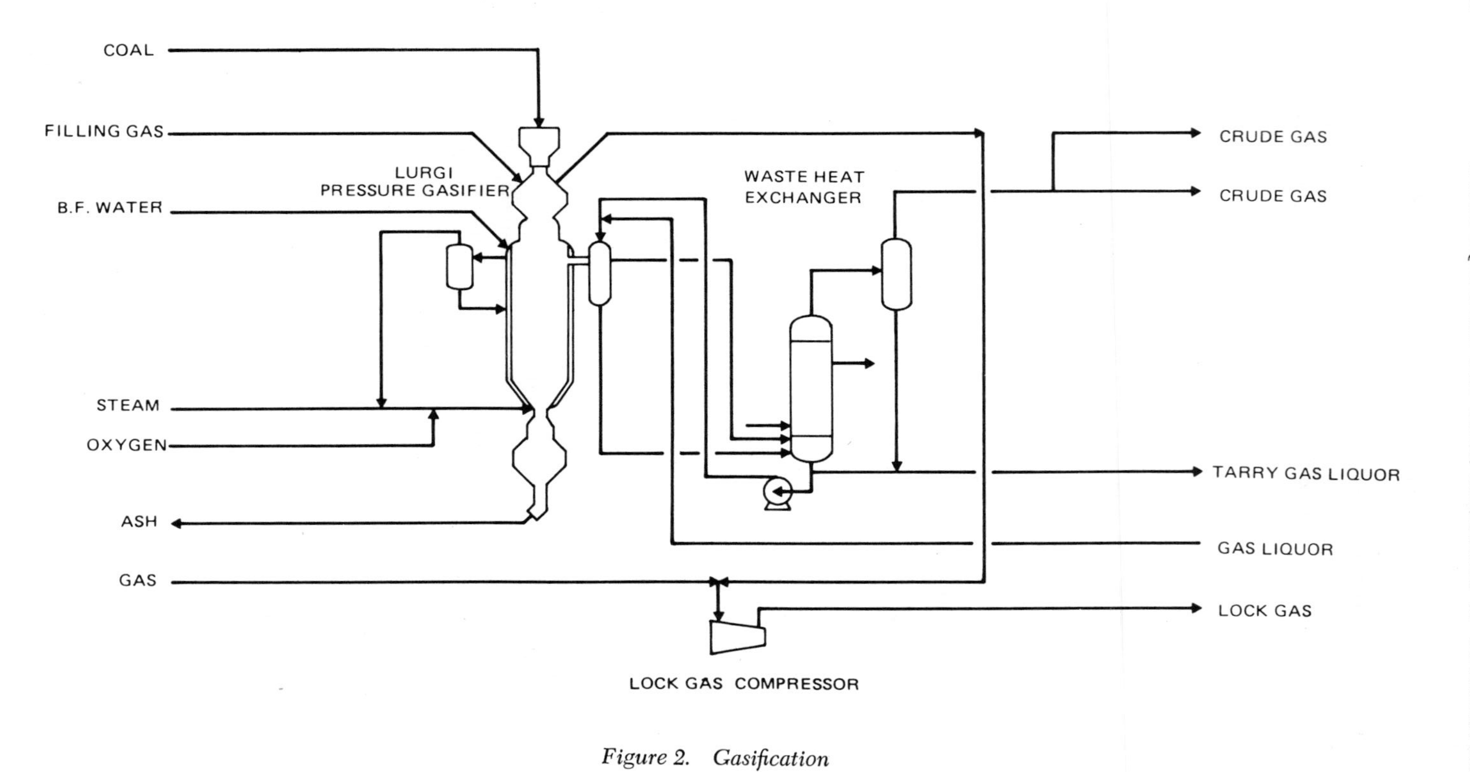

Figure 2. Gasification

tion scheme used by Lurgi for the proposed ANG Goal Casification Company Plant (2).

The gas leaving the reactor is quenched through a venturi nozzle with a mixture of recirculated water, tar, and light oil to about 800°F. The gas is then cooled by passing up through the tubes of a vertical heat exchanger which generates 100 psi steam on the shell side. (It is worth noting, in passing, that coal gasifiers consume large amounts of steam,--something over two tons of steam for each ton of coal.) The condensate from this exchanger, a mixture of tar, oil, and water, washes the tubes in countercurrent flow and keeps them from fouling. Part of the condensate is recycled to the venturi scrubber (Figure 2).

The gas from this exchanger is split into two streams. One stream goes to a second countercurrent, vertical heat exchanger which generates 60 psi steam, then to the two shift converters, where the carbon monoxide in the stream is almost entirely converted to carbon dioxide, in order to adjust the ratio of carbon oxides to hydrogen for subsequent methanation (Figure 3). The second stream bypasses the shift converters, and goes to a series of four vertical exchangers which generate 60 psi steam and 20 psi steam, preheat boiler feedwater, and finally cool the gas with cooling tower water. The gas from the shift converters is cooled through two boiler feedwater exchangers in series, is cooled in an air cooler, and finally, is cooled against cooling tower water, and is compressed to join the bypassed stream. In this processing scheme, the gas is countercurrently scrubbed with mixed-phase condensate five times at successively lower temperatures. This repeated contacting, together with the venturi scrub at the reactor outlet, removes almost all the particulates in the raw gas from the gasifier, and does it in a way that keeps the exchanger tubes from fouling with tar. The condensates from the high temperature exchangers go to "tarry gas liquor" treatment, and the condensate from the low temperature exchangers go to "oily gas liquor" treatment (Figure 4).

At this point, the gas is essentially at ambient temperature, and the condensate contains all the tar and particulates, all the ammonia and phenol, most of the water and oil, and some of the hydrogen sulfide and carbon dioxide. The gas is refrigerated and treated in a Rectisol unit to remove hydrogen sulfide. Rectisol is Lurgi's tradename for a proprietary process using refrigerated methanol as a solvent. Methanol is an excellent solvent for hydrogen sulfide, carbonyl sulfide, carbon disulfide, carbon dioxide, hydrocarbons, and water. It dissolves hydrogen sulfide preferentially to carbon dioxide, and it is possible to provide a concentrated hydrogen sulfide stream from a Rectisol unit. In the design shown in Figure 5, the hydrogen sulfide, with some carbon dioxide, is removed first, then the gas after methanation is scrubbed to remove the remaining carbon dioxide. Light oil, hydrogen cyanide, and water are all recovered in the Rectisol unit.

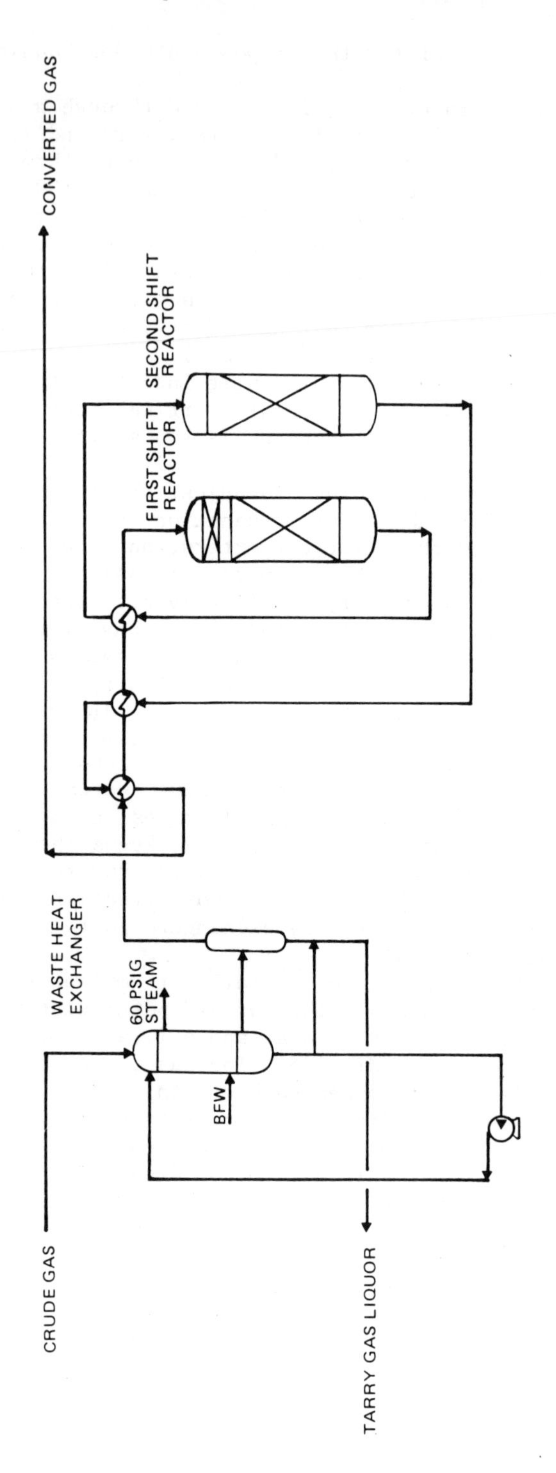

Figure 3. Shift conversion

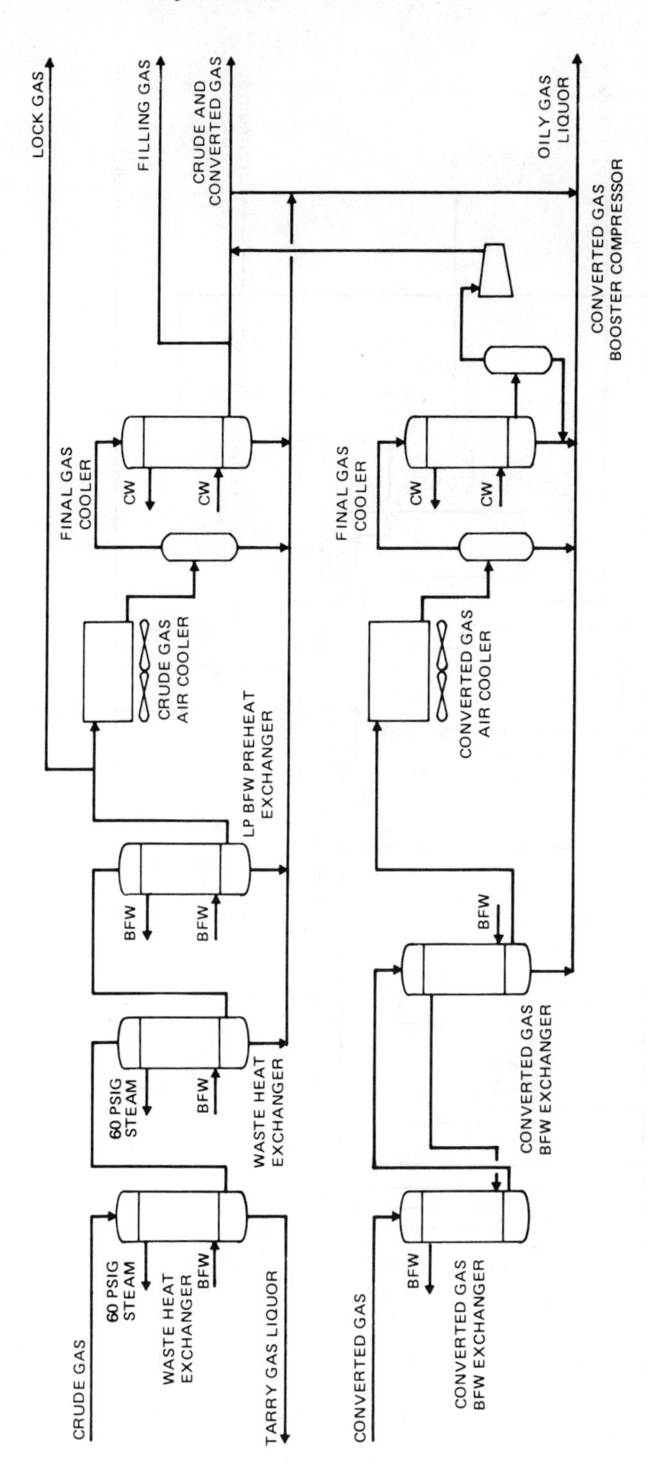

Figure 4. Gas cooling

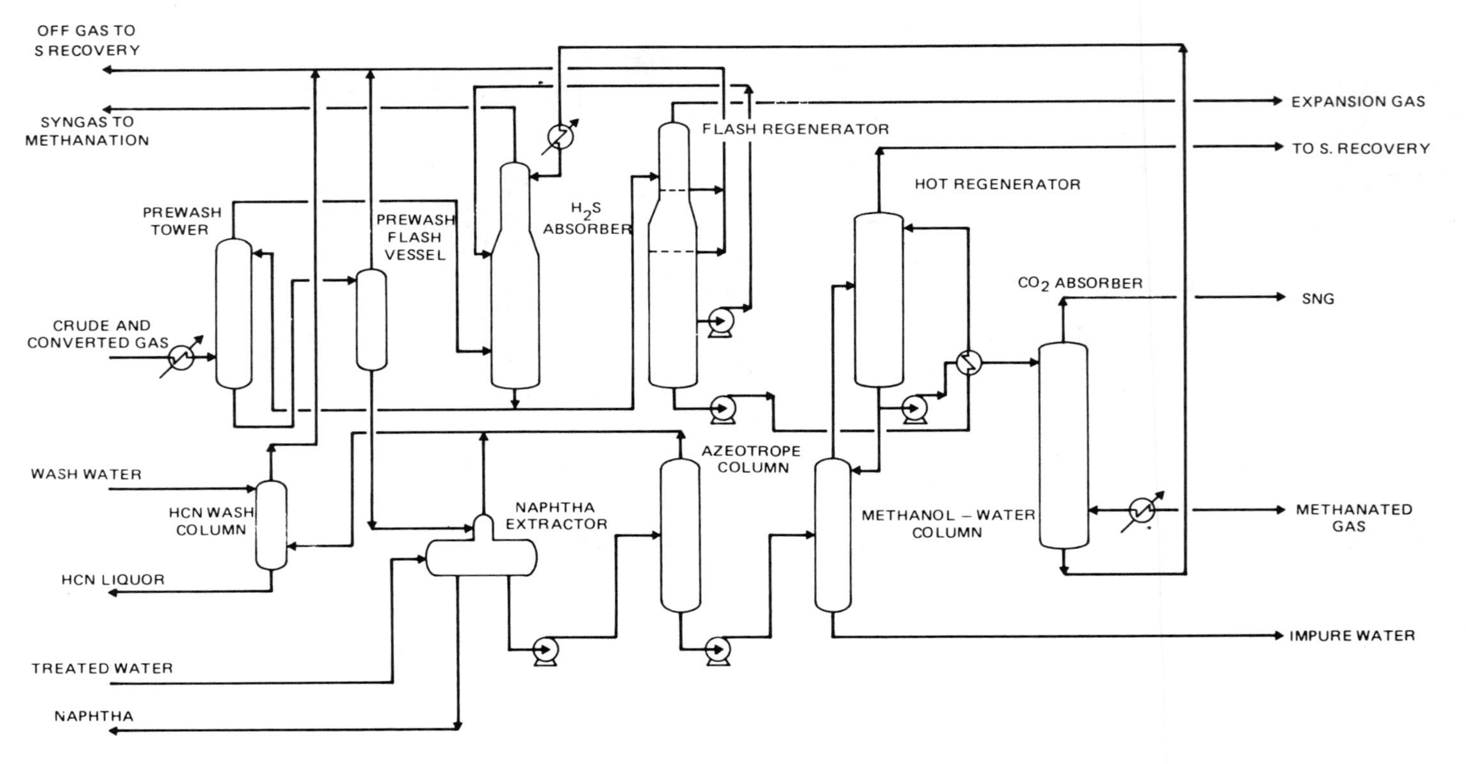

Figure 5. Rectisol unit

The condensates are combined, dropped in pressure to flash off dissolved gases, and separated by settling into tar (heavier than water), tar oil (lighter than water), and contaminated water (Figure 6). The water is treated in another Lurgi proprietary process, the Phenosolvan process, to remove phenols and ammonia (Figure 7).

The various sulfur streams are processed through a Claus plant for treatment of the more concentrated hydrogen sulfide streams. In the Claus plant, part of the H_2S is burned to SO_2, which is reacted with the residual H_2S to make elemental sulfur according to the reaction

$$SO_2 + 2H_2S = 3S + 2H_2O$$

Since the tail gas from the Claus plant still contains more sulfur than EPA regulations permit, the tail gas is treated through a proprietary IFP process for cleanup.

The low sulfur concentrations are processed in a Stretford unit, which also produces elemental sulfur. All these sulfur removal facilities are shown in Figure 8.

It can be seen that cleaning up the crude synthesis gas from a Lurgi gasifier is a complicated operation. Part or all of the cost of this cleanup is offset by the value of the recovered by-products,--ammonia, phenols, hydrogen cyanide, aromatics, tar oil, tar, and sulfur. The gas cleanup in a Winkler or Koopers-Totzek gasifier is considerably simpler.

Winkler. The Winkler reactor is a fluidized bed, as shown in Figure 9. The coal is pulverized before being fed to the bed, and the fluidizing medium is steam and oxygen, or steam and air. The reactor has no internal moving parts, is quite simple, and has a well-established reputation for reliability. The reactor must operate in the non-slagging mode to keep the bed fluidized, and so is limited to coals with reasonably high ash fusion temperatures. Since the operation is not countercurrent, the gas would leave the reactor at reaction temperature, but an internal heat exchanger in the gas space above the bed removes some heat and drops the gas temperature. The chief disadvantage of the Winkler reactor is its atmospheric pressure operation, but the licensor is working on a pressure modification.

Because the coal is brought rapidly to reaction temperature and the pyrolysis products are exposed to the full temperature, the Winkler reactor produces no tars or phenols, and very little methane. A typical gas composition from bituminous coal is (dry and sulfur-free):

CO_2	21 volume %
CO	33
H_2	41
CH_4	3
N_2, Ar, etc.	2

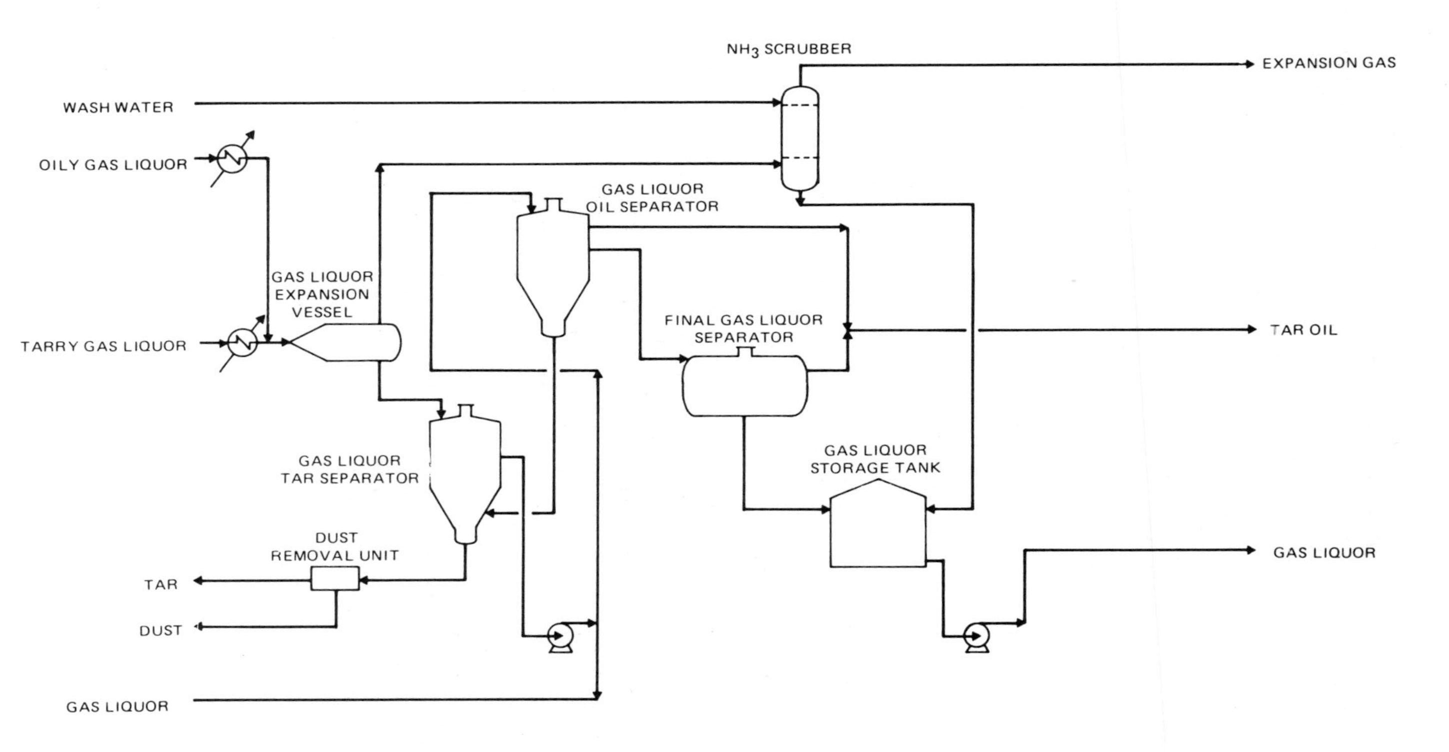

Figure 6. Gas liquor separation

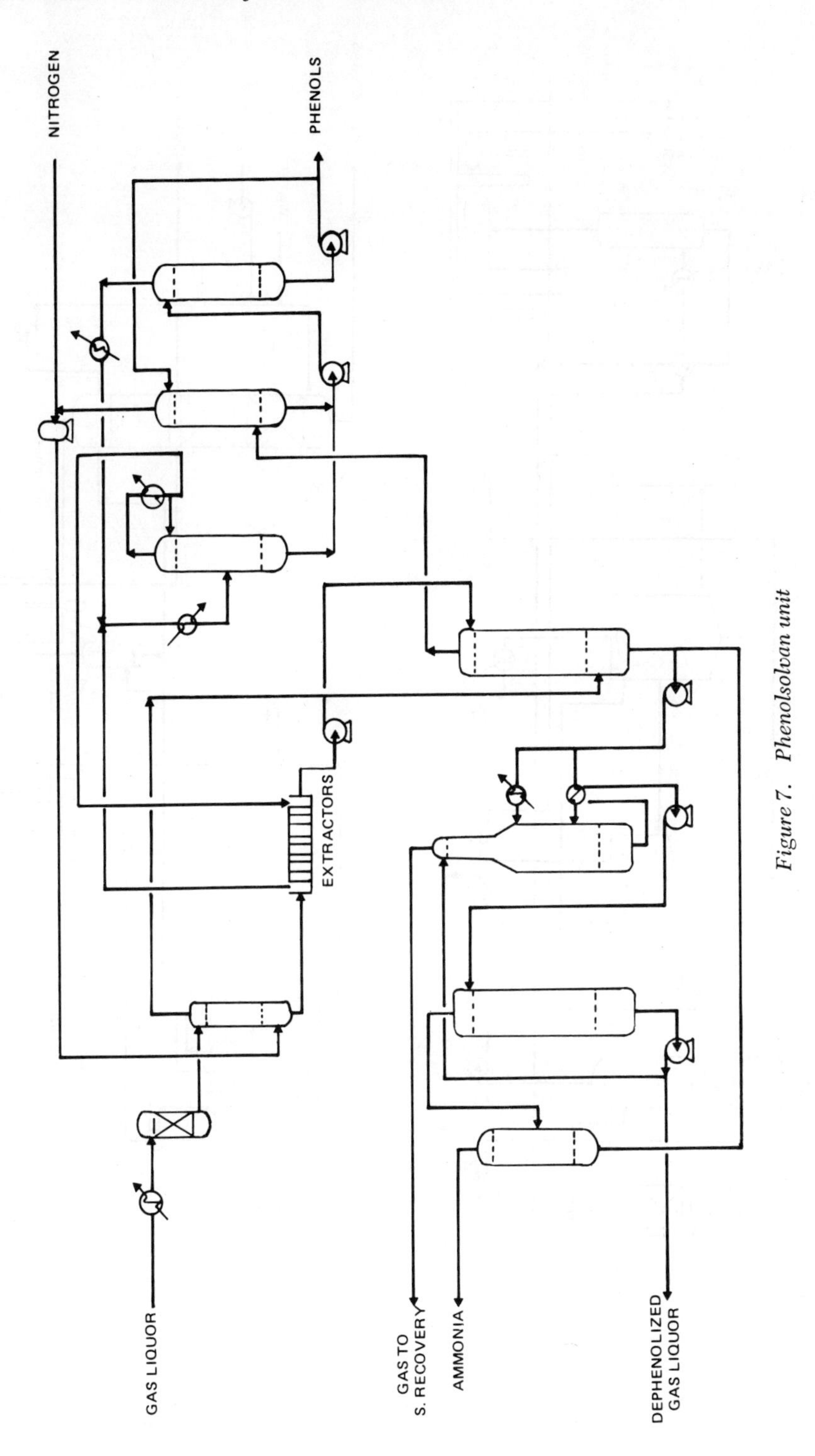

Figure 7. Phenolsolvan unit

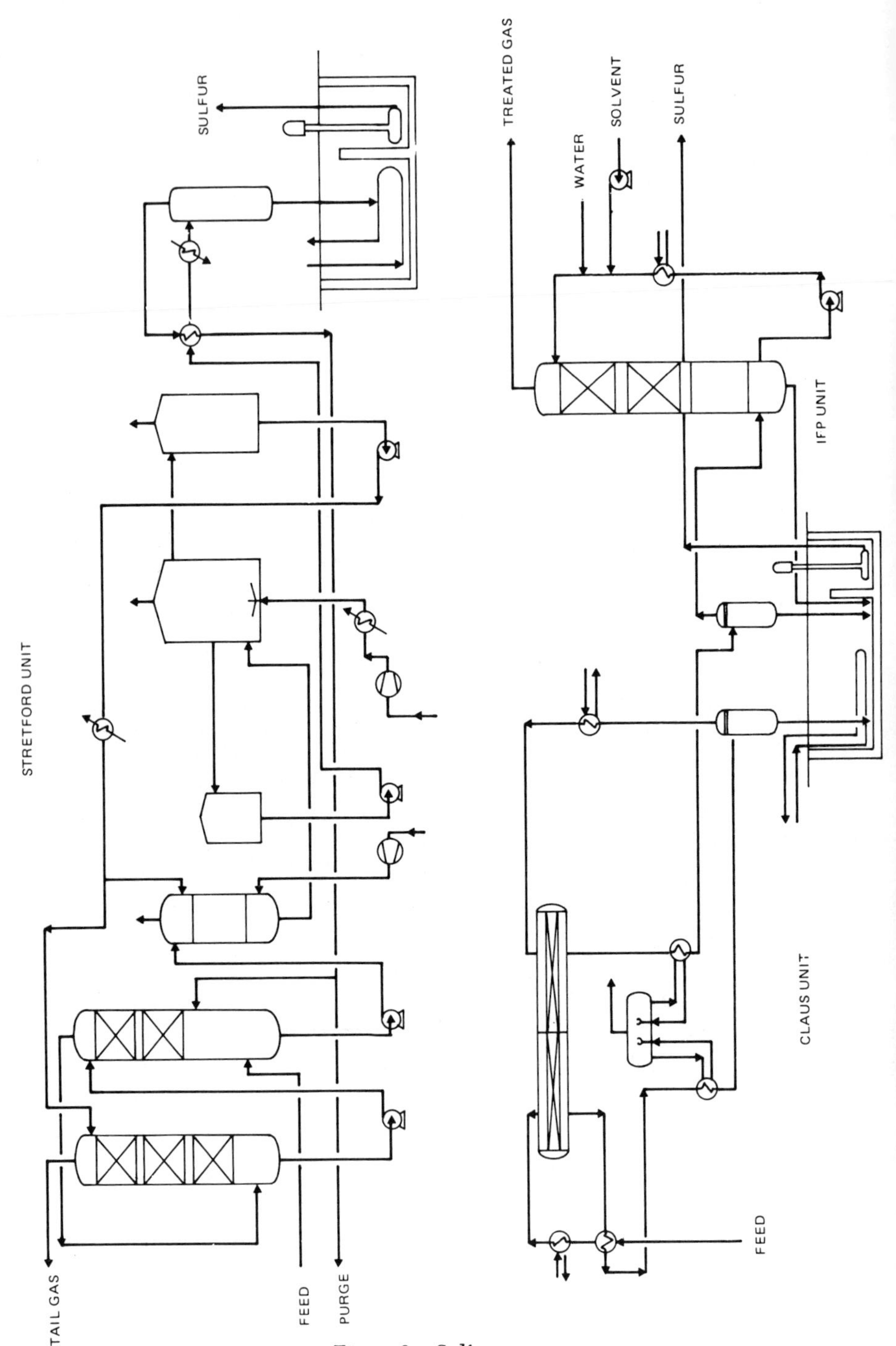

Figure 8. Sulfur recovery

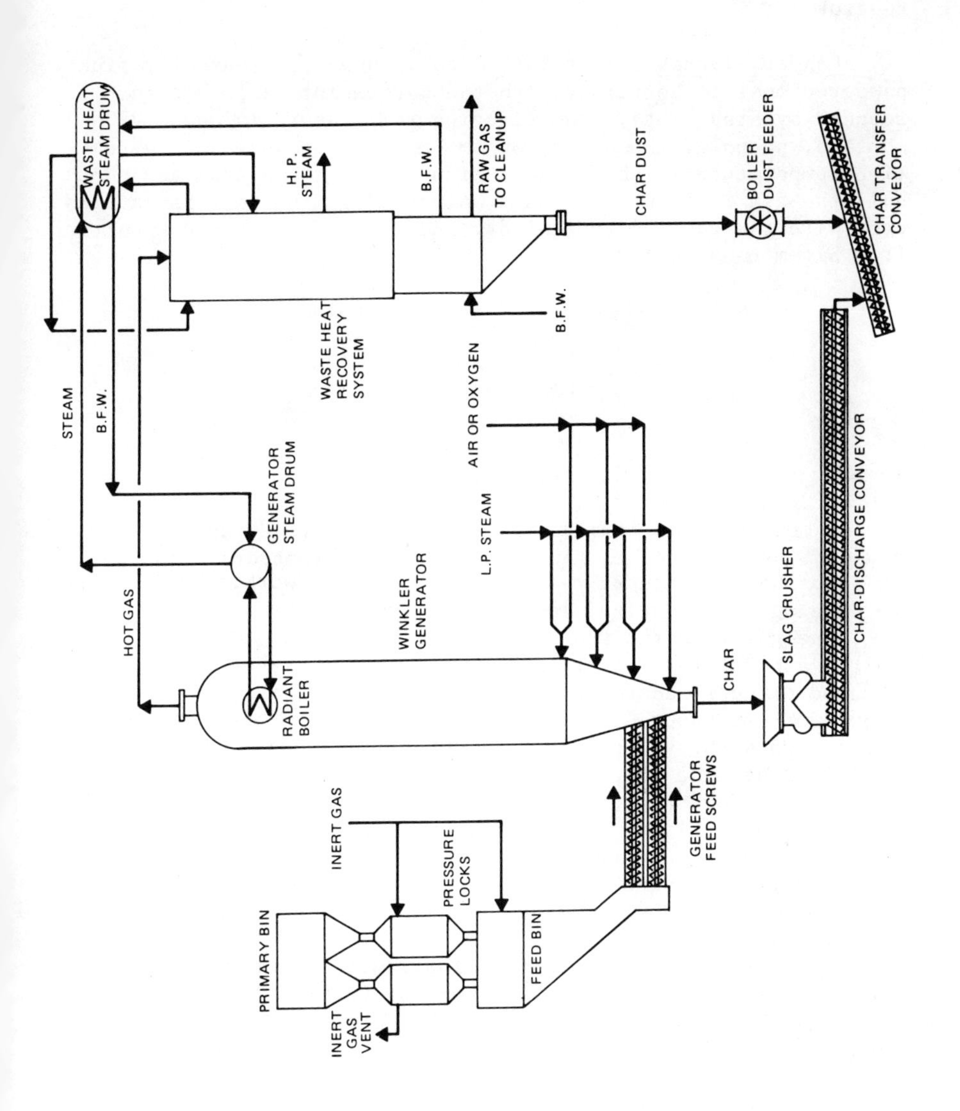

Figure 9. Winkler gasifier

Most of the ash leaves the reactor with the gas, so that particulate removal is a more difficult problem than with the Lurgi reactor.

Koppers-Totzek. In this reactor, shown in Figure 10, finely powdered coal is entrained with the oxygen into a burner surrounded by steam jets. The reaction temperature is about 2700°F, no tars, phenols, aromatics, or ammonia are produced. The reaction temperature is above the fusion point of the ash, and about half the ash is tapped off as a molten slag; the rest is carried over with the gas. A typical dry and sulfur-free gas composition from bituminous coal is

CO_2	12 volume %
CO	53
H_2	33
CH_4	0.2
N_2, Ar, etc.	1.5

The higher reaction temperature results in more CO and less CO_2 and CH_4, as compared to the Lurgi or Winker reactors. The K-T is an atmospheric pressure reactor, but Shell is working with Koppers to develop a pressurized model.

The gas from the reactor is quenched enough to solidify the molten ash droplets before they reach the waste-heat boiler, mounted directly above the gasifier. A spray washer cools the gas from 500°F to about 95°F after the boiler. At this temperature, the gas goes through two Thiessen disintegrators, which mechanically agitate the gas with water to remove more particulates. A demister follows the disintegrators and, if the gas is to be compressed, the demister is followed by electrostatic precipitators.

The gas is then treated to remove sulfur, using any of a number of different processes, including dry iron oxide, Sulfinol, or Rectisol. It is then shift converted and carbon dioxide is removed (Figure 11).

Obviously, the gas purification required for Winkler or Koppers-Totzek is simpler than that for Lurgi, although the problem of particulate removal is worse. The condensates from these processes must still be treated to remove contaminants.

New Gasification Processes

There are about a dozen gasification processes under development in this country and in Europe. A number of them have reached the stage of large pilot plants. The one that is probably closest to commercialization is the Texaco partial oxidation process. Texaco and Shell have both licensed partial oxidation processes for use with a variety of petroleum feeds since the late 1940's, and over 250 gasifiers have been installed, largely for making

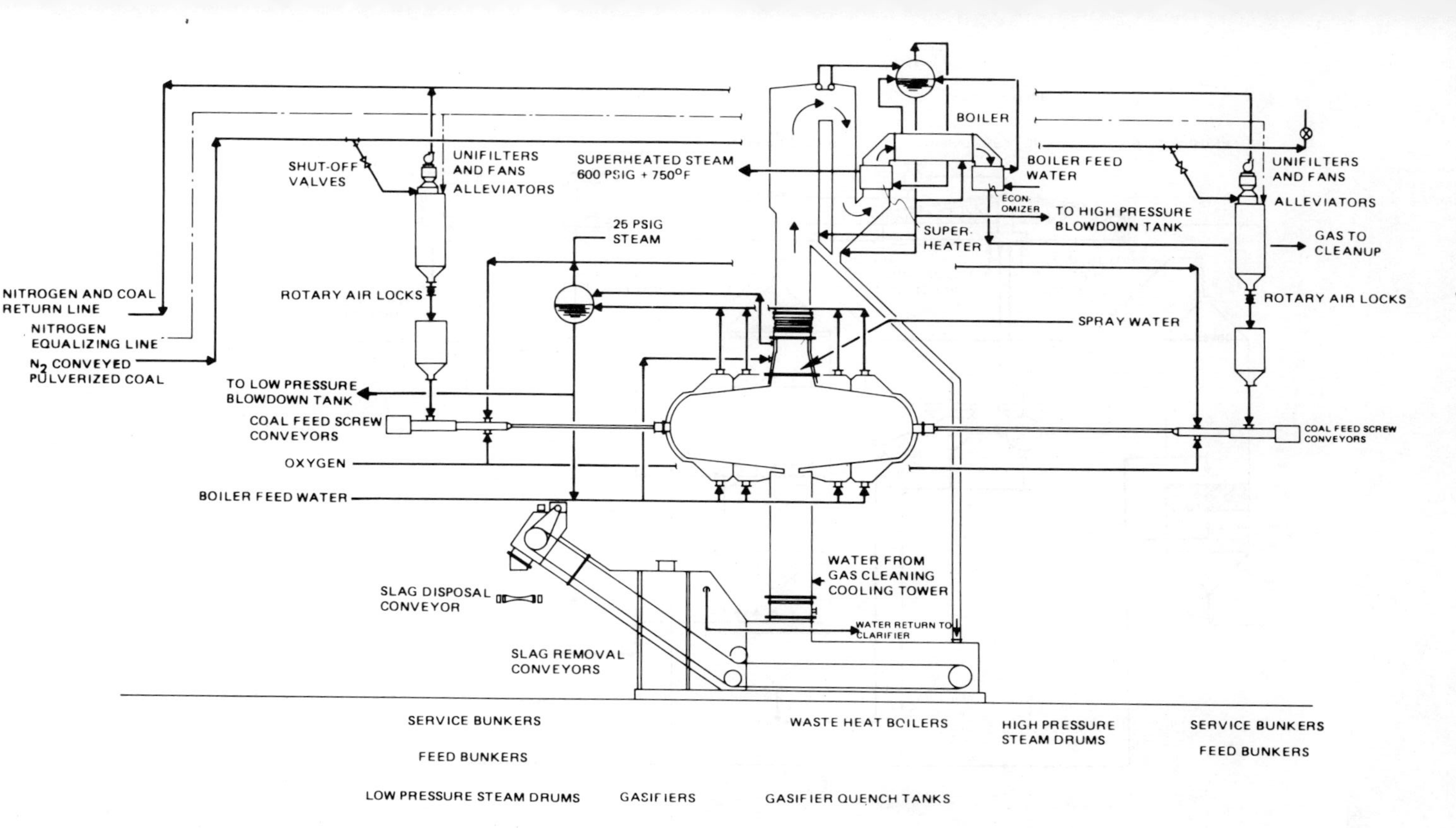

Figure 10. Koppers-Totzek gasifier

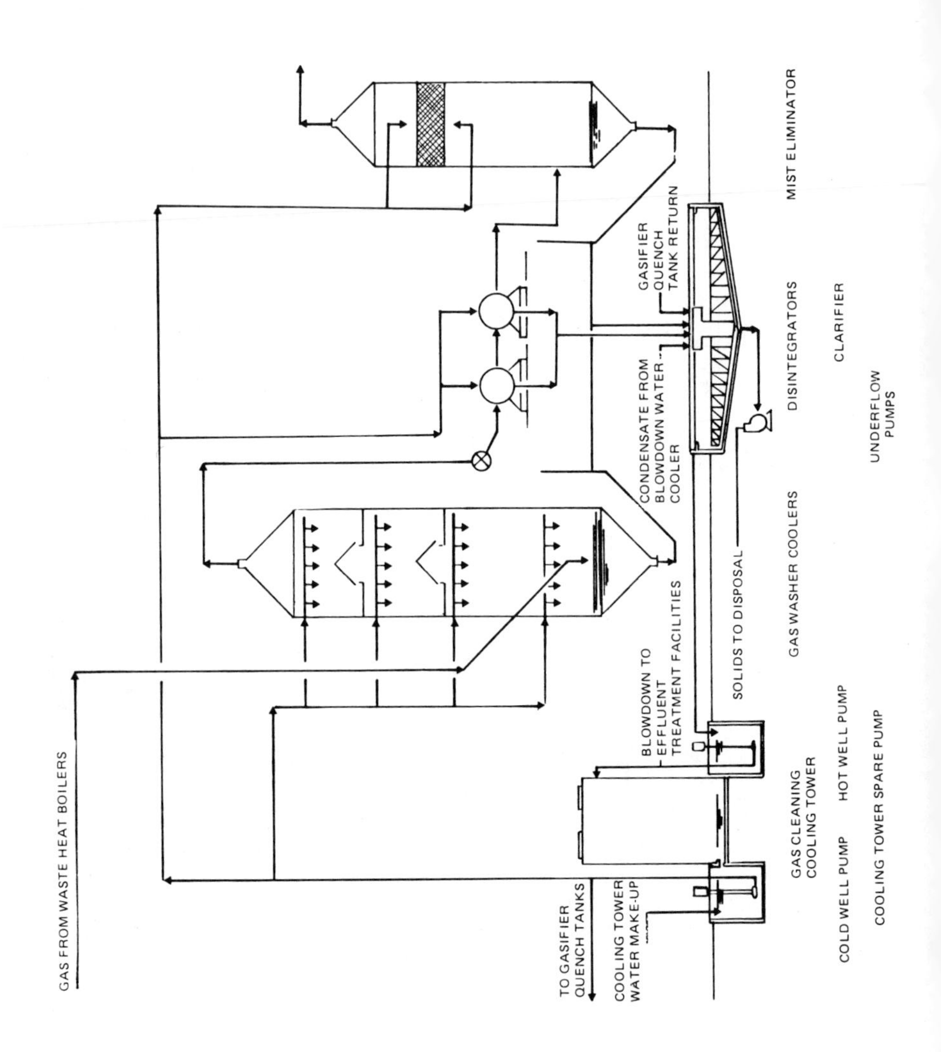

Figure 11. Koppers-Totzek gas cleanup system

hydrogen as ammonia plant feed. The process has been particularly useful for gasifying high sulfur, high metallic content resids. Recently, Texaco has tested through the pilot plant stage, and has announced the availability for licensing of a modification to gasify coal (3).

The other processes under development, the main ones being Hygas, CO_2-Acceptor, Synthane, Bi-Gas, COED, COGAS, Atgas-Patgas, Molten Salt, B&W, and Exxon Catalytic, are designed with three main objectives in mind:

to operate under pressure

to create modules large enough to supply gas for synthetic fuels production (a moderate size synthetic gas plant would require 20 Lurgi gasifiers, or 10 K-T gasifiers)

to maximize production of methane, and to minimize production of other byproducts, such as oils, tar, phenols, hydrogen cyanide, etc.

The advantages of pressure operation are that it eliminates the need to compress the gas as much, and that it increases the amount of methane in the product from the gasifier, thus reducing the subsequent load on the methanation facilities. The advantages of pressure operation are most pronounced for production of high-BTU gas, but are also considerable for production of methanol, ammonia, and hydrogen for subsequent hydrogenation. The equilibrium of the methane-forming reactions is better under pressure, but of more importance is the fact that the heat load (that is, the amount of oxygen required) is reduced if methane is made directly rather than carbon monoxide and hydrogen.

Shale Oil

A rich oil shale from the Piceance Creek Basin runs 30 or 35 gallons per ton (Fisher assay), and lies, generally, deep in the formation,--from 200 to 2000 feet deep. To supply a small refinery of 100,000 barrels per day capacity, 140,000 tons per day of this rich shale must be mined. This is a large operation, much bigger than the largest underground coal mines in the country, and larger than any single open-pit operation in the western states. The only larger operation on this continent is the Syncrude tar sands project in northern Alberta, at 250,000 tons per day. Furthermore, the spent shale from such a mine amounts to about 125,000 tons per day, and is a fine, dry dust occupying more volume than the original oil shale. This dust has to be disposed of somehow, and to keep it from blowing all over the State of Colorado, it has to be soaked with water,--in an arid country. Since 90% of the shale in the Basin lies below a reasonable level for strip mining, and since shaft-and-tunnel mining is considerably more expensive than

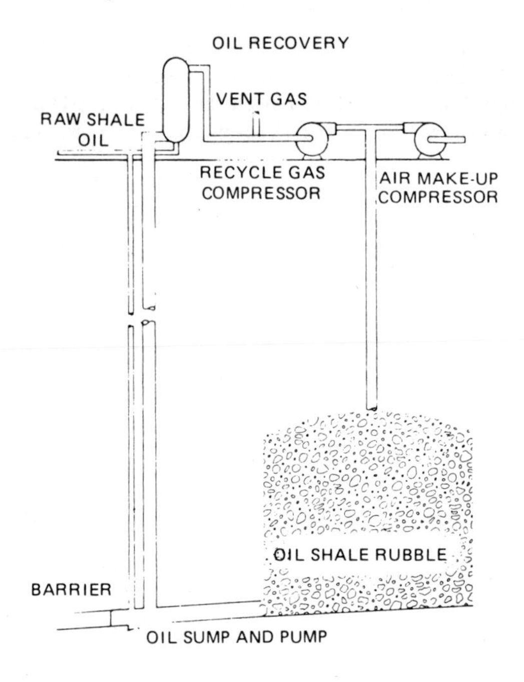

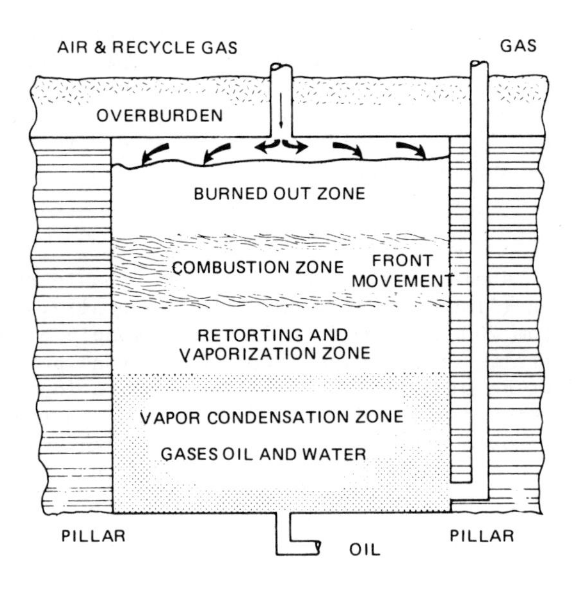

Figure 12. Oxy oil shale process retort operation

open-pit mining, there is a real incentive for an in-situ retorting operation which eliminates the necessity to mine the shale at all.

A number of concepts for in-situ retorting have been proposed. The best developed concept is shown in Figure 12. A chamber is mined out, either in the rock above or below the formation, or in the upper or lower oil shale strata, or both, and shafts are bored into the shale. Large amounts of conventional explosives are detonated throughout the shale to break it up. The formation is thus rendered porous enough to permit the flow of gas through it, while the intact surrounding shale walls confine the gas and pyrolysis products. Each such retort is, typically, 300 x 300 x 500 feet.

In one concept, air is pumped down through the formation while the upper surface is heated to ignition temperature with gas or propane burners. Once ignition starts, the process is self-sustaining. The hot gases in front of the flame heat the shale to retorting temperature and pyrolyze it, leaving a carbonaceous residue which supports combustion when the flame reaches it. The hot spent shale behind the flame front preheats the air, and the cold unretorted shale in front of the retorting zone cools the combustion gases and (at least initially) serves to condense the liquid products, which are collected at the bottom of the retort and pumped to the surface. An idealized temperature gradient for an active retort is shown in Figure 13. To maintain continuous production, several retorts are in different stages of development at any given time, with one being mined, another being rubblized, and a third in operation. In practice, the hot gases from a retort in which the flame front is approaching the bottom would most likely be directed up through a second completed retort waiting to be fired, so that the liquid products can be condensed without using cooling water.

The problem with this scheme is that the gas from such a retort has a low heating value,--probably around 50 BTU per cubic foot, and possibly approaching the lower flammable limit. There is an enormous amount of this gas, and it cannot be discarded because:

- it is contaminated with hydrogen sulfide and ammonia
- it represents an appreciable fraction of the heating value of the oil shale. For example, in a good-sized operation, the fuel oil equivalent of the retort gas could be 20,000 barrels per day.

To increase the heating value of the gas, and to reduce its volume, the gas can be recirculated from the retort through a heater and back to the retort. In this case, no actual combustion occurs in the retort, and the carbon residue is left behind. Part of the gas make is used to fire the heater; the remainder is available for steam generation, or for sale. Obviously, the addition of the heaters increases the capital cost, but the increased cost may be offset by the increased value of the high BTU gas. All of the gas

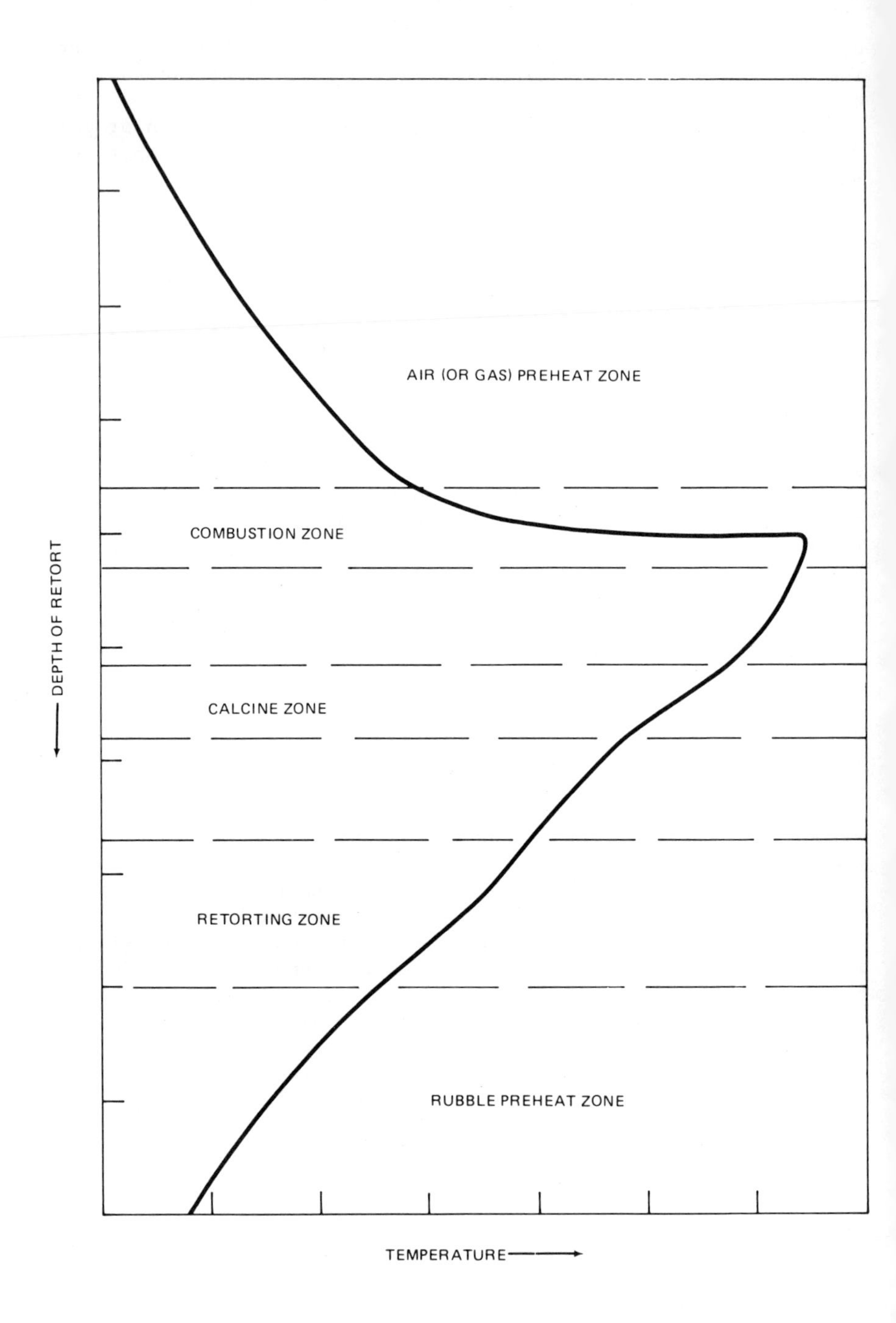

Figure 13. Idealized temperature gradient for in-situ retorting

must be treated for sulfur and ammonia removal; and the water produced from the retort must be treated before discharge.

The rock matrix of western oil shale is largely carbonate, and during retorting it is partly decomposed. In some cases, it may be necessary to scrub the carbon dioxide from the gas to increase its heating value. The base rock of eastern oil shales is silicate.

The shale oil produced is a viscous material that required hydrogenation to eliminate sulfur, nitrogen, and oxygen, to reduce its viscosity, and to improve its stability. The hydrogenation will be done at the production site to improve the properties of the shale oil before pipelining to a refinery. Hydrogenated shale oil is a reasonable substitute for crude oil, and still requires refining.

What are the Data Needs?

Three areas where more thermodynamic data are needed are free energies and enthalpies of formation; high temperature and high pressure region, particularly with polar mixtures; and the cryogenic region for certain mixtures.

Free Energies and Enthalpies of Formation. Coal is the basis for several of the synfuels, and probably will be the most important one domestically. Coals vary tremendously,--from lignite to anthracite; and in all of its forms coal is more reactive than graphite, which is the thermodynamic standard state for carbon. In view of the very large differences in reactivity, the industry will need better heats and free energies of formation, probably as differences between graphite and carbon in the coal, to predict reliable chemical equilibria and heat requirements.

A first step is a tabulation of such data for a representative variety of coals. A second step is the correlation of these data against some easily measured properties of coal, so that the free energy and enthalpy of formation can be predicted for a new coal from some simple laboratory measurements.

The same type of formation is needed for shales, since it can be anticipated that shales from various sources will vary considerably. The problem here is complicated by the need to distinguish between organic and inorganic carbon in shales. A further complication, for both coals and shales, is that the reactivity of the carbon varies as gasification proceeds and that for accurate work the prediction of this variation is necessary.

High-Temperature, High-Pressure Region. The gasification processes are distinguished from the more familiar petroleum refining processes by much higher temperatures, and by the fact that mixtures containing polar compounds are involved. Processing pressures are no higher than many encountered in petroleum work, but the different types of compounds handled add an extra complexity to the high pressure operations.

Very few refinery operations operate about 800°F. The coal gasification processes, on the other hand, start at about 1300°F and some operate up to nearly 3000°F. Many of the new processes operate at pressures of about 1000 psia. Reliable zero pressure enthalpies up to these temperatures are available for most of the materials found in gasifiers, but the mixing rules and pressure corrections which the trade has been using for nonpolar mixtures are probably inadequate.

Another complication is that the systems found in gasifiers are reacting systems at high temperatures. In addition to the heterogeneous reactions, the homogeneous gas phase has many possible reactions such as:

$$CO + H_2O = CO_2 + H_2$$

$$CH_4 + H_2O = CO + 3H_2$$

$$2NH_3 = N_2 + 3H_2$$

and these reactions are catalyzed by the ash components and are promoted by the large area solid surfaces of the coke. For some designs, it may be necessary to predict the chemical equilibria in these systems.

Vapor-liquid equilibrium predictions in these systems are particularly interesting, and will require some different techniques than the ones that the petroleum industry has used. At the high pressures of the second generation gasifiers, water starts to condense as the gas is cooled to about 500°F, and continues to condense down to the lowest temperature, probably around 100°F. The condensate contains appreciable quantities of hydrogen sulfide and carbon dioxide, probably all of the ammonia, possibly phenols and cyanides. Depending on the type of gasifier, a hydrocarbon phase may also condense,--ranging from tar to benzene. In some unlikely circumstances, two hydrocarbon phases may separate, one lighter and one heavier than water. The mutual solubility of the phases is affected by the dissolved components, all of which are soluble in all the phases. And the equilibrium calculations are further complicated by reactions in the liquid water phase, between the acidic and basic components.

The Lurgi gas cleanup system, shown in Figures 2, 3, and 4, is a good example of the problems involved. Each of the five counter-current exchangers represents a series of complicated, simultaneous equilibrium and heat transfer calculations for a polar mixture, with a three- and possibly four-phase system. (If the solid fines are considered, the system is four, possibly five phases; but the solid phase, which most likely stays with the tar, is generally neglected.) Neither the available enthalpy data, nor the available equilibrium correlations, are really adequate for such mixtures, and the problems would be worse if the pressures were higher, as they may be in the future. This is not to say that

Lurgi cannot design such a system; obviously they have, the plants so designed do work. Either Lurgi uses proprietary data, or they design from past experience; in either case, their techniques are not available to the technical public. Nor is it known how much safety is built into their designs.

Cryogenic Region. At the low end of the temperature scale, there are other data needs. Here, the designer needs reliable data for a small number of simple systems. And he needs data on solid-liquid phase relations for a few materials, notably carbon dioxide, hydrogen sulfide, and the higher hydrocarbons. The chief area where these data (or correlations) are required has been for liquified natural gas plants. It would appear that the available data should be adequate, since there is a lot of information on the light hydrocarbons; but, if the reader has not done it, he will be unpleasantly surprised to find how much the specific heat of methane gas at low temperatures varies between various prediction methods. He will also find that it is difficult to predict where carbon dioxide crystallizes out of the liquid hydrocarbon phase as natural gas is cooled. Since deep refrigeration is expensive, cryogenic processing requires accurate correlations.

Exxon has recently published a paper on their new catalytic gasification process, in which coal impregnated with potassium carbonate is gasified with steam and oxygen at a low temperature (1200-1300°F) and a pressure of about 500 psi (4). Under these conditions, the production of methane is maximized, and the overall reaction is almost isenthalpic. The carbon dioxide and methane are separated from the carbon monoxide and hydrogen, which are recycled to the reactor. The separation of methane from carbon monoxide and hydrogen is cryogenic and, for good economy, requires reliable enthalpy and equilibrium data.

Almost any cryogenic separation design becomes, at some point, a balance between heat exchange costs and compression costs. And the optimum (minimum total cost) usually occurs at exchanger temperature differences of only a few degrees, sometimes only a fraction of a degree. To design exchangers to these close approaches requires very accurate equilibria data, and good enthalpy data. If the plant uses a mixed refrigerant, the need is more pronounced.

How Good Do the Data Need to Be?

For the high temperature polar mixtures at high pressures, the need is primarily for prediction methods that can be fairly sloppy, but even more important is some knowledge of the degree of uncertainty in the method. The design engineer faced with calculating the enthalpy of a mixture of CO, CO_2, H_2, and CH_4 with 50% steam at 1000 psia and 1500°F will probably take the four gases at zero pressure and 1500°F, use some generalized pressure correction to raise the mixture to 500 psia, and add the enthalpy of pure steam at 500 psia and 1500°F. Or, he may take all five components at zero pressure (1 psia for the steam because he is using steam tables) and 1500°F, and use the generalized pressure correction to raise the mixture to 1000 psia. If he does both, he will have two different answers. If he uses different pressure corrections, he will have more different answers. If he is knowledgable, he will worry a little about the fact that the pressure correction is based on light hydrocarbons, or on air. If he checks, he will find that the effect of pressure on the enthalpy of steam, from his steam tables, is badly predicted by the generalized correlation. What he needs is some method which tells him that his calculated enthalpy has a reasonable probability of being off by, say, ± 10%. Given that uncertainty, he can design enough safety factor into the unit to take care of it. His problem occurs when he believes his calculated enthalpy and it is really off by 10%.

Later, as the synfuel industry becomes more sophisticated, better accuracy will be needed. The heat flows around a large gasifier plant are immense, and much of that heat is supplied from expensive oxygen reacting with the coal. A high-BTU syngas plant producing 250 million standard cubic feet per day of gas also produces about 15,000 tons per day of carbon dioxide, equivalent to a heat of combustion of 5 billion BTU per hour. To avoid large and expensive safety factors in the design of such plants, accurate enthalpy methods will be needed.

For cryogenic separations, industry already has the approximate methods. What is needed now is accurate data and methods to reduce the cost of unnecessary safety factors on expensive deep refrigeration. The industry also needs techniques to predict solid phase formation.

When Are the Data Needed?

The petroleum industry and the natural gas processing industry operated for about 40 years before any real attempt to develop data and correlations was made. They used crude approximations, such as Raoult's and Dalton's Laws, because they were good enough for the processing that the industry was doing at that time. As processing became more complex, better data were needed and were developed,--a trend that still continues.

The synthetic fuel industry cannot undergo a comparable incubation period for its data requirements for two main reasons:

1. The synthetic fuels are intended to augment petroleum fuels that are presently produced by sophisticated processing techniques based on adequate thermodynamic data.
2. Synthetic fuel plants will be horrendously expensive, and there is a large economic incentive to provide data good enough to eliminate expensive safety factors from the design.

Literature Cited

1. Musser, William N., and John H. Humphrey, "In-Situ Combustion of Michigan Oil Shale: Current Field Studies", Eleventh Intersociety Energy Conversion Engineering Conference, page 341 (1976).
2. "Joint Application of Michigan-Wisconsin Pipeline Company and ANG Coal Gasification Company for Certificates of Public Convenience and Necessity", Docket No. CP75-278 before the Federal Power Commission, Volume 1 (1975).
3. Crouch, W. G., and R. D. Klapatch, "Solids Gasification for Gas Turbine Fuel: 100 and 300 BTU Gas", Eleventh Intersociety Energy Conversion Engineering Conference, page 268 (1976).
4. Epperly, W. R., and Siegel, H. M., "Catalytic Coal Gasification for SNG Production", Eleventh Intersociety Energy Conversion Engineering Conference, page 249 (1976).

Discussion

C. A. ECKERT

Two papers given in this session assessed the severe problems encountered by scientists and engineers in dealing with processing under unusual conditions. Mike Modell's paper stressed the difficulties thermodynamicists have in dealing with multicomponent systems in the critical region. He reviewed the criteria for criticality in multicomponent systems and developed the use of Legendre transformations to find a stable point on the spinodal surface. Much of the discussion following the paper centered about whether one could use the same type of approach to determine the binodal surface--that is, in a practical sense, find the composition of coexisting phases. Some comments on this problem were as follows:

H. Ted Davis, University of Minnesota, "The spinodal conditions are useful in constructing the binary and ternary phase diagrams. Meijering (1) has used the spinodal extensively to locate critical points for regular solutions. Once the critical point is located, then a simple numerical technique can be used to march along a binodal to construct the binodal curve. Overlapping binodals can then be used to locate three phases in equilibrium, where such exist. Such a process is being carried out for liquid phase diagrams by Jeff Kolstad working with C. E. Scriven and me at Minnesota."

John S. Rowlinson, University of Oxford, United Kingdom, "I would like to make two points:

1. By concentrating on the spinodal surface, as Modell and Reid's elegant transformations do, one runs the risk of overlooking other kinds of behavior on critical surfaces. For example, the spinodal curve may be outside one of the binodal curves. This does happen with the ternary diagram for which G^E is quadratic in composition, a system which was analyzed fully by Meijering (1). Here, there are three tricritical points, and it is the presence of these, which are singularities not envisaged in Gibb's treatment, which would invalidate the use of the spinodal alone as a sole criterion of critical behavior.

2. The Gibbs (and related) treatments assume that the exten-

sive thermodynamic functions form an analytic surface $U = f(V,S,N_i)$ around the critical point. This assumption conflicts with the known existence of "weak" singularities at this point (e.g., $C_V \rightarrow \infty$). In a mixture these singularities can give rise to complications not, I think, encompassed in Modell's treatment. Thus, R. B. Griffiths and Wheeler (2) have suggested that there are "anomalies" on a gas-liquid critical curve for a binary mixture, not only at the critical azeotrope but also at points where the critical curve passes through the extremum with respect to changes of pressure or temperature."

The second paper of the evening given by Howard Hipkin of Bechtel Corporation complimented the first in that it stressed, from a much more practical point of view, specific needs that will be encountered in the near future by industrial designers dealing with methods for energy recovery from fossil fuels. He discussed what synthetic fuels might be, and how they might be made, and from what we know about such processes now predicted what they indeed might be. He stressed the need for heats of formation and Gibbs energies of formations at higher temperatures, especially for coal and oil shale; for high temperature and pressure data especially for polar mixtures; and the need for calculational methods for handling such data. As one example he held forth the spectre to an analyst of a five-phase system emerging from a Lurgi gasifier.

Rather extensive discussion involving a number of individuals followed dealing with the imminent needs for new energy sources with small likelihood of it being satisfied by nuclear, solar, or geothermal power. However, severe problems exist in the utilization of coal or oil shale in terms of high capital requirements coupled with the environmental restrictions on emissions from such plants. One quite considerable dilemma that becomes apparent is that the production of energy from syntehtic fuels with our current technology would only be practical at current capital costs if the selling price of energy went up. However, it is quite evident that capital costs are linked to energy costs.

Thus, the consensus of the discussion was that better data are needed at higher temperatures and pressures, and in the initial stages, even rather "sloppy" data would be more useful than what is now available. This will require better materials and may lead to new techniques such as, for example, supercritical separation processes and higher temperature processes. Certainly the methods we now have for estimating such properties will prove inadequate, and new and better methods will certainly be required. As illustrated so graphically by the first talk of the evening these will undoubtedly be more difficult to develop and apply than current methods. However, the general concensus was that it is quite clear that the energy crisis will soon be upon us in a much more serious sense than the general public appreciates, and we as scientists and engineers must begin now to seek solutions in terms of new data at more extreme conditions and the thermodynamic framework within which to apply them.

References

1. Meijering, J. L., Phillips Res. Rep. (1950) 5, 333; (1951) 6, 183.
2. Griffiths, R. B. and Wheeler, J. C. Phys. Rev. (1970) 2, 1047.

21

A Group Contribution Molecular Model for Liquids and Solutions Composed of the Groups CH_3, CH_2, OH, and CO

T. NITTA, E. A. TUREK, R. A. GREENKORN, and K. C. CHAO

Purdue University, West Lafayette, IN 47907

Group interaction models have achieved remarkable success in the description of activity coefficients of nonelectrolyte solutions. Notable in this development are the pioneering work by Pierotti, Deal and Derr (1), Wilson and Deal (2), and subsequent contributions by Scheller (3), Ratcliff and Chao (4), Derr and Deal (5), and Fredenslund, Jones, and Prausnitz (6).

Nitta et. al. (7) extended the group interaction model to thermodynamic properties of pure polar and non-polar liquids and their solutions, including energy of vaporization, pvT relations, excess properties and activity coefficients. The model is based on the cell theory with a cell partition function derived from the Carnahan-Starling equation of state for hard spheres. The lattice energy is made up of group interaction contributions.

An important advantage of the model by Nitta et. al. is its applicability over a wide temperature range. The same group parameters used in the same equations have been found to give good results at conditions for which the cell model is known to be applicable--where the liquid is not "expanded", the reduced density is greater than two and the temperature is not much above the normal boiling point. It is not necessary to have different sets of group parameter values for different temperatures.

Nitta et. al. (7) presented the properties of the groups CH_3, CH_2, OH, and CO and their interactions. Comparisons of the model and experimental data were made for a number of pure liquids and their solutions.

Additional comparisons of solution properties with the model calculated values are presented here to cover the gamut of mixtures made up of the given groups from the non-polar/non-polar, through non-polar/polar, to polar/polar.

Figure 1 shows the predicted activity coefficients of n-hexane in solution with n-dodecane compared to experimental data by Broensted and Koefoed (8). The same agreement is obtained between our model and experimental data from the same source on the mixtures of other n-alkanes.

Figure 2 shows the predicted activity coefficients in the system ethanol/n-octane at 75C in comparison with experimental

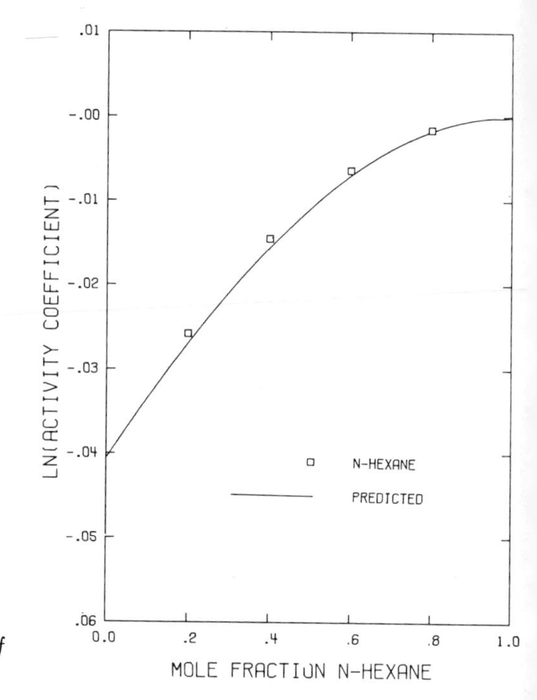

Figure 1. Activity coefficients of n-C_6 in n-C_{12} at 20°C

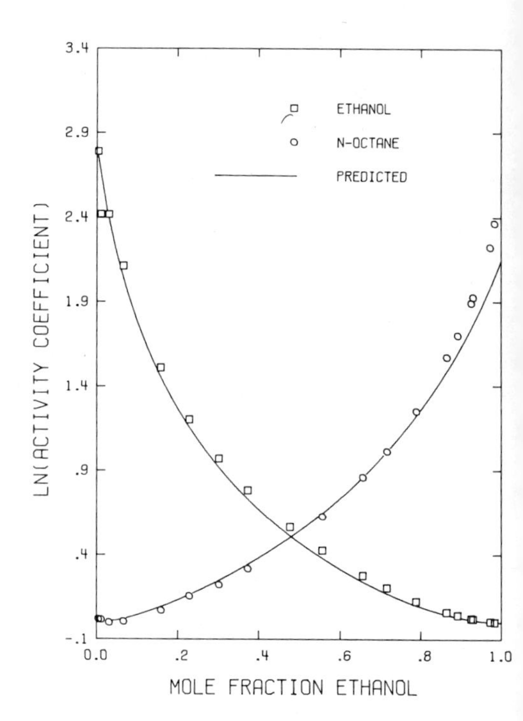

Figure 2. Activity coefficients in ethanol–n-octane at 75°C

data by Boublikova and Lu (9). The agreement is the same as that previously reported by Nitta et. al. (7) for the same system at 45C with data from the same source. There is a remarkable change in activity coefficients in this system in the temperature interval of 30C. Thus the infinite dilution value of ethanol is reduced by a factor of about two while that of n-octane is reduced by only about 10% with this temperature increase. This remarkable change is quantitatively described by the model.

Figure 3 shows the predicted activity coefficients in n-butanol/n-heptane at 50C in comparison with the experimental data of Aristovich et. al. (10). The agreement that is obtained here for the high alcohol is about the same as the previously reported results (7) for the lower alcohols.

Figure 4 shows the predicted excess enthalpy of n-butanol/n-heptane at 15 and 55C in comparison with experimental data by Nguyen and Ratcliff (11). The agreement is not quite quantitative and deviations up to 30 cal/g-mole are observed for some compositions.

Figure 5 through 8 show the activity coefficients in the system n-hexane/2-propanone at four temperatures 45, 20, -5, and -25C. Experimental data are from Schaefer and Rall (12) at 45 and -20C, and from Rall and Schaefer (13) at 20 and -25C. The variation of the activity coefficients with temperature appears to be quantitatively described by our model. The 45C isotherm was used in the development of the properties of the CO group and the model is therefore in a sense fitted to this isotherm. But the other isotherms were not used in the development of the model, and the calculations for them are of a predictive nature.

Figure 9 shows the activity coefficients in the system 2-propanal/n-hexanol (14). The molecular interactions in this polar/polar mixture is complex leading to an apparent maximum in the figure. The existence of the maximum is correctly predicted by our model, but the calculated values seem to vary too rapidly at small concentrations of acetone. There also seems to be considerable uncertainty and scattering of the experimental data in the same range.

Acknowledgment

This work was supported by National Science Foundation through grants GK-16573 and ENG76-09190. D. W. Arnold assisted in the calculations.

Abstract

The group contribution molecular model for liquids and solutions developed by Nitta et. al. is applied to properties of liquid solutions made up of the groups CH_3, CH_2, OH, and CO, and the results are compared with experimental data. A wide range of molecular species in mixtures is included over a wide temperature range.

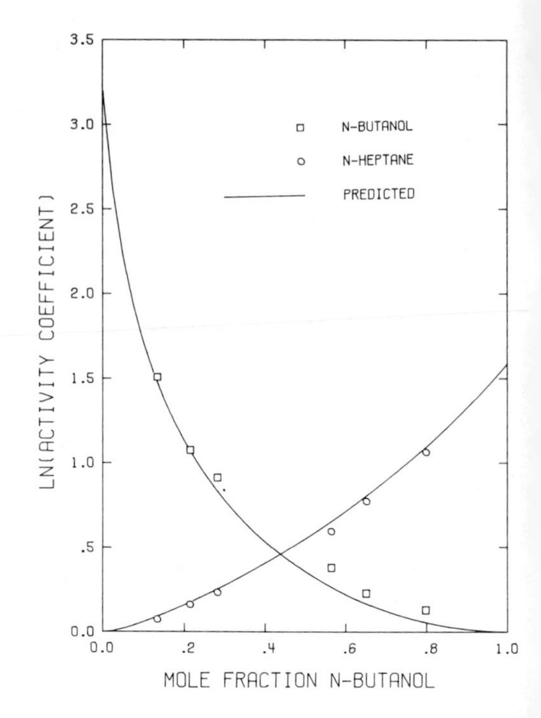

Figure 3. Activity coefficients in n-*butanol*–n-*heptane at 50°C*

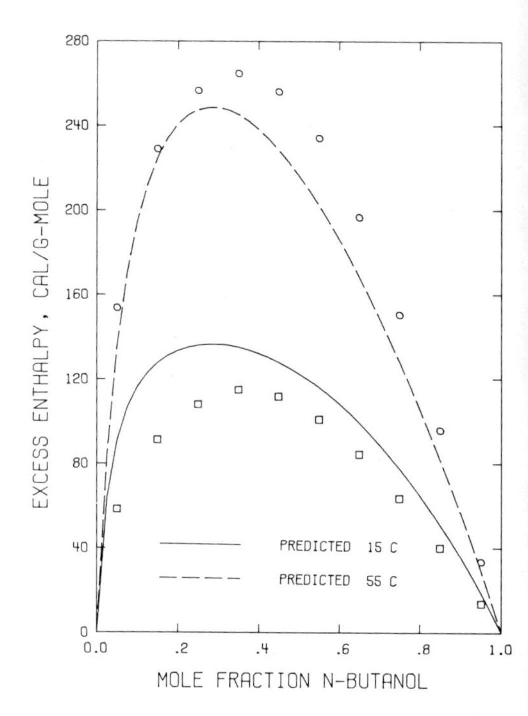

Figure 4. Excess enthalpy of n-*butanol–n-heptane*

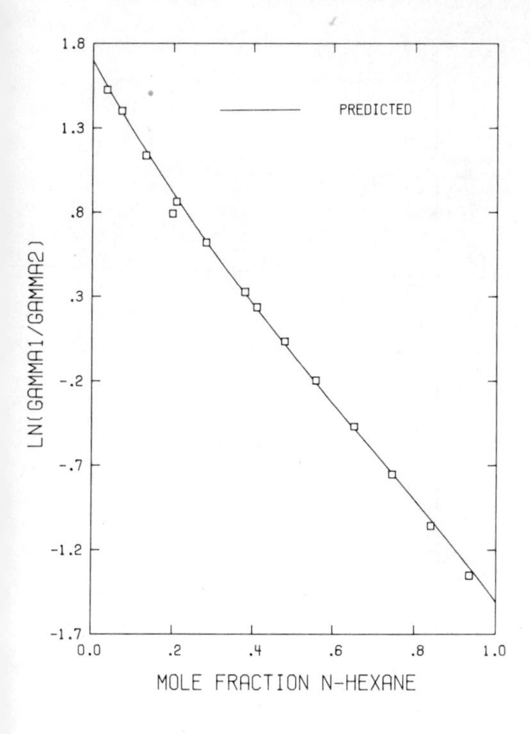

Figure 5. Activity coefficients in n-hexane–2-propanone at 45°C

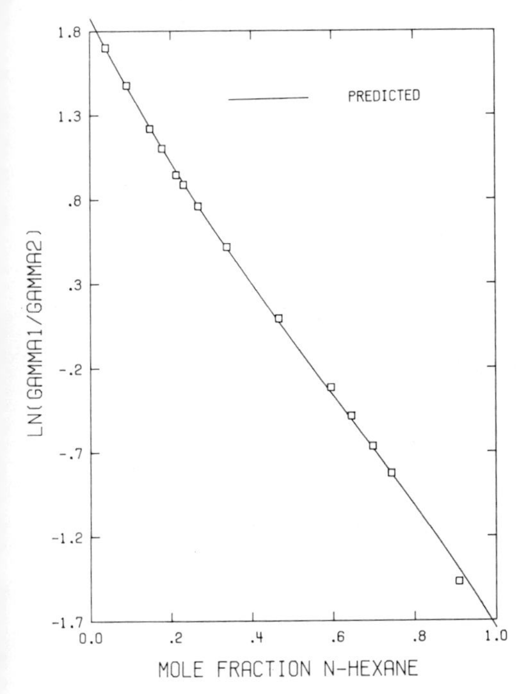

Figure 6. Activity coefficients in n-hexane–2-propanone at 20°C

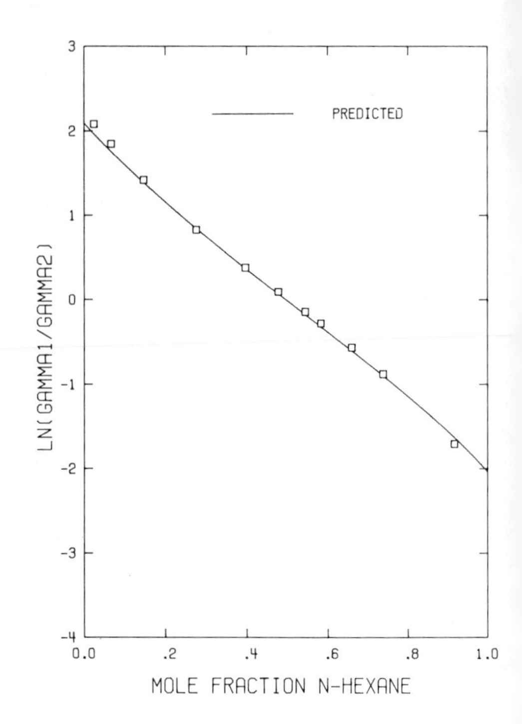

Figure 7. Activity coefficients in n-hexane–2-propanone at 5°C

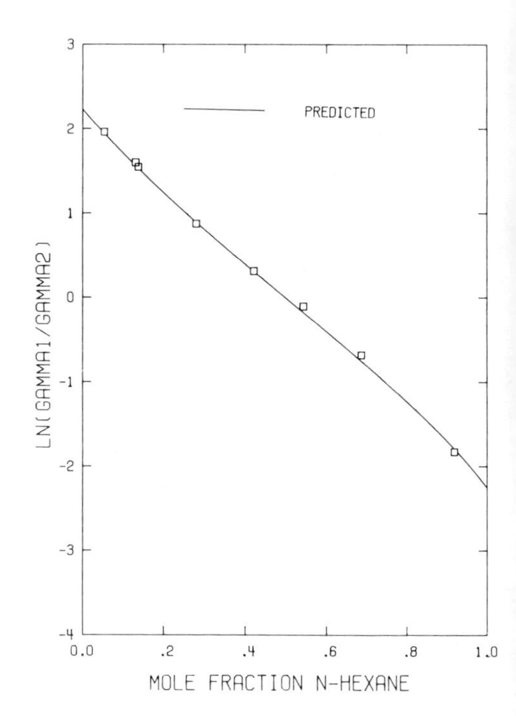

Figure 8. Activity coefficients in n-hexane–2-propanone at 20°C

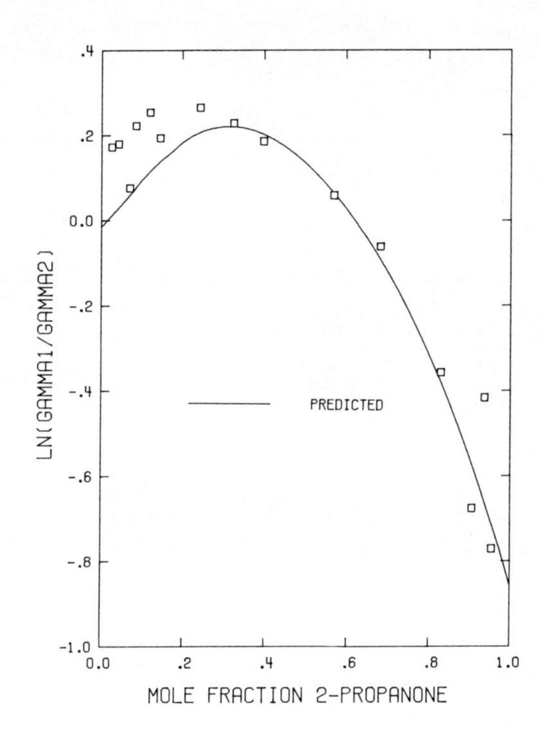

Figure 9. Activity coefficients in 2-propanone– n-*hexanol at 1 atm*

Literature Cited

1. Pierotti, G. J., Deal, C. H., and Derr, E. L., Ind. Eng. Chem. (1959) 51, 95.
2. Wilson, G. M., and Deal, C. H., Ind. Eng. Chem. Fundamen., (1962) 1, 20.
3. Scheller, W. A., Ind. Eng. Chem. Fundamen. (1965) 4, 459.
4. Ratcliff, G. A., and Chao, K. C., Canad. J. Chem. Eng. (1969) 47, 148.
5. Derr, E. L. and Deal, C. H., Distillation 1969, Sec. 3, p. 37, Brighton, England: Intern. Conf. Distillation, Sept. 1969.
6. Fredenslund, Aa., Jones, R. L., and Prausnitz, J. M., AIChE J. (1975) 21, 1086.
7. Nitta, T., Turek, E. A., Greenkorn, R. A., and Chao, K. C., AIChE J. (1977) 23, 144.
8. Broensted, J. N., and Koefoed, J., Selskab. Mat. Psy. Medd. (1946) 22, No. 17, 1.
9. Boublikova, L., and Lu, B. C. -Y., J. Appl. Chem. (1969) 19, 89.
10. Aristovich, V. Y., Morachevskii, A. G. and Sabylin, I. I., J. Appl. Chem. USSR (1965) 38, 2633.
11. Nguyen, T. H. and Ratcliff, G. A., J. Chem. Eng. Data (1975) 20, 252.
12. Schaefer, K. and Rall, W., Z. Elektrochem. (1958) 62, 1090.
13. Rall, W., and Schaefer, K., Z. Elektrochem. (1959) 63, 1019.
14. Rao, P. R., Chiranjivi, C., and Dasarao, C. J., J. Appl. Chem. (1967) 17, 118.

22

Areas of Research on Activity Coefficients from Group Contributions at the Thermochemical Institute

GRANT M. WILSON

Brigham Young University, Provo, UT 84602

The concept of calculating the thermodynamic properties of mixtures from group contributions has been a dream of workers in solution thermodynamics for some time. Significant progress in this direction has been made since the discovery of a size effect contribution to the excess free energy of mixing. This effect was recognized in 1941 by both Flory and Huggins in their independent publications of the now well known Flory-Huggins equation (1,2). By subtracting the size effect from the excess free energy of mixing, it has now been found that the residual free energy can nearly be correlated in terms of group interactions. Some initial work in this area was reported by Wilson and Deal (3) in 1962; and there have been several subsequent papers by Derr and Deal, (4) Scheller, (6) and Fredenslund et al. (7) on this subject. Probably the most significant publications to date are the "Analytical Solutions of Groups" model (ASOG) by Derr and Deal, (4) and the UNIFAC method of Prausnitz (7) et al.

There are still a number of problems and unresolved areas to be studied before the group contribution model can be applied rigorously and dependably. This paper summarizes several areas of activity under study at the Thermochemical Institute in attempts to improve prediction accuracy of the group contribution method. This paper also identifies problem areas which need further study in the future.

Areas of Research Activity at the Thermochemical Institute

Areas of research activity on group contribution methods at the Thermochemical Institute at Brigham Young University can be categorized into the following areas.

(a) The effect of void spaces between molecules on the excess free energy of mixing.

(b) The estimation of the infinite dilution activity coefficient of a component in a binary mixture knowing the infinite dilution activity coefficient of the other component.

(c) The measurement of binary vapor-liquid equilibrium data for modeling purposes in the development of generalized group interaction models.

(d) The estimation of pure compound properties from a knowledge of group activity coefficients in the group environment of the pure compound.

(e) The development of equations of state from group contribution methods assuming void spaces as an additional component in a fluid.

Each of these areas are described in subsequent paragraphs of this section. A subsequent section deals with problem areas and future work.

The Effect of Void Spaces. The Flory-Huggins equation (1,2) was derived assuming no vacant sites in a liquid mixture. Presumably there would be an additional effect on the excess free energy of mixing due to void spaces between molecules which is not accounted for in the Flory-Huggins equation. This is apparent from the fact that there is no size effect on the excess free energy of mixing of ideal-gas molecules. Thus the effect of void spaces varies the Flory-Huggins effect so that it becomes zero for an ideal-gas mixture. A paper on this subject was published in 1973 (8) from the Thermochemical Institute in which the effect of void spaces was estimated by assuming void spaces as an additional component of a mixture. The number of moles of void spaces present was calculated by assuming the molar volume of void spaces to be equal to the volume of one mole of lattice sites occupied by methylene groups or 15.5 cc, and the volume of molecules was assumed to be a factor of 1.57 times the volume calculated from group volumes given by Bondi (9). The resulting equation for the activity-coefficient contribution is as follows.

$$\ell n\ \gamma_i^{MF-H} = \ell n \frac{v_i}{v} + 1 - \frac{v_i}{v} + \frac{bv_i - vb_i}{vb_H} + \frac{v_i - b_i}{b_H} \ell n \left[\left(\frac{v-b}{v}\right)\left(\frac{v_i}{v_i - b_i}\right) \right] \quad (1)$$

where v = molar volume of mixture

v_i = molar volume of pure component i

b = volume of one mole of molecules in mixture

b_i = volume of one mole of molecules of component i

b_H = volume of one mole of void spaces (= 15.5cc).

The effect of this correction to the Flory-Huggins equation for selected binary mixtures is shown in Table 1 where equation (1) is compared with the unmodified Flory-Huggins equation. This comparison shows that for the cases selected the effect of void spaces is to produce a lower activity coefficient than does size effect alone. This equation would therefore have a significant effect on group contribution correlations because a larger

Table 1

INFINITE-DILUTION ACTIVITY COEFFICIENTS FROM MODIFIED FLORY-HUGGINS EQUATION*

Solute	Solvent	γ_∞	
		Mod. F-H	F-H
n-Pentane	Methyl ethyl ketone	0.866	0.968
n-Heptane	Methyl ethyl ketone	0.859	0.870
n-Decane	Methyl ethyl ketone	0.662	0.676
n-Heneicosane	Methyl ethyl ketone	0.127	0.164
n-Octacosane	Methyl ethyl ketone	0.0443	0.0453
n-Butane	Furfural	0.361	0.983
n-Pentane	Furfural	0.459	0.943
n-Heptane	Furfural	0.538	0.825
n-Hexadecane	Furfural	0.271	0.283
n-Triacontane	Furfural	0.0209	0.0255
n-Butane	Ethanol	0.797	0.843
n-Pentane	Ethanol	0.744	0.749
n-Heptane	Ethanol	0.544	0.558
n-Decane	Ethanol	0.269	0.326
n-Hexadecane	Ethanol	0.0463	0.0910
n-Eiscosane	Ethanol	0.0152	0.0334
n-Heptane	Ethylene glycol	0.341	0.523
n-Decane	Ethylene glycol	0.238	0.293
n-Hexadecane	Ethylene glycol	0.0725	0.0760

*Temperature, 25°C. Lattice parameters assumed in the calculations are: $b_H = 15.5\ cm^3$, $b_i = 1.57b_{Bondi}$.

correction would be made than is now made with the Flory-Huggins equation. The effect of varying the volume of one mole of void spaces can be seen from equation (1) where b_H appears in the denominator of the last two terms. Thus a larger value for b_H would reduce the effect of void spaces. The choice of 15.5 cc/mole was arbitrary, and other values could be used. There doesn't appear to be a good method for determining the correct value. The value used here is consistent with the lattice model of Flory and Huggins where we have assumed that one mole of lattice sites is equivalent to 15.5cc.

Infinite Dilution Activity Coefficient Estimation of One Binary Component from the Other. Two methods have been published recently by Bruin and Prausnitz (10) and by Tassios (11) on one-parameter activity-coefficient equations. In principle the single parameter in these equations could be adjusted to fit one known activity coefficient of a binary mixture. The infinite dilution activity coefficient of the other component could then be calculated from the derived parameter. Unfortunately the equation proposed by Bruin and Prausnitz is nearly equivalent to predicting $\ln \gamma_2^\infty/\ln\gamma_1^\infty$ ratios from volume ratios or by assuming ratios of unity depending on the system. This assumption is rather poor as is shown in Figure 1 (12) where the $\ln \gamma_2^\infty/\ln_1^\infty$ for hydrocarbons over 2-butanone or ethanol are close to unity, but they vary with the carbon number of the paraffin. The assumption of ratios proportional to volume ratios is way too big and the assumption of ratios of unity is generally too small. The method of Tassios accounts for differences in molecular size but the method does not appear to be adequate.

The Thermochemical Institute has recently published a paper (12) which in principle is similar to the one-parameter methods of Bruin and Prausnitz and of Tassios. The method provides a means based on group contributions whereby the infinite dilution activity coefficient of one component of a binary mixture can be estimated from the infinite dilution activity coefficient of the other component. This result then produces two infinite dilution activity coefficients which can be fitted to any 2-parameter activity coefficient equation to predict activity coefficients at finite concentrations. The method assumes that the ratio of $\ln \gamma_2^\infty/\ln\gamma_1^\infty$ after correction for size effect and void space effect, can be calculated from group contributions according to the following equation:

$$\frac{\ln\gamma_2^{\infty G}}{\ln\gamma_1^{\infty G}} = \frac{(\Sigma_i n_i \nu_i S_i)_2}{\Sigma_i n_i \nu_i S_i)_1} \quad (2)$$

where $\gamma^{\infty}G$ = infinite dilution activity coefficient due to group interactions

n_i = number of groups of type i in molecule

ν_i = number of lattice sites occupied by group i

S_i = group contribution weighing factor applied to group i.

By this method, the parameters S_i are found to be given by the following equation:

$$S_i = \exp(\frac{\Delta h_i}{Z\nu_i RT}) \quad (3)$$

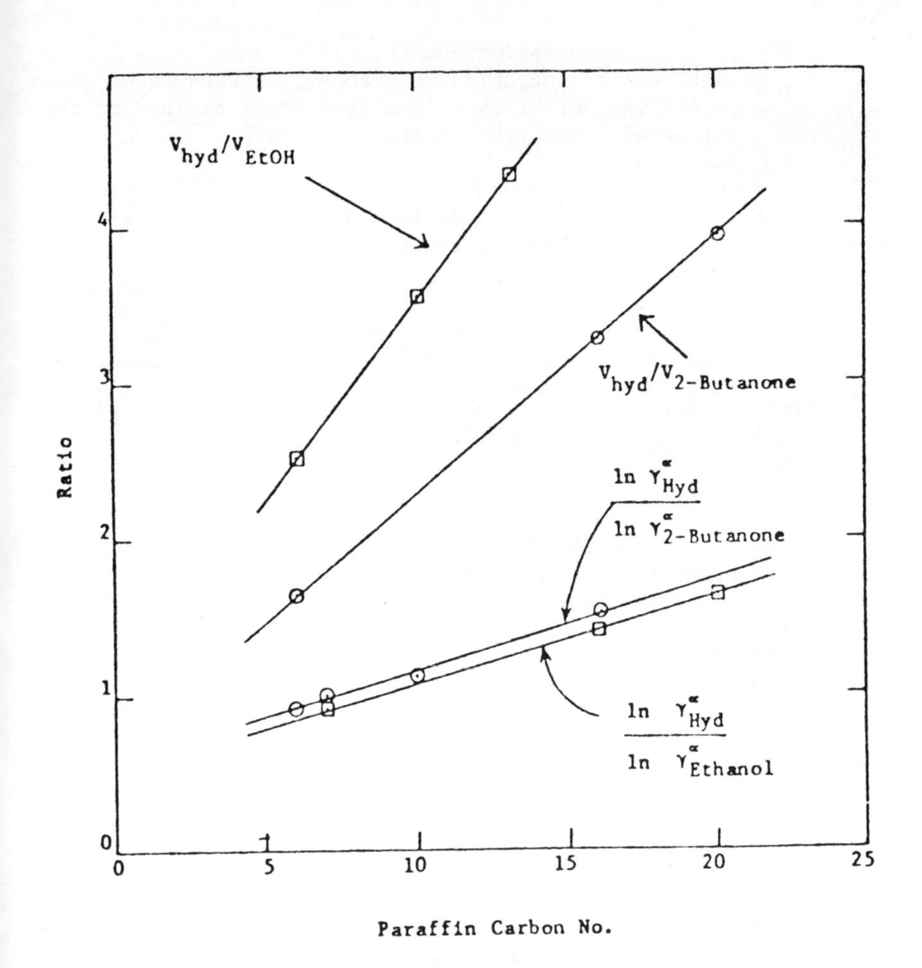

Figure 1. Effect of size and polarity on the ratio of the logarithm of infinite dilution activity coefficients, paraffin–ethanol and paraffin–2-butanone systems at 60°C

where Δh_i = group interaction energy parameter approximated as the incremental change in latent heat per group in a molecule

Z = lattice coordination number; 6.3 assumed

ν_i = lattice sites occupied by group i; assumed to be given by the Van der Waals volume from Bondi divided by the volume of a methylene group

R = gas constant

T = absolute temperature

Table 2 gives a summary of derived group contribution parameters. The method has been tested against 130 binary systems with the following results.

Type of Mixture	Average Error in γ_2^∞, %
Mixtures without -OH groups	11
Mixtures containing alcohols	25
Water-alcohol mixtures	32
Water-non-alcohol mixtures	59

We believe the method should prove to be quite useful because it applies to branched and ring compounds as well as straight-chain compounds. In many cases it is possible to measure infinite dilution activity coefficients of a volatile compound in a less volatile compound by gas chromatographic retention time. This method would then provide a means for estimating the activity coefficient of the less volatile component in the more volatile component; thus providing the necessary data required for modeling multicomponent distillation conditions.

Experimental Modeling for Group Interaction Correlations. Experimental methods have been developed at the Thermochemical Institute which permit the rapid determination of isothermal binary vapor liquid equilibrium data. The method is based on measurement of total pressure versus total composition of a mixture in a vapor-liquid equilibrium apparatus. Because no analyses are made of the equilibrium phases, one can study large numbers of binary mixtures in a short period of time. Wayne Eng (13) has recently completed the study of 17 binary mixtures involving thiol compounds with other compounds involving 14 other functional groups. This work represents a significant contribution because it provides modeling data for the prediction of activity coefficients based on group contribution methods for just about any component of a mixture containing thiols with other compounds. Table 3 gives a summary of measured infinite dilution activity coefficients derived from these data. Wayne Eng also used the vapor-liquid data to derive UNIFAC parameters (13) in the model published by Prausnitz, et al. (7). We believe the method could be used to measure data required to "fill in" gaps of a group

Table 2

PREDICTED AND EMPIRICAL PARAMETERS FOR GROUP WEIGHTING FACTORS

Group	Sites per group, ν	Interaction energy[a] °K $\Delta H/\nu R$ Predicted	Empirical[b]
$-CH_3$	1.37	688	1150
$-CH_2-$	1.02	503	50
$-\overset{\mid}{\underset{\mid}{C}}-H$	.68	-132	-132
$-\overset{\mid}{\underset{\mid}{C}}-$	.33	-2230	(-2230)
CH_4	1.71	627	(627)
$=C{<}^{H}$ ar	.81	848	750
$=CH_2$ par	1.19	731	731
$=CH_2{<}^{H}$ par	.85	650	650
$=C=$	.70	1210	(1210)
phenyl	4.58	737	737
$\equiv CH$	1.16	1010	1010
$\equiv C-$	.81	888	(888)
$-F$	.60	--	688
$-Cl$	1.22	1057	750
$-Br$	1.46	845	(845)
$-OH$	.80	4720	4720
H_2O	.98	5360	3400
$-C{\lneq}^{O}$	1.17	1678	2800
$-O-$	.37	1908	2500
$-C{\lneq}^{O}O-$	1.44	--	3300
CO_2	1.64	1697	1900
$-C\equiv N$	1.47	2160	3000
$-N{<}$	.43	--	(2210)
$-NH_2$	1.05	2210	(2210)
$-NO_2$	1.68	--	1900
H_2S	1.88	1287	1250
CS_2	3.00	1140	1140
HCl	1.65	1280	(1280)

[a] $\ell n\ s_i = \frac{\Delta h_i}{Z\nu_i RT}$, Z = 6.3

[b] Numbers in parenthesis have not been tested

Table 3

CLASSIFICATION OF BINARY SYSTEMS ACCORDING TO THE SOLUTION BEHAVIOR OF THE THIOL COMPOUNDS, BASED ON INFINITE DILUTION ACTIVITY COEFFICIENTS

Behavior	A	B	γ_A^∞	γ_B^∞
Slight Positive Deviations	Methanethiol	Benzene	1.065	1.142
	Methanethiol	N,N-Dimethyl formamide	1.076	1.395
	Methanethiol	Methylamine	1.141	1.141
	Methanethiol	Methyl chloride	1.170	1.144
	Methanethiol	Dimethyl ether	1.171	1.171
	Methanethiol	Acetone	1.237	1.342
Moderate Positive Deviations	Methanethiol	n-Decane	1.404	2.727
	Methanethiol	Dimethyl sulfoxide	1.532	2.363
	Methanethiol	Methyl formate	1.624	1.790
	Methanethiol	Acetonitrile	1.990	2.528
	Methanethiol	Propane	2.463	2.984
	1-Dodecanethiol	Propane	3.042	1.253
	Methanethiol	Methanol	4.245	9.047
	1-Butanethiol	Propane	4.584	1.924
	Benzenethiol	Propane	10.558	4.678
Large Positive Deviations	1,2-Ethane-dithiol	Propane[a]	26.691	10.555
Negative Deviations	Methanethiol	Propionaldehyde[b]	0.0509	0.0887

[a]Liquid-liquid immiscibility occurs.

[b]Chemical reaction occurs.

interaction table or to verify interactions based on limited data. Proposals are being circulated to industrial firms at the present time for further measurement of modeling data.

The Estimation of Pure Compound Properties from a Knowledge of Group Activity Coefficients. The vapor pressure and other thermodynamic properties of pure compounds have been found by the author to be related to the activity coefficients of groups in the group environment of the pure compound. In a paper given by the author in 1970, (14) it was shown that the vapor pressure of pure compounds can be related to the properties of the constituent groups according to the following equation:

$$\ln(P^o\phi) = \ln P_t - \ln J + J - 1 + \Sigma_i \nu_i (G_i^{og}/RT) + \Sigma_i \nu_i \ln \Gamma_i \qquad (4)$$

where $P_t = -T\left(\frac{\partial P}{\partial T}\right)_V$

J = number of lattice groups in a molecule

ν_i = number of groups of type i in the molecule

G_i^{og} = ideal free energy of interaction of group of type i

Γ_i = activity coefficient of group i

ϕ = fugacity coefficient in vapor

Since $\ln P_t$ is nearly constant for a homologous series a straight line should be obtained if the left hand value of the following equation is plotted versus carbon number in a homologous series.

$$\ln (P^o\phi) + \ln J - \Sigma_i \nu_i \ln \Gamma_i = \ln P_t + J - 1 + \Sigma_i \nu_i (G_i^{og}/RT) \qquad (5)$$

This plot is shown to be straight in two cases tested for paraffins and alcohols shown in Figure 2 at 60°C. Group activity coefficients are derived from data on methanol-hexane, and the $\ln J$ term comes from a size effect on the entropy of vaporization derived in the same paper. Figure 2 shows that generalized group contribution methods could be applied to the prediction of pure component vapor pressures because the lines in Figure 2 are parallel. The separation between the lines represents an —OH group contribution and the slope represents a $-CH_2-$ group contribution. Similar plots for other homologous compounds could provide the necessary parameters for a generalized model.

Equations of State from Group Contribution Methods. The concept of void spaces in a fluid as an additional component provides a means for deriving equations of state from existing activity coefficient equations. Thus if one derives an activity coefficient equation based on group contributions, then an analogous equation of state can also be derived. This method was

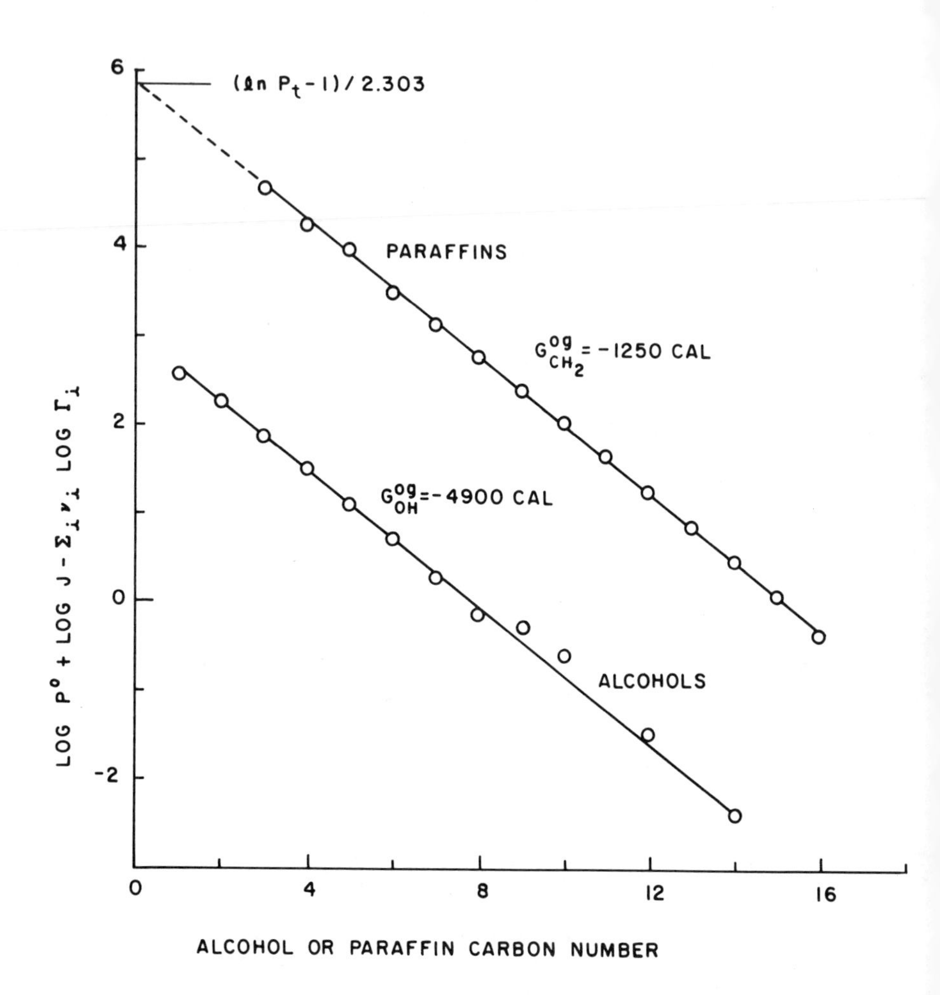

Figure 2. Alcohol and paraffin vapor pressure after correction for polymerization and nonideal group interactions. T = *60°C, p° (mm Hg),* P_t = *2600 atm. Slope of the paraffin line is —0.387. Separation between the paraffin line and the alcohol line is —2.82.*

applied by John Cunningham (15) in his thesis on the PFGC (parameters from group contributions) equation of state. Basically the method involves the assumption of a molar volume b_H for void spaces in a fluid; then the number of moles of void spaces is calculated as the total volume minus the Van der Waals volume of the molecules divided by b_H according to the following equation.

$$n_H = \frac{V - \Sigma_i n_i b_i}{b_H} \quad (6)$$

Additionally the Helmholtz free energy is assumed to have the same analytical form as the Gibbs free energy. The substitution of equation (6) into the Helmholtz free energy equation introduces a volume dependence which can then be differentiated to give an equation of state according to the following equation.

$$P = -\left(\frac{\partial A}{\partial V}\right)_{T,n} \quad (7)$$

The PFGC equation assumes the Wilson equation for group activity coefficients and a modified Flory-Huggins equation for molecular size effects. The resulting equation of state is therefore consistent with these activity coefficient equations in the liquid phase, and it converts to the ideal gas equation at infinite volume. This is a tremendous advantage over other equations of state because we can take advantage of what we have already learned about activity coefficients in equation-of-state-development studies. The PFGC equation was used to model interactions of methyl and methylene groups with the following groups.

CH_4

$-\overset{|}{C}H-$

$-\overset{|}{\underset{|}{C}}-$

$-OH$

$-CHO$

$-\overset{O}{\overset{\|}{C}}-$

$-\overset{O}{\overset{\|}{C}}-O-$

H_2S

H_2O

CO_2

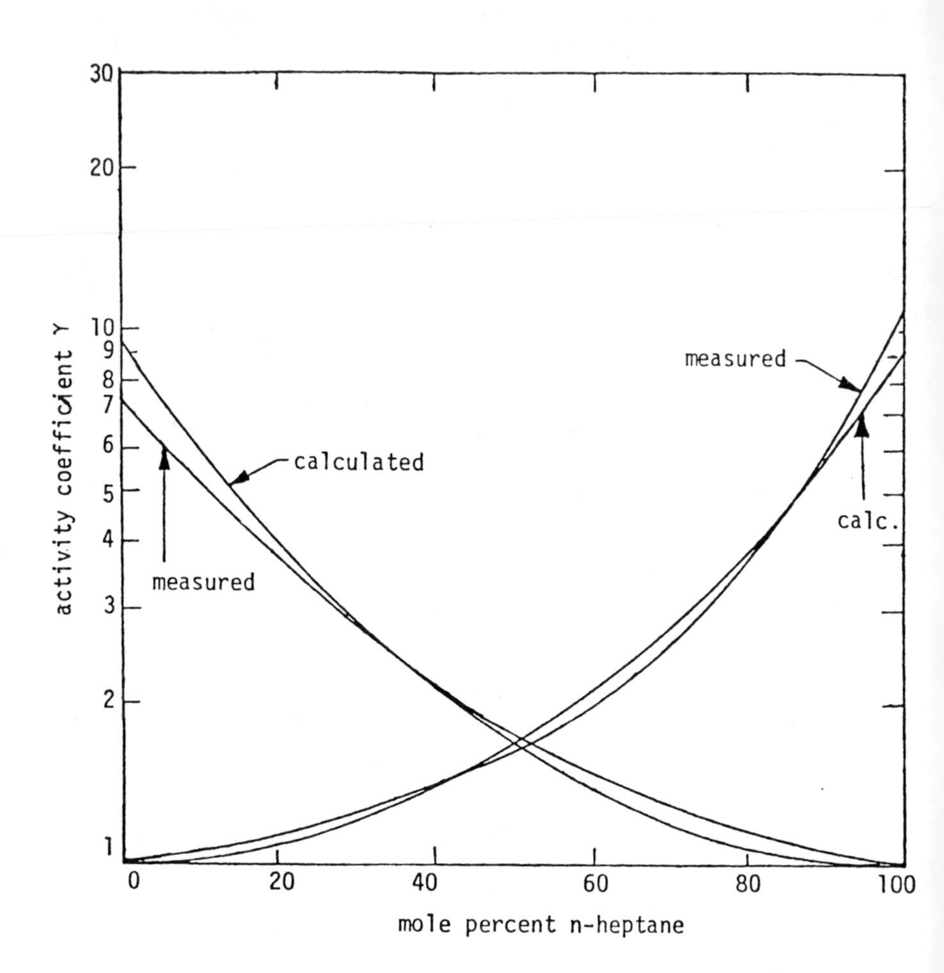

Figure 3. Comparison of measured [C. P. Smyth, E. W. Engel, J. Amer. Chem. Soc. **51**, *2660 (1929)] and calculated activity coefficients for* n-*pentane–ethanol mixtures at 158°F*

The same equation was used to simultaneously calculate vapor phase non-idealities and was successful in correlating light hydrocarbon, CO_2, and H_2S K-values; and it was also found capable of predicting infinite dilution activity coefficients of various homologous series in other compounds. We consider this work as very significant because it makes possible the prediction of phase equilibria in regions where activity coefficient methods fail such as the following:

Components above their critical point temperature.

Mixtures at high pressures where retrograde condensation phenomena occur.

The equation appears suitable for both polar and non-polar compounds, and represents the assymetric nature of activity coefficients. These two capabilities are problem areas with other equations of state. Figures 3 and 4 are examples of the representation of activity coefficients by the PFGC equation of state for n-heptane-ethanol at 158^oF and n-pentane-acetone at 212^oF. Both figures show good representation of the activity coefficient curves.

We believe that equation of state models such as this will provide better activity coefficient prediction methods based on groups than existing group contribution methods because the equation of state is capable of taking all physical factors into account such as the effect of void space in a liquid on group interactions. However, more work needs to be done before this will occur.

Problem Areas

Areas regarding the prediction of activity coefficients from group contributions which at present appear unresolved include the following:

(a) What is the effect of steric hindrance effects in isomers on group interactions? Shielding effects change the liquid environment and the magnitude of $\ell n\ \Gamma_i$ for a given group. Is the effect predictable?

(b) Can account be satisfactorily made for induction effects due to other substituent groups in a molecule?

(c) What is the effect of positive or negative volumes of mixing on group contributions? Is this effect accounted for?

(d) What is the effect of cohesive pressure in a liquid on component activity coefficients? Are changes in cohesive pressure accounted for in the group contribution method?

(e) What is the effect of void spaces on group interactions? Do they vary as a function of the amount of void space between molecules?

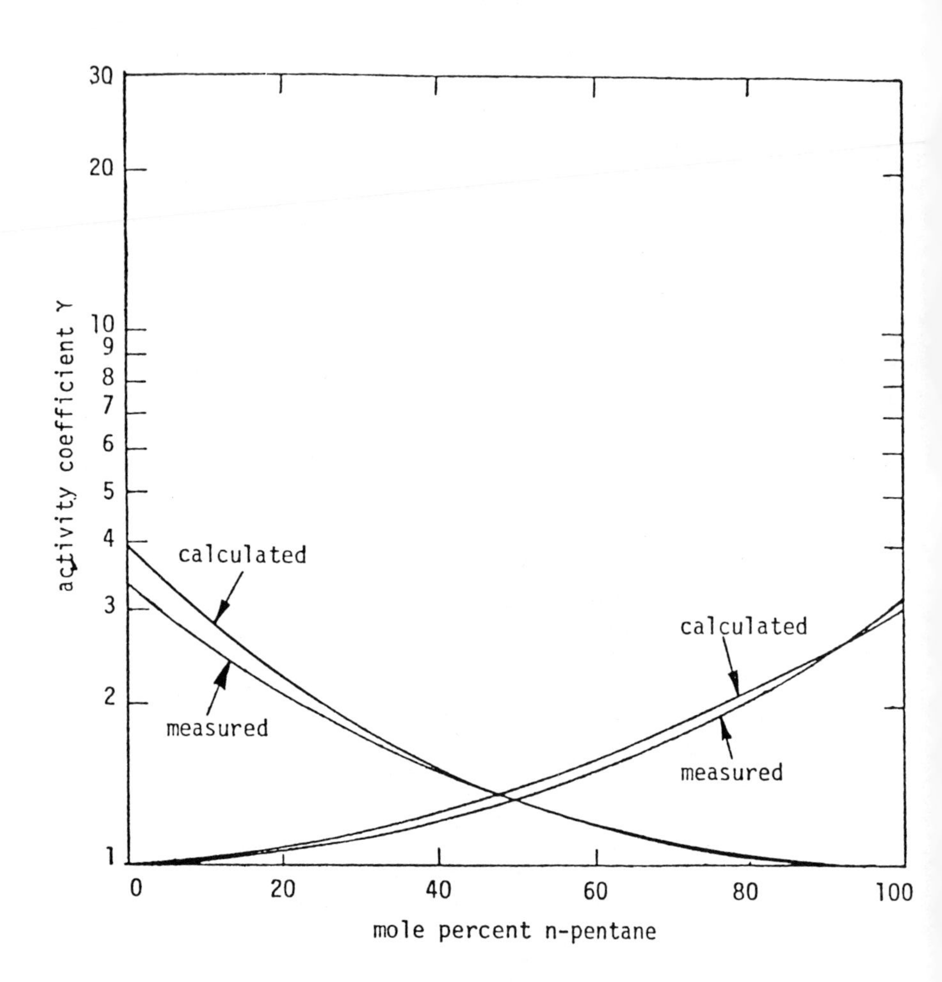

Figure 4. Comparison of measured (new lab data) and calculated activity coefficients for n-*pentane–acetone mixtures, at 212°F*

Answers to some of these questions are partly answered in our work and other work in the literature, but more work needs to be done to resolve these problems. Of particular importance is the prediction of activity coefficients of branched compounds versus straight chain compounds. This effect appears to be accounted for in the UNIFAC model, but more work needs to be done to demonstrate that this is the case.

Abstract

This paper summarizes various research areas studied at the Thermochemical Institute regarding the prediction of activity coefficients from group contributions as follows.

(a) The effect of void spaces between molecules on the excess free energy of mixing.
(b) The estimation of the infinite dilution activity coefficient of a component in a binary mixture knowing the infinite dilution activity coefficient of the other component.
(c) The measurement of binary vapor-liquid equilibrium data for modeling purposes in the development of generalized group interaction models.
(d) The estimation of pure compound properties from a knowledge of group activity coefficients.
(e) The development of equations of state from group contribution methods assuming void spaces as an additional component in a fluid.

Also problem areas for further research are identified as follows.

(a) What is the effect of steric hindrance in isomers on group interactions?
(b) Can induction effects due to other substituent groups be predicted?
(c) Are volume of mixing effects accounted for?
(d) Are changes in cohesive pressure in a liquid accounted for?
(e) What is the effect of void spaces between molecules on group interactions?

Bibliography

1. Flory, P. J. J. Chem. Phys. (1941) 9, 660; 10, 51 (1942).
2. Huggins, M. L. J. Phys. Chem. (1941) 9, 440; Ann. N. Y. Acad. Sci. 43, 1 (1942).
3. Wilson, G. M. and Deal, C. H. Ind. Eng. Chem. Fund., (1962), 1, 20.
4. Derr, E. L. and Deal, C. H. I. Chem. E. Symposium Series No. 32, p. 3-40 (1969: Instn. Chem. Engrs., London).
5. Ronc, M. and Ratcliff, G. A. Can. J. Chem. Eng., (1971), 49, 825.
6. Scheller, W. A. I. & E.C. Fund. (1965), 4, 459.

7. Fredenslund, A., Jones, R. L., and Prausnitz, J. M. AIChE J. (1975) 21, 1086.
8. Wilson, G. M. Adv. in Cry. Eng. (1973) 18, 264.
9. Bondi, A. J. Phys. Chem. (1964) 68, 441.
10. Bruin, S. and Prausnitz, J. M. Ind. Eng. Chem. Process Design Develop. (1971) 10, 562.
11. Tassios, D. AIChE J. (1971) 17, 1367.
12. Wilson, G. M. AIChE Symposium Series No. 140, Vol. 70, 120 (1974).
13. Eng, W. W. Y., Masters Thesis, Dept. of Chemical Engineering, Brigham Young University (1976).
14. Wilson, G. M. Paper no. 34d of 63rd Annual Meeting of AIChE, Chicago, Ill., Nov. 29-Dec. 3, 1970.
15. Cunningham, J. R. Masters Thesis, Chemical Engineering Dept. Brigham Young University, 1974.

23

Application of the Group Contribution Approach to Equilibrium Problems in Simulation of a Complex Chemical Process

DAVID A. PALMER

Amoco Chemicals Corp., Naperville, IL 60540

I have been asked to be part of this panel for the purpose of giving you a description of the use of group contribution methods in Amoco Chemicals. I would like to do this in the framework of a specific process design problem which needed to be solved.

We were given the objective of developing a simulation program capable of giving the material and energy balances for a liquid phase phthalic anhydride plant. The unit operations include: a simultaneous reaction and phase equilibrium, recovery operations, three phase flashes, and purification towers. The process simulator is called CHIPS which means Chemical Information Processing System.

The obstacles encountered in development of this simulation were the following:

a. We had at least 14 components including gases, water, ortho-xylene, phthalic acid, phthalic anhydride, and reaction intermediates.
b. There was a chemical equilibrium between phthalic anhydride, phthalic acid, and water.
c. We had almost no mixture data.
d. There were great variations in temperature, pressure, and composition.

The equilibrium constants for the gases were determined by using Henry's constants and the unsymmetric activity coefficient convention as per the following equation:

$$K_i = \frac{H_{i,j}\,\gamma_{i,r}}{\phi_i P} \exp \frac{v_{i,r}(P^o - P_k^*) + v_i(P - P^o)}{RT}$$

Because the solubility of the gases was very low, the activity coefficient for the gas in the reference solvent ($\gamma_{i,r}$) was set equal to unity.

Henry's constants for gas solubility in water were obtained from the literature. The Yen-McKetta correlation was used to predict Henry's constants for gases in the organic solvents. Partial molar volumes for the pressure correction factor were estimated from literature values. Over the range of pressures encountered, the pressure correction affects the equilibrium constant by only two percent.

The equation for subcritical component equilibrium is:

$$K_i = \frac{\gamma_i \phi_{i,P_i^*} P_i^* \exp \frac{V_i^L (P - P_i^*)}{RT}}{\phi_{i,P} P}$$

The information required for this equation includes:

- Vapor pressure constants
- The critical properties for the Redlich-Kwong equation of state which was used for calculating the gas phase fugacity coefficients.
- The activity coefficients for calculating Van Laar parameters within the framework of the Chien-Null multicomponent activity coefficient equation.

The most difficult problem in this development was the estimation of activity coefficient parameters. Inasmuch as we knew practically nothing about each of the minor mixtures, we used the group contribution approach. At the time, the only approach which was sufficiently developed was the ASOG method developed by the session chairman, Carl Deal, and E. L. Derr. Activity coefficients of group interactions are correlated with the Wilson equation, and a Flory-Huggins expression is used to account for size effects. We calculated the Wilson parameters from experimental data on molecules having groups characteristic of those molecules which we were specifically interested in.

The data sources which we had for development of these group contributions were as follows:

Cyclohexane + acetic acid
Acetic acid + water
Toluene + acetic acid
Cyclohexane + acetic anhydride
Acetic acid + benzene
Acetic acid + acetic anhydride

Using the group contributions from these six systems, the vapor-liquid equilibrium for 26 new systems were predicted, thus vividly illustrating the way in which a small amount of experimental data can be vastly expanded using the group contribution method.

Now I would like to illustrate the accuracy of some of the predictions which we made. In Figure 1 we show the activity coefficient predictions for the propionic acid/water system and the formic acid/water system. Neither of these systems was used in the data base and yet the predictions are quite accurate. It will be

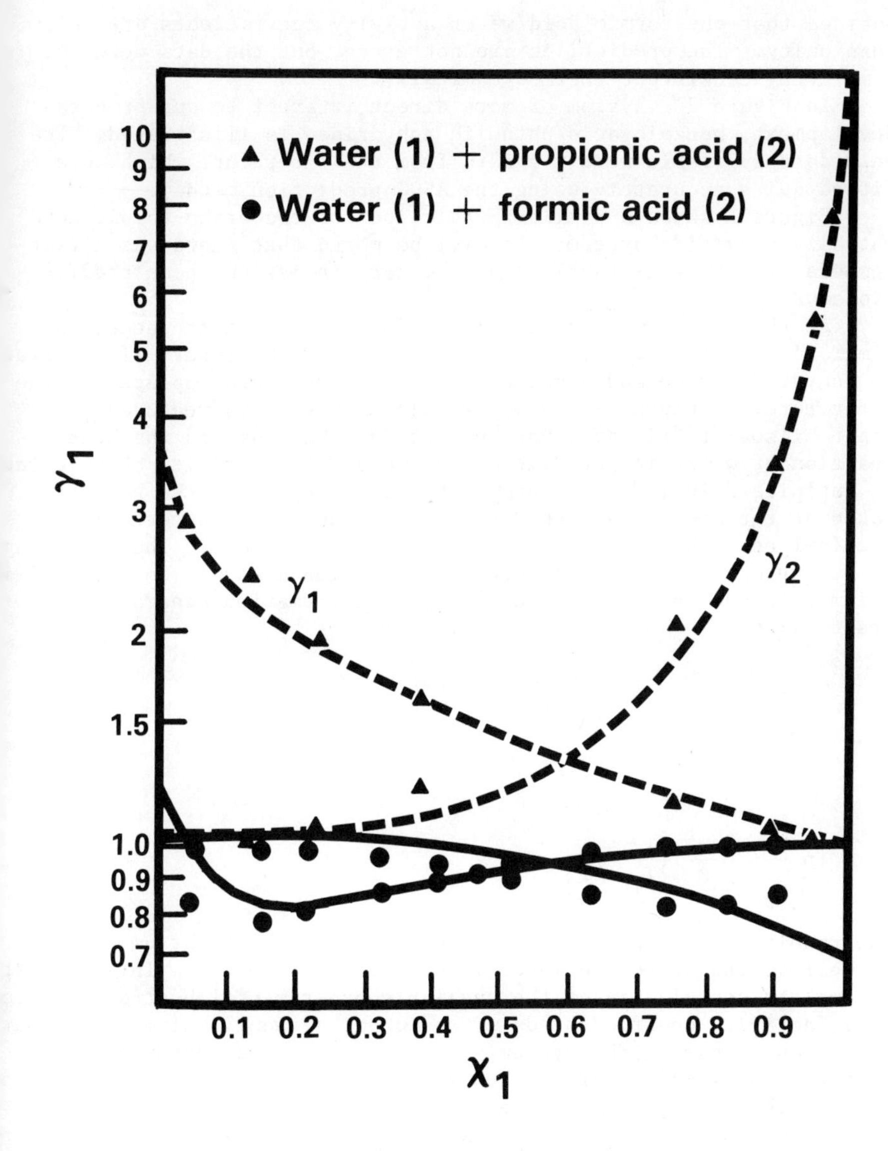

Figure 1. ASOG activity coefficients for propionic acid (2) + water (1) and formic acid (2) + water (1)

noticed that the formic acid/water activity coefficients are less than unity. The predictions are not exact, but the data were not completely consistent thermodynamically.

In Figure 2 a system of more direct interest to our process development, benzoic acid/phthalic anhydride, is illustrated. The only data available were x-y data from the literature which were fitted quite accurately using the ASOG prediction method.

Figure 3 shows a similar prediction for the ortho-toluic acid/phthalic anhydride system. It will be noted that there was a fair temperature range extrapolation necessary in making these predictions.

At this point we were still missing a key element needed in simulation of the overall system and that was the interaction betwee an anhydride group and a water group. No such data appeared in the literature. Fortunately, we were able to have data measured by Grant Wilson of Brigham Young University. He measured the heat of reaction of water in phthalic anhydride and the activity coefficient at infinite dilution of water in phthalic anhydride. We had data taken at the Amoco Research Center on the anhydride/water/acid chemical equilibrium, using Laser-Raman spectroscopy. The following two equations were used in doing a regression analysis on the equili brium data and the activity coefficient information generated by Grant Wilson.

$$\tilde{K} = \frac{x_{PAN}\, x_{W}}{x_{OA}} \cdot \frac{\gamma_{PAN}\, \gamma_{W}}{\gamma_{OA}}$$

$$\ell n \frac{\tilde{K}_1}{\tilde{K}_2} = \frac{\Delta H^r}{R} \left(\frac{1}{T_1} - \frac{1}{T_2}\right)$$

The result of the analysis was a base chemical equilibrium constant as well as the anhydride/water interaction parameters. This permitt a complete correlation of the available experimental heat of reactio data, activity coefficient data, and chemical equilibrium data, that we had available on the system. It also permitted checking of chemical equilibrium information at other temperatures which were no part of the original data base.

We now had all the information necessary to calculate the equilibrium in the reactor. The calculation mode is illustrated in the following sketch.

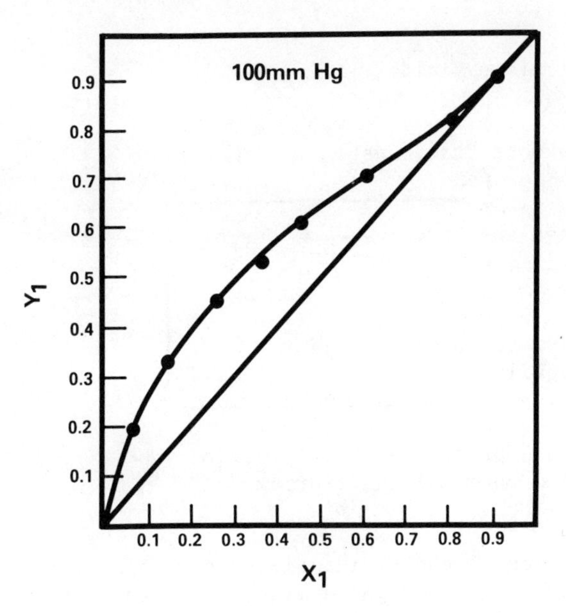

Figure 2. ASOG prediction of vapor–liquid equilibrium for benzoic acid (1) + phthalic anhydride (2) at 100 mm Hg

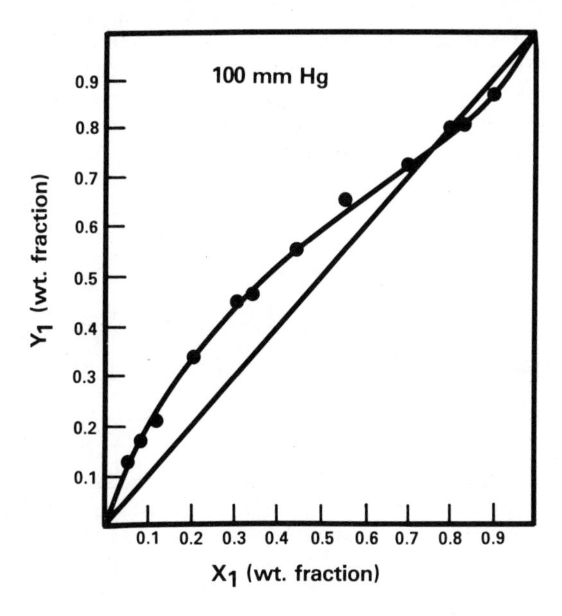

Figure 3. ASOG prediction of vapor–liquid equilibrium for o-toluic acid (1) and phthalic anhydride (2) at 100 mm Hg

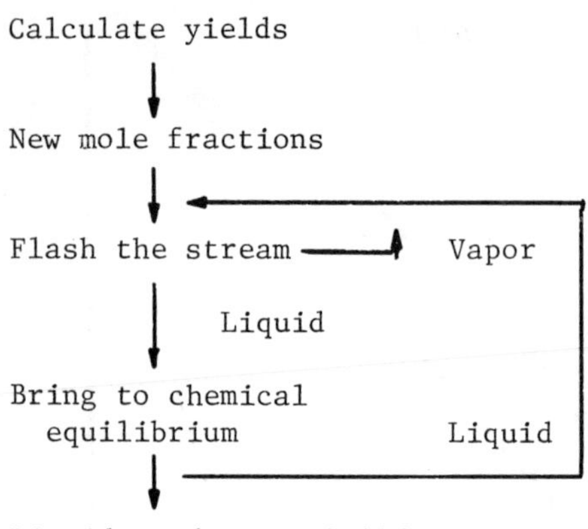

After calculation of the yields giving new mole fractions, the strea is flashed giving a vapor and a liquid, and the latter is brought to chemical equilibrium. Then the liquid and vapor are mixed, computationally, and flashed once more. This procedure is repeated a numbe of times until there is convergence, giving a liquid product at the bubble point in chemical equilibrium, and a vapor stream in equilibrium with it. After all of these calculations, we were very concerned whether we would, in fact, be able to simulate the reactor operation accurately. Therefore, experimental data were taken in th pilot plant. The following table gives the comparison of experiment and predicted K-values for water and ortho-xylene in a mixture of phthalic anhydride and reaction intermediates. The agreement is excellent.

Comparison of Predicted With Experimental Reactor K-Values

	Predicted	Experimental
Water	1.66	1.67, 1.80
O-xylene	0.32	0.32, 0.36

We have also used the UNIFAC method of predicting group interactions in various mixtures and have been able to correlate the tie lines in partially miscible ternary systems. It is necessary, however, to fit the interaction parameters which most strongly affect the partially miscible binary. We are very pleased with the development of a large number of group contribution parameters for the UNIFAC method. It is finding considerable use in Amoco Chemicals.

Conclusions

1. We find that existing predictive methods can be used for design of chemical plants with complex phase equilibrium problems, which operate at low to intermediate pressure in the absence of hydrogen. This assumes that the necessary group interaction parameters can be obtained.

2. Group contributions are the only practical means of estimating equilibrium when there are a large number of polar components.

3. The ASOG and UNIFAC methods can be used for correlation of systems in which some, but not all, of the necessary data are available.

4. We believe that UNIFAC has an edge over ASOG because the parameters have less temperature dependence. For those of us outside of Shell, more parameters are available.

Recommendations

1. We favor an experimental program which would have industrial and/or government funding to study unmeasured group interactions.

2. We are very much interested in expansion of the UNIFAC parameter data base.

3. We believe that the plan should be publicized so that university labs will contribute needed data to fill in gaps in the interaction parameter grid.

4. It would be helpful if there were data center involvement for data screening and dissemination of the data base to others developing UNIFAC parameters and developing other correlations.

5. Future group contribution theories should be based on the same data and tested on the same systems as UNIFAC, thus permitting a rapid comparison of predictive accuracy and capabilities.

Nomenclature

ΔH^r = heat of reaction
K_i = vapor=liquid equilibrium constant (y/x)
$\tilde{K}$ = chemical equilibrium constant (mole fraction)
P^* = vapor pressure
P = system pressure

P^o = reference pressure
T = temperature, ^oKelvin
R = gas constant
V_i^L = liquid molar volume
x = liquid mole fraction
y = vapor mole fraction
γ_i = activity coefficient for symmetric or unsymmetric convention referred to system pressure
$\phi_{i,p}$ = gas phase fugacity coefficient at system conditions
$\phi_{i,p}^*$ = gas phase fugacity coefficient at vapor pressure of component

24

Recent Developments in the UNIFAC Method for Calculating Activity Coefficients from Group Contributions

J. M. PRAUSNITZ

University of California, Berkeley, CA 94720

The form of the semi-theoretical UNIQUAC equation suggested by Abrams (10) is particularly useful for establishing a group-contribution method to estimate activity coefficients in liquid solutions of nonelectrolytes. This method was established by Fredenslund and Jones in 1975 (2).

As suggested earlier by Deal, Derr and Wilson in their ASOG method (3), the activity coefficient of a component in a liquid mixture consists of two parts: a configurational part and a residual part. The first of these requires only pure-component data reflecting the sizes and surface areas of the molecules as calculated from van der Waals radii. The second requires group-group interaction parameters; these are obtained from reduction of binary phase-equilibrium data.

The UNIFAC correlation presented in 1975 has now been significantly enlarged thanks to extensive data reduction by Fredenslund, Michelsen and Rasmussen at the Technical University of Denmark and Gmehling and Onken at the University of Dortmund, Germany. Each group-group interaction is characterized by two parameters and a new table of these parameters (comprising 25 functional groups) has been prepared. Because experimental binary data are often unreliable or unavailable, this table is not complete. Of a total possible 600 parameters, the present table contains 344.

In binary data reduction, extensive use was made of the literature compilation prepared by Hála and coworkers (4). All data were first checked for thermodynamic consistency; only consistent data were used for determining interaction parameters. Vapor-phase corrections through fugacity coefficients were included in data reduction. The fugacity coefficients were calculated using the correlation of Hayden and O'Connell (5).

Since the number of different functional groups is much smaller than the number of different molecules, a group-contribution method provides a highly efficient technique for estimating activity coefficients for a very large number of binary and multicomponent mixtures, including many for which no experimental data are at hand. The accuracy of prediction is not high but, especially for mixtures containing polar components, it is better than that of other methods (using only molecular instead of group parameters). The accuracy achieved by UNIFAC is sufficient for many practical engineering calculations, especially those concerned with preliminary design.

The new UNIFAC table has been submitted for publication. Preprints and computer programs are available from the author or from Dr. Aa. Fredenslund (Instituttet for Kemiteknik, Tekniske Hojskole, Bygning 229, DK 2800 Lyngby, Denmark).

Further progress in UNIFAC is strongly dependent on new experimental data for binary mixtures, particularly those which have received no previous study.

To obtain UNIFAC parameters, it is possible to use activity coefficients at infinite dilution, provided that these are measured accurately. Thanks to modern instrumentation, it is now possible to make such measurements with relative ease and speed. As shown by Charles Eckert and coworkers (Univ. of Illinois, Urbana), activity coefficients at infinite dilution can be obtained efficiently and accurately, using differential ebulliometry for systems where the relative volatility is in the range (approximately) 0.1 to 10. Gas-liquid chromatography should be used if the relative volatility falls outside that range.

To obtain additional UNIFAC parameters, new measurements are needed especially for mixtures containing aldehydes, carboxylic acids, glycols and other bi (or tri) functional hydrocarbon derivatives. Some experimental efforts toward that end are now in progress at several laboratories. It is clear, however, that appreciably more experimental work will be needed to extend UNIFAC's range of application.

At present, T. Oishi (Berkeley) is working on application of UNIFAC to polymer solutions. It may also be possible to use UNIFAC for correlating gas solubilities but considerable modification will be required to achieve that goal.

Literature Cited

1. Abrams, D. S. and Prausnitz, J. M., AIChE Journal (1975) 21, 116.
2. Fredenslund, Aa., Jones, R. L. and Prausnitz, J. M., AIChE Journal (1975) 21, 1086.
3. Derr, E. L. and Deal, C. H., Intl. Chem. Eng. Symp. Ser. No. 32 (Instn. Chem. Engrs., London) 3:40 (1969).
4. Wichterle, I., Linek, J. and Hála, "Vapor Liquid Equilibrium Data Bibliography," Elsevier (1973); Supplement (1976).
5. Hayden, J. G. and O'Connell, J. P., Ind. Eng. Chem. Process Des. Dev. (1975) 14, 209.

25

The Industrial Data Bank: Utopia vs. the Real World

EVAN BUCK

Chemicals and Plastics Division, Union Carbide Corp., South Charleston, WV 25303

It is a truism that the techniques used in today's engineering calculations require accurate basic thermodynamic and physical property data. Computer developments over the last 20 years have fostered the creation of quite sophisticated design methods; however, it is also clear that the results of even the most correct mathematical model are no better than the basic data that go into it.

Having established the need for reliable property information, one must decide (among other things) in what form this mass of data should be contained. In other words, once the data have been collected, what is the optimum way to make this information available to those who wish to use it?

We have found that one of the best ways of accomplishing this task is via the computerized data bank. There are a number of advantages to this approach. First of all, updating is very easy since all of the numbers are in one place. This also guarantees that everyone uses the same data. The computer programs that utilize this information can directly access the data bank, thus eliminating the time and effort to transcribe the data to computer input sheets, as well as copying errors. And last but not least, the computerized data bank can be made quite secure with the judicious use of backup tapes.

Let us now consider the industrial data bank: what is in it, and what are its attributes. Suppose we have the admittedly not very likely situation of a good sized chemical company whose management is rather loose with its purse strings, and so authorizes the creation of the "world's best" data bank. What would it look like?

As a start, our perfect data bank should have data for all of the common chemical compounds used throughout the industry. In addition, the compounds peculiar to this particular company should also be included: raw materials, products, by-products, intermediates, catalysts, etc. The total would probably number several thousand chemicals.

The properties in this data bank would consist of all the quantities used by all of the various elements of our mythical industrial firm, from the basic research chemist, through the design engineer, and even including the marketing personnel who publish property values in sales brochures. This information would be divided into two broad categories, pure component and mixture. Within each category there would be several different types of data:

(1) Quantities which are independent of temperature, pressure, and composition are contained in the data bank as single values; for example, molecular weight, critical constants, and binary interaction parameters for an equation of state.

(2) Properties whose values depend on temperature, pressure, and composition, such as vapor pressure and activity coefficients, are represented by the constant coefficients of a correlation for that property. In many cases it would be desirable to include the parameters for more than one correlation for a given property, particularly in the mixture category.

(3) In some instances, it may be decided to include single values of type (2) properties at some specified conditions; for example, the heat of vaporization at the normal boiling point. In these cases, the reference conditions must also be in the data bank.

(4) The correlations of the type (2) properties are often applicable only within certain temperature, pressure, and composition ranges. Thus, the upper and lower limits of applicability must also be included.

(5) It is extremely important that every value in the data bank be referenced so that users can determine the source of information. Therefore, it is necessary to include a data source code which would refer to a thorough bibliography.

(6) Some sort of data quality index should be included for each property which would indicate how much reliability may be placed on the data bank value. A $\pm$ error might also be incorporated.

Our super data bank should also have the following attributes:

(a) Each compound must be uniquely identified within the data bank. For retrieval purposes, a computerized synonym file would have to be established to link the data bank identifier with the various possible compound names.

(b) All properties should have a precise, universal definition. "Modified" properties, such as the acentric factor used in the Chao-Seader fugacity correlation, should not be put into the data bank.

(c) The property values in the data bank should be the "best" available. Thus, all included quantities are subject to updating at any time.

(d) All data bank properties which are not dimensionless should be in a consistent set of units. In addition, all values for a given compound must be consistent with each other. For example, the four critical constants would be related by $P_cV_c = Z_cRT_c$.

(e) Finally, the data bank should be completely filled. Quantities which cannot be derived from reliable literature sources should be determined from experimental measurements.

We could specify other equally desirable attributes, but these are probably enough to establish the concept.

Now, of course, such an idealized industrial data bank could never exist. Any company whose management would allow even the attempt to produce this utopian data bank would certainly not be in business very long. However, by making a few compromises, it is possible to develop a practical and useful computerized data bank. As a case in point, there is the one which has been established by the Union Carbide Chemicals and Plastics Division.

This file contains data for about 1000 compounds embodied in approximately 130 properties, which include single values, appropriate reference conditions,correlation parameters and limits, and data source codes. However, these are all for the pure component, and there is only one representation for each property.

As far as the attributes are concerned, each compound is uniquely identified, and there also exists a synonym file so that users do not have to use the unique data bank identifier to obtain information about that compound. Incidentally, the basis for this synonym system is the Chemical Abstracts registry number.

The Carbide data bank does have a precise definition for each included property and is being continually updated as new data become available. The dimensional units employed throughout are basic International System metric (SI), and property consistency for each compound is required. For example, the normal boiling point is in fact predicted by the vapor pressure equation, as represented by its constant parameters, and that vapor pressure equation curve does end at the critical point contained in the data bank.

This data bank does have a number of blank spaces; as a matter of fact, it is only about half full. These holes are filled only on an as needed basis, as are compound and property additions. The data bank was created about seven years ago with an initial loading of about 800 compounds and 75 properties; as mentioned earlier, it now has over 1000 compounds and 130 properties.

It also should be pointed out that there really is not just one Carbide data bank; there are actually several. The basic ground rules are that each data bank contains only properties which are generally attainable for the compounds in that file, and that each property is contained in only one data bank. The data bank just described contains process design engineering properties; there is also an environmental property data bank containing such quantities as biological oxygen demand and median tolerance limits for aquatic life. A mechanical design property data bank is currently being created. Of course, some of the data files are quite small. For example, there is a 26 compound data bank for B-W-R constants.

In conclusion, it should be stated that this computerized data bank system has proved to be quite successful at Carbide. The two major selling points have been that everyone can conveniently use the same value for a given property, and that these quantities have been selected by those most qualified to do the choosing. Although perhaps somewhat modest in scope when compared to the grandiose scheme first proposed in this paper, these data banks have been attainable at a reasonable cost, and more importantly, they are widely accepted and used throughout the corporation.

26

Centralized Service for Thermophysical Data in Germany—DSD: Dechema Data Service

HELMUT KNAPP

Institut für Thermodynamik, Technische Universitat, Berlin, West Germany

This paper comprises a review:

of the organization (history, management, financing, by-laws)
of the present capacity (structure, contents, transmittal)
of the future development (improvement, expansion)
of DSD
the thermodynamic problems
the methods of measuring, calculating and predicting thermodynamic data
the data banks owned by private industry are not discussed.

Two essentially different kinds of problems arise when setting up a centralized data bank:

1. Organizational: who takes the initiative, who contributes at what price, who does the daily work, who is entitled to use it, who determines the cost of the data service, who is responsible for financial management, who controls, checks, develops the information.

2. Scientific:

   a. cybernetic: what is the best system for searching, finding, storing and retrieving data, how is information most practically received, transmitted, displayed.

   b. thermodynamic: what is the best, i.e., the most accurate, most universal and fastest method to calculate thermodynamic data.

Although very different specific capabilities are required for the fulfillment of the organizational or scientific task, both jobs seem to be equally difficult and great patience, dedication and diplomacy as well as scientific knowledge and experience are required to be successful. It is obvious that the problems involved are so numerous and so complex that no single company, university or institution can hope to solve them alone.

Organization: History and Management

Based on the experience of the Dechema institution in serving the chemical industry and based on a functioning program package developed by the Uhde engineering Company at Dortmund, a joint project was started several years ago in which interested chemical production and engineering companies, such as BASF, Bayer, Höchst, Lurgi, Uhde as well as a number of institutes of German Universities are cooperating under the management of Dechema.

Financing. Beginning 1977 subsidies are expected from the Federal government's ministry of Research and Technology in order to finance the work done at the Universities.

By-laws. A data development group (EG-SDC) has been established and by-laws have been issued and signed defining:

1. the activities of the group, such as collecting, critically testing and storing basic data, developing and improving methods of computation and integrating programs into the compiler.

2. the duties of the members such as supplying substantial contributions to Dechema who will coordinate, administrate, distribute information to the EG-SDC members and offer data-services to the public.

3. rights of the members, such as being allowed to use all results of the group without loosing the title to original contributions, however not being allowed to pass on information to third parties.

Present Capacity: Structure

The data bank offers two services:

1. The retrieval system (SDR) stores and delivers published information.

2. The Data Calculation System, the so-called Data Compiler (SDC), which was purchased from Uhde Company, Dortmund, Germany computes thermo-physical properties of pure components or mixtures at a desired design point (temperature, pressure and composition) or within a desired design range.

The SDR System must be capable of recovering specified information from its storage, answering questions such as, where to find certain properties of a certain component or which publication describes certain methods of calculating certain properties.

The documentation of literature is accomplished by recording briefs of the contents of publications using key words alphabetically listed in a thesaurus. Typical indexing words: activity coefficient, BWR equation, calculation, Cooper-Goldfrank method, diagram, mixture, multicomponent system, phase-equilibrium etc.

About 70 periodicals and journals are regularly surveyed and extracted. Publications on the following subjects are being registered: thermodynamic properties such as pvT-data, specific heats, enthalpies, entropies, free energies, excess properties, phase equilibria, solubilities, interaction parameters, transport properties such as viscosity, thermal conductivity, diffusivity, electrolytic conductance, ionic mobilities, surface and interfacial tension, molar refraction, parachor, dielectricity, dipole moments etc.

Even if the SDR would be complete and up to date it is highly improbable that literature data will be available for any desired data point. In most cases it is therefore necessary to calculate a "new" set of data by suitable methods of inter- or extrapolation, by using generalized correlations, by using incremental techniques etc.

The data bank therefore contains the SDC which can answer individual questions by calculating thermophysical properties for any specified state based on a limited number of basic data using proven correlations.

The SDC program package contains:

1. Evaluated basic data for the gaseous and liquid states of more than 350 pure substances in an array with 400 labeled registers assigned to each substance.

2. Programs (33 methods) for approximating missing basic data, e.g., by an incremental technique.

3. Programs (18 methods) for calculating properties of pure compounds.

4. Programs (50 methods) for calculating properties of mixtures with as many as 50 components e.g. density, specific heat capacity, compressibility, enthalpy, heat of combustion, heat of condensation, heat conductivity, viscosity, surface tension, pseudocritical properties.

   In addition the following data can be delivered for each individual component at the specified state: critical properties molar mass, heat of vaporization, vapor pressure, normal boiling point, ideal and real K-values, activity coefficients, specific heat, enthalpy heat conductivity, viscosity, heat of combustion, diffusivity.

5. A flash subprogram (using 3 different methods: ideal, Chao-Seader, van Laar activity coefficients) provides information about the state of the mixture (liquid, vapor or two phase).

The data compiler can be used in combination with computer programs for the process design of rectification columns, heat exchangers, compressors etc. Here direct access is necessary and for users who constantly use the data compiler it is practical and economical to purchase the entire package.

The data compiler can also be used to answer individual questions in daily routine work, either by calculating conditions for one single point or by plotting diagrams for a entire range of temperature, pressure or composition. Communication with the DSD is possible by telex, telephone, mail or--for external users with in-house terminals--by direct access.

In the original system used by Uhde the dialogue between user and data compiler is controlled by the message control program in TCAM (Tele Communication Access Method). The monitor (in ASSEMBLER) connects the terminal to the programs when processing begins, communication with the data program (in FORTRAN) passes through a TCAM-FORTRAN interface (in ASSEMBLER).

At Dechema the dialogue between user and data compiler is accomplished by a FORTRAN IV - MASK program. The inquiry is entered at the display terminal. The answer is temporarily stored on a disk file and can be displayed on the terminal screen or printed out. This system is shown in block diagram in Figure 1.

Future Development

The information offered by SDR and SDC is neither complete with respect to quantity nor perfect with respect to quality. The information stored in SDR is continuously implemented and therefore growing.

The data stored in SDC are exchanged against better and critically tested data, the programs in SDC are replaced or implemented by more accurate, more flexible, more general programs. The development group EG-SDC is responsible for the improvement and extension of SDC.

The information contained in SDR and SDC is specifically selected for the need of the process engineer and therefore limited. SDC is intended to assist the engineer in his daily routine work.

Contribution of the Institute of Thermodynamics, TUB

The institute of Thermodynamics at the Technical University of Berlin participates as a member of the development group. It is engaged in various research projects:

Thermodynamic properties of mixtures are measured (vapor-liquid and solid-liquid equilibria, gas solubility, Joule Thomson coefficient, isobaric heat capacity and isothermal throttling effect) and correlated.

Transport properties (diffusivity and viscosity) are measured in gaseous and liquid mixtures.

Separation processes (permeation, absorption and desorption) are investigated.

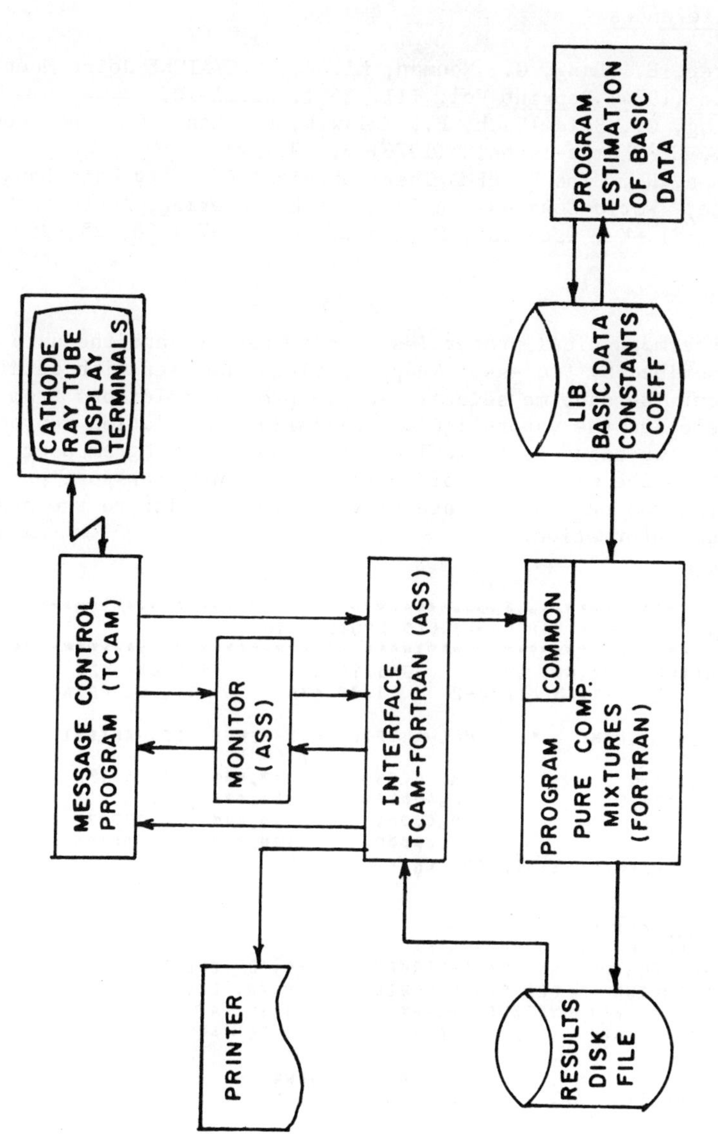

Figure 1. Block diagram of the Dechema Data Service System

The institute supplies to SDC additional calculation methods and programs especially for the application in low temperature engineering.

General References

1. Futterer, E., Lang, G., Neuman, Kl.OK., GVC/AIChE Joint Meeting, München (1974) Reprint Vol. III, E6-1, pp. 1-10.
2. Ostertag, G., Seidelbach, F., Kulawik, P., Lang, G., and Neuman, Kl.-K., CZ-Chemie-Technik (1973) 5, 191-197.
3. Eckerman, R., "The DECHEMA Thermophysical Property Data Bank and Service," World Congress on Chemical Engineering, Amsterdam (1976
4. Neuman, Kl.-K., Ostertag, G., Chem. Ind. (1976) 28, 253-255.

Editor's Note

The thermophysical properties program package and the data bank system described by Professor Knapp provides the user with detailed information. He sent me several example problem solutions from which I selected the vapor-liquid equilibrium calculation reproduced on the next two pages as typical of the results a program user would obtain. Note the wide range of thermodynamic and transport propertie data for the two phases that are provided in addition to the phase equilibrium information.

```
*****************************************************************************
    EXAMPLE PROBLEM: VAPOUR-LIQUID EQUILIBRIUM
*****************************************************************************
    TEMPERATUR:   70.00  CEL      343.15 K         158.00  DEG F
    DRUCK:         1.03  KP/CM2     1.00 ATM        14.69  PSIA

    KODE        NAME           MOL-ANTEIL  MOL-MENGE  GEW-ANTEIL   GEW-MENG
                                             KMOL                      KG
      91  ACETON                0.200000    20.000    0.367132    1161.56
      63  METHANOL              0.400000    40.000    0.405098    1281.68
      10  WASSER                0.400000    40.000    0.227771     720.64
    SUMME                       1.000000   100.000    1.000000    3163.88
    MITTLERE MOLMASSE: 31.639  KG/KMOL


    ZWEIPHASENGEBIET:
    (REALE K-WERTE NACH VAN-LAAR)
    SIEDETEMPERATUR  BEI SYSTEMDRUCK          66.72  CEL
    TAUTEMPERATUR    BEI SYSTEMDRUCK          79.79  CEL
    SIEDEDRUCK  BEI SYSTEMTEMPERATUR           1.16  AT
    TAUDRUCK    BEI SYSTEMTEMPERATUR           0.76  AT

    DAMPF            37.53  MOL%   45.32  GEW%
    FLUESSIGKEIT     62.47  MOL%   54.68  GEW%
```

Figure 2

```
**********************************************************************
   STOFFDATEN DER FLUESSIGPHASE
**********************************************************************

   KODE        NAME           MOL-ANTEIL  MOL-MENGE  GEW-ANTEIL  GEW-MENGE
                                             KMOL                    KG
     91  ACETON               0.113603      7.097    0.238237     412.16
     63  METHANOL             0.365552     22.836    0.422938     731.70
     10  WASSER               0.520846     32.537    0.338825     586.18
   SUMME                      1.000000     62.469    1.000000    1730.03
   MITTLERE MOLMASSE         27.69    G/MOL

   GEMISCHDATEN DER FLUESSIGPHASE
   ==============================


   DICHTE
    NACH BWR                 39.9998    MOL/L          1.10777  KG/L
    VOLUMENMITTEL            28.9131    MOL/L          0.80073  KG/L
    NACH YEN-WOOD            30.6048    MOL/L          0.84758  KG/L

   SPEZ.WAERME
    AUS FLUESS.KOEFF.        21.0638    CAL/(MOL.K)    0.7606   CAL/(G.K)

   ENTHALPIE
    AUS FLUESS.KOEFF.     1416.87       CAL/MOL        50.94     CAL/G
      NACH YENALEX      RED.DRUCK KLEINER ALS 0.01


   HEIZWERT
    GEW.ANT.MAESSIG    108824.75        CAL/MOL      3929.49     CAL/G

   VERDAMPFUNGSWAERME
    MOL.ANT.MAESSIG     9107.55         CAL/MOL       328.86     CAL/G

   WAERMELEITFAEHIGKEIT
    GEW.ANT.MAESSIG           0.2876    KCAL/(M.HR.K)

   ZAEHIGKEIT
    MOLANT.MAESSIG            0.35475   CP

   OBERFLAECHENSPANNUNG
    NACH SUGDEN AUS
    PARACHOR                 25.87      DYN/CM
    NACH SUGDEN AUS
    EINZEL-OB.FL.SP.         25.70      DYN/CM

   PSEUDOKRITISCHE GROESSEN                      REDUZIERTE GROESSEN

    DRUCK                   148.305     ATM          0.0067
    TEMPERATUR              582.44      K            0.5892
    MOLVOLUMEN                0.096     L/MOL

   VORGESCHLAGENE NENNWEITEN

      NW                GESCHWINDIGKEITEN       DRUCKVERLUST/ROHRLAENGE
      40   MM               0.477     M/S       0.507E-01      AT/100 M
      25   MM               1.222     M/S       0.484          AT/100 M
```

Figure 2. Continued

```
**************************************************************************************
    STOFFDATEN DER DAMPF- BZW. GASPHASE
**************************************************************************************

    KODE          NAME              MOL-ANTEIL   MOL-MENGE   GEW-ANTEIL   GEW-MENGE
                                                  KMOL                       KG
      91   ACETON                    0.343803     12.903      0.522652      749.40
      63   METHANOL                  0.457337     17.164      0.383572      549.98
      10   WASSER                    0.198858      7.463      0.093776      134.46
    SUMME                            0.999999     37.531      1.000000     1433.85
    MITTLERE MOLMASSE               38.20    G/MOL

    GEMISCHDATEN DER DAMPF- BZW. GASPHASE
    =====================================


    DICHTE
     NACH BWR                      0.0355    MOL/L              0.00136   KG/L
***  NACH BEATT.BRIDGM.            0.0359    MOL/L              0.00137   KG/L
     ALS IDEALGAS                  0.0355    MOL/L              0.00136   KG/L

    SPEZ.WAERME
     AUS GAS-KOEFF.               13.5971    CAL/(MOL.K)        0.3559    CAL/(G.K)
     AUS GAS-KOEFF.
     UND ABWEICHUNG
     NACH BWR                     13.6984    CAL/(MOL.K)        0.3586    CAL/(G.K)

    ADIABATENEXPONENT              1.1892

    KOMPRESSIBILITAETSFAKTOR
     N. BEATT.BRIDGM. --->   0.99022    NACH BWR --->   1.00109

    ENTHALPIE
     AUS GAS-KOEFF.
     (H=0 BEI 0 CEL )           893.50       CAL/MOL          23.39        CAL/G
     AUS GAS-KOEFF.
     PLUS KOND.WAERME
     BEI 0 CEL                10043.76       CAL/MOL         262.90        CAL/G
     AUS FLUESS.KOEFF.
     PLUS KOND.WAERME          9865.34       CAL/MOL         258.23        CAL/G
     AUS GAS-KOEFF.
     UND ABWEICHUNG
     NACH BWR                 10043.30       CAL/MOL         262.89        CAL/G
     NACH YEN-ALEX.            9967.47       CAL/MOL         260.90        CAL/G
```

Figure 3

```
HEIZWERT
 GEW.ANT.MAESSIG    212775.31       CAL/MOL        5569.46       CAL/G

KONDENSATIONSWAERME
 MOL.ANT.MAESSIG      8240.45       CAL/MOL         215.70       CAL/G

WAERMELEITFAEHIGKEIT
 MOLANT.MAESSIG             0.0157   KCAL/(M.HR.K)
 N.LINDSAY U.BROMLEY        0.0149   KCAL/(M.HR.K)
 N.STIEL U.THODOS           0.0149   KCAL/(M.HR.K)

ZAEHIGKEIT
 MOLANT.MAESSIG             0.01008  CP
 NACH LEE                   0.00998  CP
 NACH WILKE                 0.01004  CP
 N.STIEL U. DEAN            0.01007  CP

PSEUDOKRITISCHE GROESSEN                              REDUZIERTE GROESSEN

 DRUCK                      74.049   ATM                 0.0135
 TEMPERATUR                538.37    K                   0.6374
 MOLVOLUMEN                  0.137   L/MOL

VORGESCHLAGENE NENNWEITEN

  NW                    GESCHWINDIGKEITEN        DRUCKVERLUST/ROHRLAENGE
 300  MM                    4.150    M/S          0.660E-03     AT/100 M
 250  MM                    5.976    M/S          0.158E-02     AT/100 M
 200  MM                    9.338    M/S          0.462E-02     AT/100 M
 150  MM                   16.601    M/S          0.184E-01     AT/100 M
```

Figure 3. Continued

27

The NSRDS and Data Banks

HOWARD J. WHITE

National Bureau of Standards, Washington, DC 20234

In the 1950's and early 1960's a number of articles appeared in the technical press about the "information explosion" and the problem attendant to it. The major problem cited was the size of the technical literature; it had grown so extensive and diverse that the individual could not keep up with all of the articles germane to his interests by the traditional browsing methods. New methods of organization and condensation were called for.

A subset of these problems involved the numerical data of science and technology. Again the size and diversity of the literature were principal problems, but here condensation or distillation was definitely needed. Thoughtful analysts recognized that, although the primary literature was the source for the results of experimental measurements, it was not a satisfactory reference base for most users. The raw data needed to be codified, standardized and put on a common basis, the conflicts raised by discordant data sets needed to be resolved, and seriously flawed data removed from use. The ideas were not really new: comprehensive handbooks had existed for the better part of one hundred years and one of the later ones, the *International Critical Tables*, had involved a measure of critical judgment as the name made apparent. However, comprehensive compilation and evaluation were called for on a scale never before envisioned.

In 1963 by executive action of the Federal Council for Science and Technology the establishment of a National Standard Reference Data System (NSRDS) to provide critically evaluated reference data for the scientific and technical communities became the Federal policy of the United States. This policy was endorsed and strengthened by the Congress with the passage of PL 90-396, the Standard Reference Data Act of 1968.

Management responsibility for the NSRDS was assigned to the National Bureau of Standards (NBS) in recognition of its long involvement with previous, less formal reference data activities arising from its responsibilities to provide reference measurements and data of many types within the United States. Program management responsibility within NBS was assigned to the Office of Standard Reference Data (OSRD).

The National Standard Reference Data System

The NSRDS exists to provide critically evaluated reference data. The data in question are quantitative numerical data arising from the well-established measurements of physics, chemistry, or engineering on materials of established composition. The data are obtained by compilation and evaluation of data from the literature. There is, in general, not an experimental-measurement component of the NSRDS, although there are necessarily intimate interactions of several types between the data evaluators and the measurement community.

The technical work of the NSRDS is carried out by a network of data centers and data projects. A data center is an organization that accepts continuing responsibility for the compilation and evaluation of the data in a given technical area. The data center may produce a variety of products over a period of time and stands ready to provide informed judgments on any question involving the data in its region of competence. More will be said on this subject later. A data project is similar to a data center in that it involves data compilation and evaluation; however, it is limited in term and scope. The staff expects to proceed to other activities after the term of the project, and one or, at most, a small number of specific products is expected.

The data centers in the NSRDS may or may not receive full or partial support from NBS. In general, the larger data centers receive support from more than one source.

The principal outlet for critically evaluated data from the NSRDS is the Journal of Physical and Chemical Reference Data published by the American Chemical Society, the American Institute of Physics, and the National Bureau of Standards. Another outlet worth mentioning is the NSRDS-NBS Series published by the Government Printing Office. There are still other outlets for reference data and for other products such as bibliographies and current-awareness services.

Data Centers and Data Banks

The assembled output of evaluated reference data of the NSRDS is a data bank (although an unautomated one at present)--one of the most extensive and carefully prepared in existence and one that will continue to increase in content and scope while the NSRDS continues to exist in its current form. However, it is not specifically a data bank of thermophysical properties for the chemical industry, nor is it any one of a dozen other specialized data banks that could be mentioned, although it would be able to make a major contribution to any of them. The NSRDS data bank might be called the fundamental, or natural, or, perhaps, archival data bank. To understand the difference between the fundamental data bank and the specialized data banks, it is useful to consider the different types in some detail. Although the specialized data banks differ in detail, most, if not all, have certain common characteristics which can be compared with those of the fundamental data bank.

The fundamental data bank responds to the internal logic of the NSRDS. Its scope is as broad as that of the NSRDS. Authors preparing texts for publication in the Journal of Physical and Chemical Reference Data or the NSRDS-NBS Series are required as matters of editorial policy to provide comprehensive bibliographies, detailed documentation of the reasoning used in arriving at their recommended values and estimations of the reliability of the recommended values. These requirements allow the critical reader to assess the technical quality and comprehensiveness of the process of critical evaluation. They also provide a sound base for future evaluation so that the older literature need not be endlessly recompiled and re-evaluated by later workers.

It is occasionally said that careful data evaluators do not resort to data "prediction" or, more properly, the conversion of data on one or more properties into data on the property desired through the use of theoretical or empirical processes. This is seldom true. In some cases, as, for example, the ideal-gas thermal functions compiled in the JANAF Tables, all of the data are predicted values. The low-pressure measurements needed to extrapolate to zero pressure are difficult to make and there is general agreement that the values produced from the use of partition functions will be more reliable than measured values. In other cases theoretical methods will be used to interpolate along temperature or pressure scales or even along concentration scales or to estimate values of collateral data. Anticipated relationships among chemically related molecules, for example, the molecules of an homologous series, may be used to smooth recommended values or set limits of reliability. For that matter, theoretical treatments are often needed to convert raw measurements to the desired (and reported) data. It is true, however, that good data evaluators will try to avoid degradation of their product below some intuitively conceived limit by overextending their estimating efforts. If the data available from direct or indirect measurements do not meet minimum standards for a certain substance, no value for that substance will be included among the recommended values.

Given adequate data, the evaluator will try to give the best and most comprehensive representation of the data possible. What such a representation consists of varies with the type of data. In some cases attempts will be made to make the data internally consistent over a range of properties and states, thus one equation may suffice to express the pvT and related thermodynamic properties of a fluid over a range of temperatures and pressures including the liquid and vapor state. Such extended coverage and internal consistency are usually achieved at the cost of some added complexity in the representation, however.

The specialized data bank is usually a mission-oriented or problem-oriented data bank. This would certainly be true of a data bank of thermophysical properties of fluids for the chemical industry. The mission-oriented data bank is limited in terms of properties and substances by the scope of the mission, although it may include properties and/or substances which are not closely related scientifically but are important to the mission. Within the field of coverage, blank entries can generally not be tolerated; and data must be estimated or, if of sufficient importance, measured by some coordinated experimental programs. The principal uses to which a mission-oriented data bank will be put are known in advance and play a role in its design and format. Simplicity and uniformity of the treatment of the properties of the various substances covered make for easy, quick use and hence are desirable traits. If uniformity of treatment is to be maintained, it is clear that the data on those substances for which the poorest and sparsest data are available control the nature of the treatment used. In some cases superior, but more complex, representations of the data might be deliberately degraded in the interests of uniformity. Speed and simplicity of use imply short condensed notation, especially if computerized handling is to be used as is usually the case.

It should not be inferred that the documentation of the evaluation process and the analysis of reliability that characterize the fundamental data bank, or the more comprehensive but, perhaps, more complex representation which is sometimes produced are not useful in the formation of a mission-oriented data base. Characteristically, this information is kept in auxiliary files. Details of the evaluation process, estimated reliability, or a formulation for a property that provides maximum precision and accuracy will be of great importance to some users. The average user will be satisfied with less detail, and the existence of the more extensive analysis is often indicated by reference flags in the main data bank. Thus, the developer of a mission-oriented data bank will use the material in the fundamental data bank but will rework it to maximize the ease with which it can be applied to the purposes in question. However, he will go to greater lengths to avoid blank entries than will the developer of a fundamental data bank, and accept the resulting degradation in technical quality as justified by the increase in utility.

There is no need to go into detail on the characteristics of a data bank of thermophysical properties of fluids for the chemical industry. Evan Buck has given an outline for one, with which, I think, most engineers would agree.

NSRDS and Data Banks for the Chemical Industry

The primary output of the NSRDS has been to, what has been called here, the fundamental, natural or archival data bank. The basic importance of this data bank has also been stressed. It does not follow that other data banks are not also of importance and interest within the NSRDS. In fact, a national data system, and the data centers which form its backbone should stand ready to make a contribution to any reference data question within their competences.

Elements of the NSRDS have developed, or helped develop, several mission-oriented data banks. Examples are a thermodynamic data bank for the design of industrial incinerators and a data bank for the study of the chemistry of the stratosphere. Experience has shown that, not only can data centers make important contributions to mission-oriented data problems, but they can do so in a timely and economical fashion.

Thus it is clear that a thermophysical data bank for fluids for the chemical industry would be a suitable subject for consideration within the NSRDS. In point of fact, there are several data-related activities already under way in NBS which could make contributions to such a data bank. These involve data evaluation, data measurement, or combinations of the two. They are outlined in a document listed as NBSIR 76-1002 and entitled "Industrial Process Data for Fluids: A Survey of Current Research at the National Bureau of Standards."

However, the most direct involvement of the NSRDS in an attempt to provide a data bank for the chemical industry is called "Project Evergreen." This is not a project, but a proposal at the present time. A detailed description of Project Evergreen would be too lengthy for the time available. The principal features of Project Evergreen are the following:

1. The project would provide evaluated data on single substances, the so-called "ubiquitous substances," those that every company with chemical interests must necessarily deal with in the course of its activities.

2. The properties would include the thermodynamic and thermophysical properties under discussion at this meeting.

3. Every effort would be made to avoid blanks in the substance-property matrix and indications of the antecedents of the data and their estimated reliability would be provided.

4. In general outline the data bank would resemble very closely that described by Evan Buck.

5. The data bank would be limited to numerical data for properties and would not include programs for process computations.

6. The development and maintenance of the data bank would be the responsibility of a data center established for the purpose.

7. The data bank would be jointly supported by the NSRDS through the Office of Standard Reference Data and industry. This support would include joint financial support, and joint technical support.

8. There would be close continuing interactions of the data center and industrial advisors in matters of selection of substances and properties, selection of data and evaluative and estimating techniques and estimation of reliability.

9. The data bank and details of the data selection and estimation procedures would be available to all and would not be kept as proprietary information. The value received from the more extended review and criticism obtained in this way and the desirability of having a commonly agreed-upon data set for the ubiquitous substances would outweigh any proprietary advantage.

Complete details on Project Evergreen are available in the project proposal.

In summary, a thermophysical data bank for the properties of fluids for the chemical industry is a suitable subject for the involvement of the NSRDS, and several steps have been taken in this direction.

Discussion

R. C. REID

Group Contribution Estimation Methods

In the presentations, it was noted that few additional group interaction parameters could now be obtained for UNIFAC as essentially all available and reliable phase equilibrium data had been used. A. Bondi suggested that other mixture property data might be of value to fill the vacancies. G. Wilson indicated that, from his experience, this was possible but only approximate values could be determined.

D. Tassios noted that an extension of group contribution estimation techniques into the area of electrolyte solutions would be useful. R. Gray also suggested an extension to poorly-defined, high-molecular weight mixtures. J. Prausnitz noted that the constituent molecules must still be known to determine the configurational portion of the activity coefficient expression although C. Deal indicated that this objection may possibly be overcome if the mixture is somehow characterized by an "overall" molecular weight (or volume) average.

Several members inquired about the expected accuracy of a group contribution method such as UNIFAC. As C. Deal pointed out, the error would vary on the type of mixture, how different it was from those used to obtain the group interaction parameters, the reliability of the original data, etc. P. Rasmussen did say that when the new UNIFAC values were completed, there would also be tables showing how these parameters could be used to "predict" phase equilibrium data for many systems. The goodness of the fits would be of value to users to indicate the expected accuracy.

There was considerable discussion on how infintely-dilute activity coefficients might best be obtained. C. Eckert (among others) advocated a direct measurement using differential ebulliometry or gas chromatography as extrapolation of γ data at finite concentrations to infinite dilution was difficult and prone to large uncertainties. D. Tassios raised the issue that we now have considerable data not at infinite dilution and a reasonable extrapolation method to low concentrations would be to plot ln (γ_1/γ_2)

vs. composition. Some felt this method would not be particularly helpful, but few other suggestions were made as to how to use finite composition data on γ to obtian γ^{∞}. One could use a liquid-phase model to fit the data and then extrapolate to x_1, $x_2 \rightarrow 0$ (S. Adler indicated Kellogg does this using a best fit to a Margules equation.). G. Wilson said that even if one could fit P - x data within 1% by a liquid model, an uncertainty of 5% in γ^{∞} could be expected.

C. Deal pointed out that rather poor extrapolations to infinite dilution apparently prevented the recognition of systematic effects that had led to successful group approaches in the past. Once it was realized that one could obtain such systematic changes in examining γ^{∞} for different systems (i.e., to extract molecular group interaction parameters), then more interest was generated in obtaining accurate values of γ^{∞}.

Several members pointed out that UNIFAC parameters were not reliable to estimate liquid-liquid equilibria. Also, C. Deal emphasized that their principal value may be in screening or exploratory studies. If certain interactions were found to be critical then additional, in-house experiments might be warranted. Others in the audience noted that UNIFAC parameters had been of real value to estimate vapor phase equilibria. There would appear to be no limitation of using binary UNIFAC parameters (or other group contributions as in ASOG), to calculate equilibria for multi-component systems.

D. Fussell noted that the Chao-Greenkorn approach became less accurate for high-molecular weight compounds. K-C. Chao agreed but felt there was still reasonable accuracy up to C_{20}.

In the Reporter's summary, there seems to be an ever-increasing interest in using some theory, coupled with molecular group inter-action parameters to calculate properties. This approach has been widely used for ideal-gas properties and is now beginning to mature for liquid-phase activity coefficients. Success can only breed other approaches. G. Wilson noted some of these (vapor pressure, equation-of-state mixture constants) and others (transport properties and surface tensions of mixtures) could be suggested.

Comments by A. Bondi:

After listening to Grant Wilson's success with the use of activity-coefficient-derived group contributions for the estimation of heats of vaporization, it seems to me that one should be able to do the following: (1) Correlate the γ^{∞} group contributions in the UNIFAC or ASOG matrix with the appropriate group contributions to ΔE_{vap} (at some suitable reference state). (2) Once such a corre-lation has been created, one should be able to approximate currently unavailable γ^{∞} group contributions from external ΔE_{vap} data of the compound pair in question.

Comments by P. Rasmussen:

Prepared by Peter Rasmussen, Technical University of Denmark

1. Selection of VLE data for the UNIFAC database.

The revised table of UNIFAC group-interaction parameters (submitted for publication in Ind. Eng. Chem. Proc. Des. Dev.) is based on vapor-liquid equilibrium data for more than 2000 binary systems. The data were selected from the large data collection at the University of Dortmund (1) and from Wichterle et al. (2).

The VLE data were tested for thermodynamic consistency by the method outlined by M. Abbott earlier at this meeting and described by Christiansen and Fredenslund (3).

The data thus tested were accepted as reliable for parameter estimation when the mean deviation between experimental and calculated values of the mole fraction in the gas phase was less than 0.01. The vapor-phase non-idealities are accounted for as shown by Hayden and O'Connell (4).

References

1. Gmehling, J. and Onken, U., "Vapor Liquid Equilibrium Data Collection," DECHEMA Chemistry Data Series, first part available in December 1976.
2. Wichterle, J., Linek, J. and Halá, E. "Vapor-Liquid Equilibrium Data Bibliography," Elsevier, Amsterdam, 1973 and Supplement I, Elsevier, 1976.
3. Christiansen, L. J. and Fredenslund, Aa., AIChE Journal (1975) 21, 49.
4. Hayden, J. G. and O'Connell, J. P., Ind. Eng. Chem. Proc. Des. Dev. (1975) 14, 209.

2. Alcohol isomers.

The question was raised by G. M. Wilson if the UNIFAC method could distinguish between 1-propanol and 2-propanol. The answer is that UNIFAC certainly can.

1-Propanol is considered to consist of two groups: 1 CH_3 and 1 CH_2CH_2OH. 2-Propanol contains also two groups: 1 CH_3 and 1 $CHOHCH_3$.

The table shows the values of the group volume R_k and the group area Q_k for group k and the group interaction parameters a_{mn} between groups m and n.

	Group k	R_k	Q_k
	CH_3	0.9011	0.848
	$(-CH_2CH_2OH$	1.8788	1.664
CCOH	$(-CHOHCH_3$	1.8780	1.660
	$(-CHOHCH_2-$	1.6513	1.352

$$a_{CH_3/CCOH} = 737.5$$

$$a_{CCOH/CH_3} = -87.93$$

It can be seen from the table that the interaction parameters for the different alcohol groups are identical and that the difference between the two propanols are accounted for by the different values of R_k's and Q_k's.

It can also be seen that UNIFAC will in general distinguish between 1-, 2-, and 3-alkanols but not between 3- and 4-alkanols.

3. Reliability of UNIFAC predictions of activity coefficients.

It has been suggested during the discussion that an "error indicator" should be published together with each UNIFAC group-interaction parameter. The idea is that such error indicators should enable the user of UNIFAC to estimate which errors to be expected in the predicted activity coefficients.

This is not possible since the errors in the predicted activity coefficients will depend so much on the mixture for which the parameters will be used. The errors in the predicted activity coefficients for ethanol will thus depend on the other components in the mixture, on the temperature and on the concentration.

A book on UNIFAC is now being written by Aa. Fredenslund, J. Gmehling and P. Rasmussen and it will hopefully be published by Elsevier Publishing Company later this year.

The book will contain extensive tables with comparisons between experimental and predicted activity coefficients for many mixtures. It will thus be possible for the user of UNIFAC to judge from these tables if the accuracy of the UNIFAC predictions are acceptable for mixtures resembling the mixture for which the user might wish to use UNIFAC.

Comments by D. T. Binns:

The British Institution of Chemical Engineers has now incorporated a new, flexible and powerful data bank and physical property point generation package into their Physical Property Data Service (PPDS).

The programs have been developed by ICI, and are based on

experience with at least three first generation banks developed in various parts of the company. It has only recently been acquired by the Institution, and should not be confused with the system used by PPDS over the last four years.

The bank is entirely controlled in the storage, editing and retrieval of data by alphanumeric English language free-format commands. It is designed to be interfaced with application programs requiring physical property values, of which a program to calculate and print a table of values is one.

Some of the main features are:

Information, like names (including synonyms) of substances, physical properties, correlations and numerical data are easily added to the bank.

There are no restriction on the type of information which can be stored by the system, but generally single values, correlation coefficients or, more rarely, tabulated data are stored.

Different correlations for a given property can be used.

For a multicomponent mixture, for properties where simple mixing rules are adequate, pure component properties may be calculated by different correlations.

Different sets of coefficients for the same substance and a given correlation can be stored by the system, and can be distinguished by stored attributes like reference, units, security password, temperature and pressure range. A minimum acceptable reliability can be specified by a user of the bank on retrieval.

The concept of 'stream type' has been introduced, whereby the physical properties of different process streams (or the same process stream under widely different conditions) in, for instance, a flow sheeting program, can be calculated by different correlations.

Data may be provided by users of an interfaced application program, and these can be used alongside data from the bank, in the physical property point generation package.

At the moment, in the Institution system, which is for sale, there are data on 400 compounds, 17 constant properties and 14 properties varying with temperature and pressure. Several equations of state are available for calculation of density, fugacity, enthalpy, etc.

The size of the system, excluding the direct access data store, end-to-end is about 700 K bytes on an IBM 370/158. Depending on the application program the system can be overlayed to around 70 K bytes.

To give some impression of the control commands, I list the following miscellany:

The program for manipulating information in the data store recognizes instructions like

```
ADD COMPONENTS
     HYDROGEN                          1
     METHANE; CH4                      6ØØ
```

```
ADD CORRELATIONS
     ANTOINE 3                                        45Ø1
     WILSON A -1, 1.Ø                                 4212
     HV TABLE -2                                      4399
     REDLICH-KWONG Ø                                  53Ø4
ADD DATA
     NAMES = 'METHANE', 'ETHANE', 'PROPANE'
     MOLWT = 16, 3Ø, 44
ADD DATA
     TRANGE = 3ØØ, 1ØØØ
     REFERENCE = 'TICHE 1976 432'
     RELIABILITY = 593
     UNITS = 'SI'
     ANTOINE ('AMMONIA') = 79.736 + 11.513Ø44, -47Ø6.3, Ø
```

There is an arithmetic decoding facility.
The Program for retrieving data recognizes instructions like:

```
NAMES FOR ORGANICS 'METHANE', 'PROPANE'
```

This defines the components in a stream type called ORGANICS. If there were only one stream type in the application program, this could be shortened to:

```
NAMES 'METHANE', 'PROPANE'
```

Physical properties and correlations can be chosen like this:

```
CRITICAL TEMPERATURE
LIQUID ENTHALPY BY EXAPI REDLICH-KWONG
LIQUID ENTHALPY BY MY CORRELATION CODE 91 USING TC, PC
LIQUID DENSITY FOR WATER BY FRANCIS L
```

Sets of correlation coefficients can be chosen like this:

```
REFERENCE = 'BLOGGS DATA'
RELIABILITY = 211
SECURITY = 'NOT AT ANY PRICE'
TRANGE = 273,673
PRANGE = 5ØØØØ, 1ØØØØØ
```

There are default values for these attributes. Data values can be assigned like this:

```
MOLWT = 16, 36.5, 92, 112
EXANT ('METHANE') = 45.465, -1506.7, Ø, Ø.Ø27724, -7429
WILSON A ('EDC', 'HCL') = 1.7282
```

Comments by Eiji Shima:

The computer program of data bank and estimation of physical properties, JUSE-EASOPP, on which I have presented a paper at a CODATA Conference held in Freiburg, West Germany, is currently used by several industrial firms in Japan. From my experience in developing the data bank, I would like to make some comments from the practical point of view.

Accuracy received first priority in featuring the program package. We have, however, received comments from industrial people, after practical application, that it is inconvenient and inefficient to link the data bank with some other systems as design packages. That is, a rather long time is required to obtain the physical property data from the storage area, e.g., on a magnetic tape.

One should be aware of two different kinds of data that are called in the practical applications; (1) highly evaluated, accurate data (which could be time-consuming or expensive to obtain); (2) easily processable data for engineering purposes (which could be less accurate but should be available for a wide range of conditions).

In connection with the group contribution procedure, one may be interested in the work of Professor Yukio Yoneda of the University of Tokyo. He recently opened his program, EROICA, for public use at the Computer Center of the University of Tokyo. His program is characterized by a data bank containing about 5000 compounds. The input formula of CHEMO notation is used to represent the molecular structure.

I would like to stress the importance of the input method for specifying the molecular structures if it is desired to carry out any kind of group contribution calculation. One might find, later on, that a functional group gives a good correlation. Then for all the compounds that have been filed in the data bank one should be able to retrieve data again with respect to that particular functional group.

Thermophysical Properties Data Banks

There was general discussion of problems associated with establishing general data banks. The political, organizational, and economic problems of establishing a common data bank system may never all be solved but the benefits of such a system appear to be great enough to provide an agenda for discussion. The established data bank systems in Europe and Japan also provide experience factors that may be useful to those organizing a data bank effort in the United States.

28

Three E's of Our Decade—Entropy, Energy, and Economics

JOSEPH J. MARTIN

University of Michigan, Ann Arbor, MI 48109

At first I was a little uncertain on the kind of salutation by which to address so distinguished a group as gathered here tonight, and then it dawned on me that today is January 20th. Thus, I can say, "Fellow Republicans." If this were not so, we would all be swinging at the Inaugural Ball on the other side of the continent. If our absence from the inauguration is not sufficient indication that we are not part of the "in" group, one might ask why we are not rubbing elbows with the celebrities playing golf this week in the Bing Crosby tournament at Pebble Beach on the other side of this peninsula. Being part of the "in" group, however, does not always guarantee the expected bid to participate in a glamorous event. In the State of Michigan, for example, there are about 3000 Democrats who thought they would be invited to the inauguration, but only half received invitations. Quoting the newspaper, "for reasons unknown many of the party faithful, veteran Democrats as well as early Carter supporters, have found their mail boxes bare of that long-awaited invitation . . . Miffed Democrats include several who supervised telephone call operations on election day. They didn't get invitations, but the phone callers they supervised did."

Seriously, please do not think that I underestimate the importance to the nation of you, the scientists and engineers who are specialists in many different areas of thermodynamics. After all, one of the hottest (or is it coldest?), topics of our times and certainly one of the major problems of the day is energy. And what is thermodynamics but the science of energy in its various forms, transfers, and manifestations. Thus, for good or bad, our field has become somewhat unknowingly the center of attention. If our nation, or the whole world for that matter, is to find solutions to the energy problem, it must be by experts in energy. Certainly, the politicians, the lawyers, the business executives, the financiers, or the labor leaders are not by themselves able to solve the energy crisis. They may be a party to actions that lead to solutions but it must be the professional, knowledgeable experts who understand the production, the utilization, and the science of energy, who will furnish the technical basis for any solution. This places the problem squarely in our laps. Through our knowledge and understanding of thermodynamics we appreciate the grim reality and magnitude of the situation, so we realize our work is cut out for us. Can we meet the challenge? I believe so.

Now, I am sure you all did not come here tonight to hear just another talk on the energy crisis. At least, I hope you find this one is different. If you wanted an overview of the problem, you could simply read a few of the many articles and reports on the subject that have been turned out by ERDA, FEA, NAS, EPRI, and a host of others. For one of the better summaries I recommend for your reading the recent article by Richard E. Balzhiser of the Electric Power Research Institute located a little north of here. Writing in Chemical Engineering magazine for Jan. 3, 1977, Dick addresses the topic, "Energy Options to the Year 2000." In order to save time, I would like to skip over Dick's thorough discussion of practically every known source of energy and note his conclusion. "Solar, geothermal, and fusion will not contribute significantly to our needs in the year 2000. Oil and gas, while still substantial, will lessen in importance over this period. Only coal and nuclear power are clearly capable of providing increased amounts of energy deemed essential in this period, but government action has frustrated, not furthered their cause." On this score Dick comments, "It . . . puzzles me, as an engineer, that so many of our elected representatives believe we can solve the problems associated with solar, geothermal, and fusion energy overnight, but will not help develop acceptable solutions for considering coal, oil and nuclear waste disposal, oil spills, strip-mining reclamation, or pollution control."

If we accept Dick's conclusions (and he certainly has spent as much time studying the problem as anyone else), we can see the basis of an energy policy. Put research and development dollars where they have the greatest potential for making the most effective impact on the problem, and this is in the areas of coal and nuclear. The only thing which could change this is the discovery of some fantastic new oil and gas fields. Meanwhile, this does not infer the complete elimination of all research in other areas. For example, energy conservation studies and techniques are worthwhile if the amounts of energy to be saved are appreciable and if the capital investment is not great. Solar energy can be capitalized upon in architecture in some cases through alternative design of buildings at little additional cost, but the broad scale capture of solar energy, other than through biomass such as agriculture or forests, is not likely. Using ocean tides or waves and the common geothermal sources outside a few isolated areas is not realistic. Windmills can only be justified in a few locations and ocean gradients are so small in a thermodynamic sense as to be totally out of the question. Lest we fall into a trap that these conclusions are too simplistic, let us examine the problem from the more basic point of view of thermodynamics.

While the newspapers and magazines speak glibly of the problems of energy and economics, they do not nention the most fundamental quantity of all which is entropy. This is only natural because they do not know what entropy is and they realize the public knows nothing about it. Since entropy is not unfamiliar to this audience, I am not constrained to build a discussion without it. You and I know that entropy is central to all of nature's processes be they animate or inanimate.

Who was responsible for the introduction of entropy into our storehouse of scientific knowledge? We all know it was Clausius, but before laying too much blame at his door for a concept that is a stumbling block for the beginner in thermodynamics, we must recognize that entropy has been with us since the beginning of time. Clausius only discovered it and perceived its significance and showed us how to use it. Entropy was not really invented any more than energy or economics. They all have been a part of man's existence since he came on this earth and will continue to the end of time.

What is entropy? In simplest terms it is a measure of the mixed-up-ness or randomness of a system, and it provides a means of getting a quantitative handle on the irreversible processes of nature. Remember that famous quotation of Clausius, "Die Entropie das Welt strebt ein Maximum zu." The entropy of the world (universe) tends toward a maximum. Of course, Clausius did not mean we as a people are becoming more mixed up, though he might have! What he meant is that ordered systems when left to their own devices tend to become more disordered or random. All natural processes are irreversible and result in an increase of entropy somewhere. Now there are two things which Clausius might have said, but did not. First, he did not discuss the rate at which the entropy of the world is increasing, because he did not know the rate and neither do we; and second, he did not point out that the energy of the world is a constant if it is an isolated system and energy and mass are treated together through $E=Mc^2$. When people say we are using up our energy, we know they are not precisely correct. We are using up the energy available to do work by converting it into energy unavailable to do work, thereby maintaining a constant amount of energy but increasing the entropy. Thus, the measure of our activities might well be in terms of available energy, or entropy just as Clausius stated.

Let's explore this point a little further. Suppose we consider the planet earth the equivalent of an isolated system. This is a reasonable assumption if the energy input from the sun happens to match exactly the energy radiated to outer space. Actually, earth scientists have been trying for years to prove that the earth is heating up or cooling down, but so far there is not complete unanimity on the conclusion. In that connection I would like to digress for a moment and tell you a true story. A few years ago I attended a conference in San Francisco sponsored by the United Nations on "Man and His Environment". So far as I could ascertain, out of about 500 attendees there was only one other engineer besides myself. There were biologists, zoologists, entomologists, anthropologists, ecologists, environmentalists, lawyers, and even politicians. I felt uneasy as the meeting progressed, and I heard repeated sneering references to technology, to engineers, to the "infernal" combustion engine, to phosphates, to DDT, to nuclear power plants, to thermal pollution, and to production and the GNP--not to mention the cheering for ZPG and legalized abortion. It was obvious that I was not among close friends, so I was intimidated to remain quiet. At one of the luncheon tables a warm but friendly exchange of ideas was taking place when it became apparent that the gentleman next to me was acknowledged by the other people at the table as an authority on the temperature of the earth. In the course of the discussion he made the unequivocal statement that in the last 50 years the earth had warmed up one degree Fahrenheit due to the burning of fossil fuels. He predicted that if we continued, we would melt the polar ice cap, raise the level of the oceans, and flood the Empire State Building. He was not implying that the heat released by the combustion would melt the ice; he was referring to the greenhouse effect of carbon dioxide in the atmosphere. You are all familiar with this effect when it warms the inside of your glassed-in automobile on a sunny cold day in the winter. Short wavelength radiation from the sun penetrates the atmospheric mantle of CO_2 and is absorbed on the surface of the earth. The earth being at a lower temperature than the sun, sends out longer wavelength radiation which is absorbed by the CO_2, thus resulting in a net energy gain to the earth and its atmosphere from the sun. Increasing the CO_2 in the atmosphere would enhance this effect and lead to the dire consequences predicted.

The authority was not a little upset when I got up enough nerve to question the statistical significance of his temperature data which he said represented average daily readings in a number of cities throughout the world. He did not regard me in a very friendly manner when I suggested the relative negligibility of one degree compared to the wide variations monthly, weekly, and even daily, and when I mentioned that the earth was covered 80% with water and no readings were taken over these vast areas. He became even more agitated when I asked whether atmospheric scientists have been able to measure CO_2 concentrations precisely enough for 50 years to prove there has been an appreciable global increase. The other people regarded my questions with disdain and coolness, as their minds seemed to be closed on the subject.

That afternoon I attended a plenary session where one of the principal speakers was an authority on aerosols and particulate matter in the atmosphere. He stated just as emphatically as the first fellow that the temperature of the earth had gone down one degree Fahrenheit in the 50 years due to a change in the earth's albedo caused by all the soot and dust we were spewing into the air with our dirty combustion and other chemical processes, resulting in more of the sun's radiation being reflected back to outer space. The audience shivered sympathetically when he predicted a return to the ice age, but I sat back relieved, comforted by the average of plus and minus one.

As an engineer and scientist I realize both authorities were partially right. Putting carbon dioxide and particulate matter into the atmosphere could cause trouble and we should not be blind to the possibility of deleterious effects, particularly locally. It is interesting that today there are scientists who are again looking at the potential problem. Just a few months ago ICSU, the International Council of Scientific Unions issued a report on Environment Issues of 1976. What caught my attention after the UN conference experience of a number of years ago was the statement that the earth could undergo major climatic changes in the next 200 years due to four to eight-fold increases of carbon dioxide in the atmosphere from fuel-burning, obviously, a resurrection of the same old hypothesis. They stated that the intensified effect of additional CO_2 could cause dramatic changes in vegetation and gradual melting of the polar caps with a resultant rise in our planet's oceans, and termed this prospect "rather alarming." They concluded that this emphasized the need for global environmental monitoring systems. This is clearly the only way to allay the contentions of the fear-mongers who predict sad consequences of man's activities without real factual evidence. Indeed, the problem with CO_2 is similar to that for ozone where some people predict the use of chlorinated hydrocarbons is destroying the ozone in the stratosphere, even though the data do not show any statistically significant decrease in ozone. Clearly more monitoring measurements are much needed if we are to know whether these gloomy predictions will actually occur.

Let us now return from our diversion to the assumption that the earth behaves as an isolated system with no net energy transfer with its surroundings. We then can say the energy of the earth is constant and the entropy is steadily increasing due to all the irreversible processes taking place on the earth, where we are neglecting the difference in wavelength of the radiation coming to and leaving the earth. We also can say that the rate of entropy increase is greater as a result of man's presence because man consciously promotes irreversible processes over and beyond those of nature. Man has developed the capability of overcoming the potential barriers that prevent spontaneous processes from taking place, with simple combustion and nuclear fission being good examples. Of course, mother nature herself occasionally overcomes potential barriers, such as by igniting a forest fire with lightning, but this is considered a natural process increasing the entropy, It can also be argued that man sometimes slows down the rate of entropy increase by retarding irreversible processes, such as by damming a river to prevent flooding and erosion or by planting a field or forest to utilize solar energy rather than just allowing it to heat up the ground. The central question here is the degree of significance of man's influence. Suppose that the natural rate of entropy increase in the planet earth is 1 billion units per year and man comes along and causes it to be 1 unit greater. Obviously, if this is truly the case, one need not worry about the effect of man's activities on an overall global scale, though localized effects might not be insignificant. On the other hand, if man's activities were shown to contribute 10 million units of entropy increase per year, we would probably have cause for concern.

Although we can calculate the entropy increase due to our own activities, we do not know how to measure the natural rate of entropy increase. In the case of the formation of carbon dioxide by combustion of fossil fuels, we do not know how much CO_2 is being formed by natural processes and is being absorbed by plant life and the oceans. What we must do then is measure the CO_2 concentration globally, as recommended by ICSU and see if we can predict the effect of any increase. If it could be shown that the earth is definitely cooling down through natural processes, it might even prove beneficial to burn more fossil fuel to enhance the greenhouse effect and prevent the earth from passing into a permanent ice age and to assist in the growth of plant life which thrives on the absorption of CO_2. Existing knowledge is inadequate to tell us whether burning carbonaceous fuels is really good or bad. In all probability quite a few years will pass before man will deliberately produce CO_2 to try to keep the earth from becoming too cold a place to live--though I can tell you that the subzero weather in Michigan the last two months causes me to wonder if that time has not come.

In the past century when sources of available energy were plentiful and seemingly inexhaustible, the measurement of man's activities was through the medium of money. This situation is not going to change and the field of economics is going to stay with us. As our backs are put to the wall, however, to have adequate supplies of available energy, it may be worthwhile to become somewhat more fundamental in our thinking. Instead of evaluating a given product or service in terms of its costs, we can give it in terms of the entropy increase or the available energy decrease required to produce it. Let us see how this works out for some common materials of construction. To make one pound of cold-rolled steel, studies have shown that it takes approximately 700 BTU for mining, 800 for the coke furnace, 10,000 for the blast furnace, 2500 for the steel furnace, 1000 for the other materials added to the steel furnace, 7000 for hot rolling, 3000 for cold rolling, and 700 for transportation. This gives a total of almost 26,000 BTU. Other studies have shown this is a little high, so that one might accept an average of 22,000 BTU. This may be contrasted with aluminum where a pound of rolled or drawn product requires about 110,000 BTU, taking into account the energy consumed in mining bauxite, processing to alumina, electrolyzing the alumina in a flouride bath, and rolling and drawing. It is assumed in this figure that about 4% of the metal is recycled scrap which reduces the energy for electrolysis. If a given product can be manufactured with a pound of steel or a pound of aluminum, the energy requirement for the aluminum will, therefore, be five times that of steel. In some applications steel and aluminum may be interchanged on a volume basis. By including the density in the calculations it takes about 6,000 BTU to make a cubic inch of steel and 11,000 for a cubic inch of aluminum, so now the energy requirement of aluminum is a little less than twice that of steel. If it is noted that steel is about 3 times as strong as aluminum on a volume basis, then steel and aluminum normally would be interchanged on a mass basis because steel is 3 times as dense as aluminum, so the five to one ratio is the more reasonable one. Of course, other factors besides energy requirement would enter a decision to use one material or another. Appearance, corrosion, atmospheric and radiation resistance, electrical conductivity, thermal conductivity, brittleness, hardness, compressive and tensile strength, elasticity, rigidity, density, flammability, formability, temperature dependence of properties, and machinability are all important factors for specific applications. The following table gives a few useful materials and their energy requirements.

ENERGY REQUIRED TO PRODUCE SOME COMMON MATERIALS

Materials	BTU/lb	BTU/cu in
Cast Iron	10,000	2,600
Glass	13,000	1,200
Steel (cold-rolled)	22,000	6,000
Plastics (average)	25,000	1,000
Rubber (milled)	37,000	2,000
Zinc	46,000	12,000
Copper (drawn or rolled)	66,000	21,000
Aluminum	110,000	11,000

Inspection of the table shows the desirability of using plastics, glass, rubber, and cast iron where the application is on a volume basis and no other factors predominate. Although not shown here prices of these materials on a mass basis correlate fairly well with the energy requirements.

Recognizing the energy required to produce the basic materials, consideration must be given to the energy required to convert these into useful goods. The latter obviously depends on whether we are making a low-technology product such as car fenders or a high-technology product such as hand-held computers. Fairly good estimates can be made for either, though we shall not go into that here.

We often hear it said that the barrier to using some new energy technology is cost, but we also can examine it from the point of view of energy requirement. For example, when oil was $3 a barrel, it was estimated we could not make synthetic oil from coal for less than $10 a barrel. Consequently, we did not rush out and build coal liquefaction plants. Today oil is up in the $14 range, but no one can show that coal liquefaction is even close to being economical because the costs of the coal, the processing equipment, and the operation have escalated with the price of oil. This indicates the value of using the fundamental yardstick of available energy or entropy. In a competitive situation we cannot expect to use an energy source that requires a high fraction of available energy or causes a large increase in entropy to produce it. This is one of the reasons that oil and natural gas displaced other sources after World War II. We thought then that natural gas was plentiful, but we know now that a shortage of natural gas will be the first real crisis, and this is going to cause great hardship until we readjust to coal or oil, or even nuclear. The energy requirements to find and produce native oil or gas are steadily increasing because of broader scale exploration, deeper drilling, off-shore platforming, greater pumping, and secondary recovery techniques. Foreign oil does not require any more energy for production and shipping (though liquifying natural gas does), but political decisions are making it less accessible.

Since there are no political decisions in using solar energy, if suitable area is available, why do we not substitute the sun for the oil or gas well? The answer is obvious, in that it takes too much energy to manufacture large volume or area solar equipment. Only an unforeseeable scientific breakthrough, such as a highly efficient photovoltaic cell will change this. A few architectural alternatives will allow us to capture sunshine at low cost, but this is somewhat limited. Even in agriculture, arboriculture, horticulture, and aquaculture it requires too much energy to "grow" energy for general use. The optimum use for these techniques is to raise food for an expanding population and to produce raw materials. In my home State of Nebraska, though, they are promoting the use of grain-based alcohol in "gasohol" for motor vehicles. An energy analysis shows this is not very practical, however.

From the Carnot relation the fraction of available energy that can be extracted from heat flowing between two temperatures depends upon the temperature difference. If this difference is small, vast quantities of heat must be handled. This will require extensive equipment which will require enormous amounts of energy for its manufacture and operation. Consequently, proposals to use ocean thermal gradients (or salt concentration gradients) seem ridiculous.

If a high fraction of energy is required to produce a unit of available energy, such as with shale, for example, the overall entropy increase is greater than if a small fraction of energy is needed for the same amount of available energy. Only in the case of solar-related energy can we improve on nature and cause the entropy increase to be no more than if man were not on earth. The process of sunshine falling on bare earth is highly irreversible. If we extract some available energy and use it, the overall entropy increase is practically the same if the energy expended to produce the equipment comes from the sun.

The only sources of available energy in the amounts needed that can be obtained by spending a small fraction of energy, are coal and nuclear. Controlled fusion is still just a pot of gold at the end of the rainbow. Uncontrolled fusion has been here for a quarter century, but we are still searching for ways to use it, as by underground fracturing or earth moving.

We will continue to measure manufactured goods, or services such as transportation, by the dollars of economics, but the immutable gold standard of our activities is available energy or entropy increase. Comparisons of alternative schemes can well use these quantities which are independent of the fluctuations in the value of the dollar.

29

Some Aspects of Henry's Constants and Unsymmetric Convention Activity Coefficients

JOHN P. O'CONNELL

University of Florida, Gainesville, FL 32611

The purpose of this talk is to show some important unrealized aspects of the liquid-phase thermodynamics for mixed systems containing some substances well above their critical temperatures. Examples are hydrogen, nitrogen, carbon monoxide, methane, etc. in mixtures with polar solvents for which no equation of state is applicable.

Thermodynamics

The fugacity of a component in a liquid can be written as function of quantities related to measurables (T, P, $\underline{x}$) in two main ways.

$$f_i^L\ (T,P,\underline{x}) = x_i\ \gamma_i\ (T,\underline{x})\ f_i^{oL}(T)\ \exp\left[\int_{P_i^o}^{P} \frac{\bar{v}_i^L\ (T,P,\underline{x})}{RT}\, dP\right]_{T,\underline{x}} \quad (1a)$$

$$= x_i\ \gamma_i^*\ (T,\underline{x})\ H_i(T,\underline{z})\ \exp\left[\int_{P_i^o}^{P} \frac{\bar{v}_i^L\ (T,P,\underline{x})}{RT}\, dP\right]_{T,\underline{x}} \quad (1b)$$

Where the activity coefficients γ_i and γ_i^* have the "symmetric" convention and "unsymmetric" convention limits respectively.

$$\lim_{x_i \to 1} \gamma_i = 1 = \lim_{x_i \to 0} \gamma_i^* \quad (2)$$

Effectively, all the change of convention does is substitute $\gamma_i^*\ H_i$ for $\gamma_i\ f_i^{oL}$ and regardless of which is used, the product must yield a correct value of f_i^L. The pure component reference state fugacity f_i^{oL} is usually calculated or extrapolated using $\phi_i^{sat}\ P_i^{sat}$ when the system temperature is below or only somewhat above the critical temperature of the component. The use of Henry's Constant, which may be a function of solvent composition, $\underline{z}$, in multicomponent solvents, requires binary system data since it often cannot be predicted with sufficient accuracy to do good design calculations. Yet, as Figure 1 shows, if data are available over a limited range of composition, the proper value of f_i^{oL} is subject to great uncertainty because correlations for γ_i are not necessarily trustworthy.

While $f_i^{oL}(T)$ is not an accessible quantity, it still can be used in thermodynamic derivations. Its dependence only on temperature is of particular significance since the following derivation reveals constraints on theories to be used relating H_i, γ_i, and γ_i^*

$$\ell n \left[f_i^{oL}(T) \right] = \ell n H_i(T,\underline{z}) + \ell n \left[\frac{\gamma_i^*\ (T,\underline{x})}{\gamma_i\ (T,\underline{x})} \right] \tag{3}$$

$$= \ell n H_i\ (T,\underline{z}) - \lim_{x_i \to 0} \ln \gamma_i\ (T,\underline{z}) \tag{4}$$

$$= \ell n\ H_i(T,\underline{z}) + \lim_{x_i \to 1} \ln\ \gamma_i^*(T,\underline{z}) \tag{5}$$

Equations (4) and (5) come from (3) when the limits of (2) are applied with the recognition that neither f_i^{oL} nor H_i depend on $\underline{x}$. Further, the $\underline{z}$ variations of $\ell n\ H_i$ and $\lim_{x_i \to 0} \ell n\ \gamma_i$ and $\lim_{x_i \to 1} \ell n\ \gamma_i^*$ must cancel in Equations (4) and (5). Finally, it is in principle possible to use a theory for g^E (to yield $\ell n\gamma_i$) or g^{E*} (to yield $\ell n\gamma_i^*$) to obtain the activity coefficient in the opposite convention.

$$\ell n\ \gamma_i^* = \ell n\ \gamma_i - \lim_{x_i \to 0} \ell n\gamma_i \tag{6}$$

Multicomponent Solvents

Single-solvent systems are simple in the sense that $\lim_{x_1 \to 0}$ implies $\lim_{x_2 \to 1}$. Yet, the $x_i \to 0$ specification is not complete for multicomponent solvents. The value of H_i is specified for a particular state associated with solvent compositions, $\underline{z}$ having

$$z_i \equiv x_j/\ (1-x_s) \tag{7}$$

$$\text{where } x_s \equiv \sum_{i \text{ solutes}} x_i$$

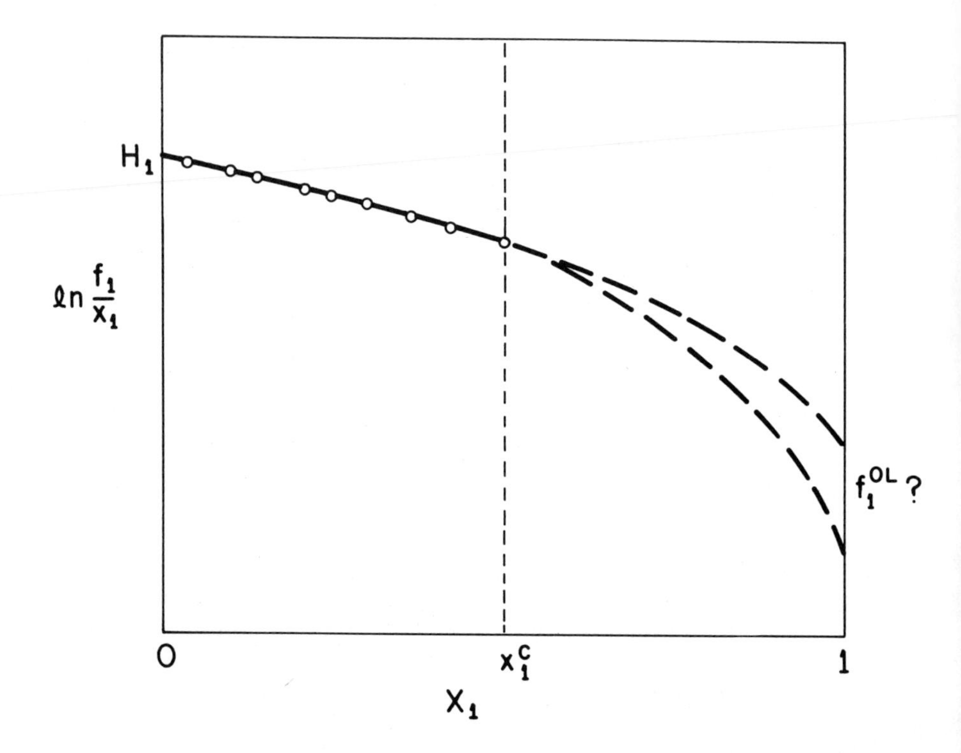

Figure 1. Fugacity of a super critical component illustrating uncertainty of extrapolation

Thus all z_j must be indicated.

Prausnitz, et. al. (1) chose a reference solvent, j=r such that H_i was the single solvent Henry's Constant

$$\lim_{x_i \to 0} H_i \equiv \lim_{x_r \to 1} H_{ir} \tag{8}$$

$$\text{and } \lim_{x_r \to 1} \gamma_i^* = 1$$

Then to be consistent, a constraint existed on γ_i^* for other pure solvents.

$$\lim_{\substack{x_j \to 1 \\ j \neq r}} \gamma_i^* H_{ir} = H_{ij} \tag{9}$$

However, as O'Connell (2) shows, the usual expressions for g^{E*} for systems such as Methane-Butane-Decane do not yield accurate values of $\lim_{\substack{x_r \to o \\ x_i \neq o}} \gamma_i^* H_i$ probably because they are insufficiently in form.

Prausnitz and Cheuh (3) chose a reference state whose solvent composition varied and the mixture H_i was "solvent-averaged" over single solvent values H_{ik}. The general formulation of this used a solvent composition variable θ_i (mole fraction, volume fraction, etc.) for $\ell n f_i^{oL}$

$$\ell n\ f_i^{oL} = \sum_{j = \text{solvents}} \frac{\theta_j}{(1-\theta_s)} \left[\ell n\ f_i^{oL} \right]_j \tag{10a}$$

$$= \sum \frac{\theta_j}{(1-\theta_s)} \left\{ \ell n\ H_{ij} + \ell n \left[\frac{\gamma_{ij}^*}{\gamma_{ij}} \right] \right\} \tag{10b}$$

This yields

$$\ell n\ \gamma_i^* H_i = \sum_{j=\text{solvents}} \frac{\theta_j}{1-\theta_s} \left[\ell n\ H_{ij} - \lim_{x_j \to 1} \ell n\ \gamma_i \right] + \ln \gamma_i \tag{12}$$

where a theory for g^E is assumed available. If this must be corrected with additional expression, g^{E*}_{corr}, to account for the unusual effects associated with supercritical components (e.g. dilation of the UNIQUAC lattice, Abrams and Prausnitz, (4)) then Equation (12) becomes

$$\ln \gamma_i^* H_i = \sum_j \frac{\theta_j}{1-\theta_s} \left[\ln H_{ij} - \lim_{x_j \to 1} \ln \gamma_i \right] + \ln\gamma_i + \ln\gamma^*_{i,corr} \tag{13}$$

where

$$\ln \gamma^*_{i,corr} = \left[\frac{\partial \left(\frac{n_t g^{E*}{}_{corr}(T,\underline{x})}{RT} \right)}{\partial n_i} \right]_{T,n_{j \neq i}} \tag{14}$$

Dilute Solutions ($\gamma_i^* \simeq 1$)

If the concentration x_s is small, Equation (13) simplifies to

$$\ln H_i - \sum_j \frac{\theta_j}{1-\theta_s} \ln H_{ij} = - \sum_j \frac{\theta_j}{1-\theta_s} \lim_{x_j \to 1} \ln \gamma_i \tag{15}$$

O'Connell (2) showed that the right-hand side is not a strong function of the species i; the difference between the mixed solvent H_i and the solvent-averaged H_{ij} depended only on the solvents, not the solute. Further, Nitta, et. al. (5) showed that if θ_j is the volume-fraction rather than the mole fraction, the rhs was not large. While no systematic investigation has been done, their plots indicate that an "area" fraction $(\theta_j \equiv x_j v_j^{2/3} / \sum_{k=solvents} x_k v_k^{2/3})$ might make it quite close to zero.

Some Cautions

Experimental evidence (Prausnitz, (6); Prausnitz and Chueh, (3)) indicates that for supercritical components the curvature of $\ln\gamma_i^*$ is different than that normally encountered with subcritical components having positive deviations from Raoult's "Law". Figure 2 shows that whereas for subcritical components

$\left(\frac{\partial^2 \ln x_1}{\partial x_1^2}\right)_{T,P} > o$, we find $\left(\frac{\partial^2 \ln \gamma_1^*}{\partial x_1^2}\right) < o$ for super critical species. Equation (6) implies they should have the same sign.

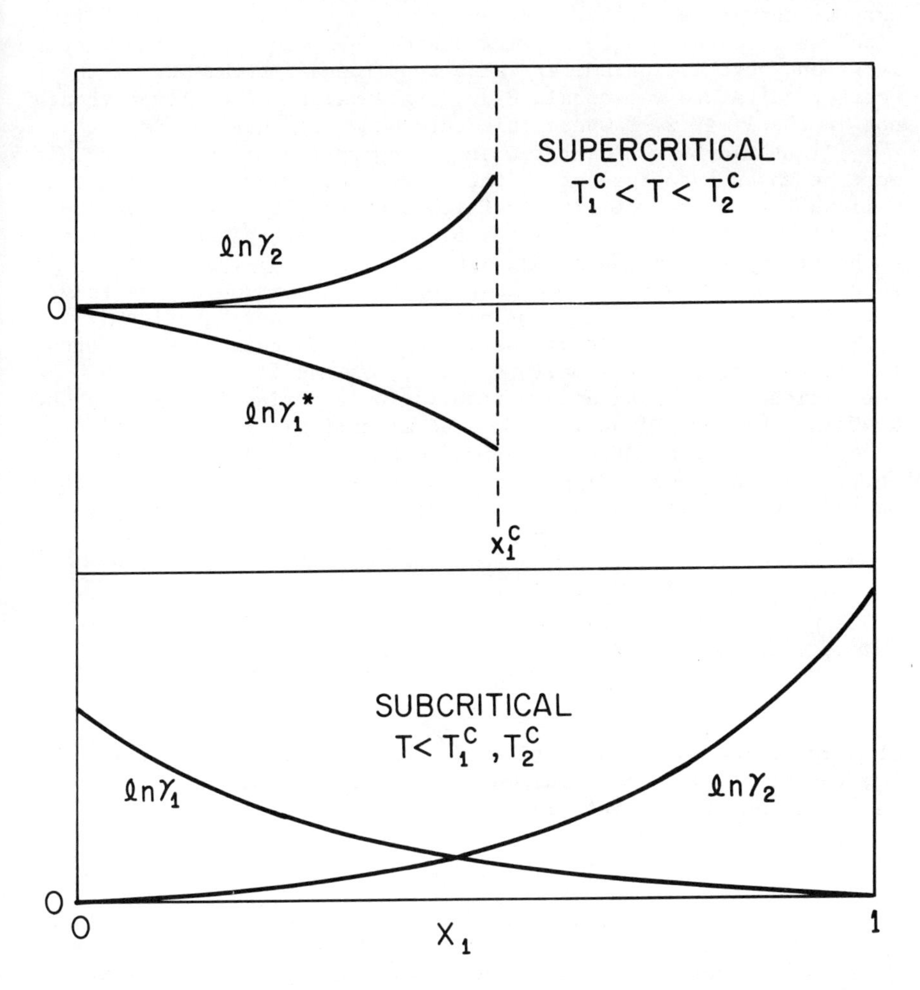

*Figure 2. Activity coefficients in binary systems illustrating difference of curvature (second derivative) of ln γ_1 and ln γ_1**

Thus, for an expression for g^E which is of the Porter, Wilson, NRTL, UNIQUAC type, there must be a correction g^{E*}_{corr} which yields the correct curvature of $\ell n\gamma^*_i$. Prausnitz and Chueh (3) added a "dilation" term to the single parameter Wohl equation. With the very large and negative parameter value they found for this term, the correct curvature was obtained at finite values of x_1 (even though not in the limit $x_1 \rightarrow 0$ where this term made no contribution).

It should be noted that when the correction term is added, it must be taken into account in the solvent activity coefficients as well as the solute activity coefficients, and it must be consistently used when Equations like (6) and (14) are pieced together as by Prausnitz and Chueh (3). In fact, Abrams, et al. (7) point out that Prausnitz and Chueh omitted a portion of the solvent activity coefficient. A safer procedure is to use Equation (13).

A fundamental question associated with calculations concerns the limits where approximations for γ^*_i, $\overline{v}^L_i$ and vapor fugacity coefficient, ϕ^V_i, break down. From Taylor's series expansion about $x_1 \rightarrow 0$ in a binary, O'Connell (8) used a statistical mechanical theory to show consistent usage would be $\overline{v}^L_1 = \overline{v}^\infty_1$ and $\ell n\ \gamma^*_1 = A_{12}\ (x_2^2 - 1)$. Higher order corrections require that both be modified. Empirically (Prausnitz, (6)), this can be used for $x_1 < 0.5x^c_1$. At the same time, the pressures and components are such that normally, ϕ^V_i can be described in this range by the second virial coefficient

$$\ell n\ \phi^V_i = \left\{ \sum_{k=1} y_k \left[2B_{ik}(T) - \sum_{j=1}^{M} y_j B_{kj}(T) \right] \right\} P/RT$$

Choosing to be more "sophisticated" in any of these quantities γ^*_i, $\overline{v}^L_i$, ϕ^V_i requires becoming more sophisticated in all. The above range is one of considerable practical significance, however.

Calculational Aspects

Two particular things should be borne in mind when computing phase equilibria within these systems. One is that convergence depends on smooth derivatives and projections. Therefore since changing correlations or conventions always yields discontinuous values as well as slopes, calculation schemes should not be changed in "midstream." It is better to stay with one form or another through a calculation and if the correlations are not applicable at the end states, redo the problem with more appropriate calculations. Do not risk becoming involved with inordinately long calculations due to discontinuities.

The other aspect is that when the temperature is to be solved for in a given problem, particularly when $\underline{x}$ is specified (bubble temperature problem) the correlation for Henry's constant must be highly accurate or poor values of the final temperature will be found. This occurs because H_i is not as sensitive to T as is

P^{sat}, for example, even though the total pressure on the system is determined primarily by this value. (For example, for N_2 or CH_4, H_i can even go through a minimum at moderate temperatures.) The calculation will attempt to adjust T to satisfy the pressure constraint and may have to go many degrees away from the correct value to account for a small error in the predicted H_i. While x_i may be adjusted as well in the dew temperature problem ($\underline{y}$ specified) so that less errors occur, my experience is that only highly accurate values of H_i will obviate this difficulty in general.

Literature Cited

1. Prausnitz, J. M., C. A. Eckert, R. V. Orye and J. P. O'Connell, Computer Calculations for Multicomponent Vapor-Liquid Equilibria," Prentice-Hall, 1967.
2. O'Connell, J. P., AIChE. J., (1971) 17, 650.
3. Prausnitz, J. M. and P. L. Chueh, "Computer Calculations for High-Pressure Vapor-Liquid Equilibria," Prentice-Hall, 1968.
4. Abrams, D. and J. M. Prausnitz, AIChE J. (1975) 21, 116.
5. T. Nitta, A. Tatsuishi, and T. Katayama, J. Chem. Eng. Japan, (1973) 6, 475.
6. Prausnitz, J. M., "Molecular Thermodynamics of Fluid-Phase Equilibria," (1969), Prentice-Hall, Chapter 10.
7. Abrams, D., F. Seneci, P. L. Chueh and J. M. Prausnitz, Ind. Eng. Chem. Fund., (1975) 14, 52.
8. O'Connell, J. P., Mol. Phys., (1971) 20, 27.

Note Added in Proof

It may be that this anomolous behavior is observed because in the evaluation of the integral in Equation (16), the integrand is assumed constant. Thus, at high concentrations, the exponential term is too large (the value of $\bar{v}_i^L$ is actually less at high pressure) and γ_i^* appears to be smaller in order to compensate.

30

Phase Equilibria in the Critical Region: An Application of the Rectilinear Diameter and "⅓ Power" Laws to Binary Mixtures

M. R. MOLDOVER and JOHN S. GALLAGHER

National Bureau of Standards, Washington, DC 20234

It is well known that binary mixtures near their critical loci have properties which seem to differ qualitatively from the corresponding properties in pure fluids. The most conspicuous difference is the occurrence of retrograde condensation in mixtures (1). Thus, as a rule, the critical point of a binary mixture of a given mole fraction, x, occurs at neither the maximum temperature nor the maximum pressure at which that mixture can separate into two phases. This apparently drastic difference between binary mixtures and pure fluids suggests that two powerful and simple rules for correlating coexisting density data in pure fluids are not applicable to binary mixtures. These two rules are the "1/3 power law", or

$$(\rho_{liquid} - \rho_{vapor})/(2\rho_c) = C_1(-t)^{1/3} \tag{1}$$

and the "law of the rectilinear diameter"

$$(\rho_{liquid} + \rho_{vapor})/(2\rho_c) = 1 + C_2 t \tag{2}$$

(Here $t = (T-T_c)/T_c$ and ρ_c is the critical density.) It is the purpose of this note to point out that there is a sense in which these two laws may in fact be applicable to mixtures.

The underlying idea is one which Griffiths and Wheeler (2) have emphasized as being particularly applicable to the critical region: The thermodynamic properties of mixtures can be expressed in simple analogy with the thermodynamic properties of pure fluids if variables such as fugacities are used as independent variables to specify the mixture under consideration. A particularly convenient variable is one used by Leung and Griffiths (3) to write down an accurate thermodynamic potential for the critical region of $He^3 - He^4$ mixtures. They introduce the variable

$$\zeta = f_1/(kf_2 + f_1) \tag{3}$$

(f_1 and f_2 are the fugacities of the first and second components, respectively and the constant k emphasizes the fact that the numerical values assigned to ζ depend upon the choice of reference state for the fugacities. Of course, physical properties do not.) In Figure 1 we have sketched the saddle shaped ρ-T-ζ surface for VLE states of CO_2 - C_2H_6 to illustrate our interpretation of the ideas of Griffiths and Wheeler. Each cross section of the ρ-T-ζ surface at constant ζ has the same functional form as the ρ-T coexistence curve for a pure fluid. All the "constants" in this function now have a smooth ζ dependence:

$$(\rho_{liquid} - \rho_{vapor})/(2\rho_c(\zeta)) = C_1(\zeta)(-t)^{1/3} \tag{4}$$

$$(\rho_{liquid} + \rho_{vapor})/(2\rho_c(\zeta)) = 1 + C_2(\zeta)\ t \tag{5}$$

The reduced temperature variable is also smoothly ζ dependent:

$$t = (T-T_c(\zeta))/T_c(\zeta) \tag{6}$$

The critical point of every mixture occurs at the maximum temperature of two phase coexistence (at constant ζ) exactly in analogy with a pure fluid. In Figure 1 the curved sheet which intersects the ρ-T-ζ saddle represents the thermodynamic states accessible to a mixture of a fixed mole fraction, x. The intersection of this sheet with the saddle is the locus VLE states accessible to a mixture of fixed composition. It is clear that the critical point on this locus is not, in general, at the maximum temperature on the locus. Thus, Figure 1 provides a pictorial representation of the origin of retrograde condensation in the Griffiths-Wheeler picture and indicates the sense in which the "1/3 power law" and the "law of rectilinear diameter" are preserved in mixtures. The remainder of this paper is devoted to an application of this point of view.

It is possible to generalize to mixtures the singularity which is known to occur in the specific heat (4) and in the vapor pressure (5). This can be done by using the P-T-ζ surface for coexisting phases instead of the P-T-x surface conventionally used. These surfaces are illustrated in Figure 2. Because the P-T-ζ surface consists of a single smooth sheet it is easy to write down an algebraic representation of the surface which has the singularity (i.e. noninteger exponent) found in the vapor pressure and which, when differentiated twice, gives a constant volume specific heat which approaches infinity at the critical point in agreement with experiment. A possible representation of the P-T-ζ surface is:

$$\frac{P}{T}\frac{T_c(\zeta)}{P_c(\zeta)} = 1 + C_3\ (-t)^{1.9} + C_4\ t + C_5\ t^2 + C_6\ t^3 \tag{7}$$

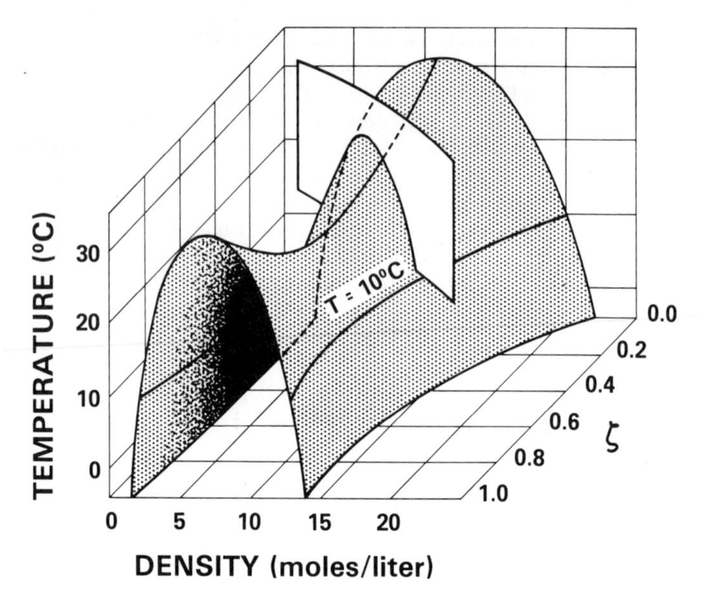

Figure 1. Vapor–liquid equilibria states near the critical locus of CO_2–C_2H_6. Values of ρ-T-ζ fall on a saddle-shaped surface. The critical locus is the heavy curve along the top of the saddle. The curved sheet intersecting this surface represents, schematically, all states of a given mole fraction x.

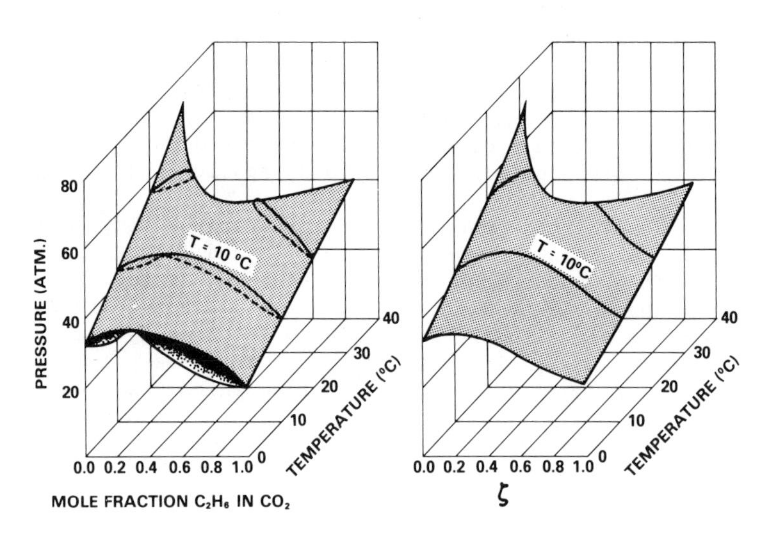

Figure 2. Vapor–liquid equilibrium states near the critical locus of CO_2–C_2H_6. Left: Values of P-T-x; *right: Values of* P-T-ζ. *Here all states with two coexisting phases are on a single surface. The azeotrope is the ridge in this surface. The critical locus is the upper boundary of this surface.*

The exponent 1.9 leads to a divergence (as $(-t)^{-0.1}$) of both the second derivative of the vapor pressure and of C_V (at constant ζ) as the critical point is approached.

Both the "1/3 power law" and the vapor pressure singularity may be combined in a thermodynamically consistent way by writing down a thermodynamic potential of the so called "scaling" type. Leung and Griffiths did so for mixtures of He^3 in He^4 and D'Arrigo et al. (6) used the Leung-Griffiths scheme in mixtures of CO_2 in C_2H_4. In each case the authors used an exponent 0.35 instead of 1/3 because such an exponent seems to agree somewhat better with a wider range of data. (These authors actually used power laws in the variable t/(1+t) which does not fit the coexisting density data well over as large a range as the power law in the variable t which we use.) In another publication (7) we have somewhat modified the Leung-Griffiths potential. In our thermodynamic potential the various constants ($C_i(\zeta)$) which vary from fluid to fluid are written in a dimensionless form where the critical parameters $T_c(\zeta)$, $P_c(\zeta)$, and $\rho_c(\zeta)$ are used as reduction factors. This is illustrated in Equations (4, 6, and 7). Upon examining the literature data for the critical region of many pure fluids we have found that except for the two helium isotopes and to a lesser extent hydrogen, the dimensionless coefficients C_i are essentially the same for the various fluids studied. (The correlation of Levelt Sengers et al. (8) was especially helpful. Our coefficient C_3 is equal to $0.7222(\overline{a})(x_o)^{-0.355}$ in their notation). The minor variation of C_i from fluid to fluid led us to try models in which $C_i(\zeta)$ is determined by linear interpolation in ζ from pure fluid data alone. The values we actually used are listed in Table I. We have also used an interpolation scheme to estimate the behavior of ζ along the critical locus. The simplest such scheme is to assume ζ is identically equal to 1-x along the critical locus itself. This assumption is exact for a mixture of ideal gases. Arguments for more general applicability of this assumption are made elsewhere. With this assumption, the only parameters that remain to be fitted to mixture data are those which specify the physical mixture critical locus: $T_c(x)$, $P_c(x)$, and $\rho_c(x)$. An alternative to obtaining these functions from data would be to "predict" them by using, for example, a van der Waals theory for mixtures. We have not tried this alternative.

We will not write down our thermodynamic potential here. (it is only one of many from which equations similar to (4), (5), and (7) may be derived. Details will be published elsewhere (7)). From our potential, it is possible to derive, by differentiation, a sequence of equations which enable one to determine isothermal sections of the coexistence surface without an iterative root-finding process and without any knowledge of the variable ζ. We now enumerate these derived equations in the same order that a computer program might execute them. It is assumed that pure fluid values for C_i are available (C_{i1} for the first component for which

Table I: Constants for Pure Components

Substance	CO_2	SF_6	C_2H_6	C_3H_8	He^3	He^4
Property (units)						
T_c (K)	304.17	318.82	305.34	369.88	3.3105	5.1884
ρ_c (mol/liter)	10.6	5.034	6.880	4.927	13.82	17.3
P_c (atm.)	72.894	37.157	48.076	41.983	1.1322	2.2434
C_1	2.009	1.980	1.924	1.947	1.289	1.425
C_2	-0.995	-0.859	-0.834	-0.861	0.05	-0.1
C_3	30.87[a]	29.98[a]	28.84[a]	29.57[a]	4.26[a]	6.586[a]
C_4	5.999	5.974	5.455	5.704	2.4174	2.9306
C_5	-26.128	-25.899	-25.942	-25.746	-3.32015	-5.1042
C_6	-5.486	-14.060	-8.130	-6.287	-.5691	-.6820
References	(9)	(11, 12)	(13)	(11, 12)	(15)	(16, 17)

a-- Value obtained by method described in this text from Table of reference (8).

$x = 0$ and C_{i2} for the second component) as well as functions specifying the critical locus $T_c(x)$, $P_c(x)$, and $\rho_c(x)$ and its derivatives dT_c/dx and dP_c/dx. Then an isotherm at temperature T is calculated by substituting successive values of ζ (say $\zeta = 0$, 0.01, 0.02 ..., 0.99, 1.0) in the following equations:

$$x_c = 1 - \zeta \tag{8}$$

$$t = (T - T_c(x_c))/T_c(x_c) \tag{9}$$

$$\frac{dC_i}{d\zeta} = C_{i2} - C_{i1} \qquad (i = 1, ..., 6) \tag{10}$$

$$C_i = \frac{dC_i}{d\zeta}\,\zeta + C_{i1} \qquad (i = 1, ..., 6) \tag{11}$$

$$\rho = \rho_c(x_c)\ (1 \pm C_1\ (-t)^{0.355} + C_2 t) \tag{12}$$

$$P = P_c(x_c)(T/T_c(x_c))(1 + C_3(-t)^{1.9} + C_4 t + C_5 t^2 + C_6 t^3) \tag{13}$$

$$\overline{Q}(\zeta,t) = -\frac{P}{RT}\frac{T_c}{P_c}\frac{d(p_c/T_c)}{dx} + \frac{P}{RT_c}\left[\frac{dC_3}{d\zeta}(-t)^{1.9} + \frac{dC_4}{d\zeta}t + \frac{dC_5}{d\zeta}t^2 + \frac{dC_6}{d\zeta}t^3\right] - \frac{P_c}{R}\frac{d(1/T_c)}{dx}(1+t)[-1.9\ C_3(-t)^{0.9} + C_4 + 2C_5 t + 3C_6 t^2] \tag{14}$$

$$x = (1-\zeta)\{1-\zeta\ (\frac{\overline{Q}(\zeta,t)}{\rho} - \frac{\overline{Q}(\zeta,0)}{\rho_c})\} \tag{15}$$

Thus, for given T, a table of equally spaced ζ values has been used to determine corresponding "experimental" variables ρ, P, and x. In these equations, R is the gas constant. If the + sign is used in Equation (13), a liquid state will be obtained; if the - sign is used, the vapor state which coexists with the liquid state will be obtained. In Equation (14) we have not written out the explicit x dependence of T_c, P_c and their derivatives.

We have studied the VLE data for four mixtues near their critical loci: He^3 - He^4, CO_2 - C_2H_6, SF_6 - C_3H_8, and C_8H_{18} - C_3H_8. In each case we used Equations (8-15) to represent the PVTx data. In each case the parameters specifying the physical critical loci were adjusted to optimize (by eyeball) the fit to the data. A convenient form for which we used to represent the physical critical loci is:

Table II: Mixture Critical Constants

These constants, which specify the location of the critical loci, are defined in Equations (16-18).

Mixture	$CO_2 - C_2H_6$	$SF_6 - C_3H_8$	$He^3 - He^4$
ρ_1 (mol/liter)	-3.152	-0.054	0.
ρ_2 (mol/liter)	-0.988	-0.466	0.
P_1 (mol/liter)	-0.2929	0.1542	1.0
P_2 (mol/liter)	-0.1537	-0.4418	0.
T_1 (mol/liter atm)	0.0075	0.006027	-0.62
T_2 (mol/liter atm)	-0.00112	0.002947	0.
T_3 (mol/liter atm)	-0.00188	0.	0.

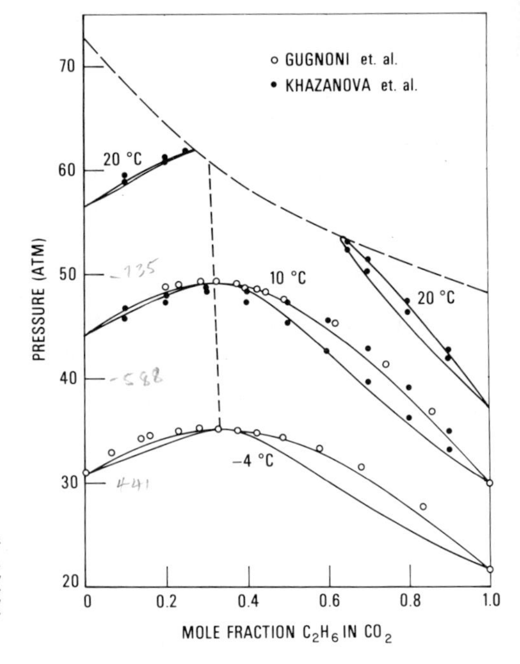

Figure 3. Vapor–liquid equilibrium in CO_2–C_2H_6. The model is shown by solid curves. The azeotrope and critical locus are shown as dashed curve.

$$\frac{1}{RT_c(x)} = \frac{1-x}{RT_{c1}} + \frac{x}{RT_{c2}} + x(1-x)\ (T_1 + (1-2x)\ T_2 + (1-2x)^2\ T_3) \quad (16)$$

$$\frac{P_c}{RT_c(x)} = \frac{(1-x)P_{c1}}{RT_{c1}} + \frac{xP_{c2}}{RT_{c2}} + x(1-x)\ (P_1 + (1-2x)\ P_2) \quad (17)$$

$$\rho_c(x) = (1-x)\rho_{c1} + x\rho_{c2} + x(1-x)(\rho_1 + (1-2x)\rho_2) \quad (18)$$

The values we used for T_1, T_2, T_3, P_1, P_2, ρ_1, and ρ_2 are listed in Table II.

We were able to obtain satisfactory representations of the VLE data for He^3 - He^4, CO_2 - C_2H_6, and SF_6 - C_3H_8. This is illustrated in Figures 3-6. We have not shown the data for the He^3 - He^4 mixtures. Its behavior is not "typical". This mixture is in some sense so "ideal" that the critical locus as a function of x is nearly straight (Table 2) and the VLE data for a mixture of a given x look almost like data for a pure substance (14). Our model fits the VLE data from the critical locus to about two thirds the critical temperature for each composition in the He^3 - He^4 system. This range of good fit is atypically large.

In Figures 3 and 4 we compare our model for CO_2 - C_2H_6 with the data of Khazanova et al. (18) and the data of Gugnoni et al (19). By adjusting the parameters which specify the critical locus we were able to obtain agreement between the model and the data which is comparable to the agreement between the two sets of data. The temperature range covered is about the same as the range for which Equations (4), (5), and (7) could be expected to fit good VLE data in pure fluids.

In Figures 5 and 6 we compare our model for SF_6 - C_3H_8 with the data of Clegg and Rowlinson (11, 12). The agreement is again quite good over a range which is comparable to the range over which the "1/3 power" law (or rather the 0.355 power law) seems to hold in pure fluids. We were unable to obtain a satisfactory representation of the data for the propane rich mixtures ($x > 0.8$) of C_3H_8 - C_8H_{18} (20) with our simple scheme (Equations (8-15)). At the moment, we suspect that in mixtures such as the latter, where $(dT_c/dx)/T_c$ becomes large, a more complete thermodynamic model may be required to describe phase equilibria accurately in the critical region. In our simple scheme we have made no reference to standard states and we have made no attempt to describe the behavior of the entropy as a function of composition along the critical line. This may not be possible in general.

We would like to thank Professor R. B. Griffiths for his continuing advice and encouragement.

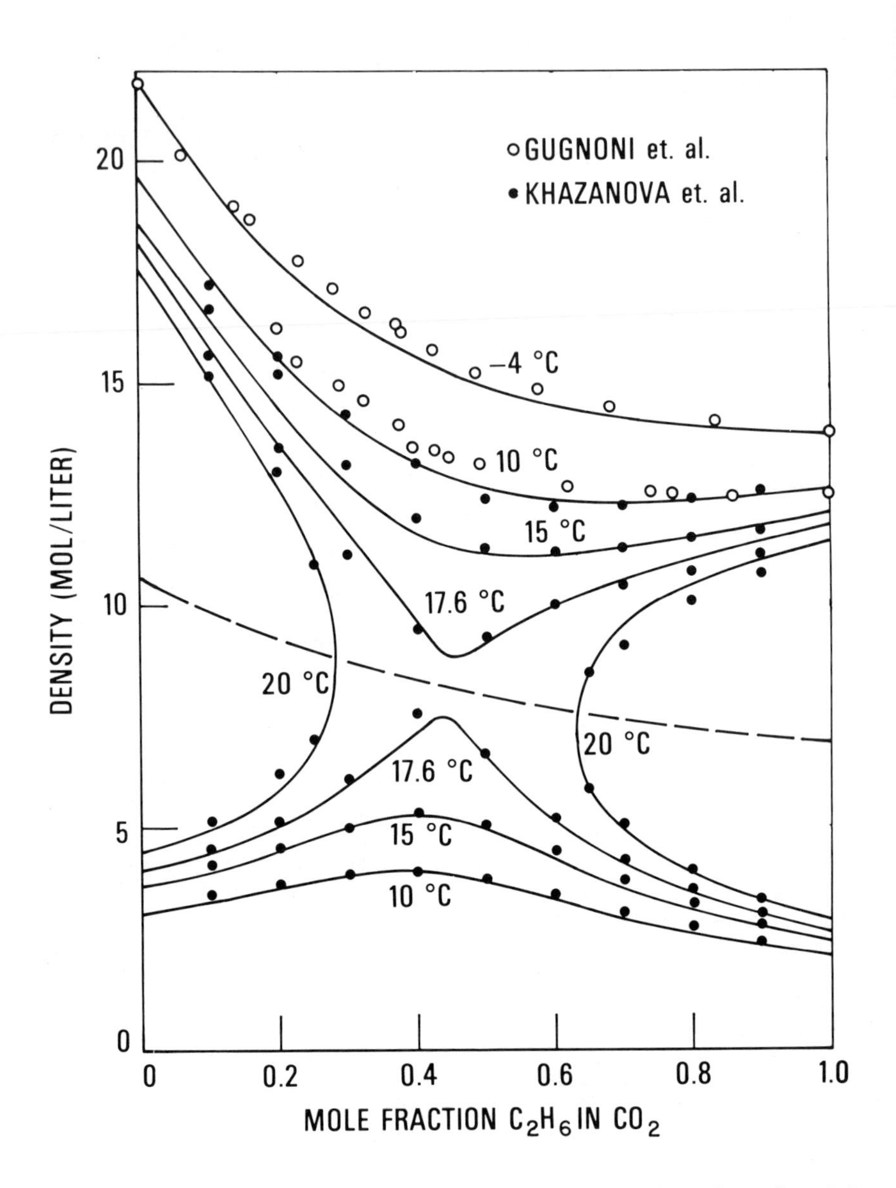

Figure 4. Vapor–liquid equilibrium in CO_2–C_2H_6. The model is shown by solid curves. The critical locus is the dashed curve.

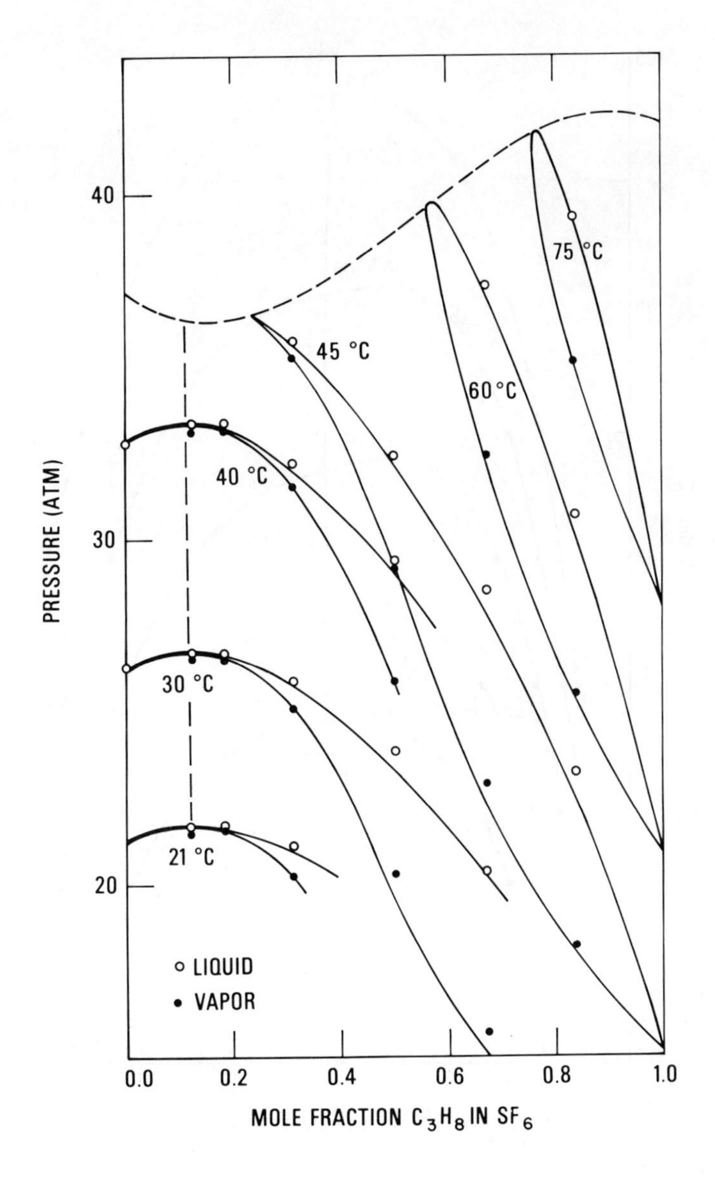

Figure 5. Vapor–liquid equilibrium in SF_6–C_3H_8. The model (solid curves) is compared to the data of Clegg and Rowlinson. The critical locus and azeotrope are indicated by dashed curves.

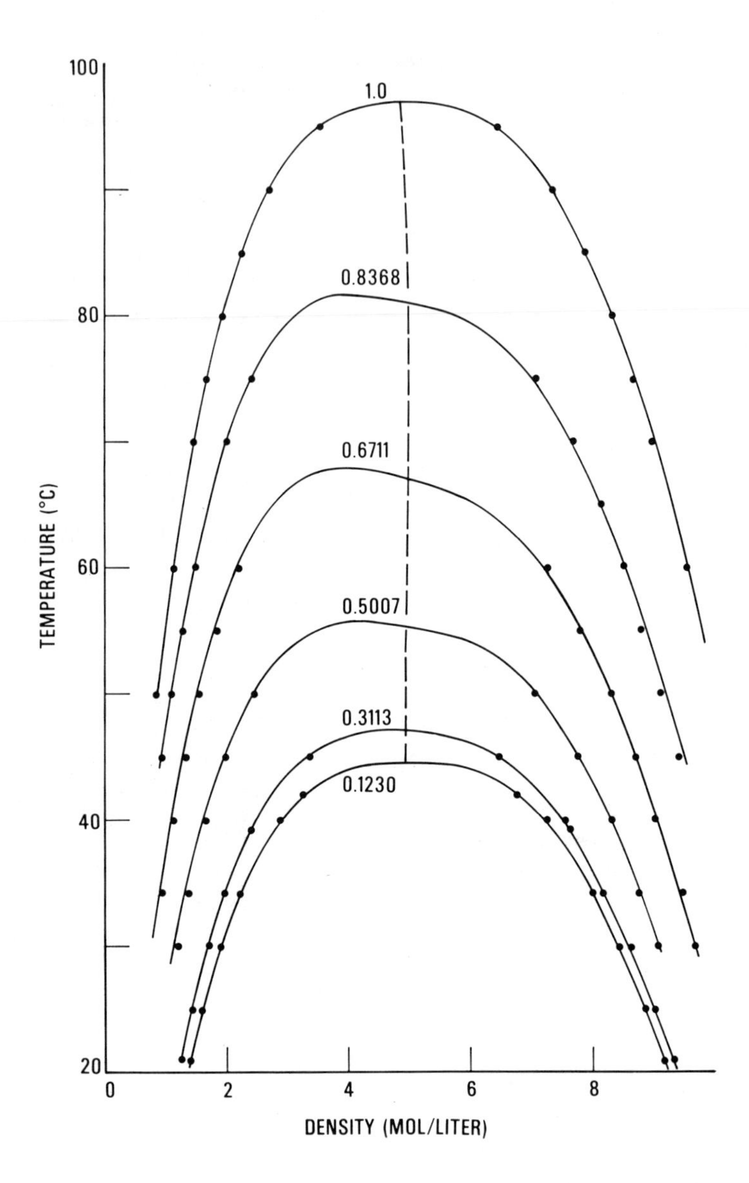

Figure 6. Vapor–liquid equilibrium in SF_6–C_3H_8. The model (solid curves) is compared with the data of Clegg and Rowlinson. The critical locus is indicated by a dashed curve.

Literature Cited

1. Kay, W. B., Accounts of Chemical Research, (1968) 1, 344.
2. Griffiths, R. B., and Wheeler, J. C., Phys. Rev. (1970) A2, 1047.
3. Leung, S. S. and Griffiths, R. B., Phys. Rev. (1973) A8, 2670.
4. Levelt, Sengers, J.M.H., "Critical Behavior in Fluids" in "High Pressure Technology and Processes," edited by I.L. Spain and J. Paauwe. Marcel Dekker (New York), in press (1977).
5. Kierstead, H. A., Phys. Rev. (1973) A7, 242; Levelt Sengers, J. M. H, and Chen, W. T., J. Chem. Phys. (1972) 56, 595.
6. D'Arrigo, G. L., L. Mistura, and P. Tartaglia, Phys. Rev. (1975) A12, 2587.
7. Moldover, M. R. and Gallagher, J. S., "Critical Points of Mixtures-An Analogy with Pure Fluids," submitted to AIChE Journal, April 1977.
8. Levelt Sengers, J.M.H., Greer, W. L. and Sengers, J. V., J. of Phys. Chem. Ref. Data (1976) 5, 1.
9. Levelt Sengers, J.M.H. and Chen, W. T., J. Chem. Phys. (1972) 56, 595.
10. Michels, A., Blaisse, B. and Michels, C., Proc. Roy. Soc. (London) (1937) A153, 201.
11. Clegg, H. P., Rowlinson, J. S. and Sutton, J. R., Faraday Soc. Trans. (1955) 51, 1327
12. Clegg, H. P. and Rowlinson, J. S., Faraday Soc. Trans. (1955) 51, 1333.
13. Douslin, D. R. and Harrison, R. H., J. Chem. Thermodynamics (1973) 5, 491.
14. Wallace, B., Jr. and Meyer, H., Phys. Rev. (1972) A5, 953; Duke University Physics Department Technical Report (1971).
15. Sherman, R. H., Sydoriak, S. G. and Roberts, T. R., J. of Research NBS (1964) 68A, 579.
16. Brickwedde, F. G., van Dijk, H., Durieux, M., Clement, J. R., Logan, J. K., J. of Research NBS (1960), 64A, 1.
17. Roach, P. R., Phys. Rev. (1968) 170, 213.
18. Khazanova, N. E., Lesnevskaya, L. S., and Zhakarova, A. B., "Liquid-Vapor Equilibrium in the System Ethane-Carbon Dioxide," Khimcheskaia Promyshlennost (1966) 44, 364.
19. Gugnoni, R. J., Eldridge, J. W., Okay, V. C. and Lee, T. J., AIChE Journal (1974) 20, 357.
20. Kay, W. B., Genco, J. and Fichtner, D. A., J. Chem. Eng. Data (1974) 19, 275.

Discussions of the Ad Hoc Sessions

A. H. LARSON

Editor's Note

The ad hoc sessions were held informally and notes were not kept during these sessions. Dr. A. H. Larson did prepare a summary of the ad hoc session held on Thursday afternoon and has consented to let us reproduce it here.

John M. Prausnitz - "New Equations of State". Large molecules, such as polymers, wiggle, rotate, and vibrate. These kinds of behavior are not considered by Redlich-Kwong type equations, but require new physics. The perturbed hard-chain theory provides an equation of state for fluids containing simple or complex molecules, covering all fluid densities--from argon to polymer, ideal gas to close-packed liquid. A partition function for large molecules is developed, including a factor for internal and external degrees of freedom. The volume is added to get the correct ideal-gas limit. The result is an equation in Helmholtz energy. There are three parameters, independent of temperature and density: a characteristic size, a characteristic energy, and the number of external degrees of freedom. Separate mixing rules are used for each constant in the equation. Flexibility is needed to meet the boundary conditions. This approach has recently been applied to mixtures as well as pure compounds.

For complex systems, including chemical equilibria, van der Waals forces and cluster formation in fluids are combined to result in analystic perturbed hard-sphere equation of state in pressure. The result correctly reduces the critical compressibility factor for more and more highly polar components. All constants in the equation can be obtained from the critical point, the normal boiling point, and the acentric factor. The equation does a fairly good job for pure compounds. The next step will be application to mixtures, such as copolymers.

Professor Prausnitz was cited for major contributions to the Conference.

Grant M. Wilson - "Industry-Sponsored Research at Universities". A proposal has been made for three cooperative research programs for measurement of vapor-liquid equilibrium data at the Thermochemical Institute at Brigham Young University:

1. Hydrogen-hydrocarbon phase equilibrium data at low temperatures, 32-500 ^{o}F.
2. Hydrogen-hydrocarbon phase equilibrium data at high temperatures, 500-900 ^{o}F.
3. Binary vapor-liquid equilibrium data for multicomponent modeling.

Copies of the proposal were available and distributed to those interested.

R. N. Maddox described Fluid Properties Research, Inc. (FPRI) facilities at Oklahoma State University for measuring liquid density, viscosity, thermal conductivity, and interfacial tension. In addition to its regular research program, FPRI carries out proprietary measurements for member companies and, on a limited basis, for others.

Lloyd L. Lee - "Perturbation Equation of State for Polar Fluids". This work with K. E. Starling, is in progress, and there are no definitive results to present yet. It is a downstream application of Rowlinson and Gubbins' work. Three approaches are under investigation: (1) Consider the total pair potential to consist of isotropic and anisotropic parts; (2) develop perturbation expansions for the Helmholtz energy; and (3) separate the multipolar interactions from the temperature--and density--dependent part of the compressibility factor, as a generalization of the three-parameter corresponding states theory.

Joseph Kestin - "Transport Properties". A formalism has been developed to calculate transport properties as well as thermodynamic properties for gaseous mixtures. The theory includes the Boltzmann equation, the Chapman-Enskog solution for rare gases with extension to mixtures, the extended law of corresponding states, the Wang-Chang-Uhlenbeck development, extended by Mason and Monchick to polyatomic gases, and intermolecular force potentials, used in calculation of functionals. A common set of functionals satisfying consistency criteria can be selected. The result showed good fit to data and prediction of properties for monatomic and simple polyatomic compounds and mixtures, including second virial coefficients, collision integrals, binary diffusion coefficients, viscosity, and thermal conductivity. Calculations can be done more accurately and more easily than data can be measured, for these kinds of systems.

John P. O'Connell - "Aspects of Henry's Law Constants and Unsymmetric Activity Coefficients". The unsymmetric convention merely substitutes $\gamma_i^* H_i$ for $\gamma_i f_i^{OL}$ and should be done only if f_i^{OL} is not available, such as for light gases or for temperatures above 1.5 times the critical temperature of the solute. A solvent -averaged $\ln f_i^{oL}(T)$ can be found, using the composition variable. This requires binary information in the form of Henry's Law constants and theory for excess Gibbs energy. Some cautions were given: (1) The theory for the excess Gibbs energy g^E and corrections by lattice dilation g^E_{corr} must yield the correct curvature for $\ln \gamma_i^*$; (2) g^E_{corr} changes γ_k as well as γ_i^*; (3) application of pressure and composition corrections must be consistent--corrections for any higher x_1 require corrections for all; (4) the same convention and correlations must be used throughout process calculations.

Paul Y. Chen - "Data Needs in Pollution Abatement". The 1977 effluent limits and the 1982 "zero discharge" EPA requirements raise many problems and needs. Some major problems are the following: (1) azeotropes, reduction to very low concentrations; (2) distribution of organics between hydrogen chloride and water; and (3) obtaining data and models quickly to meet the imposed requirements. Suggestions were requested. Group contribution methods may be useful for screening purposes and for troubleshooting.

Summaries

Academic View

ROBERT N. MADDOX

In the beginning, I think that we should thank the conference co-chairmen, Dr. Sandler and Dr. Storvick, and the members of the organizing committee for a very interesting program. Presentations were educational and enlightening and I am sure that all in attendance benefited from the lectures and discussions.

Most of my comments will be of a general nature rather than specific. However, there is one observation that I feel needs to be made, particularly in view of some of the comments presented. This concerns the fact that, while the Chao-Seader may not have been the perfect equation of state for predicting thermodynamic properties, it does have one major distinction--it was the first equation of state--thermodynamic property correlation developed solely and intentionally for use on a digital computer. As a matter of fact, for a short time it came as close to being an "industry standard" in the gas processing industry as any equation of state has been or is likely to be.

Most of the speakers at the conference addressed themselves to the subject of vapor-liquid equilibrium in systems with two phases (vapor, liquid) and, for the most part ideal systems. One must wonder if equilibria constants or K values actually dominate to this extent. If so where are the enthalpies, entropies, heats of solutions, heats of reaction, etc.

I was pleased to hear many of the speakers mention the need for experimental data. In the thermodynamics area, as well as others, there is certainly need to check theoretical predictions against reliable experimental values. In addition, as an experimentalist, I was made to feel wanted and almost personally addressed when the speakers said "There is need for these data."

The discussions of the conference, both formal and informal, seemed to indicate three general areas where additional information is needed:

1. There is need for additional theoretical development and experimental data on non-ideal systems. This is particularly true for mixtures, polar materials, those

that associate-disassociate and those containing more than two phases.

2. A particular concern among petroleum people is predictive correlations for the properties of so called "undefined components." These require something other than chemical composition as a means of predicting properties and behavior. Currently available techniques have proven to be unsatisfactory for these materials. This is an area in which the theorist, perhaps, can put forth new suggestions and new ideas as to properties or characteristics that can be used for predicting properties and behavior of these materials. The properties used as a basis for these techniques must be easily and readily measured accurately.

3. The quantity of available experimental data for non-hydrocarbon components and systems needs to be expanded. The data should be of good quality and collected with the latest and best available techniques.

If the proceedings of this conference are indicative of any trend, I would say that it would involve research, development and utilization of the group contribution technique for correlation and prediction of thermophysical properties. This conference, in a sense, brought this technique to respectability. I predict that much valuable work in this area will be accomplished in the next ten years.

Industrial View

A. H. LARSON

We have had an exciting and stimulating week. For me it has been a magnificent experience. What I would like to give you is a personal view of the impact of the Conference with an orientation toward industrial needs.

Breadth of Conference

I'd like to talk briefly about the breadth of the Conference as I see it, the dialogue that has been established between the various interests represented here, and the new developments expressed; and mention some properties that were not discussed, indicate the needs that were emphasized, and comment on some future directions that may result from this Conference.

The Conference was divided about evenly between industrial and academic participants: 65 from academia, 62 from industry, and 9 from government and other institutions. What strikes me more than this division of participants is the variety of interests that were represented, all the way from the theorists to those primarily concerned with correlation and estimation techniques, to experimentalists interested in obtaining data, to the industrialists, who are concerned about getting the engineering work done, the process designs and the decisions that have to be made on an engineering basis.

The title of the Conference indicates to me the breadth of these interests and reflects and establishes usefulness as the purpose of the Conference. Notice that the title includes "estimation and correlation techniques" which must be based at least to some degree on theory but also include empirical correlations. Chemical industry is the bottom line and that's where it all matters. Whatever is developed or considered or thought about that doesn't see an eventual application in industry somehow doesn't fit. The title thus reflects both the breadth of interests and the idea that we want to bridge this variety of interests and establish a meaningful dialogue and interaction between us.

Dialogue Established

The dialogue was established, as I see it, in three main areas. First in the formal presentation and reviews, which I thought did a magnificent job in developing the historical background in a number of areas, indicating the contrast between the academic and industrial approach. They showed also that there are some real differences in language between the theorists and the engineers in industry. For example, there are very few people working in industry who speak daily about spinodals, criticality, partition functions, congruence, or shape factors, and perhaps most of the theoreticians don't employ in their normal vocabulary such terms as process flow diagrams, tertiary recovery, or heat exchanger duty. So we have some elements of language that are not common among all of us, but we have a good deal that is in common; and these aspects that are in common are capatilized on in this Conference.

The contrast between the academic and industrial approach was shown most vividly between the two reviews in Phase Equilibria. The first emphasized the development of correlations and the second, providing process engineers with the tools they need. Now if you think about it, these two have very much in common; in fact, one follows from the other. To the extent that they mean the same thing in your mind, we have established a common dialogue.

In addition to the formal presentations and reviews, there were a number of ad hoc sessions which I think were significant and meaningful. There were meetings of task groups for CODATA and for the Manufacturing Chemists Association, discussions of various papers, and some additional presentations which contributed greatly.

Finally, this dialogue was established perhaps most meaningfully for each of us in the personal conversations and interactions we have had. In fact, I consider that to be the most valuable part of this Conference. The formal presentations served mainly as a stimulus for this personal dialogue.

New Developments

There were a large number of developments in progress discussed and some of these are very significant for industry. The group contribution estimation methods for activity coefficients I see as a significant tool that has great potential and will be usable by process engineers in the immediate future. But it also emphasizes an outstanding need for more and better experimental data to achieve further progress.

Another new development is that of thermophysical property data banks, some of which have been established in various parts of the world: Dechema, PPDS, the Japanese JUSE project, and discussions on a new project in this country--Project Evergreen. These also show promise of real help to industry by providing better and more convenient data and by avoiding a good deal of duplication of effort.

Also treated by many speakers was the development of new equations of state. One example that I would like to indicate was mentioned by John Prausnitz in an ad hoc session: the perturbed hard-sphere equation, in which all the constants for each component are obtained from the critical point, the normal boiling point, and the acentric factor. For complex molecules for which there are very little data, this is the kind of equation of state in which industry is really interested. Also, as has been mentioned in the Conference, there has been a noticeable shift in the past few years in emphasis from computer programs to data as the key factor in the thermodynamics business. But we should not minimize the computer's role in this activity and, in fact, as Bob Maddox pointed out, the role of the computer is becoming even more significant.

Properties Not Discussed

There are a few properties that were not discussed and for which estimation methods and correlation methods are important, especially for complex molecules where a single equation of state cannot be conveniently used for both vapor and liquid phases. Separate correlations are required for vapor pressure, latent heat of vaporization, liquid enthalpy, and perhaps liquid heat capacity. In fact, in the past few years there have been some developments reported in the literature in these areas but not mentioned in this Conference. These properties are indeed important to process engineers. It leaves me a little uneasy to think that the lack of mention here may indicate that no further work needs to be done in these areas.

Needs Emphasized

As to the needs that have been emphasized, I think the primary one is that of data measurements. Thermodynamics is not magic and it cannot extract something out of nothing, so accurate data are needed for correlation work, as tests for theories, and also for the complex processes and designs and decisions that must be made in industry. We have very complex molecules and systems to consider. In many cases theory and correlations do not represent them adequately and engineers must resort to measured data.

Arrangements for getting data measured, however, were not mentioned very much. They were discussed in a proposal that Grant Wilson of Brigham Young University has outlined but I don't recall any other concrete way in which arrangements for measurements were discussed.

Another need is that correlations that are developed must be usable by process engineers in design work. Often a published correlation makes only a small incremental improvement over some existing correlation. But to be really useful for process engineers as emphasized by Terry Krowlikowski it must be a widely applicable, reasonably accurate, and simple method that will solve easily on a computer. That's a very tall order for a single correlation to achieve. Yet this is the kind of compromise that industry really needs.

The final need is that of urgency. Theory contributes primarily long-range; there may be a ten-year lag, or more, in making some theoretical developments available to engineers. With our current problems of diminishing energy resources and pollution control requirements, changes are needed in the immediate future. Coal liquefaction and gasification and processing under cryogenic conditions require correlations and data for extreme ranges of temperature and pressure. Pollution control deals with parts-per-million concentrations where the thermodynamics may be difficult and data are lacking. It took forty years to establish a good data base for the petroleum industry. We can't affort to wait anywhere near that long for the data needed to solve these energy and pollution problems.

Future Directions

For future directions, we are going to be involved with more complex molecules and systems, for both theoretical and experimental work, and hopefully be cured of "argonitis". We will also be operating in shorter time scales to satisfy needs. More use will be made of computers to handle data and make calculations. As John Prausnitz expressed it, the purpose of the Conference was for each of us to be a little wiser at the end of the week than before. This wisdom will be demonstrated by what we do differently as a result of having been here.

INDEX

D

E

F

I

J

K

L

N

T

U

V